MODELING AND ANALYSIS OF DYNAMIC SYSTEMS

MODELING AND ANALYSIS OF DYNAMIC SYSTEMS

Charles M. Close and Dean K. Frederick
RENSSELAER POLYTECHNIC INSTITUTE

HOUGHTON MIFFLIN COMPANY   Boston

Dallas   Geneva, Illinois   Hopewell, New Jersey
Palo Alto   London

To our children:

Scott, Kim, and Doug Close

Steven, David, Ann, and Charles Frederick

Printed in the United States of America
Library of Congress Catalog Card Number: 77-74421
ISBN: 0-395-25040-4

CONTENTS ————————————————————

PREFACE

This book is intended for a first course in dynamic systems and is suitable for all engineering students regardless of discipline. It has been used at Rensselaer Polytechnic Institute for a one-semester course taken by every engineering student in his or her sophomore or junior year. While providing an exposure to dynamic systems for those whose interests lie in other areas, the material covered in this course serves as a basis for subsequent courses such as circuits and electronics, chemical process control, feedback systems, linear systems, vehicular dynamics and control, nuclear reactor control, biocontrol systems, systems physiology, and introduction to public systems. Because it covers such general topics as state variables, linearization of nonlinear models, numerical solution for the response, transfer functions, and feedback, this book can also be used for a general dynamic-systems course by students who have already completed a course in an area such as electrical circuits, machine dynamics, or chemical process dynamics. The rationale for the book is summarized in Section 1.1.

We assume that the reader has had differential and integral calculus and basic college physics, including mechanics and electrical phenomena. A course in differential equations should be taken at least concurrently. We have kept the mathematical level somewhere between the degree of rigor required for a mathematics book and the degree of expediency that raises inconsistencies for the discerning student and that often results in concepts that must be unlearned later. For example, the impulse has been treated in a manner that is consistent with distribution theory but that is no more difficult to grasp than the usual approach taken in introductory engineering books.

The organization of the book is indicated in the table of contents, and its scope and objectives are discussed in Section 1.5. Among the distinguishing features are the following:

1. A wide variety of physical systems is included, with one or more chapters on mechanical, electrical, electromechanical, thermal, and hydraulic systems. Each type is modeled in terms of its own fundamental laws and nomenclature.

2. The formulation of state-variable equations is included from the beginning, along with the more traditional input-output differential equation. We do not introduce matrix notation until the next-to-the-last chapter, however.

3. The technique for finding linearized models in terms of incremental variables is developed early in the book and used in a number of subsequent chapters. Incremental variables are also used to model linear systems with time-varying parameters.

4. The significance of the various components of a system's response— including the zero-input response, the zero-state response, and mode functions—is given more attention than is usual for a book at this level.

5. The numerical solution for the response of a system model is done in a way that does not emphasize a particular computer language. Numerical comparisons of the responses of nonlinear and linearized models are made, and problems requiring a programmable calculator or digital computer for their solution are included.

6. The use of Laplace transform techniques is deferred until after the introduction of time-domain and numerical solutions. The relationships of concepts such as the transfer function, poles and zeros, and frequency response to the time-domain solutions are emphasized.

7. The concluding chapter presents five case studies drawn from a variety of systems, including a sociological system.

We considered including discrete-time systems, but we did not do so because the length of the book would have become excessive. For this same reason, and because they are better suited for more advanced books, we do not discuss distributed and stochastic systems.

The majority of the material can be covered in a one-semester course, but the book can also be used as the basis for a year-long course. For schools on the quarter system, most of the material could be included in a two-quarter sequence. If necessary, any of Chapters 10, 11, or 14 to 18 can be omitted or abbreviated without loss of continuity.

In its various versions, the manuscript has been used at Rensselaer since 1973 by more than 2500 students. There are 130 examples and 420 problems. A solutions manual containing solutions to each of the problems is available.

ACKNOWLEDGMENT

For their continued assistance over a period of several years, we are deeply indebted to Doris Davis, who typed all the preliminary drafts and most of the final copy, and to Joseph H. Smith, Jr., who provided the administrative assistance that is essential for a project of this type from its inception. Colleagues who provided information and helpful comments on specific portions of the manuscript are Professors Henry Sneck, Peter Lashmet, and Pitu Mirchandani. As we were preparing this manuscript for publication, several people read it in its various stages and offered constructive criticism. We are grateful to Richard A. Baker, Washington State University, Norman H. Beachley, University of Wisconsin, James S. Demetry, Worcester Polytechnic Institute, and John Hadjilogiou, Florida Institute of Technology. To Richard A. Baker, we owe additional thanks for detailed criticism of draft chapters. Rensselaer students Bosco Leung, James Odenthal, Ron Ragonnetti, Paul Taylor, and Andy Yin assisted in preparing solutions to the homework problems. Harold Tomlinson, Jr., prepared the computer solutions. We appreciate the support and encouragement of our department chairman, Lester Gerhardt. Marita Jadlos, Charles Frederick, and Phyllis Frederick made major contributions to the final stages of manuscript preparation, and it would have been impossible to meet the deadline without their assistance. Finally, we sincerely appreciate the support of our families throughout the duration of this project.

C. M. C.
D. K. F.

CHAPTER 1 ————————————————————————
INTRODUCTION

In this chapter we present the rationale for the book, define several terms that will be used throughout, and describe various types of systems. The chapter concludes with a description of the particular types of systems to be considered and a summary of the techniques that the reader should be able to apply after completing the book.

1.1 RATIONALE

The importance of understanding and being able to determine the dynamic response of physical systems has long been recognized. It has been traditional in engineering education to have separate courses in dynamic mechanical systems, circuit theory, chemical-process dynamics, and other areas. Such courses develop techniques of modeling, analysis, and design for the particular type of physical systems relevant to that specific discipline, even though many of the techniques taught in these courses have much in

1

common. This approach tends to reinforce the student's view of such courses as isolated entities with little in common and to foster reluctance to apply what has been learned in one course to a new situation.

Another justification for considering a wide variety of different types of systems in an introductory book is that the majority of systems of practical interest contain components of more than one type. In the design of electronic circuits, for example, attention must be given to mechanical structure and to dissipation of the heat generated. Hydraulic motors and pneumatic process controllers are other examples of useful combinations of different types of elements. Furthermore, the techniques in this book can be applied not only to pneumatic, acoustical, and other traditional areas but also to systems that are quite different, such as sociological, physiological, economic, and transportation systems.

Because of the universal need for engineers to understand dynamic systems and because there is a common methodology applicable to such systems regardless of their physical origin, it makes sense to present them all together. This book considers both the problem of obtaining a mathematical description of a physical system and the various analysis techniques that are widely used.

1.2 ANALYSIS OF DYNAMIC SYSTEMS

Since the most frequent key word in the text is likely to be "system," it is appropriate to define it at the outset. A *system* is any collection of interacting elements for which there are cause-and-effect relationships among the variables. This definition is necessarily general, because it must encompass a broad range of systems. The important feature of the definition is that it tells us we must take interactions among the variables into account in system modeling and analysis, rather than treating individual elements separately.

Our study will be devoted to *dynamic systems*, for which the variables are time-dependent. In nearly all our examples, not only will the excitations and responses vary with time but at any instant the derivatives of one or more variables will depend on the values of the system variables at that instant. The system's response will normally depend on initial conditions, such as stored energy, in addition to any external excitations.

In the process of analyzing a system, two tasks must be performed: modeling the system and solving for the model's response. The combination of these steps is referred to as *system analysis*.

A *mathematical model*, or *model* for short, is a description of a system in terms of equations. The basis for constructing a model of a system is the physical laws (e.g., the conservation of energy and Newton's laws) that the system elements and their interconnections are known to obey.

The type of model sought will depend on both the objective of the engineer and the tools for analysis. If a pencil-and-paper analysis with parameters expressed in literal rather than numerical form is to be performed, a relatively simple model will be needed. To achieve this simplicity, the engineer should be prepared to neglect elements that do not play a dominant role in the system.

On the other hand, if a computer is available for carrying out simulations of specific cases with parameters expressed in numerical form, a comprehensive mathematical model that includes descriptions of both primary and secondary effects might be appropriate. The important notion is that a variety of mathematical models are possible for a system, and the engineer must be prepared to decide what form and complexity are most consistent with the objectives and the available resources.

The process of using the mathematical model to determine certain features of the system's cause-and-effect relationships is referred to as *solving the model*. For example, the responses to specific excitations may be desired for a range of parameter values, as guides in selecting design values for those parameters. As described in the discussion of modeling, this phase may include the analytical solution of simple models and the computer solution of more complex ones.

The type of equation involved in the model has a strong influence on the extent to which analytical methods can be used. For example, nonlinear differential equations can seldom be solved in closed form, and the solution of partial differential equations is far more laborious than that of ordinary differential equations. Computers can be used to generate the responses to specific numerical cases for complex models. However, the use of a computer for the solution of a complex model is not without its limitations. Models used for computer studies should be chosen with the approximations encountered in numerical integration in mind and should be relatively insensitive to system parameters whose values are uncertain or subject to change. Furthermore, it may be difficult to generalize results based only on computer solutions that must be run for specific parameter values, excitations, and initial conditions.

The engineer must not forget that the model being analyzed is only an approximate mathematical description of the system and is not the physical system itself. Conclusions based on equations that required a variety of assumptions and simplifications in their development may or may not apply to the actual system. Unfortunately, the more faithful a model is in describing the actual system, the more difficult it is to obtain general results.

One procedure is to use a simple model for analytical results and design and then to use a different model to verify the design by means of computer simulation. In very complex systems, it may be feasible to incorporate actual hardware components into the simulation as they become available, thereby eliminating the corresponding parts of the mathematical model.

1.3 CLASSIFICATION OF VARIABLES

A system is often represented by a box (traditionally called a "black box"), as shown in Figure 1.1. The system may have several *inputs*, or *excitations*, each of which is a function of time. Typical inputs are a force applied to a mass, a voltage source applied to an electrical circuit, and a heat source applied to a vessel filled with a liquid. In general discussions that are not related to specific systems, we shall use the symbols $u_1(t), u_2(t), \ldots, u_m(t)$ to denote the m inputs, shown by the arrows directed into the box.

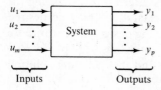

Figure 1.1 Black-box representation of a system.

Outputs are variables that are to be calculated or measured. Typical outputs are the velocity of a mass, the voltage across a resistor, and the rate at which a liquid flows through a pipe. The p outputs are represented in Figure 1.1 by the arrows pointing away from the box representing the system. They are denoted by the symbols $y_1(t), y_2(t), \ldots, y_p(t)$. There is a cause-and-effect relationship between the outputs and inputs. To calculate any one of the outputs for all $t \geq t_0$, we must know the inputs for $t \geq t_0$ and also the accumulated effect of any previous inputs. One approach to constructing a mathematical model is to find equations that relate the outputs directly to the inputs by eliminating all the other variables that are internal to the system. If we are interested only in the input-output relationships, eliminating extraneous variables may seem appealing. However, potentially important aspects of the system's behavior may be lost by deleting information from the model.

Another modeling technique is to introduce a set of *state variables*, which generally differs from the set of outputs but which may include one or more of them. The state variables must be chosen so that a knowledge of their values at any reference time t_0 and a knowledge of the inputs for all $t \geq t_0$ is sufficient to determine the outputs and state variables for all $t \geq t_0$. An additional requirement is that the state variables must be independent, meaning that it must not be possible to express one state variable as an algebraic function of the others. This approach is particularly convenient for working with multi-input, multioutput systems and for obtaining computer solutions. In Figure 1.2 the representation of the system has been modified to include the state variables denoted by the symbols $q_1(t), q_2(t), ..., q_n(t)$ within the box. The state variables can account for the important aspects of the system's behavior, regardless of the choice of output variables. Equations for the outputs can then be written as algebraic functions of the state variables, inputs, and time.

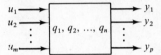

Figure 1.2 General system representation showing inputs, state variables, and outputs.

Whenever it is appropriate to indicate units for the variables and para-meters, we shall use the International System of Units (abbreviated SI, from the French *Systeme International d'Unités*). A list of the units used in this book appears in Appendix A.

1.4 CLASSIFICATION OF SYSTEMS

Systems are grouped according to the types of equations that are used in their mathematical models. Examples are partial differential equations with time-varying coefficients, ordinary differential equations with constant coefficients, and difference equations. In this section we define and briefly discuss ways of classifying the models, and in the next section we indicate those categories that will be treated in this book. The classifications that we use are listed in Table 1.1.

SPATIAL CHARACTERISTICS

A *distributed system* does not have a finite number of points at which state variables can be defined. In contrast, a *lumped system* can be described by a finite number of state variables.

Table 1.1 Criteria for classifying systems

Criterion	Classification
Spatial characteristics	Lumped
	Distributed
Continuity of the time variable	Continuous
	Discrete-time
	Hybrid
Quantization of the dependent variable	Nonquantized
	Quantized
Parameter variation	Fixed
	Time-varying
Superposition property	Linear
	Nonlinear

To illustrate these two types of systems, consider the flexible shaft shown in Figure 1.3(a) with one end embedded in a wall and with a torque applied to the other end. The angle through which a point on the surface of the shaft is twisted depends on both its distance from the wall and the applied torque. Hence the shaft is inherently distributed and would be modeled by a partial differential equation. However, if we are only interested in the angle of twist at the right end of the shaft, we may account for the flexibility of the shaft by a rotational spring constant K and represent the effect of the distributed mass by the single moment of inertia J. Making these approximations results in the lumped system shown in Figure 1.3(b), which has the important property that its model is an ordinary differential equation. Because ordinary differential equations are far easier to solve than partial differential equations, converting from a distributed system to a lumped approximation is often essential if the resulting model is to be solved with the resources available.

Another example of a distributed system is an inductor that consists of a wire wound around a core, as shown in Figure 1.4(a). If an electrical

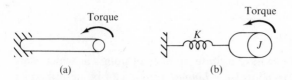

(a) (b)

Figure 1.3 A torsional shaft and its lumped approximation.

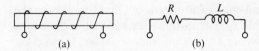

(a) (b)

Figure 1.4 An inductor and its lumped approximation.

excitation is applied across the terminals of the coil, then different values of voltage would exist at all points along the coil, characteristic of a distributed system. To develop a lumped circuit whose behavior as calculated at the terminals closely approximates that of the distributed device, we might account for the resistance of the wire by a lumped resistance R and for the inductive effect related to the magnetic field by a single inductance L. The resulting lumped circuit is shown in Figure 1.4(b). Note that in these two examples (although not in all cases), the two elements in the lumped model do not correspond to separate physical parts of the actual system. The stiffness and moment of inertia of the flexible shaft cannot be separated into two physical pieces, nor can the resistance and inductance of the coil.

CONTINUITY OF THE TIME VARIABLE

A second basis for classifying dynamic systems is the independent-variable time. A *continuous system* is one for which the inputs, state variables, and outputs are defined over some continuous range of time (although the signals may have discontinuities in their waveshapes and not be continuous functions in the mathematical sense). A *discrete-time system* has variables that are determined at distinct instants of time and that are either not defined or not of interest between those instants. Continuous systems are described by differential equations and discrete-time systems by difference equations.

Examples of the variables associated with continuous and discrete-time systems are shown in Figure 1.5. In fact, the discrete-time variable $f_2(kT)$

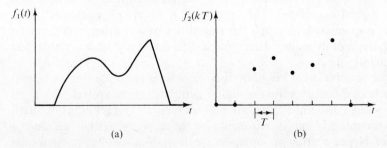

Figure 1.5 Sample variables. (a) Continuous. (b) Discrete-time.

shown in Figure 1.5(b) is the sequence of numbers obtained by taking the values of the continuous variable $f_1(t)$ at instants separated by T units of time. Hence $f_2(kT) = f_1(t)|_{t=kT}$ where k takes on integer values. In practice, a discrete-time variable may be composed of pulses of very short duration (much less than T) or numbers that reside in digital circuitry. In either case, the variable is assumed to be represented by a sequence of numbers, as indicated by the dots in Figure 1.5(b). There is no requirement that their spacing with respect to time be uniform, although this is often the case.

If a system contains both discrete-time and continuous subsystems, it is referred to as a *hybrid system*. Many modern control and communication systems contain a digital computer as a subsystem. In such cases, those variables that are associated with the computer are discrete in time, while variables elsewhere in the system are continuous. In such systems, sampling equipment is used to form discrete-time versions of continuous variables, and signal-reconstruction equipment is used to generate continuous variables from discrete-time ones.

QUANTIZATION OF THE DEPENDENT VARIABLE

In addition to possible restrictions on the values of the independent-variable time, the system variables may be restricted to certain distinct values. If within some finite range a variable may take on only a finite number of different values, it is said to be *quantized*. A variable that may have any value within some continuous range is *nonquantized*. Quantized variables may arise naturally, or they may be created by rounding or truncating the values of a nonquantized variable to the nearest quantization level.

The variables shown in Figure 1.6(a) and Figure 1.6(b) are both non-quantized, whereas those in the remaining parts of the figure are quantized. Although the variable in Figure 1.6(b) is restricted to the interval $-1 \le f_b \le 1$, it is nonquantized because it can take on a continuous set of values within that interval. The variable f_c shown in Figure 1.6(c) is restricted to the two values 0 and 1 and is representative of the signals found in devices that perform logical operations. The variable f_d is restricted to integer values and thus is quantized, although there need not be any other restriction on the magnitude of its values. The variable f_e is a discrete-time variable that is also quantized.

Because variables that are both discrete in time and quantized in amplitude, such as f_e, occur within a digital computer, they are referred to as *digital variables*. In contrast, the variable f_a in Figure 1.6(a) is both continuous and nonquantized and is representative of a signal within an analog computer. Hence continuous, nonquantized variables are often referred to as *analog variables*.

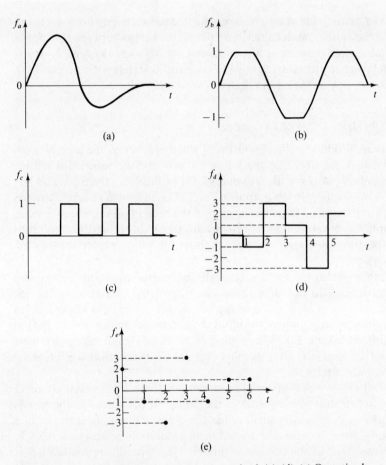

Figure 1.6 Sample variables. (a), (b) Nonquantized. (c), (d), (e) Quantized.

PARAMETER VARIATION

Systems may be classified according to properties of their parameters as well as of their variables. *Time-varying systems* are ones whose characteristics (e.g., the value of a mass or a resistance) change with time. Element values may change because of environmental factors such as temperature and radiation. Other examples of time-varying elements are the mass of a rocket, which decreases as fuel is burned, and the inductance of a coil, which increases as an iron slug is inserted into the core. In the differential equations describing time-varying systems, some of the coefficients will be functions of time. Delaying the input to a time-varying system will affect the size and shape of the response.

For *fixed* or *time-invariant systems*, whose characteristics do not change with time, the system model describing the relationships between the inputs, state variables, and outputs is independent of time. If such a system is initially at rest, delaying the input by t_d units of time will just delay the output by t_d units, without any change in its size or waveshape.

SUPERPOSITION PROPERTY

A system can also be classified in terms of whether it obeys the *superposition property*, which requires that the following two tests be satisfied when the system is initially at rest with zero energy. (1) Multiplying the inputs by any constant α must multiply the outputs by α. (2) The response to several inputs applied simultaneously must be the sum of the individual responses to each input applied separately. *Linear systems* are those satisfying the conditions for superposition; *nonlinear systems* are those for which superposition does not hold.

For a linear system, the coefficients in the differential equations comprising the system model do not depend on the size of the excitation, while for nonlinear systems at least some of the coefficients do. For a linear system initially at rest, multiplying all the inputs by a constant multiplies the output by the same constant. Likewise, replacing all the original inputs by their derivatives (or integrals) gives outputs that are the derivatives (or integrals) of the original outputs.

Nearly all systems are inherently nonlinear if there are no restrictions at all placed on the allowable values of the inputs. If the values of the inputs are confined to a sufficiently small range, the originally nonlinear model of a system may often be replaced by a linear model whose response closely approximates that of the nonlinear model. This type of approximation is desirable because analytical solutions to linear models are more easily obtained.

In other applications, the nonlinear nature of an element may be an essential feature of the system and should not be avoided in the model. Examples are mechanical valves or electrical diodes designed to give completely different types of response for positive and negative inputs. Devices used to produce constant-amplitude oscillations generally have the amplitude of the response determined by nonlinear elements in the system.

To illustrate the difference between linear and nonlinear models, consider a system with the single input $u(t)$ and the single output $y(t)$. If the input and output are related by the differential equation

$$a_1 \frac{dy}{dt} + a_0\, y(t) = b_0 u(t)$$

where a_0, a_1, and b_0 may be functions of time but do not depend on $u(t)$ or $y(t)$ in any way, then the system is linear. However, if one or more of the coefficients is a function of the input or output, as in

$$\frac{dy}{dt} + u(t)\,y(t) = u(t)$$

and

$$\frac{dy}{dt} + |y(t)|\,y(t) = u(t)$$

then the system is nonlinear.

1.5 SCOPE AND OBJECTIVES

This book is restricted to lumped, continuous, nonquantized systems that may be described by sets of ordinary differential equations. Because well-developed analytical techniques are available for solving linear ordinary differential equations with constant coefficients, we shall emphasize such methods. The majority of our examples will involve systems that are both fixed and linear. A method for approximating a time-varying or nonlinear system by a fixed linear model will be developed. For time-varying or non-linear systems that cannot be approximated by a fixed linear model, one can resort to computer solutions.

We list as *objectives* the following things that the reader should be able to do after completing the book. These objectives are grouped in the two general categories of modeling and solving for the response.

After completing this book, the reader should be able to do the following for dynamic systems composed of mechanical, electrical, thermal, and hydraulic components.

1. Given a description of the system, construct a simplified version using idealized elements and define a suitable set of variables.

2. Use the appropriate element and interconnection laws to obtain a mathematical model generally consisting of ordinary differential equations.

3. If the model is nonlinear, determine the equilibrium conditions and, where appropriate, obtain a linearized model in terms of incremental variables.

4. Arrange the equations comprising the model in a form suitable for solution. Construct simulation·diagrams and block diagrams and simplify the latter.

When a linear mathematical model has been determined or is given, the reader should be able to do the following:

1. For a first- or second-order system, solve directly for the time-domain response without transforming the functions of time into functions of other variables.

2. For a model of moderate order (of order four or less), use the Laplace transform to
    a. find the complete time response.
    b. determine the transfer function and its poles and zeros.
    c. analyze stability and, where appropriate, evaluate time constants, damping ratios, and undamped natural frequencies.

3. Find from the transformed expression the steady-state response to a constant or sinusoidal input without requiring a general solution.

In addition to using these analytical methods for obtaining the model's response, the reader should be able to use a calculator or computer to obtain the time response of a linear or nonlinear model in numerical form.

This book investigates all the foregoing procedures in detail. First we illustrate the basic modeling approaches and introduce state variables in the context of mechanical systems. We next consider the procedure for approximating a nonlinear model by a linear model. Then there are several chapters about the analytical and computer solutions of mathematical models, followed by discussion of electrical circuits and systems containing electrical and mechanical parts that are coupled together. Analytical techniques based on transforming functions of time into functions of a different variable are developed and applied to the solution of fixed linear systems. This material is followed by separate chapters on thermal and hydraulic systems. An introductory treatment of feedback systems is followed by a chapter on the use of matrix techniques for modeling multi-input, multioutput linear systems. The book concludes with several case studies drawn from a variety of engineering disciplines.

CHAPTER 2
TRANSLATIONAL MECHANICAL SYSTEMS

The modeling techniques for translational and rotational mechanical systems are discussed in this and the next two chapters. Procedures for solving the mathematical models are developed in later chapters.

After introducing the variables to be used, we discuss the laws for the individual elements, the laws governing the interconnections of the elements, and the use of free-body diagrams as an aid in formulating the equations of the model. In the following chapter, we show how the mathematical model can take the form of a set of state-variable equations or a differential equation that relates the output directly to the input.

2.1 VARIABLES

The symbols for the basic variables used to describe the dynamic behavior of translational mechanical systems are

  x, displacement in meters (m)

  v, velocity in meters per second (m/s)

  a, acceleration in meters per second per second (m/s^2)

  f, force in newtons (N)

All these variables are functions of time. In general, however, we shall add
a t in parentheses immediately after the symbol only when it denotes an
input or when we find it useful for clarity or emphasis.

Displacements are measured with respect to some reference condition,
which is often the equilibrium position of the body or point in question.
Displacements and velocities are measured with respect to an inertial
reference frame, which is defined to be one for which Newton's laws hold.
For ordinary systems on or near the surface of the earth, the earth's surface
is a very close approximation to an inertial reference frame, so it is the one
we use. The symbol for acceleration will in nearly all cases be replaced by
the derivative of the velocity or the second derivative of the displacement.

Two conventions used to define displacements are illustrated in Figure
2.1(a) and Figure 2.1(b). In Figure 2.1(a), the variable x represents the displacement of the left side of the body from the fixed vertical wall, whereas in
Figure 2.1(b), the reference position corresponding to $x = 0$ is not specifically shown. Generally, the reference position will correspond to a condition
of equilibrium for which the system inputs are constants and in which the
net force on the body being considered is zero. Figure 2.1(c) and Figure
2.1(d) indicate two methods of defining a velocity. All points on the body

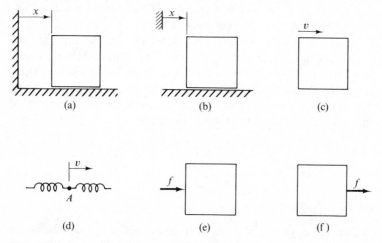

Figure 2.1 Conventions for designating variables.

in Figure 2.1(c) must move with the same velocity, so there is no possible ambiguity about which point has the velocity v. In Figure 2.1(d), the vertical line at the base of the arrow indicates that v is the velocity of the point labeled A. Forces can be represented by arrows pointing either into or away from a body, as depicted in Figure 2.1(e) and Figure 2.1(f), which are equivalent to one another.

Remember that the arrows only indicate an assumed positive sense for the displacement, velocity, or force being considered and by themselves do not imply anything about the actual direction of the motion or of the force at a given instant. If, for example, in Figure 2.1(e) and Figure 2.1(f), the force acting on the body is $f(t) = \sin t$, the force acts to the right for $0 < t < \pi$ and to the left for $\pi < t < 2\pi$, and it continues to change direction every π seconds. Note that an alternative way of describing the identical situation is to draw the arrow pointing to the left and then write $f(t) = -\sin t$. Reversing a reference arrow is equivalent to reversing the sign of the algebraic expression associated with it. There is no unique way of choosing reference directions on a diagram, but the equations must be consistent with whatever choice is made for the arrows.

Reference arrows for the displacement, velocity, and acceleration of a given point are invariably drawn in the same direction, so that the equations

$$v = \frac{dx}{dt}$$

and

$$a = \frac{dv}{dt} = \frac{d^2x}{dt^2}$$

can be used. With this understanding, a reference arrow for acceleration is not shown explicitly on the diagrams, and for the same reason only the reference arrow for either the displacement or the velocity of a point (but not both) is shown in many examples.

Variables in addition to those defined at the beginning of this section are

w, energy in joules (J)

p, power in watts (W)

where 1 joule $= 1$ newton-meter and 1 watt $= 1$ joule per second. Because the arrows defining the positive senses of the velocity and force are in the same direction, the power supplied to the mass in Figure 2.2(a) and to the spring in Figure 2.2(b) is

$$p = fv \tag{1}$$

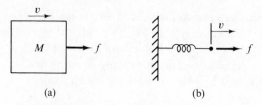

(a)   (b)

Figure 2.2 Reference arrows for (1).

Since power is defined to be the rate at which energy is supplied or dissipated, it follows that

$$p = \frac{dw}{dt} \tag{2}$$

and the energy supplied between time t_0 and t_1 is

$$\int_{t_0}^{t_1} p(t)\,dt$$

If $w(t_0)$ denotes the energy supplied up to the time t_0, then the total energy supplied up to any later time t is

$$w(t) = w(t_0) + \int_{t_0}^{t} p(\lambda)\,d\lambda \tag{3}$$

In the last integrand, t has been replaced by the dummy variable λ in order to avoid confusion between the upper limit and the variable of integration.

2.2 ELEMENT LAWS

Physical devices are represented by one or more idealized elements that obey laws involving the variables associated with the elements. As we mentioned in Chapter 1, some degree of approximation is required in selecting the elements to represent a device, and the behavior of the combined elements may not correspond exactly to the behavior of the device. The elements that we include in translational systems are mass, friction, stiffness, and the lever. The element laws for the first three relate the external force to the acceleration, velocity, or displacement associated with the element. For the lever, a pair of equations is required.

MASS

Figure 2.2(a) shows a *mass M*, which has units of kilograms (kg), subjected to a force f. Newton's second law states that the sum of the forces acting on a body is equal to the time rate of change of the momentum:

$$\frac{d}{dt}(Mv) = f \tag{4}$$

We shall restrict our attention to constant masses, neglect relativistic effects, and let v be the absolute velocity with respect to a fixed reference. The momentum and force are really vector quantities, but in this chapter the mass is constrained to move in a single direction, allowing us to write scalar equations.

For a constant mass, (4) can be written as

$$M\frac{dv}{dt} = f \tag{5}$$

Hence a mass can be modeled by an algebraic relationship between the acceleration dv/dt and the external force f. For (5) to hold, the positive senses of both dv/dt and f must be the same, since the force will cause the velocity to increase in the direction in which the force is acting.

Energy in a mass is stored as kinetic energy if the mass is in motion and as potential energy if the mass has a vertical displacement relative to its reference position. The kinetic energy is

$$w_k = \tfrac{1}{2}Mv^2 \tag{6}$$

and the potential energy, assuming a uniform gravitational field, is

$$w_p = Mgh \tag{7}$$

where g is the gravitational constant (approximately 9.807 m/s^2 at the surface of the earth) and h is the height of the mass above its reference position. In order to determine the response of a dynamic system containing a mass for $t \geq t_0$, we must know its initial velocity $v(t_0)$ and, if vertical motion is possible, its initial height $h(t_0)$.

FRICTION

Forces that are algebraic functions of the relative velocity between two points are modeled by friction elements. A mass sliding on an oil film having laminar flow, as depicted in Figure 2.3, is subject to *viscous friction* and obeys

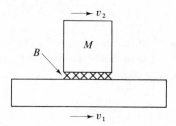

Figure 2.3 Friction described by (8) with $\Delta v = v_2 - v_1$.

the linear relationship

$$f = B \, \Delta v \tag{8}$$

where B has units of newton-seconds per meter $(\text{N} \cdot \text{s/m})$ and where $\Delta v = v_2 - v_1$. The direction of a frictional force will be such as to oppose the motion of the mass. For (8) to apply to Figure 2.3, the force f exerted on the mass by the oil film is to the left. (By Newton's third law, the mass exerts an equal force f to the right on the oil film.) The friction coefficient B is proportional to the contact area and the viscosity of the oil and inversely proportional to the thickness of the film.

Viscous friction may also be used to model a dashpot, such as the shock absorbers on an automobile. As indicated in Figure 2.4(a), a piston moves through an oil-filled cylinder, and there are small holes in the face of the piston through which the oil passes as the parts move relative to each other. The symbol often used for a dashpot is shown in Figure 2.4(b). Many dashpot devices involve high rates of fluid flow through the orifices and have nonlinear characteristics. If the flow is laminar, then the element is again described by (8). If the lower block in Figure 2.3 or the cylinder of the dashpot in Figure 2.4(a) is stationary, then $v_1 = 0$ and the element law reduces to $f = Bv_2$.

If the dashpot or oil film is assumed to be massless and if the accelerations

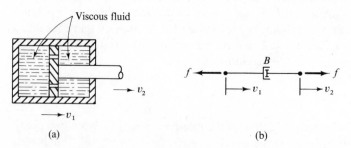

(a)

(b)

Figure 2.4 A dashpot and its symbol.

are to remain finite, then when a force f is applied to one side, a retarding force of equal magnitude must be exerted on the other side (either by a wall or by some other component), as shown in Figure 2.4(b), again with $f = B(v_2 - v_1)$. This means that in the system shown in Figure 2.5, the force f is transmitted through the dashpot and exerted directly on the mass M.

Viscous friction described by (8) is a linear element, for which the plot of f vs Δv is a straight line passing through the origin, as shown in Figure 2.6(a). Examples of friction that obey nonlinear relationships are *dry friction* and *drag friction*. The former is modeled by a force that is independent of the magnitude of the relative velocity, as indicated in Figure 2.6(b), and which can be described by the equation

$$f = \begin{cases} -A & \text{for } \Delta v < 0 \\ A & \text{for } \Delta v > 0 \end{cases}$$

Drag friction is caused by resistance to a body moving through a fluid (e.g., wind resistance) and can often be described by an equation of the form $f = D|\Delta v|\,\Delta v$, as depicted in Figure 2.6(c). Various other nonlinearities may be encountered in friction elements.

The power dissipated by friction is the product of the force exerted and the relative velocity of the two ends of the element. This power is immediately converted to heat and thus cannot be returned to the rest of the mechanical system at a later time. Because of this, we do not usually need to know the initial velocities of the friction elements in order to solve the model of a system.

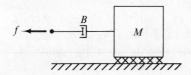

Figure 2.5 Force transmitted through a dashpot.

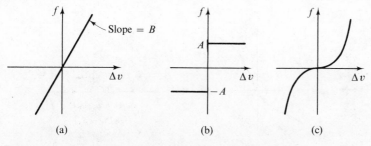

(a) (b) (c)

Figure 2.6 Friction characteristics. (a) Linear. (b) Dry. (c) Drag.

STIFFNESS

Any mechanical element that undergoes a change in shape when subjected to a force can be characterized by a *stiffness element*, provided only that an algebraic relationship exists between the elongation and the force. The most common stiffness element is the spring, although most mechanical elements undergo some deflection when stressed. For the spring sketched in Figure 2.7(a), we define d_0 to be the length of the spring when no force is applied and x to be the elongation caused by the force f. Then the total length at any instant is $d(t) = d_0 + x$, and the stiffness property refers to the algebraic relationship between x and f, as depicted in Figure 2.7(b). Since x has been defined as an elongation and the plot shows that f and x always have the same sign, it follows that the positive sense of f must be to the right in Figure 2.7(a); that is, f represents a tensile rather than a compressive force. For a linear spring, the curve in Figure 2.7(b) is a straight line and $f = Kx$, where K is a constant with units of newtons per meter (N/m).

Figure 2.7(c) shows a spring whose ends are displaced by the amounts x_1 and x_2 relative to their respective reference positions. If $x_1 = x_2 = 0$ corresponds to a condition when no force is applied to the spring, then the elongation at any instant is $x_2 - x_1$. For a linear spring,

$$f = K \, \Delta x \tag{9}$$

where $\Delta x = x_2 - x_1$. For small elongations of a structural shaft, K is proportional to the cross-sectional area and Young's modulus and inversely proportional to the length.

When a force f is applied to one side of a stiffness element that is assumed to have no mass, a force equal in magnitude but of opposite direction must be exerted on the other side. Thus for the system shown in Figure 2.8, the force f passes through the first spring and is exerted directly on the mass M. Of course all physical devices have some mass, but to obtain a lumped model we assume either that it is negligible or that it is represented by a separate element.

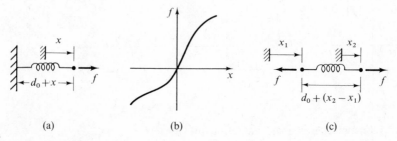

(a)   (b)   (c)

Figure 2.7 Characteristics of a spring.

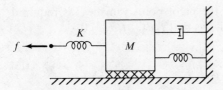

Figure 2.8 Force transmitted through a spring.

Potential energy is stored in a spring that has been stretched or compressed, and for a linear spring that energy is given by

$$w_p = \tfrac{1}{2}K(\Delta x)^2 \tag{10}$$

This energy may be returned to the rest of the mechanical system at some time in the future. Therefore, the initial elongation $\Delta x(t_0)$ is one of the initial conditions we need in order to find the complete response of a system.

THE LEVER

An *ideal lever* is assumed to be a rigid bar pivoted at a point and having no mass, no friction, no momentum, and no stored energy. In all our examples, the pivot point will be fixed. If the magnitude of the angle of rotation is small (say less than 0.25 rad), the motion of the ends can be considered strictly translational. In Figure 2.9, let θ denote the angular displacement of the lever from the horizontal position. For a rigid lever with a fixed pivot, the displacements of the ends are given by $x_1 \simeq d_1\theta$ and $x_2 \simeq d_2\theta$, where θ is in radians. Thus for small displacements we have

$$x_2 = \left(\frac{d_2}{d_1}\right)x_1 \tag{11}$$

and, by differentiating the above equation, we find

$$v_2 = \left(\frac{d_2}{d_1}\right)v_1 \tag{12}$$

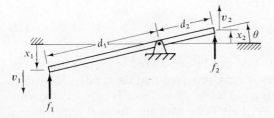

Figure 2.9 The lever.

Since the sum of the moments about the pivot point vanishes, as required by the assumed absence of mass, it follows that $f_2 d_2 - f_1 d_1 = 0$, or

$$f_2 = \left(\frac{d_1}{d_2}\right) f_1 \tag{13}$$

The pivot exerts a downward force of $f_1 + f_2$ on the lever, but this does not enter into the derivation of (13), since this force exerts no moment about the pivot. The lever differs from the mass, friction, and stiffness elements in that the algebraic relationships given by (11) through (13) involve pairs of the same types of variables: two displacements, two velocities, or two forces.

2.3 INTERCONNECTION LAWS

Having identified the individual elements in translational systems and given equations describing their behavior, we next present the laws that describe the manner in which the elements are interconnected. These include D'Alembert's law, the law of reaction forces, and the law for displacement variables.

D'ALEMBERT'S LAW

This law is just a restatement of Newton's second law governing the rate of change of momentum. For a constant mass, we can write

$$\sum_i (f_{ext})_i = M \frac{dv}{dt} \tag{14}$$

where the summation over the index i includes all the external forces $(f_{ext})_i$ acting on the body. The forces and velocity are in general vector quantities, but they can be treated as scalars provided that the motion is constrained to be in a fixed direction. Rewriting (14) as

$$\sum_i (f_{ext})_i - M \frac{dv}{dt} = 0 \tag{15}$$

suggests that the mass in question can be considered to be in equilibrium, i.e., that the sum of the forces is zero, provided that the term $-M \, dv/dt$ is thought of as an additional force. This fictitious force is called the *inertial* or *D'Alembert force*, and including it along with the external forces allows us to write the force equation as one of equilibrium, namely

$$\sum_i f_i = 0 \tag{16}$$

The minus sign associated with the inertial force in (15) indicates that when $dv/dt > 0$, the force acts in the negative direction.

In addition to applying (16) to a mass, we can apply it to any point in the system, such as the junction between components. Since a junction is considered massless, the inertial force will be zero in such a case.

THE LAW OF REACTION FORCES

In order to relate the forces exerted by the elements of friction, stiffness, and levers to the forces acting on a mass or junction point, we need Newton's third law regarding reaction forces. Accompanying any force of one element on another, there will be a *reaction force* on the first element of equal magnitude and opposite direction.

In Figure 2.10(a), for example, let f_K denote the force exerted by the mass on the right end of the spring, with the positive sense defined to the right. Newton's third law tells us that there acts on the mass a reaction force f_K of equal magnitude with its positive sense to the left, as indicated in Figure 2.10(b). Likewise, at the left end, the fixed surface exerts a force f_K on the spring with the positive sense to the left, while the spring exerts an equal and opposite force on the surface.

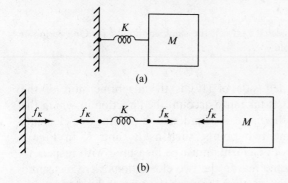

(a)

(b)

Figure 2.10 Example of reaction forces.

THE LAW FOR DISPLACEMENTS

If the ends of two elements are connected, those ends are forced to move with the same displacement and velocity. For example, since the dashpot and spring in Figure 2.11(a) are both connected between the wall and the mass, the right ends of both elements have the same displacement x and move with the same velocity v. In Figure 2.11(b), where B_2 and K are connected between two moving masses, the elongation of both elements is $x_2 - x_1$. An equivalent statement is that if we go from M_1 to M_2 and record the elongation of the dashpot B_2 and then return to M_1 and subtract the elongation of the spring K, the result is zero. In effect, we are saying that the difference between the displacements of any two points is the same regardless of which elements we are examining between those points. This statement is really a consequence of our being able to uniquely define points in space.

The discussion in the previous paragraph can be summarized by saying that at any instant the algebraic sum of the elongations around any closed path is zero; that is,

$$\sum_i \Delta x_i = 0 \quad \text{around any closed path} \tag{17}$$

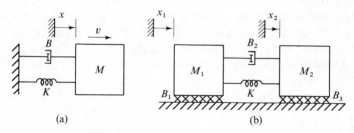

(a)   (b)

Figure 2.11 Two elements connected between the same endpoints. (a) One endpoint fixed. (b) Both endpoints movable.

It is understood that the left side of (17) is the algebraic sum of the elongations Δx_i with signs that take into account the direction in which the path is being traversed. Furthermore, it is understood that for two elements connected between the same two points, such as B_2 and K in Figure 2.11(b), the elongations of both elements must be measured with respect to the same references. If for some reason the two elongations were measured with respect to different references, then the algebraic sum of the elongations around the closed path would be a constant but not zero.

In Figure 2.12, let x_1 and x_2 denote displacements measured with respect to reference positions that correspond to a single equilibrium condition of the system. Then the respective elongations of B_1, B_2, and K are x_1, $x_2 - x_1$, and x_2. When the elongations are summed going from the fixed surface to the mass by way of the friction elements B_1 and B_2 and going back by way

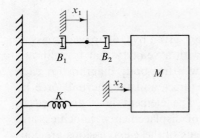

Figure 2.12 Illustration for the displacement law.

of the spring K, (17) gives

$$x_1 + (x_2 - x_1) - x_2 = 0$$

This equation can be regarded as a justification for the statement that the elongation of B_2 is $x_2 - x_1$ and that an additional symbol for this elongation is not needed.

In the analysis of mechanical systems, (17) is normally used implicitly and automatically in the process of labeling the system diagram. For example, we use the same symbol for two elongations that are forced to be equal by the element interconnections and avoid using different symbols for displacements that are known to be identical.

Differentiating (17) would lead to a similar equation in terms of relative velocities. However, we shall use only a single symbol for the velocities of two points that are constrained to move together. It will therefore not be necessary to invoke (17) in a formal way.

2.4 OBTAINING THE SYSTEM MODEL

The system model must incorporate both the element laws and the interconnection laws. The element laws involve displacements, velocities, and accelerations. Since the acceleration of a point is the derivative of the velocity, which in turn is the derivative of the displacement, we could write all the element laws in terms of x and its derivatives or in terms of x, v, and dv/dt. It is important to indicate the assumed positive directions for displacements, velocities, and accelerations. Since we shall always choose the assumed positive directions for a, v, and x to be the same, it will not be necessary to indicate all three positive directions on the diagram. Throughout the book, dots over the variables are used to denote derivatives with respect to time. For example, $\dot{x} = dx/dt$ and $\ddot{y} = d^2y/dt^2$.

FREE-BODY DIAGRAMS

We normally need to apply D'Alembert's law, given by (16), to each mass or junction point in the system that moves with a velocity that is unknown beforehand. To do so, it is useful to draw a free-body diagram for each such mass or point, showing all external forces and the inertial force by arrows that define their positive senses. The element laws are used to express all forces except inputs in terms of displacements, velocities, and accelerations. We must be sure that the signs of these expressions are consistent with the directions of the reference arrows. After the free-body diagram is completed, (16) can be applied by summing the forces indicated on the diagram, again taking into account their assumed positive senses. Normally all forces must be added as vectors, but in our examples the forces in the free-body diagram will be collinear and can be summed by scalar equations.

EXAMPLE 2.1   Draw the free-body diagram and apply D'Alembert's law for the system shown in Figure 2.13(a). The mass is assumed to move horizontally on a frictionless support (not shown in the figure), and the spring and dashpot are linear.

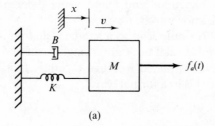

(a)

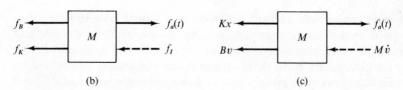

(b)   (c)

Figure 2.13 (a) Translational system for Example 2.1. (b) Free-body diagram. (c) Free-body diagram including element laws.

Solution   The free-body diagram for the mass is shown in Figure 2.13(b). The two vertical forces on the mass, i.e., the weight Mg and the upward force exerted by the frictionless support, have been omitted because these forces are perpendicular to the direction of motion. The horizontal forces, which are included in the free-body diagram, are

f_K,   the force exerted by the spring

f_B,   the force exerted by the dashpot

f_I,   the inertial force

$f_a(t)$, the applied force

The choice of directions for the arrows representing f_K, f_B, and f_I is arbitrary and will not affect the final result. However, the expressions for these individual forces must agree with the choice of arrows. The use of a dashed arrow for the inertial force f_I emphasizes that it is not a physical force like the other three.

We next use the element laws to express the forces f_K, f_B, and f_I in terms of the element values K, B, and M and the system variables x and v. In Figure 2.13(a), the positive direction of x and v is defined to be to the right, so the spring is stretched when x is positive and compressed when x is negative. If the spring undergoes an elongation x, then there must be a tensile force Kx on the right end of the spring directed to the right and a reaction force $f_K = Kx$ on the mass directed to the left. Since the arrow for f_K does point to the left in Figure 2.13(b), we may relabel this force as Kx in Figure 2.13(c). Note that if x is negative at some instant of time, the spring will be compressed and will exert a force to the right on the mass. Under these conditions, Kx will be negative and the free-body diagram will show a negative force on the mass to the left, which is equivalent to a positive force to the right. Although the result is the same either way, it is more common to assume that all displacements are in the assumed positive directions when determining the proper expressions for the forces.

Similarly, when the right end of the dashpot moves to the right with velocity v, a force $f_B = Bv$ is exerted on the mass to the left. Finally, because of (15), the inertial force $f_I = M\dot{v}$ must have its positive direction opposite to that of dv/dt. After trying a few examples, the reader should be able to draw a free-body diagram such as the one in Figure 2.13(c) without first having to show explicitly the diagram in Figure 2.13(b).

D'Alembert's law can now be applied to the free-body diagram in Figure 2.13(c), with due regard for the assumed arrow directions. If forces acting to the right are regarded as positive, the law yields

$$f_a(t) - (M\dot{v} + Bv + Kx) = 0$$

or

$$f_a(t) - (M\ddot{x} + B\dot{x} + Kx) = 0$$

EXAMPLE 2.2   Draw the free-body diagrams and write D'Alembert's law for the system shown in Figure 2.14(a).

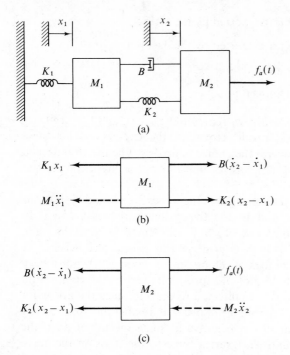

(a)

(b)

(c)

Figure 2.14 (a) Translational system for Example 2.2. (b), (c) Free-body diagrams.

Solution   Since there are two masses that can move with different unknown velocities, a separate free-body diagram should be drawn for each one, and this is done in Figure 2.14(b) and Figure 2.14(c). In Figure 2.14(b), the forces $K_1 x_1$ and $M_1 \ddot{x}_1$ are similar to those in Example 2.1. As indicated in our earlier discussion of displacements, the net elongation of the spring and dashpot connecting the two masses is $x_2 - x_1$. Hence a positive value of $x_2 - x_1$ results in a reaction force by the spring to the right on M_1 and to the left on M_2, as indicated in the figure. Of course, the force on either free-body diagram could be labeled $K_2(x_1 - x_2)$ provided that the corresponding reference arrow were reversed. For a positive value of $\dot{x}_2 - \dot{x}_1$, the reaction force of the middle dashpot is to the right on M_1 and to the left on M_2. As always, the inertial forces $M_1 \ddot{x}_1$ and $M_2 \ddot{x}_2$ are opposite to the positive directions of the accelerations.

Summing the forces on each free-body diagram separately and taking into account the directions of the reference arrows gives the following pair of differential equations:

$$B(\dot{x}_2 - \dot{x}_1) + K_2(x_2 - x_1) - M_1\ddot{x}_1 - K_1 x_1 = 0$$

$$f_a(t) - M_2\ddot{x}_2 - B(\dot{x}_2 - \dot{x}_1) - K_2(x_2 - x_1) = 0$$

Rearranging, we have

$$M_1\ddot{x}_1 + B\dot{x}_1 + (K_1 + K_2)x_1 - B\dot{x}_2 - K_2 x_2 = 0$$

$$M_2\ddot{x}_2 + B\dot{x}_2 + K_2 x_2 - B\dot{x}_1 - K_2 x_1 = f_a(t)$$

(18)

The equations in (18) constitute a pair of coupled second-order differential equations. In the next chapter, we shall discuss two alternative methods of presenting the information contained in such a set of equations.

EXAMPLE 2.3   Draw the free-body diagram, including all forces, and find the differential equation describing the system shown in Figure 2.15(a).

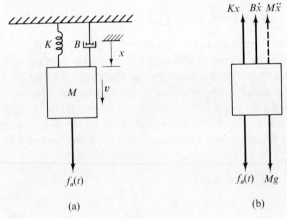

Figure 2.15 (a) Translational system with vertical motion. (b) Free-body diagram.

Solution   Assume that x is the displacement from the position corresponding to a spring that is neither stretched nor compressed. The gravitational force on the mass is Mg, and we include it in the free-body diagram shown in Figure 2.15(b) because the mass moves vertically. By summing the forces on the free-body diagram, we obtain

$$M\ddot{x} + B\dot{x} + Kx = f_a(t) + Mg$$

(19)

Suppose that the applied force $f_a(t)$ is zero and that the mass is not moving.

Then $x = x_0$, where x_0 is the constant displacement caused by the gravitational force. Since $\dot{x}_0 = \ddot{x}_0 = 0$, the above differential equation reduces to the algebraic equation

$$Kx_0 = Mg \tag{20}$$

This can also be seen directly from the free-body diagram by noting that all but two of the five forces vanish under these conditions.

We now reconsider the case where $f_a(t)$ is nonzero and where the mass is moving. Let

$$x = x_0 + z \tag{21}$$

This equation defines z as the displacement caused by the input $f_a(t)$, namely the additional displacement beyond that resulting from the constant weight Mg. Substituting (21) into (19) and again noting that $\dot{x}_0 = \ddot{x}_0 = 0$, we have

$$M\ddot{z} + B\dot{z} + K(x_0 + z) = f_a(t) + Mg$$

or, by using (20),

$$M\ddot{z} + B\dot{z} + Kz = f_a(t) \tag{22}$$

Comparison of (19) and (22) indicates that the gravitational force Mg can be ignored when drawing the free-body diagram and when writing the system equation, provided that the displacement is defined to be the displacement from the static position corresponding to no inputs except gravity.

This conclusion is valid for all cases where masses are suspended vertically by one or more linear springs. Under static-equilibrium conditions, there are no inertial or friction forces, while the force exerted by the spring has the form

$$f_K(t) = K(x_0 + z)$$

$$= Kx_0 + Kz$$

which is just the superposition of the static force caused by gravity and the force caused by additional inputs. For nonlinear springs, however, this conclusion is not valid because superposition does not hold (see Problems 2.18 and 2.19 for examples).

Normally, a new symbol such as z is not introduced in problems involving gravitational forces. Instead, the symbol x can be redefined to be the additional displacement from the static position.

EXAMPLE 2.4   A situation similar to that in the previous example is illustrated in Figure 2.16(a), which has two linear springs with constants K_1 and K_2 connected between the wall and the mass M. The mass is constrained to move horizontally, and the unstretched lengths of the springs are d_1 and d_2. Develop the equations describing the system.

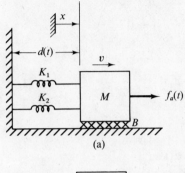

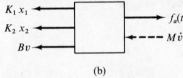

Figure 2.16 (a) Translational system for Example 2.4. (b) Free-body diagram.

Solution If x_1 and x_2 denote the elongations of the springs K_1 and K_2 with respect to their unstretched positions, then the total distance from the wall to the mass is

$$d(t) = d_1 + x_1 = d_2 + x_2 \qquad (23)$$

so

$$x_2 = x_1 + d_1 - d_2 \qquad (24)$$

Note that the velocity v of the mass can be expressed as $v = \dot{x}_1$ and also as $v = \dot{x}_2$. The free-body diagram for the mass is shown in Figure 2.16(b). With $v = \dot{x}_1$, summing the forces gives

$$M\ddot{x}_1 + B\dot{x}_1 + K_1 x_1 + K_2 x_2 = f_a(t)$$

Since x_1 and x_2 are not independent, we can use (24) to write

$$M\ddot{x}_1 + B\dot{x}_1 + K_1 x_1 + K_2(x_1 + d_1 - d_2) = f_a(t) \qquad (25)$$

In order to obtain more insight into the behavior of the two springs, consider the case where $f_a(t)$ is zero and where the mass is motionless in a position of static equilibrium. We use an extra subscript zero to denote the static displacements x_{1_0} and x_{2_0} and the static length d_0. Since $\dot{x}_{1_0} = \ddot{x}_{1_0} = 0$,

$$K_1 x_{1_0} + K_2(x_{1_0} + d_1 - d_2) = 0$$

or

$$x_{1_0} = \frac{K_2(d_2 - d_1)}{K_1 + K_2}$$

In a similar way, we can show that

$$x_{2_0} = \frac{K_1(d_1 - d_2)}{K_1 + K_2}$$

Now

$$d_0 = d_1 + x_{1_0}$$

$$= d_1 + \frac{K_2(d_2 - d_1)}{K_1 + K_2}$$

$$= \frac{K_1 d_1 + K_2 d_2}{K_1 + K_2} \tag{26}$$

We can come to the same conclusion by considering

$$d_0 = d_2 + x_{2_0}$$

$$= d_2 + \frac{K_1(d_1 - d_2)}{K_1 + K_2}$$

$$= \frac{K_1 d_1 + K_2 d_2}{K_1 + K_2}$$

Note that if $d_1 = d_2$, then $d_0 = d_1 = d_2$ as expected. The quantity d_0 is the length of each spring when $f_a(t) = 0$, $v = 0$, and $\dot{v} = 0$. It is reasonable to define the variable x as the distance of the mass beyond d_0 when a force input is applied, i.e., as the displacement with respect to the reference position given by (26). Then

$$x = d(t) - d_0 \tag{27}$$

Now consider the quantity $K_1 x + K_2 x$, which by (23), (26), and (27) can be written as

$$K_1 x + K_2 x = (K_1 + K_2)(d - d_0)$$

$$= (K_1 + K_2)\left[d_1 + x_1 - \frac{K_1 d_1 + K_2 d_2}{K_1 + K_2} \right]$$

$$= K_1 x_1 + K_2 x_1 + K_1 d_1 + K_2 d_1 - K_1 d_1 - K_2 d_2$$

$$= K_1 x_1 + K_2 (x_1 + d_1 - d_2)$$

By comparing this result with (25), we see that $K_1 x + K_2 x$ is, as expected, the net force exerted by the two springs on the mass. Thus, with x defined by (27), D'Alembert's law gives

$$M\ddot{x} + B\dot{x} + (K_1 + K_2)x = f_a(t)$$

Again in this example, the displacement variable associated with the springs is not the elongation of either spring from its unstretched position. As long as the springs are linear, the form of the equations for a mechanical system will be the same no matter what reference length is used for the springs in the definition of their elongation variable.

PROBLEMS

2.1 The position of the mass M in Figure P2.1 is shown for $x = 2$ m and $y = 5$ m. When $x = 0$, the two springs are neither stretched nor compressed.

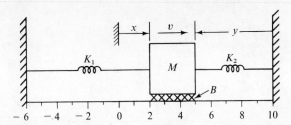

Figure P2.1

a. Give the magnitudes and directions of the forces exerted on the mass by each spring when $x = 2$ m and when $x = -2$ m.

b. Draw the complete free-body diagram and write the differential equation describing the system in terms of the variable x.

c. Find the acceleration of the mass if $x = 2$ m and $v = 3$ m/s.

d. Repeat part c when $x = -2$ m and $v = 3$ m/s.

e. Write the algebraic equation relating x and y.

f. Write the differential equation describing the system in terms of the variable y.

2.2 Applying (1), (6), (8), and (10) to the system shown in Figure 2.13(a), write an equation describing the fact that the power supplied by the force input equals the power dissipated in the friction element plus the time rate of change of the energy stored in the mass and spring. Simplify the equation and show that it reduces to the answer in Example 2.1.

2.3 Write an expression for the power supplied to the ideal lever in Figure 2.9 by the forces f_1 and f_2. Since the lever has no mass, friction, or stored energy, the power supplied must be zero. Express this fact by an equation and show that (12) and (13) satisfy the equation.

2.4 Draw the free-body diagrams and apply D'Alembert's law to the system shown in Figure 2.12.

2.5 The mechanical system shown in Figure P2.5 is driven by the applied force $f_a(t)$. When $x_1 = x_2 = 0$, the springs are neither stretched nor compressed.

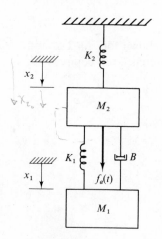

Figure P2.5

a. Draw the free-body diagrams and write the differential equations of motion for the two masses in terms of x_1 and x_2.

b. Find x_{1_0} and x_{2_0}, the constant displacements of the masses caused by the gravitational forces when $f_a(t) = 0$ and when the system is in static equilibrium.

c. Rewrite the system equations in terms of z_1 and z_2, the relative displacements of the masses with respect to the static-equilibrium positions found in part b.

2.6 For the system shown in Figure P2.6, draw the free-body diagram for each mass and write the differential equations describing the system.

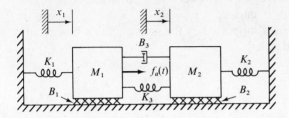

Figure P2.6

2.7 Repeat Problem 2.6 for the system shown in Figure P2.7.

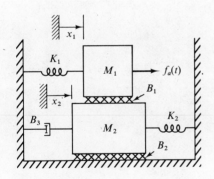

Figure P2.7

2.8 Repeat Problem 2.6 for the system shown in Figure P2.8.

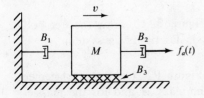

Figure P2.8

2.9 When $x_1 = x_2 = 0$ for the system shown in Figure P2.9, the springs are neither stretched nor compressed. Draw the free-body diagram for each mass and write the differential equations describing the system for each of the following cases.

   a. Write the equations in terms of x_1 and x_2.

   b. Write the equations in terms of x_1 and the elongation of K_2.

   c. Write the equations in terms of x_1 and the elongation of K_3.

   d. Write the equations in terms of z_1 and z_2, the relative displacements of M_1 and M_2 measured with respect to their static-equilibrium positions. When $f_a(t) = 0$ and $z_1 = z_2 = \dot{z}_1 = \dot{z}_2 = 0$, the system is motionless with the masses supported by the springs. Show how z_1 and z_2 are related to x_1 and x_2.

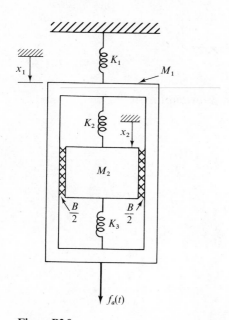

Figure P2.9

2.10 For each of the systems shown in Figure P2.10, draw the free-body diagrams and write the differential equations describing the system.

2.11 The input for the system shown in Figure P2.11 is the displacement $x_3(t)$. Draw the free-body diagrams for the mass M and for the massless point A. Write the differential equations describing the system.

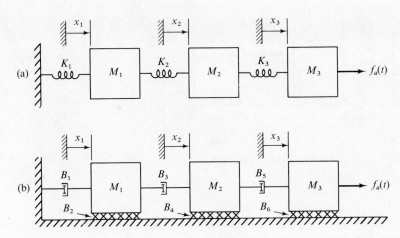

Figure P2.10

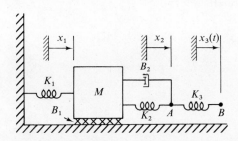

Figure P2.11

2.12 Write the differential equations describing the system shown in Figure P2.11 when the input is the force $f_a(t)$ applied to point B. The positive sense of $f_a(t)$ is to the right.

2.13 In the mechanical system shown in Figure P2.13, the spring forces are zero when $x_1 = x_2 = x_3 = 0$.

    a. Draw free-body diagrams and write a pair of coupled differential equations that govern the motion when $x_3(t) = 0$, i.e., when the base is stationary.

    b. Repeat the problem when $x_3(t)$ is the input and $f_a(t) = 0$.

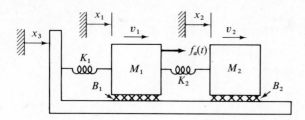

Figure P2.13

2.14 The mass M in Figure P2.14 is supported by two identical springs. Draw the free-body diagram, including the gravitational force. When $f_a(t) = 0$ and when the mass remains motionless, the springs are compressed 0.1 m compared with their free lengths. Find the value of K when $M = 10$ kg. Using these values for M and K and $B = 0.5$ N·s/m, write the differential equation describing the system in numerical form.

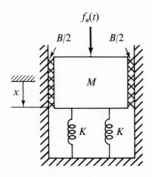

Figure P2.14

2.15 All the springs in Figure P2.15 are identical, each with spring constant K. The spring forces are zero when $x_1 = x_2 = x_3 = 0$. Draw the free-body diagrams, including the gravitational forces, and write the differential equations describing the system. Give the constant elongation of each spring caused by the gravitational forces when the masses are stationary.

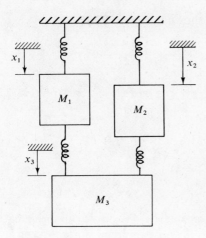

Figure P2.15

2.16 a. Repeat Problem 2.15 for the system shown in Figure P2.16.

b. Rewrite the system equations in terms of z_1 and z_2, the relative displacements of M_1 and M_2 measured with respect to their static-equilibrium positions.

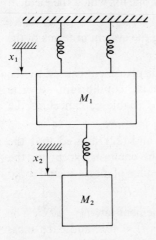

Figure P2.16

2.17 Repeat Problem 2.15 for the system shown in Figure P2.17.

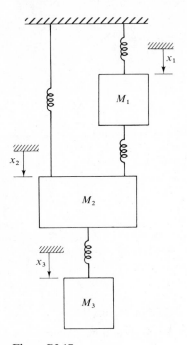

Figure P2.17

2.18 Replace the linear spring in Figure 2.15(a) by one for which the element law is $f_K = x^3$.

a. Write the differential equation describing the system in terms of the displacement x.

b. Let $x = x_0 + z$ where x_0 is the constant displacement caused by the gravitational force when the system is in static equilibrium. Rewrite the differential equation in terms of the variable z, canceling the gravitational term.

c. For a linear spring, comparison of (19) and (22) showed that the differential equation in z was the same as the one in x except for the deletion of the Mg term. Compare the results of parts a and b to determine whether that is the case in this problem.

2.19 The lower spring in Figure P2.16 has the element law $f_K = 10x^3$. The element law for each of the remaining springs is $f_K = 40x$, and each mass is 2 kg. When $x_1 = x_2 = 0$, the springs are neither stretched nor compressed.

a. Write the differential equations describing the system in terms of x_1 and x_2.

b. Find the constant elongation of each spring caused by the gravitational forces when the masses remain motionless.

c. Rewrite the system's differential equations in terms of z_1 and z_2, the relative displacements of the masses with respect to their static-equilibrium positions. Compare the results to the equations obtained in part a.

2.20 In Figure P2.20, a steel wire that moves over a pulley connects K_2 and M_2. The pulley is assumed to have no inertia and no friction. Any spring constant that is associated with the wire is included in K_2. When $x_1 = x_2 = 0$, the springs are neither stretched nor compressed.

a. Draw free-body diagrams for M_1 and M_2 and write the system equations in terms of x_1 and x_2.

b. Give expressions for x_{1_0} and x_{2_0}, the constant displacements caused by the gravitational forces when the masses are stationary and when $f_a(t) = 0$.

c. Rewrite and simplify the system equations in terms of the relative displacements z_1 and z_2, where $x_1 = x_{1_0} + z_1$ and $x_2 = x_{2_0} + z_2$.

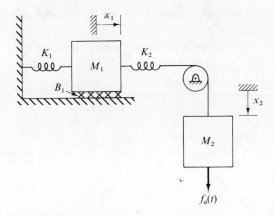

Figure P2.20

2.21 The mass M in Figure P2.21 moves on an inclined plane. The value of the viscous-friction parameter B depends on the component of the gravitational force on M that is perpendicular to the inclined surface, and hence on the angle θ. This is indicated in the diagram by the symbol $B(\theta)$.

Draw a free-body diagram, including the effect of the gravitational force, and write the differential equation describing the system in terms of x and the angle θ. To what does this equation reduce when (1) $\theta = 0$, and (2) $\theta = \pi/2$ rad?

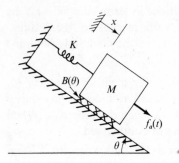

Figure P2.21

2.22 The three masses in Figure P2.22 are identical. The pulley is assumed to have no inertia and no friction. As in Problem 2.21, the values of the parameters B_1 and B_2 depend on the angle θ.

a. Draw free-body diagrams and write the equations describing the system. To what do the equations reduce when (1) $\theta = 0$, and (2) $\theta = \pi/2$ rad?

b. If the masses are stationary with no inputs applied except for the gravitational forces, for what value of θ will the spring K_1 be neither stretched nor compressed?

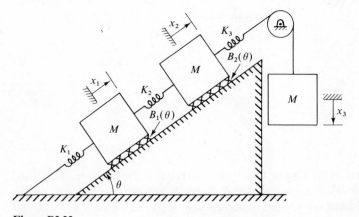

Figure P2.22

2.23 The two springs in Figure P2.23 are coupled by an ideal lever. For small values of the angular displacement θ from the vertical position, the translational displacements at the top and bottom of the lever are $a\theta$ and $b\theta$, respectively.

a. Write an expression in terms of the quantities labeled on the diagram for the forces exerted on the top and bottom of the lever. Use (13) to relate the two forces to each other.

b. Use the results of part a to express θ as an algebraic function of x_1 and x_2.

c. Draw free-body diagrams for M_1 and M_2 and write the corresponding equations. Write the system model as a pair of coupled differential equations involving only the variables x_1, x_2, and $f_a(t)$.

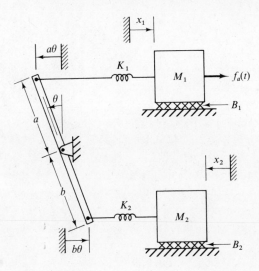

Figure P2.23

CHAPTER 3
STANDARD FORMS FOR SYSTEM MODELS

In Chapter 2, we introduced the element and interconnection laws for translational systems and the procedure for drawing free-body diagrams and applying D'Alembert's law. These are essential steps regardless of the final form of the equations.

Not only should the complete mathematical model contain as many independent equations as unknown variables, it should also have a form that is convenient for its solution. The two most common forms for the model are discussed, illustrated, and compared in this chapter. Although translational mechanical systems are used as examples, the same techniques are applied in subsequent chapters to other types of systems.

3.1 STATE-VARIABLE EQUATIONS

One procedure for formulating the system model is to begin by selecting a set of *state variables*. This set of variables must completely describe the effect of the past history of the system on its response in the future. Then knowing the values of the state variables at a reference time t_0 and the values of the inputs for all $t \geq t_0$ is sufficient for evaluating the state variables and outputs for all $t \geq t_0$. It is further understood that the state variables must be independent; that is, it must be impossible to express any state variable as an algebraic function of the remaining state variables and the inputs.

Although the choice of state variables is not unique, the state variables for mechanical systems are usually related to the energy stored in each of the system's energy-storing elements. Since any energy initially stored in these elements can affect the response of the system at a later time, one state variable will normally be associated with each of the energy-storing elements. Generally, this will adequately summarize the effect of the past history of the system. In some systems, the number of state variables is different from the number of energy-storing elements because a particular interconnection of elements causes redundant variables or because there is need for a state variable not related to the storage of energy.

After making an appropriate choice of state variables, we must find a set of first-order differential equations having the form

$$
\begin{aligned}
\dot{q}_1 &= f_1(q_1, q_2, ..., q_n, u_1, u_2, ..., u_m, t) \\
\dot{q}_2 &= f_2(q_1, q_2, ..., q_n, u_1, u_2, ..., u_m, t) \\
&\ \vdots \\
\dot{q}_n &= f_n(q_1, q_2, ..., q_n, u_1, u_2, ..., u_m, t)
\end{aligned}
\tag{1}
$$

where the q_i are the state variables, the $u_j(t)$ are the inputs, t denotes time, and the f_i are algebraic functions.* Thus we must express the derivative of each of the n state variables as an algebraic function whose arguments can be only the state variables, the inputs, and possibly time.

The outputs can be expressed as algebraic functions whose arguments are the same as in (1). For a system with p outputs, these equations will have the general form

$$
\begin{aligned}
y_1 &= g_1(q_1, q_2, ..., q_n, u_1, u_2, ..., u_m, t) \\
y_2 &= g_2(q_1, q_2, ..., q_n, u_1, u_2, ..., u_m, t) \\
&\ \vdots \\
y_p &= g_p(q_1, q_2, ..., q_n, u_1, u_2, ..., u_m, t)
\end{aligned}
\tag{2}
$$

An output variable may be identical to a state variable, so that (2) may

---

* Note that the use of the symbol f in this very general context is not related to the use of f to denote forces in mechanical systems.

contain one or more equations such as $y_1 = q_1$. When all the outputs are identical to some of the state variables, or when the outputs of interest are not specified, (2) is often omitted.

In this chapter, we shall deal almost entirely with linear systems. Then (1) and (2) reduce to

$$\dot{q}_1 = a_{11} q_1 + a_{12} q_2 + \cdots + a_{1n} q_n + b_{11} u_1 + b_{12} u_2 + \cdots + b_{1m} u_m$$
$$\dot{q}_2 = a_{21} q_1 + a_{22} q_2 + \cdots + a_{2n} q_n + b_{21} u_1 + b_{22} u_2 + \cdots + b_{2m} u_m$$
$$\vdots$$
$$\dot{q}_n = a_{n1} q_1 + a_{n2} q_2 + \cdots + a_{nn} q_n + b_{n1} u_1 + b_{n2} u_2 + \cdots + b_{nm} u_m$$

(3)

and

$$y_1 = c_{11} q_1 + c_{12} q_2 + \cdots + c_{1n} q_n + d_{11} u_1 + d_{12} u_2 + \cdots + d_{1m} u_m$$
$$y_2 = c_{21} q_1 + c_{22} q_2 + \cdots + c_{2n} q_n + d_{21} u_1 + d_{22} u_2 + \cdots + d_{2m} u_m$$
$$\vdots$$
$$y_p = c_{p1} q_1 + c_{p2} q_2 + \cdots + c_{pn} q_n + d_{p1} u_1 + d_{p2} u_2 + \cdots + d_{pm} u_m$$

(4)

For time-varying linear systems, some of the coefficients in (3) and (4) will be functions of time, but for fixed linear systems all the coefficients will be constants.

Solving (1) or (3) for $t \geq t_0$ requires knowledge of the inputs for $t \geq t_0$ and also of the initial values of the state variables, namely $q_1(t_0), \ldots, q_n(t_0)$. In general, we cannot solve the individual state-variable equations separately but must solve them as a group. For example, the equation $\dot{q}_1 = -q_1 + q_2$ cannot be solved for q_1 unless another equation exists that can be solved for q_2.

Since the energy stored in translational mechanical systems must be associated with the masses or springs, it is logical to consider as possible state variables the velocities of the masses and the elongations of the springs. In most problems, we can express the elongations of the springs in terms of the displacements of the masses, or vice versa.

In the remainder of this section, a variety of examples will illustrate the technique of deriving the mathematical model in state-variable form. The general approach is as follows:

1. Identify the state variables and write those state-variable equations that do not require a free-body diagram, e.g., equations of the form $\dot{x} = v$.

2. Draw free-body diagrams for each independent mass and junction point that can move with an unknown motion. Sum the forces on each free-body diagram separately to obtain a set of differential equations.

3. Manipulate the equations into state-variable form. For each of the state variables, there must be an equation that expresses its derivative as an algebraic function of the state variables, inputs, and possibly time.

4. When necessary, express the output variables as algebraic functions of the state variables, inputs, and possibly time.

EXAMPLE 3.1   Find the state-variable equations for the system shown in Figure 3.1(a), which is identical to Figure 2.13(a).

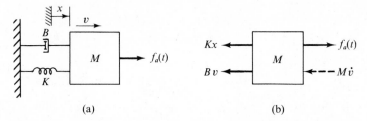

(a)   (b)

Figure 3.1 (a) Translational system for Example 3.1. (b) Free-body diagram.

Solution   We choose as state variables the elongation x of the spring, which is related to its potential energy, and the velocity v of the mass, which is related to its kinetic energy. With this choice, we can write the first of the two state-variable equations by inspection, namely

$$\dot{x} = v$$

Summing the forces on the free-body diagram, which is shown in Figure 3.1(b), gives

$$M\dot{v} + Bv + Kx = f_a(t)$$

Solving this equation for $\dot{v}$ results in the second state-variable equation, where the right side is a function of only the state variables x and v and the input $f_a(t)$, as required. Thus the state-variable equations are

$$\dot{x} = v$$

$$\dot{v} = \frac{1}{M}[-Kx - Bv + f_a(t)]$$

EXAMPLE 3.2   Find the state-variable equations for the system shown in Figure 3.2(a), which is identical to Figure 2.14(a). Then modify the equations for the case where the linear stiffness and friction elements characterized by K_1 and B are replaced by nonlinear elements.

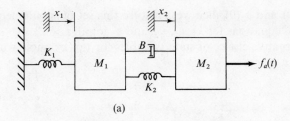

(a)

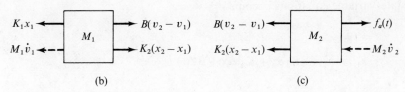

(b)   (c)

Figure 3.2 (a) Translational system for Example 3.2. (b), (c) Free-body diagrams.

Solution An appropriate choice of state variables is x_1, v_1, x_2, and v_2, since we can express the velocity of each mass and the elongation of each spring in terms of these four variables and since none of these variables can be expressed in terms of the other three. Since $\dot{x}_1 = v_1$ and $\dot{x}_2 = v_2$, two of the four state-variable equations are available immediately.

The free-body diagrams for the two masses are shown in Figure 3.2(b) and Figure 3.2(c), with all forces labeled in terms of the state variables and the input. By D'Alembert's law,

$$M_1 \dot{v}_1 + K_1 x_1 - K_2(x_2 - x_1) - B(v_2 - v_1) = 0$$

$$M_2 \dot{v}_2 + K_2(x_2 - x_1) + B(v_2 - v_1) = f_a(t)$$

which may be solved for $\dot{v}_1$ and $\dot{v}_2$, respectively. The state-variable equations are

$$\dot{x}_1 = v_1$$

$$\dot{v}_1 = \frac{1}{M_1}\left[-Bv_1 - (K_1 + K_2)x_1 + Bv_2 + K_2 x_2\right]$$

$$\dot{x}_2 = v_2$$

$$\dot{v}_2 = \frac{1}{M_2}\left[Bv_1 + K_2 x_1 - Bv_2 - K_2 x_2 + f_a(t)\right]$$

If we know the element values, the input $f_a(t)$ for $t \geq 0$, and also $x_1(0)$,

$v_1(0)$, $x_2(0)$, and $v_2(0)$, then we can solve this set of simultaneous first-order differential equations for x_1, v_1, x_2, and v_2 for all $t > 0$.

An alternative choice of state variables for this example would be x_1, v_1, Δx, and Δv, where

$$\Delta x = x_2 - x_1$$

$$\Delta v = v_2 - v_1$$

The quantities Δx and Δv represent the elongation of the spring K_2 and the velocity difference associated with the dashpot, respectively. For this choice, the state-variable equations become

$$\dot{x}_1 = v_1$$

$$\dot{v}_1 = \frac{1}{M_1}(-K_1 x_1 + K_2 \Delta x + B \Delta v)$$

$$\dot{\Delta x} = \Delta v$$

$$\dot{\Delta v} = \frac{K_1}{M_1} x_1 - \left(\frac{M_1 + M_2}{M_1 M_2}\right)(K_2 \Delta x + B \Delta v) + \frac{1}{M_2} f_a(t)$$

The forces exerted by the linear spring K_1 and the linear dashpot are $K_1 x_1$ and $B \Delta v$, respectively. Now assume that these two elements are replaced by nonlinear elements. The expression relating the spring force f_{K_1} to the displacement x_1 is denoted by $f_{K_1}(x_1)$, while the force on the dashpot is denoted by $f_B(\Delta v)$. The revised free-body diagrams are shown in Figure 3.3, and the resulting state-variable equations are

$$\dot{x}_1 = v_1$$

$$\dot{v}_1 = \frac{1}{M}[-f_{K_1}(x_1) + K_2 \Delta x + f_B(\Delta v)]$$

$$\dot{\Delta x} = \Delta v$$

$$\dot{\Delta v} = \frac{1}{M_1} f_{K_1}(x_1) - \left(\frac{M_1 + M_2}{M_1 M_2}\right)[K_2 \Delta x + f_B(\Delta v)] + \frac{1}{M_2} f_a(t)$$

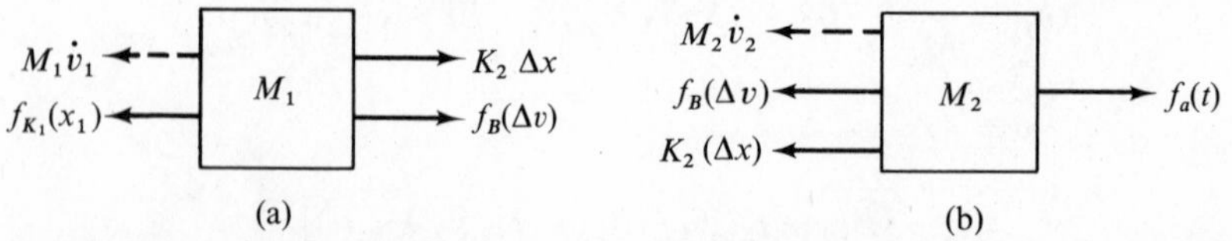

Figure 3.3 Free-body diagrams for Example 3.2 when K_1 and B are replaced by nonlinear elements.

which are identical to the previous set of state-variable equations except for the contributions of the two nonlinear elements. Although the free-body diagrams and the state-variable equations are no more complex than those for the linear case, solving the differential equations analytically would be much more difficult, if not impossible.

EXAMPLE 3.3   Write the state-variable equations for the system shown in Figure 3.4(a), which is identical to Figure 2.16(a). Then repeat the example for the two systems shown in Figure 3.5.

Solution   Let x denote the displacement of the mass from the static-equilibrium position corresponding to $f_a(t) = 0$, $v = 0$, and $\dot{v} = 0$, as discussed in Example 2.4. The free-body diagram is shown in Figure 3.4(b), and the state-variable equations are

$$\dot{x} = v$$

$$\dot{v} = \frac{1}{M}[-(K_1 + K_2)x - Bv + f_a(t)]$$

(5)

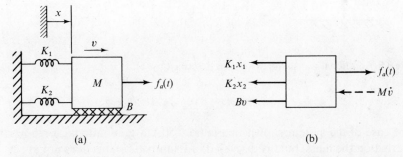

(a)   (b)

Figure 3.4 (a) Translational system with fewer state variables than energy-storing elements. (b) Free-body diagram.

Note that in this example we need only the two state variables x and v, even though there are three energy-storing elements. This is because the displacement x is sufficient to determine the energy stored in each of the springs. Since the elongations of the two springs are not independent of one another, as required by the definition of state variables, they cannot both be state variables.

Each of the systems shown in Figure 3.5 is also described by the state-variable equations in (5). Again there are three energy-storing elements, but again the displacements associated with K_1 and K_2 are not independent.

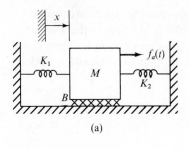

(a)

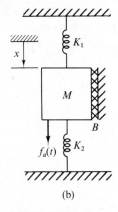

(b)

Figure 3.5 Additional translational systems described by (5).

In the case of the system shown in Figure 3.5(b), a gravitational force Mg is exerted on the mass, but (as discussed in Example 2.3) this does not affect the form of the equations.

EXAMPLE 3.4   The system shown in Figure 3.6(a) contains a massless junction that can move with a velocity different from and not directly proportional to the velocity of the mass M. If $f_a(t)$ denotes the force input, find the state-variable equations describing the system.

Solution   A satisfactory choice of state variables is x_1, v_1, and x_2, since these three variables determine the elongations of the springs and the velocity of the mass. One of the three state-variable equations is $\dot{x}_1 = v_1$.

To obtain the other two equations, we draw free-body diagrams for both the mass and the junction point, as shown in Figure 3.6(b) and Figure 3.6(c). Because the point A is massless, no inertial force is present in its free-body

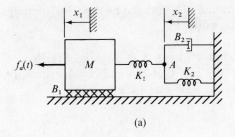

(a)

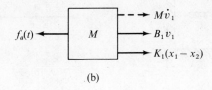

(b)

$$K_1(x_1 - x_2) \longleftarrow \overset{A}{\bullet} \longrightarrow B_2 v_2 + K_2 x_2$$

(c)

Figure 3.6 (a) Translational system containing a massless junction. (b), (c) Free-body diagrams.

diagram. Summing the forces shown in these diagrams gives

$$M\dot{v}_1 + B_1 v_1 + K_1(x_1 - x_2) = f_a(t) \tag{6a}$$

$$B_2 \dot{x}_2 + K_2 x_2 + K_1(x_2 - x_1) = 0 \tag{6b}$$

Solving (6) for $\dot{v}_1$ and $\dot{x}_2$, we arrive at the state-variable equations

$$\dot{x}_1 = v_1 \tag{7a}$$

$$\dot{v}_1 = \frac{1}{M}[-K_1 x_1 - B_1 v_1 + K_1 x_2 + f_a(t)] \tag{7b}$$

$$\dot{x}_2 = \frac{1}{B_2}[K_1 x_1 - (K_1 + K_2)x_2] \tag{7c}$$

If the dashpot is removed from the system, the motion of the massless junction denoted as point A will be directly proportional to that of the mass, and the system will have only two state variables. Setting $B_2 = 0$ in (7c), however, which corresponds to removing the dashpot, leads to division by zero, an invalid mathematical operation. The proper method for treating this case is discussed in the following example.

EXAMPLE 3.5   Show that the motion of point A in Figure 3.7(a) is not independent of that of the mass M, but that x_1 and x_2 are directly proportional to one another. Derive the state-variable equations.

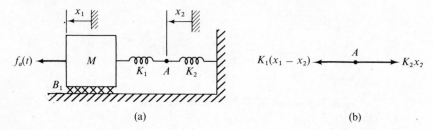

(a)   (b)

Figure 3.7 (a) Translational system with fewer state variables than energy-storing elements. (b) Free-body diagram for the massless point.

Solution   The free-body diagram for the mass M remains the same as shown in Figure 3.6(b). However, the diagram for the massless point reduces to that shown in Figure 3.7(b), which in turn leads to the algebraic equation

$$K_2 x_2 + K_1(x_2 - x_1) = 0$$

Solving this equation for x_2 in terms of x_1 gives

$$x_2 = \left(\frac{K_1}{K_1 + K_2}\right) x_1$$

Thus x_1 and x_2 are proportional to one another and cannot both be state variables. If we choose x_1 and v_1 as the state variables, we know that $\dot{x}_1 = v_1$. We can find a suitable expression for $\dot{v}_1$ by substituting the foregoing expression for x_2 into (6a). The resulting state-variable model is

$$\dot{x}_1 = v_1$$

$$\dot{v}_1 = \frac{1}{M}\left[-\left(\frac{K_1 K_2}{K_1 + K_2}\right) x_1 - B_1 v_1 + f_a(t)\right]$$

(8)

As required, the variables appearing on the right sides of these equations are either state variables or the input.

SERIES AND PARALLEL COMBINATIONS

The state-variable equations for the system shown in Figure 3.7(a) were given by (8), which we can rewrite as

$$\dot{x}_1 = v_1$$

$$\dot{v}_1 = \frac{1}{M}[-K_{eq}x_1 - B_1 v_1 + f_a(t)]$$

where

$$K_{eq} = \frac{K_1 K_2}{K_1 + K_2} \tag{9}$$

Thus the state-variable equations are the same as for a system in which the two springs are combined into a single element whose spring constant is given by (9).

Two springs or dashpots are said to be in *series* if they are connected at only one end of each element and if there is no other element connected to their common junction, as shown in Figure 3.7(a) and Figure 3.8(a).

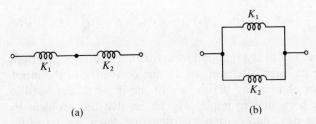

$$K_1 \qquad K_2$$

$$(a) \qquad\qquad\qquad (b)$$

Figure 3.8 Combinations of springs. (a) Series connection. (b) Parallel connection.

Springs or dashpots are connected in *parallel* if respective pairs of ends are joined so that the elements have the same elongation, as shown in Figure 3.8(b). From Example 2.4, the equivalent spring constant for the two linear springs in parallel is $K_1 + K_2$.

EXAMPLE 3.6   An external force is applied to the mass M_1 shown in Figure 3.9(a) such that the displacement $x_1(t)$ of the mass is a prescribed function of time. Derive the mathematical model for the system, treating $x_1(t)$ as the input.

Solution   Since we know the motion of M_1 in advance, there is no need to draw a free-body diagram for M_1 (unless we wish to determine the external

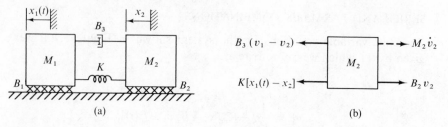

(a)   (b)

Figure 3.9 (a) Translational system with displacement input. (b) Free-body diagram.

force that must be applied to M_1 to achieve the prescribed displacement). The free-body diagram for M_2 is given in Figure 3.9(b), and the corresponding force equation is

$$M_2 \dot{v}_2 + (B_2 + B_3) v_2 + K x_2 = B_3 \dot{x}_1 + K x_1(t)$$

We may choose x_2 and v_2 as the state variables, write $\dot{x}_2 = v_2$, and solve the force equation for $\dot{v}_2$, getting

$$\dot{x}_2 = v_2 \tag{10a}$$

$$\dot{v}_2 = \frac{1}{M_2} \left[-K x_2 - (B_2 + B_3) v_2 + B_3 \dot{x}_1 + K x_1(t) \right] \tag{10b}$$

Equation (10b) does not fit the required form for state-variable equations given by (3) because its right side contains $\dot{x}_1$, the derivative of the input. A general method for eliminating derivatives of the input is discussed in Chapter 6. For this particular example, we can resolve the problem by replacing v_2 with the new state variable q such that the only derivative appearing in (10b) is $\dot{q}$. To determine how q must be defined, we rewrite (10b) with the derivative of the input moved to the left side, getting

$$\dot{v}_2 - \frac{B_3}{M_2} \dot{x}_1 = \frac{1}{M_2} \left[-K x_2 - (B_2 + B_3) v_2 + K x_1(t) \right]$$

If we select the new state variable q as

$$q = v_2 - \frac{B_3}{M_2} x_1(t)$$

then the old state variable v_2 is given by

$$v_2 = q + \frac{B_3}{M_2} x_1(t) \tag{11}$$

We can rewrite (10) in terms of x_2 and q as

$$\dot{x}_2 = q + \frac{B_3}{M_2}x_1(t)$$

$$\dot{q} = \frac{1}{M_2}\left\{-Kx_2 - (B_2 + B_3)q + \left[K - \frac{B_3}{M_2}(B_2 + B_3)\right]x_1(t)\right\}$$

which is in state-variable form. If the output of the system is v_2, then (11) is the output equation and fits the standard form described in (4).

EXAMPLE 3.7   Find the state-variable equations for the system shown in Figure 3.10(a). Also find the output equation when the output is defined to

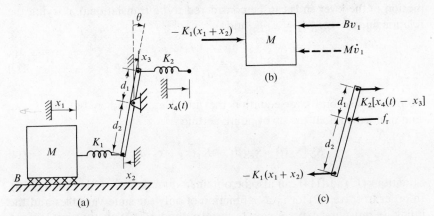

Figure 3.10 (a) Translational system containing a lever. (b), (c) Free-body diagrams.

be the force exerted on the pivot by the lever. The input is the displacement $x_4(t)$ of the right end of the spring K_2 and affects the mass M through the lever. The lever has a fixed pivot and is assumed to be massless yet rigid. Its angular rotation θ is so small that only horizontal motion need be considered. In a practical situation, the springs K_1 and K_2 might represent the stiffness of the lever and of associated linkages that have a certain degree of flexibility.

Solution   The displacements x_2 and x_3 are directly proportional to the angle θ and hence to one another. Furthermore, the two springs appear to form a series combination similar to the one shown in Figure 3.8(a), since the lever has no mass. Hence we can express x_2, x_3, and θ as algebraic functions of x_1 and $x_4(t)$. Thus we will select only x_1 and v_1 as state variables, with $x_4(t)$ being the input. By inspection, we determine that one of the required state-variable equations is $\dot{x}_1 = v_1$.

The next step is to draw free-body diagrams for the mass M and the lever, as shown in Figure 3.10(b) and Figure 3.10(c). We must pay particular attention to the signs of the force arrows and to the expressions for the elongations of the springs. For example, the elongation of spring K_1 is $-(x_1 + x_2)$ because of the manner in which the displacements have been defined. Summing the forces on the mass M yields

$$M\dot{v}_1 + Bv_1 + K_1(x_1 + x_2) = 0 \tag{12}$$

The forces on the lever are those exerted by the springs and the reaction force f_r of the pivot. Because the lever's angle of rotation θ is small, the motion of the lever ends can be considered to be translational, obeying the relationships $\theta = x_2/d_2 = x_3/d_1$ and

$$x_3 = \frac{d_1}{d_2} x_2 \tag{13}$$

To obtain a second lever equation that involves x_1 and x_2 but not f_r, we sum moments about the pivot point, getting

$$K_2[x_4(t) - x_3]d_1 - K_1(x_1 + x_2)d_2 = 0 \tag{14}$$

Equations (13) and (14) can also be obtained directly from (2.11)* and (2.13). In order to solve (12) for $\dot{v}_1$ as a function of only the state variables and the input, we must first express x_2 in terms of x_1 and $x_4(t)$. Substituting (13) into (14) and solving for x_2, we obtain the algebraic expression

$$x_2 = \frac{(d_1/d_2) K_2 x_4(t) - K_1 x_1}{K_1 + (d_1/d_2)^2 K_2} \tag{15}$$

We find the second of the two state-variable equations by substituting (15) into (12) and rearranging terms so as to solve for $\dot{v}_1$. Doing this, we find that the state-variable model is

$$\dot{x}_1 = v_1$$

$$\dot{v}_1 = -\frac{1}{M}\left[Bv_1 + \alpha K_1 x_1 + \alpha\left(\frac{d_2}{d_1}\right) K_1 x_4(t) \right] \tag{16}$$

---

* This denotes Equation (11) in Chapter 2.

where

$$\alpha = \frac{1}{1 + \dfrac{K_1}{K_2}\left(\dfrac{d_2}{d_1}\right)^2} \tag{17}$$

To develop the output equation expressing f_r as an algebraic function of the state variables and the input, we first sum the forces on the lever in Figure 3.10(c) to obtain

$$f_r = K_2[x_4(t) - x_3] + K_1(x_1 + x_2) \tag{18}$$

Substituting (13) and (15) into (18) gives

$$f_r = \left[K_1 + \left(K_2 - \frac{d_2}{d_1}K_1\right)\left(\frac{d_2 K_1}{\alpha d_1 K_2}\right)\right]x_1 + \left[K_2 + \frac{\left(\dfrac{d_2}{d_1}\right)K_1 - K_2}{\alpha}\right]x_4(t)$$

which has the desired form.

It is instructive to consider the special case where $d_1 = d_2$, which corresponds to having the lever pivoted at its midpoint. From (17), $\alpha K_1 = K_1 K_2/(K_1 + K_2)$, which is the equivalent spring constant for a series connection of the two springs, as in (9). Furthermore, the reader can readily verify that (16) reduces to the equations describing the system shown in Figure 3.11(a), in which the massless junction A replaces the lever. In turn,

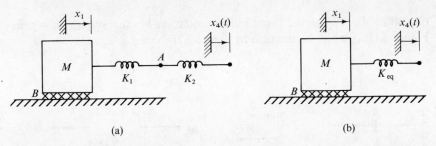

(a)   (b)

Figure 3.11 Systems equivalent to Figure 3.10(a) when $d_1 = d_2$.

this system is equivalent to that shown in Figure 3.11(b), where a single spring with the coefficient $K_{eq} = K_1 K_2/(K_1 + K_2)$ replaces the series spring connection.

3.2 INPUT-OUTPUT EQUATIONS

In this section, we develop the system models in the form of input-output differential equations by eliminating all variables except the inputs and outputs and their derivatives. For a system with one input $u(t)$ and one output y, the input-output equation has the general form

$$a_n y^{(n)} + \cdots + a_2 \ddot{y} + a_1 \dot{y} + a_0 y = b_m u^{(m)} + \cdots + b_1 \dot{u} + b_0 u(t) \quad (19)$$

where $y^{(n)} = d^n y/dt^n$ and $u^{(m)} = d^m u/dt^m$, and where for systems of practical interest $m \leq n$. For fixed linear systems, all the coefficients in (19) are constants. In order to solve such an equation for y for all $t \geq t_0$, we need to know not only the input $u(t)$ for $t \geq t_0$ but also the n initial conditions $y(t_0)$, $\dot{y}(t_0)$, $\ddot{y}(t_0)$, ..., $y^{(n-1)}(t_0)$. Finding these initial conditions may be a difficult task.

For systems with more than one input, the right side of (19) will include additional input terms. If there are several outputs, we need a separate equation similar to (19) for each output. For example, the pair of equations

$$\ddot{y}_1 + 2\dot{y}_1 + 2y_1 = 3\dot{u}_1 + 2u_1(t) + u_2(t) + 3\dot{u}_3$$

$$\ddot{y}_2 + 2\dot{y}_2 + 2y_2 = u_1(t) + 2u_2(t) + u_3(t)$$

corresponds to a system with two outputs and three inputs. In the general case, each of the input-output equations involves only one unknown variable and its derivatives. Thus, unlike state-variable equations, each equation can be solved independently of the others.

EXAMPLE 3.8 Write the input-output equation for the system shown in Figure 3.12(a), which is identical to Figure 3.1(a).

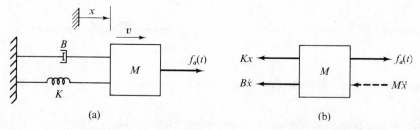

(a)   (b)

Figure 3.12 (a) Translational system for Example 3.8. (b) Free-body diagram.

Solution   In Example 3.1, the state-variable equations were found to be

$$\dot{x} = v$$

$$\dot{v} = \frac{1}{M}[-Kx - Bv + f_a(t)]$$

We can combine these equations by replacing v and $\dot{v}$ in the second equation by $\dot{x}$ and $\ddot{x}$, respectively, to obtain

$$M\ddot{x} + B\dot{x} + Kx = f_a(t)$$

The direct approach, which does not make use of the state-variable model, would be to label all the forces on the free-body diagram except $f_a(t)$ in terms of x and its derivatives. This is done in Figure 3.12(b). Then the input-output equation we have just derived follows directly from the diagram.

EXAMPLE 3.9   Find the input-output equation relating the output x_1 to the input $f_a(t)$ for the system shown in Figure 3.13(a), which is the same as that shown in Figure 3.6(a).

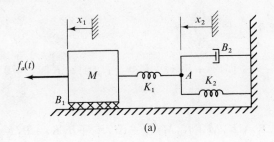

(a)

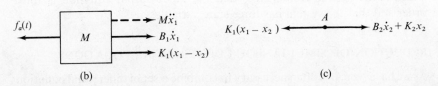

(b)   (c)

Figure 3.13 (a) Translational system containing a massless junction. (b), (c) Free-body diagrams.

Solution The free-body diagrams shown in Figure 3.13(b) and Figure 3.13(c) are labeled in terms of the displacements and their derivatives. By D'Alembert's law,

$$M\ddot{x}_1 + B_1\dot{x}_1 + K_1 x_1 - K_1 x_2 = f_a(t) \tag{20a}$$

$$B_2\dot{x}_2 + (K_1 + K_2)x_2 - K_1 x_1 = 0 \tag{20b}$$

To obtain a single differential equation relating x_1 to $f_a(t)$ from this pair of coupled equations, we must eliminate x_2 and $\dot{x}_2$. If one of the equations contains an unwanted variable but none of its derivatives, we can solve for it in terms of the remaining variables and their derivatives. Then we can eliminate the unwanted variable from the model by substitution. Thus we rewrite (20a) as

$$x_2 = \frac{1}{K_1}[M\ddot{x}_1 + B_1\dot{x}_1 + K_1 x_1 - f_a(t)]$$

and, by differentiating once, we obtain

$$\dot{x}_2 = \frac{1}{K_1}(M\dddot{x}_1 + B_1\ddot{x}_1 + K_1\dot{x}_1 - \dot{f}_a)$$

Substituting these expressions for x_2 and $\dot{x}_2$ into (20b) gives

$$\frac{B_2}{K_1}(M\dddot{x}_1 + B_1\ddot{x}_1 + K_1\dot{x}_1 - \dot{f}_a)$$

$$+ \frac{K_1 + K_2}{K_1}[M\ddot{x}_1 + B_1\dot{x}_1 + K_1 x_1 - f_a(t)] - K_1 x_1 = 0$$

or

$$MB_2\dddot{x}_1 + (B_1 B_2 + K_1 M + K_2 M)\ddot{x}_1 + (B_2 K_1 + B_1 K_1 + B_1 K_2)\dot{x}_1$$

$$+ K_1 K_2 x_1 = B_2\dot{f}_a + (K_1 + K_2)f_a(t) \tag{21}$$

which is the desired result. Note that the input-output equation is third-order and that the system has three state variables.

REDUCTION OF SIMULTANEOUS DIFFERENTIAL EQUATIONS

As we have seen, it is often necessary to combine a set of differential equations involving more than one dependent variable into a single differential equation with a single dependent variable. We did this in Example 3.9 by straight-forward substitution. However, when it is not obvious how to eliminate the unwanted variable easily, the following procedure is recommended (see Wylie).*

Let p denote the differentiation operator d/dt such that $py = \dot{y}$, $p^2 y = \ddot{y}$, etc. Then, for example,

$$(p+2)y = \dot{y} + 2y$$

$$[(p+1)(p+2)]y = (p^2 + 3p + 2)y$$

$$= \ddot{y} + 3\dot{y} + 2y$$

$$(a_n p^n + \cdots + a_2 p^2 + a_1 p + a_0)y = a_n y^{(n)} + \cdots + a_2 \ddot{y} + a_1 \dot{y} + a_0 y$$

where $y^{(k)} = d^k y/dt^k$ for any positive integer value of k. Remember that p must operate on the variable or expression that follows it and that it is not a variable or algebraic quantity itself.

Suppose that we have the pair of equations

$$\dot{y}_1 + 2y_1 + y_2 = 3u(t) \tag{22}$$

$$2\dot{y}_1 + 5y_1 - 2\dot{y}_2 + 2y_2 = 0$$

and want to find a single differential equation involving only the variables y_2 and $u(t)$. In terms of the p operator, we can rewrite (22) as

$$(p+2)y_1 + y_2 = 3u(t)$$

$$(2p+5)y_1 + (-2p+2)y_2 = 0$$

If we treat these equations as if they were algebraic equations, even though they are not, we can eliminate y_1 by solving the first equation algebraically for y_1 and substituting the result into the second equation. If we do this, the result is

$$y_2 = \left(\frac{6p+15}{2p^2 + 4p + 1} \right) u(t)$$

This equation as it stands does not make sense in terms of our definition of the p operator, so we rewrite it as

$$(2p^2 + 4p + 1)y_2 = (6p + 15)u(t) \tag{23}$$

To return to a differential equation relating y_2 and $u(t)$, we observe that (23) is the operator form of

$$2\ddot{y}_2 + 4\dot{y}_2 + y_2 = 6\dot{u} + 15u(t)$$

* Complete references are given in the Reference section at the back of the book.

which is indeed the correct input-output equation. This algebraic procedure provides a useful means of manipulating sets of differential equations with constant coefficients. The reader may consult Wylie or Section 14.2 of Desoer and Kuh for further discussion and examples.

EXAMPLE 3.10   For the system in Figure 3.14(a), which is identical to Figure 3.2(a), find the input-output equation relating x_1 and $f_a(t)$.

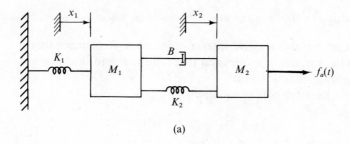

(a)

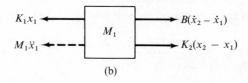

(b)

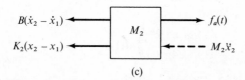

(c)

Figure 3.14 (a) Translational system for Example 3.10. (b), (c) Free-body diagrams.

Solution   From the free-body diagrams in Figure 3.14(b) and Figure 3.14(c), we have the pair of simultaneous second-order differential equations

$$M_1 \ddot{x}_1 + B\dot{x}_1 + (K_1 + K_2)x_1 - B\dot{x}_2 - K_2 x_2 = 0$$

$$-B\dot{x}_1 - K_2 x_1 + M_2 \ddot{x}_2 + B\dot{x}_2 + K_2 x_2 = f_a(t)$$

In terms of the p operator, these equations become

$$[p^2 M_1 + pB + (K_1 + K_2)] x_1 - (pB + K_2) x_2 = 0$$
$$-(pB + K_2) x_1 + (p^2 M_2 + pB + K_2) x_2 = f_a(t)$$

If we combine this pair of operator equations algebraically to eliminate x_2, we find that

$$\{M_1 M_2 p^4 + (M_1 + M_2) B p^3 + [M_1 K_2 + M_2(K_1 + K_2)] p^2 + BK_1 p$$
$$+ K_1 K_2\} x_1 = (pB + K_2) f_a(t)$$

which is the operator form of the differential equation

$$M_1 M_2 x_1^{(iv)} + (M_1 + M_2) B x_1^{(iii)} + [M_1 K_2 + M_2(K_1 + K_2)] \ddot{x}_1 + BK_1 \dot{x}_1$$
$$+ K_1 K_2 x_1 = B\dot{f}_a + K_2 f_a(t)$$

The symbols $x^{(iv)}$ and $x^{(iii)}$ are defined as $x^{(iv)} = d^4 x/dt^4$ and $x^{(iii)} = d^3 x/dt^3$. As expected for a system that has four state variables, the input-output differential equation is of order four.

COMPARISON WITH THE STATE-VARIABLE METHOD

The order of the input-output differential equation describing a system is usually the same as the number of state variables. When one or more of the state variables have no effect on the output, the order of the input-output equation is less than the number of state variables.

For a first-order system, both forms of the system model involve a single first-order differential equation and are essentially identical. For higher-order systems they are quite different. We must solve a set of n first-order differential equations in state-variable form as a group, and we must know the initial value of each state variable to solve the set of n equations. An input-output equation of order n contains only one dependent variable, but we need to know the initial values of that variable and its first $n-1$ derivatives. In practice, finding the input-output equation and the associated initial conditions may require more effort than finding the information needed for a state-variable solution.

Using the state-variable equations has significant computational advantages when a computer solution is to be found, as is discussed in Chapter 7. In fact, standard methods for solving a high-order, possibly nonlinear input-output differential equation on a computer usually require decomposition into a set of simultaneous first-order equations anyway. The analytical solution of input-output equations is considered in Chapter 8.

State-variable equations are particularly convenient for complex multi-input, multi-output systems. They are often written in matrix form and, in addition to their computational advantages, they can be used to obtain considerable insight into system behavior. The state-variable concept has formed the basis for many of the recent theoretical developments in system analysis. A number of these topics are introduced in Chapter 17.

PROBLEMS

3.1 a. Find a set of state-variable equations for the system shown in Figure P2.5.

b. Find the input-output differential equation relating x_2 to $f_a(t)$.

3.2 Repeat Problem 3.1 for the system shown in Figure P2.6.

3.3 Repeat Problem 3.1 for the system shown in Figure P2.7.

3.4 a. Find a set of state-variable equations for the system shown in Figure P2.9.

b. Find the input-output differential equation relating x_2 to $f_a(t)$ and to the gravitational forces.

c. Find the input-output differential equation relating the elongation of K_3 to $f_a(t)$ and to the gravitational forces.

d. Write the input-output differential equation when the output is z_2, the relative displacement of M_2 from its static-equilibrium position.

3.5 a. Find a set of state-variable equations for the system shown in Figure P2.10(a).

b. Find the input-output differential equation when the output is x_1.

c. Repeat part b when the output is x_3.

3.6 a. Find a set of state-variable equations for the system shown in Figure P2.10(b).

b. Find the input-output differential equation when the output is $v_3 = \dot{x}_3$.

c. Repeat part b when the output is x_3.

3.7 a. Find a set of state-variable equations for the system shown in Figure P2.11 when the input is the displacement $x_3(t)$.

b. Find the input-output differential equation when the output is x_1.

3.8 Repeat Problem 3.7 when the input is a force $f_a(t)$ applied to point B. The positive sense of $f_a(t)$ is to the right.

3.9 a. Find a set of state-variable equations for the system shown in Figure P2.13 when the inputs are $x_3(t)$ and $f_a(t)$.

b. Find the input-output differential equation relating x_1 to $x_3(t)$ and $f_a(t)$.

3.10 Find a set of state-variable equations for the system shown in Figure P2.15.

3.11 a. Find a set of state-variable equations for the system shown in Figure P2.16.

b. Find the input-output differential equation relating x_2 to the gravitational forces, where for $x_1 = x_2 = 0$ the springs are neither stretched nor compressed.

c. Write the input-output differential equation when the output is z_2, the relative displacement of M_2 with respect to its static-equilibrium position.

3.12 Find a set of state-variable equations for the system shown in Figure P2.17.

3.13 a. Find a set of state-variable equations for the system shown in Figure P2.20.

b. Find the input-output equation when the output is x_1.

3.14 Repeat Problem 3.13 for the system shown in Figure P2.22, where the inputs are the gravitational forces.

3.15 a. Find a set of state-variable equations for the system shown in Figure P2.23.

b. Find the input-output differential equation when the output is x_1.

c. Repeat part b when the output is x_2.

d. Repeat part b when the output is θ. Compare the result to the results for parts b and c.

3.16 Three masses are connected by a cable that is always in tension but that cannot stretch, as shown in Figure P3.16. The cable moves over pulleys that have negligible inertia and friction.

a. Select a set of state variables and write the model in state-variable form.

b. Assume the cable is able to stretch. Sketch an appropriate lumped-element approximation and write the corresponding state-variable equations.

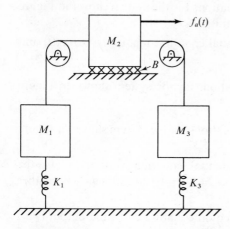

Figure P3.16

3.17 Figure P3.17 shows a mass suspended by three springs that are connected at the massless junction A.

a. Draw the free-body diagrams, including the effect of the gravitational force, and find a single differential equation relating $\ddot{x}_M$ and x_M.

b. If z_M denotes the relative displacement of the mass with respect to its static-equilibrium position, find a single differential equation relating $\ddot{z}_M$ and z_M. Identify the spring constant of the single spring that would be equivalent to the three-spring combination shown. (*Hint:* See Example 3.5.)

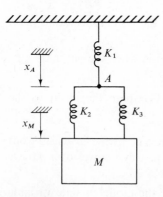

Figure P3.17

3.18 Replace each combination of dashpots in Figure P3.18 by a single equivalent dashpot. The equivalent dashpot must have the same relationship between the force and velocities at its ends as the original combination.

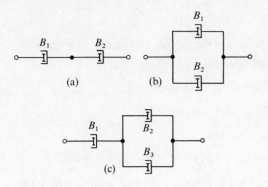

Figure P3.18

3.19 a. Write the equations describing the series combination of elements shown in Figure P3.19(a). Define all displacement and velocity variables and show their assumed positive senses on the diagram.

b. Find expressions for K_{eq} and B_{eq} in Figure P3.19(b) such that the motion of the ends of the combination is the same as that in Figure P3.19(a).

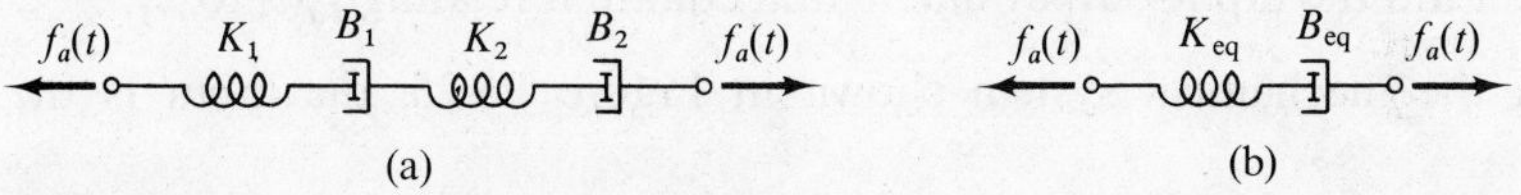

$$f_a(t) \quad K_1 \quad B_1 \quad K_2 \quad B_2 \quad f_a(t) \qquad\qquad f_a(t) \quad K_{eq} \quad B_{eq} \quad f_a(t)$$

$$\text{(a)} \qquad\qquad\qquad\qquad \text{(b)}$$

Figure P3.19

3.20 Use the p operator to derive (21) from (20).

3.21 A linear dynamic system with input $u(t)$, output y, and state variables x_1 and x_2 is characterized by the equations

$$\dot{x}_1 + 2\dot{x}_2 = 3x_1 + 4x_2 - 5u(t)$$

$$\dot{x}_1 - \dot{x}_2 = 2x_1 + x_2 + u(t)$$

$$y = \dot{x}_1 + 2x_2$$

a. Find the state-variable equations and output equation in the form of (3) and (4).

b. Find the input-output differential equation.

3.22 The following pair of equations corresponds to a linear dynamic system having $f_a(t)$ as its input and y as its output.

$$\ddot{y} + 4\dot{y} + 2y = x$$

$$\dot{x} + x + y = f_a(t)$$

    a. Derive the state-variable form of the model. Define any new symbols.

    b. Derive the input-output differential equation.

3.23 The model for a certain dynamic system is

$$\dot{x}_1 = -3x_1 + 2x_2^2 + u_1(t) + 2\dot{u}_2$$

$$\dot{x}_2 = 2x_1 + x_2 + \dot{u}_1$$

$$y = x_1 - x_2 + u_2(t)$$

where x_1 and x_2 are state variables, $u_1(t)$ and $u_2(t)$ the inputs, and y the output.

    a. Is the system linear? Is the system time-invariant?

    b. Rewrite the model in the form of (1) and (2) by avoiding derivatives on the right side of the equations. Define any new symbols.

3.24 a. Find the state-variable equations for the system shown in Figure 3.10(a) when the force input $f_a(t)$ is applied to the right side of the spring K_2 with its positive sense to the right.

    b. Find the input-output differential equation relating $f_a(t)$ to x_1.

3.25 In the mechanical system shown in Figure P3.25, the input is the

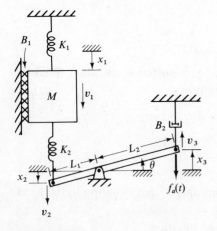

Figure P3.25

applied force $f_a(t)$. The lever is ideal and is horizontal when the system is in equilibrium with $f_a(t) = 0$ and M_1 supported by the spring K_1. The lever angle θ remains small.

    a. Taking x_1, v_1, and θ as state variables, write the state-variable equations.

    b. Write the algebraic output equation for the reaction force of the pivot on the lever, taking the positive sense as upward.

3.26 For the system shown in Figure P3.26, the angular motion of the ideal lever from the vertical position is small, so the motion of the top and midpoint can be regarded as horizontal.

    a. Select a suitable set of state variables and write the corresponding state-variable equations.

    b. Write an algebraic output equation for the support force on the lever, taking the positive sense to the right.

    c. Find the input-output differential equation when the output is the velocity of the mass.

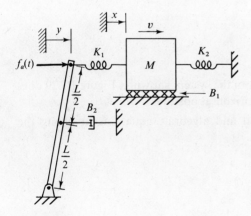

Figure P3.26

3.27 Modify Figure P3.26 by removing the force $f_a(t)$ and replacing the fixed reference at the right side of K_2 by the displacement input $x_a(t)$, with its positive sense to the right. Then repeat Problem 3.26.

3.28 For the system shown in Figure P3.28, the angular displacements θ_1 and θ_2 of the ideal levers are small and the translational elements may be assumed to move horizontally.

a. Select state variables and write the corresponding state-variable equations.

b. Write an algebraic output equation for the support force on the pivot of the right-hand lever, taking the positive sense to the left.

c. Repeat part b for the left-hand lever.

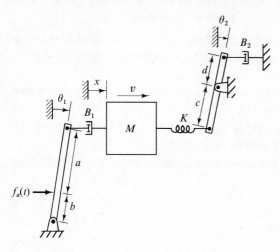

Figure P3.28

3.29 Assume that the middle lever of the system shown in Figure P3.29 does not deviate significantly from a horizontal position.

a. Write a set of differential and algebraic equations involving the variables x_1, θ_1, θ_2, and $f_a(t)$.

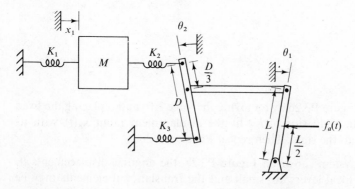

Figure P3.29

b. Explain why x_1 and $v_1 = \dot{x}_1$ are a suitable set of state variables and indicate the steps required to obtain the corresponding state-variable equations.

c. Write an algebraic output equation for the support force on the pivot of the right-hand lever, taking the positive sense to the left.

3.30 Starting with (22), find a single differential equation involving only the variables y_2 and $u(t)$ by subtracting twice the first equation from the second and then substituting the result into either of the original equations. Compare the answer with that obtained using the p operator.

CHAPTER 4
ROTATIONAL MECHANICAL SYSTEMS

Having introduced techniques for developing mathematical models of dynamic systems and having illustrated their use with translational mechanical systems, we now apply these techniques to rotational mechanical systems. As might be expected, the transition from translational to rotational systems requires little in the way of new concepts. The chapter will close with several examples of combined translational and rotational systems.

4.1 VARIABLES

For rotational mechanical systems, the symbols used for the variables are

θ, angular displacement in radians (rad)

ω, angular velocity in radians per second (rad/s)

α, angular acceleration in radians per second per second (rad/s^2)

τ, torque in newton-meters (N·m)

all of which are functions of time. Angular displacements are measured with respect to some specified reference angle in an inertial reference frame, often the equilibrium orientation of the body or point in question. We shall always choose the reference arrows for the angular displacement, velocity, and acceleration of a body to be in the same direction, so that the relationships

$$\omega = \dot{\theta}$$

$$\alpha = \dot{\omega} = \ddot{\theta}$$

hold. The conventions used are illustrated in Figure 4.1, where τ denotes an external torque applied to the rotating body by means of some unspecified mechanism, e.g., by a gear on the supporting shaft. Because of the convention that the assumed positive directions for θ, ω, and α are the same, it is not necessary to show all three reference arrows explicitly.

The power supplied to the rotating body in Figure 4.1 is

$$p = \tau \omega \tag{1}$$

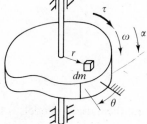

Figure 4.1 Conventions for designating rotational variables.

The power is the derivative of the energy w, and the energy supplied to the body up to time t is

$$w(t) = w(t_0) + \int_{t_0}^{t} p(\lambda) \, d\lambda$$

4.2 ELEMENT LAWS

The elements used to represent physical devices in rotational systems are moment of inertia, friction, stiffness, and gears. We shall see that moment of inertia is analogous to mass for translational systems and that gears are the counterpart of a lever. We shall restrict our consideration to elements that rotate about fixed axes.

MOMENT OF INERTIA

If Newton's second law is applied to the differential mass element dm in Figure 4.1 and the result integrated over the entire body, the result is

$$\frac{d}{dt}(J\omega) = \tau \tag{2}$$

where τ denotes the net torque applied about the fixed axis of rotation. The symbol J denotes the *moment of inertia* in kilogram-meters^2 (kg·m^2). We can obtain it by carrying out the integration of $r^2\,dm$ over the entire body. The product $J\omega$ is the angular momentum of the body.

We consider only nonrelativistic systems and constant moments of inertia, so (2) reduces to

$$J\dot{\omega} = \tau \tag{3}$$

where $\dot{\omega}$ is the angular acceleration. As is the case for a mass having translational motion, a rotating body can store energy in both kinetic and potential forms. The kinetic energy is

$$w_k = \tfrac{1}{2}J\omega^2 \tag{4}$$

and for a uniform gravitational field the potential energy is

$$w_p = Mgh \tag{5}$$

where M is the mass, g the gravitational constant, and h the height of the center of mass above its reference position. If the fixed axis of rotation is vertical or passes through the center of mass, there is no change in the potential energy as the body rotates, and (5) is not needed. To find the complete response of a dynamic system containing a rotating body, we must know its initial angular velocity $\omega(t_0)$. If its potential energy can vary or if we want to find $\theta(t)$, then we must also know $\theta(t_0)$.

FRICTION

A *rotational friction* element is one for which there is an algebraic relationship between the torque and the relative angular velocity between two surfaces. *Rotational viscous friction* arises when two rotating bodies are separated by a film of oil. Figure 4.2(a) shows two concentric rotating

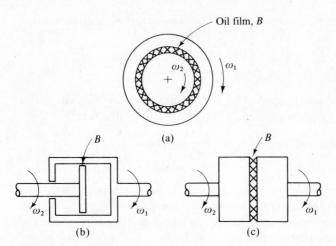

Figure 4.2 Rotational devices characterized by viscous friction.

cylinders separated by a thin film of oil, where the angular velocities of the cylinders are ω_1 and ω_2 and the relative angular velocity is $\Delta\omega = \omega_2 - \omega_1$. The torque

$$\tau = B\,\Delta\omega \tag{6}$$

will be exerted on each cylinder, in directions that tend to reduce the relative angular velocity $\Delta\omega$. Hence, the positive sense of the frictional torque must be counterclockwise on the inner cylinder and clockwise on the outer cylinder. The friction coefficient B has units of newton-meter-seconds. Note that the same symbol is used for translational viscous friction, where it has units of newton-seconds per meter.

Equation (6) also applies to a rotational dashpot that might be used in modeling a fluid drive system, such as shown in Figure 4.2(b) or Figure 4.2(c). The inertia of the parts is assumed to be negligible or else is accounted for in the mathematical model by separate moments of inertia. If the rotational friction element is assumed to have no inertia, then when a torque τ is applied to one side, a torque of equal magnitude but opposite direction must be exerted on the other side (either by a wall, the air, or some other component), as shown in Figure 4.3(a), where $\tau = B(\omega_2 - \omega_1)$. Thus in Figure 4.3(b), the torque τ passes through the first friction element and is exerted directly on the moment of inertia J.

Other types of friction, such as the damping vanes shown in Figure 4.4(a), may exert a retarding torque that is not directly proportional to the angular velocity, but they may be described by a curve of τ vs $\Delta\omega$, as shown

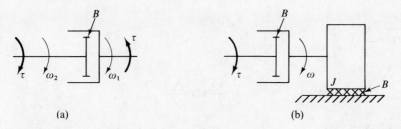

(a)   (b)

Figure 4.3 (a) Dashpot with negligible inertia. (b) Torque transmitted through a friction element.

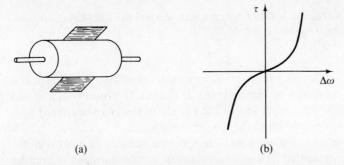

(a)   (b)

Figure 4.4. (a) Rotor with damping vanes. (b) Nonlinear friction characteristic.

in Figure 4.4(b). For a linear element, the curve must be a straight line passing through the origin. The power supplied to the friction element, $\tau \, \Delta\omega$, is immediately lost to the mechanical system in the form of heat.

STIFFNESS

Rotational stiffness is usually associated with a torsional spring, such as the mainspring of a clock, or with a relatively thin structural shaft. It is an element for which there is an algebraic relationship between τ and θ, and it is generally represented as shown in Figure 4.5(a) and Figure 4.5(b). The angle θ is the relative angular displacement of the ends of the spring from the positions corresponding to no applied torque. If both ends of the device can move as shown in Figure 4.5(c), then the element law relates τ and $\Delta\theta$, where $\Delta\theta = \theta_2 - \theta_1$. For a linear torsional spring,

$$\tau = K \, \Delta\theta \tag{7}$$

where K is a constant with units of newton-meters (N·m), in contrast to

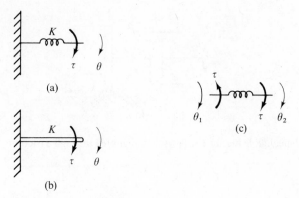

Figure 4.5 (a), (b) Rotational stiffness elements with one end fixed. (c) Rotational stiffness element with $\Delta\theta = \theta_2 - \theta_1$.

newtons per meter for the parameter K in translational systems. For a thin structural shaft, as shown in Figure 4.5(b), K is directly proportional to the shear modulus of the material and to the square of the cross-sectional area and is inversely proportional to the length of the shaft.

Since we assume that the moment of inertia of a stiffness element is either negligible or is represented by a separate element, the torques exerted on the two ends of a stiffness element must be equal in magnitude and opposite in direction, as was indicated in Figure 4.5(c). Thus for the system shown in Figure 4.6, the applied torque τ passes through the first torsional spring and is exerted directly on the body having moment of inertia J.

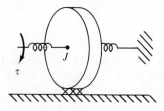

Figure 4.6 Torque transmitted through a torsional spring.

Potential energy is stored in a twisted torsional spring and can affect the response of the system at later times. For a linear spring, the potential energy is

$$w_p = \tfrac{1}{2}K(\Delta\theta)^2 \tag{8}$$

The relative angular displacement at time t_0 is one of the initial conditions we need in order to find the response of a system for $t \geq t_0$.

GEARS

A pair of gears, as shown in Figure 4.7, is the rotational analog to the lever. In order to develop the basic geometric and torque relationships, we shall assume *ideal gears*, which have no moment of inertia, no stored energy, no friction, and a perfect meshing of the teeth. Any inertia or bearing friction in an actual pair of gears can be represented by separate lumped elements in the free-body diagrams.

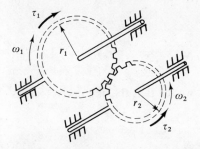

Figure 4.7 A pair of gears.

The relative size of the two gears results in a proportionality constant for the angular displacements, angular velocities, and transmitted torques of the respective shafts. For purposes of analysis, it will be convenient to visualize the pair of ideal gears as two circles, shown in Figure 4.8(a), that are tangent at the contact point and rotate without slipping. Since the spacing between teeth must be equal for each gear in a pair, the radii of the gears are proportional to the number of teeth. Thus if r and n denote the radius

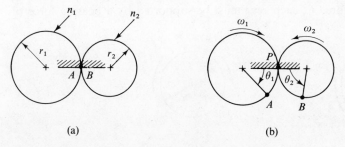

(a)   (b)

Figure 4.8 Ideal gears. (a) Reference position. (b) After rotation.

and number of teeth, respectively,

$$\frac{r_2}{r_1} = \frac{n_2}{n_1} = N \tag{9}$$

where N is called the *gear ratio*.

Let points A and B in Figure 4.8(a) denote points on the circles that are in contact with each other at some reference time t_0. At some later time, points A and B will have moved to the positions shown in Figure 4.8(b), where θ_1 and θ_2 denote the respective angular displacements from their original positions. Because the arc lengths PA and PB must be equal,

$$r_1 \theta_1 = r_2 \theta_2 \tag{10}$$

which we can rewrite as

$$\frac{\theta_1}{\theta_2} = \frac{r_2}{r_1} = N \tag{11}$$

By differentiating (10) with respect to time, we see that the angular velocities are also related by the gear ratio, namely

$$\frac{\omega_1}{\omega_2} = \frac{r_2}{r_1} = N \tag{12}$$

Note that the positive directions of θ_1 and θ_2, and likewise of ω_1 and ω_2, are taken in opposite directions in the figure. Otherwise a minus sign would be introduced into (10), (11), and (12).

We can derive the torque relationship for a pair of gears by drawing a free-body diagram for each gear, as shown in Figure 4.9. The external torques applied to the gear shafts are denoted by τ_1 and τ_2. The force exerted on each gear at the point of contact by its mate is f_c. By the law of reaction forces, the arrows must be in opposite directions for the two gears. The corresponding torques $r_1 f_c$ and $r_2 f_c$ are shown on the diagram. In addition to the contact force f_c, each gear must be supported by a bearing force of

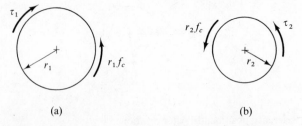

(a)   (b)

Figure 4.9 Free-body diagrams for a pair of ideal gears.

equal magnitude and opposite direction, because the gears have no translational motion. However, since the bearing support forces act through the center of the gear, they do not contribute to the torque and hence have been omitted from the figure. Because the gears have no inertia, the sum of the torques on each of the gears must be zero. Thus, from Figure 4.9,

$$f_c r_1 - \tau_1 = 0$$
$$f_c r_2 + \tau_2 = 0 \tag{13}$$

Eliminating the contact force f_c, the value of which is seldom of interest, we obtain

$$\frac{\tau_2}{\tau_1} = -\frac{r_2}{r_1} = -N \tag{14}$$

The minus sign in (14) should be expected, because both τ_1 and τ_2 in Figure 4.9 are shown as driving torques; that is, the reference arrows indicate that positive values of τ_1 and τ_2 both tend to make the gears move in the positive direction. Since the gears are assumed to have no inertia, the two torques must actually be in opposite directions, and either τ_1 or τ_2 must be negative at any instant.

An alternative derivation of (14) makes use of the fact that the power supplied to the first gear is $p_1 = \tau_1 \omega_1$ and the power supplied to the second gear is $p_2 = \tau_2 \omega_2$. Since no energy can be stored in the ideal gears and since no power can be dissipated as heat because of the assumed absence of friction, conservation of energy requires that $p_1 + p_2 = 0$. Thus

$$\tau_1 \omega_1 + \tau_2 \omega_2 = 0$$

or

$$\frac{\tau_2}{\tau_1} = -\frac{\omega_1}{\omega_2} = -N$$

which agrees with (14).

4.3 INTERCONNECTION LAWS

The interconnection laws for rotational systems involve laws for torques and angular displacements that are analogous to (2.16) and (2.17) for translational systems. The law governing reaction torques has an important modification when compared to the one governing reaction forces.

D'ALEMBERT'S LAW

For a body with constant moment of inertia rotating about a fixed axis, we can write (3) as

$$\sum_i (\tau_{\text{ext}})_i - J\dot{\omega} = 0 \tag{15}$$

where the summation over i includes all the torques acting on the body. Like the translational version of D'Alembert's law, the term $-J\dot{\omega}$ can be considered an *inertial torque*. When it is included along with all the other torques acting on the body, (15) reduces to the form appropriate for a body in equilibrium, namely

$$\sum_i \tau_i = 0 \tag{16}$$

In the application of (16), the torque $J\dot{\omega}$ is directed opposite to the positive sense of θ, ω, and α. D'Alembert's law can also be applied to a junction point that has no moment of inertia, in which case the $J\dot{\omega}$ term vanishes.

THE LAW OF REACTION TORQUES

For bodies that are rotating about the same axis, any torque exerted by one element on another is accompanied by a *reaction torque* of equal magnitude and opposite direction on the first element. In Figure 4.10, for example, a counterclockwise torque $K_1 \theta_1$ exerted by the shaft K_1 on the right disk is accompanied by a clockwise torque $K_1 \theta_1$ exerted by the disk on the shaft. However, for bodies not rotating about the same axis, the magnitudes of the two torques are not necessarily equal. For the pair of gears shown in

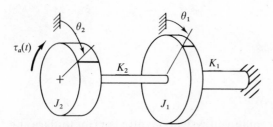

Figure 4.10 Rotational system to illustrate the laws for reaction torques and angular displacements.

Figure 4.7, Figure 4.8, and Figure 4.9, the contact forces where the gears mesh are equal and opposite, but because the gears have different radii, the torque exerted by the first gear on the second has a different magnitude than the torque exerted by the second on the first.

THE LAW FOR ANGULAR DISPLACEMENTS

When examining the interconnection of elements in a system, we automatically use the fact that we can express the motions of some of the elements in terms of the motions of other elements. In Figure 4.10, the angular velocity of the left end of the shaft K_1 is identical to the angular velocity of the disk labeled J_1. In the same figure, suppose that the reference marks on the rims are at the top of the two disks when no torque is applied, but make the angles θ_1 and θ_2 with the vertical reference when the torque $\tau_a(t)$ is applied to the left disk. Then the net angular displacement for the shaft K_2 with respect to its unstressed condition is $\theta_2 - \theta_1$.

One way to summarize these facts in the form of a general interconnection law is to say that at any instant the algebraic sum of the angular-displacement differences around any closed path is zero. The equation

$$\sum_i \Delta\theta_i = 0 \qquad \text{around any closed path} \tag{17}$$

is analogous to (2.17) for translational systems. It is understood that the signs in the summation take into account the direction in which the path is being traversed and that all measurements are with respect to reference positions that correspond to a single equilibrium condition for the system. If in Figure 4.10, for example, the relative angular displacement differences are summed going from the vertical reference to disk 1, then to disk 2, and finally back to the vertical reference, (17) gives

$$(\theta_1 - 0) + (\theta_2 - \theta_1) + (0 - \theta_2) = 0$$

Although this is a rather trivial result, it does provide a formal basis for stating that the relative angular displacements for the shafts K_1 and K_2 are θ_1 and $\theta_2 - \theta_1$, respectively. If for some reason the references for the angular displacements did not correspond to a single equilibrium condition, then the summation in (17) would be a constant other than zero.

In formulating the system equations, we normally use (17) automatically to reduce the number of displacement variables that must be shown on the diagram. Equation (17) could be differentiated to yield a comparable equation governing the relative velocities.

4.4 OBTAINING THE SYSTEM MODEL

The methods for using the element and interconnection laws to develop an appropriate mathematical model for a rotational system are the same as those discussed in Chapter 2 and Chapter 3 for translational systems. We indicate the assumed positive directions for the variables, with the assumed senses for the angular displacement, velocity, and acceleration of a body chosen to be the same. We automatically use (17) to avoid introducing more symbols than are necessary to describe the motion. For each mass or junction point whose motion is unknown beforehand, we normally draw a free-body diagram showing all torques including the inertial torque. We express all the torques except inputs in terms of angular displacements, velocities, or accelerations by means of the element laws. Then we apply D'Alembert's law, as given in (16), to each free-body diagram.

If we seek a set of state-variable equations, we identify the state variables and write those equations, such as $\dot{\theta} = \omega$, that do not require a free-body diagram. We write the equations obtained from the free-body diagrams by D'Alembert's law in terms of the state variables and inputs and manipulate them into the standard form of (3.1) or (3.3). Each of the state-variable equations should express the derivative of a different state variable as an algebraic function of the state variables, inputs, and time. For any output of the system that is not one of the state variables, we need a separate algebraic equation.

If, on the other hand, we seek an input-output differential equation, we normally write the equations from the free-body diagram in terms of only angular displacements or only angular velocities. Then we combine the equations to eliminate all variables except the input and output, which can be done by the p operator described in Section 3.2. For single-input, single-output systems, the equation will have the form of (3.19).

We shall now consider a number of examples involving free-body diagrams, state-variable equations, and input-output equations for rotational systems. Although these examples involve linear friction and stiffness elements, it is a simple matter to replace the appropriate terms by the non-linear element laws when we encounter a nonlinear system. For example, a linear frictional torque $B\omega$ would become $\tau_B(\omega)$, which is a single-valued algebraic function that describes the nonlinear relationship.

EXAMPLE 4.1   Derive the mathematical model for the rotational system shown in Figure 4.11(a), giving the result in both state-variable and input-output form.

Solution   The first step is to draw a free-body diagram of the disk. This is done in Figure 4.11(b), where the left face of the disk is shown. The torques

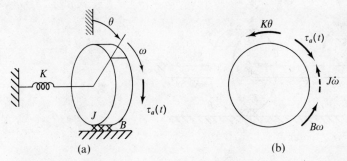

Figure 4.11 (a) Rotational system for Example 4.1. (b) Free-body diagram.

acting on the disk are the spring torque $K\theta$, the viscous-frictional torque $B\omega$, and the applied torque $\tau_a(t)$. In addition, the inertial torque $J\dot\omega$ is indicated by the dashed arrow. By D'Alembert's law,

$$J\dot\omega + B\omega + K\theta = \tau_a(t) \tag{18}$$

The normal choices for the state variables are θ and ω, which are related to the energy stored in the rotational spring and in the disk, respectively. One state-variable equation is $\dot\theta = \omega$, and the other can be found by solving (18) for $\dot\omega$. Thus

$$\dot\theta = \omega$$
$$\dot\omega = \frac{1}{J}[-K\theta - B\omega + \tau_a(t)] \tag{19}$$

For the input-output equation with θ designated as the output, we merely rewrite (18) with all terms on the left side expressed in terms of θ and its derivatives. Then the desired result is

$$J\ddot\theta + B\dot\theta + K\theta = \tau_a(t)$$

If the output is ω, we differentiate this equation term by term and then replace $\dot\theta$ by ω. The result is

$$J\ddot\omega + B\dot\omega + K\omega = \dot\tau_a$$

EXAMPLE 4.2 Write the state-variable equations for the system shown in Figure 4.12(a). Also find the input-output differential equation when the output is θ_2. The two shafts are assumed to be flexible, with spring constants K_1 and K_2. The two disks, with moments of inertia J_1 and J_2, are supported by bearings whose friction is negligible compared with the viscous-friction elements denoted by the coefficients B_1 and B_2. The reference positions for

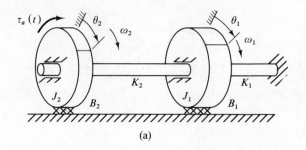

(a)

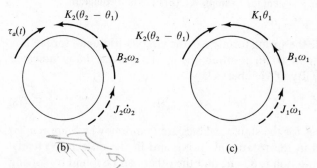

(b)   (c)

Figure 4.12 (a) Rotational system for Example 4.2. (b), (c) Free-body diagrams.

θ_1 and θ_2 are the positions of the reference marks on the rims of the disks when the system contains no stored energy.

Solution   Since the system has two inertia elements with independent angular velocities and two shafts with independent angular displacements, four state variables are required. The logical choices are θ_1, θ_2, ω_1, and ω_2, since they reflect the potential energy stored in each of the springs and the kinetic energy stored in each disk.

The resulting free-body diagrams are shown in Figure 4.12(b) and Figure 4.12(c), where only the torque $K_2(\theta_2 - \theta_1)$, which is the reaction torque on the left disk by the shaft connecting the two disks, should require an explanation. The corresponding reference arrow has arbitrarily been drawn counterclockwise in Figure 4.12(b), implying that the torque is being treated as a retarding torque on disk 2. The actual torque will act in the counterclockwise direction only when $\theta_2 > \theta_1$, so the torque must be labeled $K_2(\theta_2 - \theta_1)$ and not $K_2(\theta_1 - \theta_2)$. By the law of reaction torques, the effect of the connecting shaft on disk 1 will be a torque $K_2(\theta_2 - \theta_1)$ with its positive sense in the clockwise direction. We can reach the same conclusion by first selecting a clockwise sense for the arrow in Figure 4.12(c), thereby treating that torque as a driving torque on disk 1, and then noting that disk 2 will tend to drive disk 1 in the positive direction only if $\theta_2 > \theta_1$. Thus the correct expression is $K_2(\theta_2 - \theta_1)$. Of course, if we had selected a counter-clockwise arrow in Figure 4.12(c), we would label the arrow either $-K_2(\theta_2 - \theta_1)$ or $K_2(\theta_1 - \theta_2)$.

For each of the free-body diagrams, the algebraic sum of the torques may be set equal to zero by D'Alembert's law, giving the pair of equations

$$J_1 \dot{\omega}_1 + B_1 \omega_1 + K_1 \theta_1 - K_2(\theta_2 - \theta_1) = 0$$

$$J_2 \dot{\omega}_2 + B_2 \omega_2 + K_2(\theta_2 - \theta_1) - \tau_a(t) = 0 \tag{20}$$

Two of the state-variable equations are $\dot{\theta}_1 = \omega_1$ and $\dot{\theta}_2 = \omega_2$, while we can find the other two by solving the two equations in (20) for $\dot{\omega}_1$ and $\dot{\omega}_2$, respectively. Thus

$$\dot{\theta}_1 = \omega_1$$

$$\dot{\omega}_1 = \frac{1}{J_1}[-(K_1 + K_2)\theta_1 - B_1\omega_1 + K_2\theta_2]$$

$$\dot{\theta}_2 = \omega_2 \tag{21}$$

$$\dot{\omega}_2 = \frac{1}{J_2}[K_2\theta_1 - K_2\theta_2 - B_2\omega_2 + \tau_a(t)]$$

To obtain an input-output equation, we rewrite (20) in terms of the angular displacements θ_1 and θ_2 to obtain

$$J_1 \ddot{\theta}_1 + B_1 \dot{\theta}_1 + K_1 \theta_1 - K_2(\theta_2 - \theta_1) = 0 \tag{22a}$$

$$J_2 \ddot{\theta}_2 + B_2 \dot{\theta}_2 + K_2(\theta_2 - \theta_1) = \tau_a(t) \tag{22b}$$

Neither of the equations in (22) can be solved separately, but we want to combine them into a single differential equation that does not contain θ_1. Since θ_1 appears in (22b) but none of its derivatives do, we rearrange that equation to solve for θ_1 as

$$\theta_1 = \frac{1}{K_2}[J_2 \ddot{\theta}_2 + B_2 \dot{\theta}_2 + K_2 \theta_2 - \tau_a(t)]$$

Substituting this result into (22a) gives

$$J_1 J_2 \theta_2^{(iv)} + (J_1 B_2 + J_2 B_1)\theta_2^{(iii)} + (J_1 K_2 + J_2 K_1 + J_2 K_2 + B_1 B_2)\ddot{\theta}_2$$

$$+ (B_1 K_2 + B_2 K_1 + B_2 K_2)\dot{\theta}_2 + K_1 K_2 \theta_2$$

$$= J_1 \ddot{\tau}_a + B_1 \dot{\tau}_a + (K_1 + K_2)\tau_a(t) \tag{23}$$

which is the desired result. Equation (23) is a fourth-order differential equation relating θ_2 and $\tau_a(t)$, in agreement with the fact that four state variables appear in (21).

EXAMPLE 4.3   Find the state-variable equations for the system shown in Figure 4.12(a) and studied in Example 4.2, but with the shaft connecting disk 1 to the wall removed.

Solution   Because there are only three energy-storing elements in the modified system, we expect that we shall need only three state variables. Two of these are chosen to be ω_1 and ω_2, which are related to the kinetic energy stored in the disks. The relative displacement of the ends of the connecting shaft is $\theta_2 - \theta_1$, which is related to the potential energy in that element. Hence we select as the third state variable

$$\Delta\theta = \theta_2 - \theta_1$$

although $\theta_1 - \theta_2$ would have been an equally good choice.

The free-body diagrams for each of the disks, with torques labeled in terms of the state variables and input, are shown in Figure 4.13. By D'Alembert's law,

$$J_1 \dot{\omega}_1 + B_1 \omega_1 - K_2 \Delta\theta = 0$$

$$J_2 \dot{\omega}_2 + B_2 \omega_2 + K_2 \Delta\theta = \tau_a(t)$$

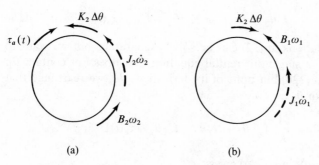

(a)   (b)

Figure 4.13 Free-body diagrams for Example 4.3.

We obtain one of the state-variable equations by noting that $\dot{\Delta\theta} = \dot{\theta}_2 - \dot{\theta}_1 = \omega_2 - \omega_1$, and we find the other two by rearranging the last two equations.

Thus the third-order state-variable model is

$$\dot{\Delta\theta} = \omega_2 - \omega_1$$

$$\dot{\omega}_1 = \frac{1}{J_1}(-B_1\omega_1 + K_2\,\Delta\theta)$$

$$\dot{\omega}_2 = \frac{1}{J_2}[-K_2\,\Delta\theta - B_2\omega_2 + \tau_a(t)]$$

Note that it is not possible to write an output equation expressing either θ_1 or θ_2 as an algebraic function of only $\Delta\theta$, ω_1, ω_2, and $\tau_a(t)$. Hence, if we wish to have either θ_1 or θ_2 as an output, three state variables are not sufficient. We could use the four state variables in Example 4.2, in which case the state-variable equations would be given by (21) with $K_1 = 0$. Alternatively, we could add either of the equations $\dot{\theta}_1 = \omega_1$ or $\dot{\theta}_2 = \omega_2$ to the above three state-variable equations. However, both of these equations could not be added without deleting the variable $\Delta\theta$, because the resulting five variables would not be independent.

EXAMPLE 4.4   The shaft supporting the disk in the system shown in Figure 4.14 is composed of two sections having spring constants K_1 and K_2. Show how to replace the two sections by an equivalent stiffness element, and derive the state-variable equations.

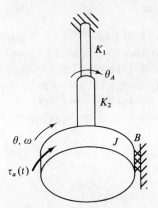

Figure 4.14 Rotational system for Example 4.4.

Solution   Free-body diagrams for each of the sections of the support shaft and for the inertia element J are shown in Figure 4.15. No inertial torques

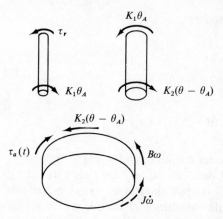

Figure 4.15 Free-body diagrams for Example 4.4.

are included on the shafts, since their moments of inertia are assumed to be negligible. The quantity τ_r is the reaction torque applied by the support on the top of the shaft. Summing the torques on each of the free-body diagrams gives

$$K_1 \theta_A - \tau_r = 0 \tag{24a}$$

$$K_2(\theta - \theta_A) - K_1 \theta_A = 0 \tag{24b}$$

$$J\dot{\omega} + B\omega + K_2(\theta - \theta_A) - \tau_a(t) = 0 \tag{24c}$$

We need the first of these equations only if we wish to find τ_r, which is not normally the case. Equations (24b) and (24c) can also be obtained by considering the free-body diagrams for just the disk and the massless junction of the two shafts. The corresponding free-body diagrams are shown in Figure 4.16. Applying D'Alembert's law to them yields (24b) and (24c).

We see from (24b) that θ_A and θ are proportional to each other. Specifically,

$$\theta_A = \left(\frac{K_2}{K_1 + K_2}\right)\theta \tag{25}$$

Substituting (25) into (24c) yields

$$J\dot{\omega} + B\omega + K_{eq}\theta = \tau_a(t) \tag{26}$$

where

$$K_{eq} = \frac{K_1 K_2}{K_1 + K_2}$$

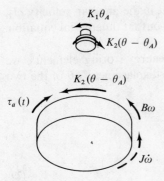

Figure 4.16 Alternative free-body diagrams for Example 4.4.

The parameter K_{eq} can be regarded as an equivalent spring constant for the combination of the two shafts. Selecting θ and ω as the state variables and using (26), we can write the state-variable model as

$$\dot{\theta} = \omega$$

$$\dot{\omega} = \frac{1}{J}[-K_{eq}\theta - B\omega + \tau_a(t)] \tag{27}$$

EXAMPLE 4.5   The system shown in Figure 4.17 consists of a moment of inertia J_1 corresponding to the rotor of a motor or a turbine, which is coupled to the moment of inertia J_2 representing a propeller. Power is transmitted through a fluid coupling with viscous-friction coefficient B and a shaft with spring constant K. A driving torque $\tau_a(t)$ is exerted on J_1, and a

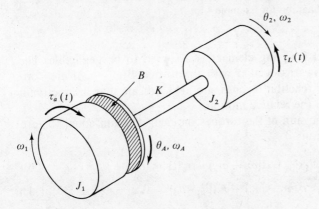

Figure 4.17 Rotational system for Example 4.5.

load torque $\tau_L(t)$ is exerted on J_2. If the output is the angular velocity ω_2, find the state-variable model and also the input-output differential equation.

Solution Since there are three independent energy-storing elements, we select as state variables ω_1, ω_2, and the relative displacement $\Delta\theta$ of the two ends of the shaft, where

$$\Delta\theta = \theta_A - \theta_2 \qquad (28)$$

Note that the equation

$$\dot{\Delta\theta} = \omega_A - \omega_2 \qquad (29)$$

is not yet a state-variable equation because of the symbol ω_A on the right side.

Next we draw the free-body diagrams for the two inertia elements and for the shaft, as shown in Figure 4.18. Note that the moment of inertia of the

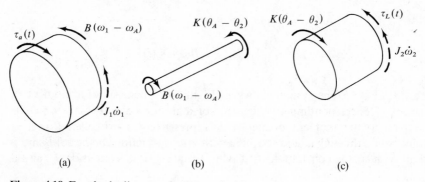

(a)   (b)   (c)

Figure 4.18 Free-body diagrams for Example 4.5.

right side of the fluid coupling element is assumed to be negligible. The directions of the arrows associated with the torque $B(\omega_1 - \omega_A)$ are consistent with the law of reaction torques and also indicate that the frictional torque tends to retard the relative motion.

Setting the algebraic sum of the torques on each diagram equal to zero yields the three equations

$$J_1\dot{\omega}_1 + B(\omega_1 - \omega_A) - \tau_a(t) = 0 \qquad (30a)$$

$$B(\omega_1 - \omega_A) - K(\theta_A - \theta_2) = 0 \qquad (30b)$$

$$J_2\dot{\omega}_2 - K(\theta_A - \theta_2) + \tau_L(t) = 0 \qquad (30c)$$

Using (28), we can rewrite (30) as

$$J_1 \dot{\omega}_1 + B(\omega_1 - \omega_A) - \tau_a(t) = 0 \tag{31a}$$

$$B(\omega_1 - \omega_A) = K \, \Delta\theta \tag{31b}$$

$$J_2 \dot{\omega}_2 - K \, \Delta\theta + \tau_L(t) = 0 \tag{31c}$$

Substituting (31b) into (31a) and repeating (31c) give

$$\begin{aligned} J_1 \dot{\omega}_1 + K \, \Delta\theta - \tau_a(t) = 0 \\ J_2 \dot{\omega}_2 - K \, \Delta\theta + \tau_L(t) = 0 \end{aligned} \tag{32}$$

Also from (31b),

$$\omega_A = \omega_1 - \frac{K}{B} \Delta\theta \tag{33}$$

Substituting (33) into (29) and rearranging (32) gives the three state-variable equations

$$\dot{\Delta\theta} = -\frac{K}{B} \Delta\theta + \omega_1 - \omega_2$$

$$\dot{\omega}_1 = \frac{1}{J_1} [-K \, \Delta\theta + \tau_a(t)] \tag{34}$$

$$\dot{\omega}_2 = \frac{1}{J_2} [K \, \Delta\theta - \tau_L(t)]$$

To obtain the input-output equation, we first rewrite (30) in terms of the angular velocities ω_1, ω_2, and ω_A and the torques $\tau_A(t)$ and $\tau_L(t)$. Differentiating (30b) and (30c) and noting that $\dot{\theta}_2 = \omega_2$ and $\dot{\theta}_A = \omega_A$, we have

$$J_1 \dot{\omega}_1 + B(\omega_1 - \omega_A) = \tau_a(t)$$

$$B(\dot{\omega}_1 - \dot{\omega}_A) - K(\omega_A - \omega_2) = 0$$

$$J_2 \ddot{\omega}_2 - K(\omega_A - \omega_2) + \dot{\tau}_L = 0$$

The reader is encouraged to apply the p-operator technique used in Example 3.10 to verify that eliminating ω_A and ω_1 from these equations gives the input-output equation

$$\dddot{\omega}_2 + \frac{K}{B} \ddot{\omega}_2 + K\left(\frac{1}{J_1} + \frac{1}{J_2}\right) \dot{\omega}_2 = \frac{K}{J_1 J_2} \tau_a(t) - \frac{1}{J_2} \ddot{\tau}_L - \frac{K}{B J_2} \dot{\tau}_L - \frac{K}{J_1 J_2} \tau_L(t) \tag{35}$$

Although this result can be viewed as a second-order differential equation in $\dot{\omega}_2$, we will need three initial conditions if we are to determine ω_2 rather than $\dot{\omega}_2$. Note that if the load torque $\tau_L(t)$ were given as an algebraic function of ω_2, as it would be in practice, ω_2 would appear in (35). Then the input-output equation would be a strictly third-order equation.

EXAMPLE 4.6 Derive the mathematical model for the gear-driven disk shown in Figure 4.19(a). The torque $\tau_a(t)$ is applied to a gear with radius r_1. The mating gear with radius r_2 is rigidly connected to the moment of inertia J, which in turn is restrained by the flexible shaft K and viscous damping B.

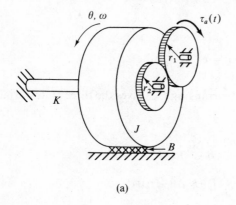

(a)

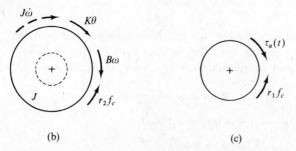

(b) (c)

Figure 4.19 (a) System for Example 4.6. (b), (c) Free-body diagrams.

Solution The free-body diagram for the disk and the gear attached to it is shown in Figure 4.19(b), and the diagram for the other gear is shown in Figure 4.19(c). The contact force where the gears mesh is denoted by f_c, and the corresponding torques are included on the diagrams. Rather than drawing a separate free-body diagram for the shaft, we show the torque $K\theta$

that it exerts on the disk in Figure 4.19(b). By D'Alembert's law,

$$J\dot{\omega} + B\omega + K\theta - r_2 f_c = 0 \tag{36a}$$

$$r_1 f_c = \tau_a(t) \tag{36b}$$

Solving (36b) for f_c and substituting the result into (36a), we have

$$J\dot{\omega} + B\omega + K\theta = N\tau_a(t)$$

where $N = r_2/r_1$. If θ and ω are chosen as the state variables, we write

$$\dot{\theta} = \omega$$

$$\dot{\omega} = \frac{1}{J}[-K\theta - B\omega + N\tau_a(t)] \tag{37}$$

Note that (19), which describes a system identical to this one except for the gears, is identical to (37) with $N = 1$. Hence, as expected, the only effect of the gears is to multiply the applied torque $\tau_a(t)$ by the gear ratio N.

EXAMPLE 4.7   Find the state-variable equations for the system shown in Figure 4.20(a), in which the pair of gears couples two similar subsystems.

Solution   Because of the two moments of inertia and the two shafts, it might appear that we could choose ω_1, ω_2, θ_1, and θ_2 as state variables. However, θ_1 and θ_2 are related by the gear ratio, as are ω_1 and ω_2. Since the state variables must be independent, either θ_1 and ω_1 or else θ_2 and ω_2 consitute a suitable set.

The free-body diagrams for each of the moments of inertia are shown in Figure 4.20(b) and Figure 4.20(c). As in Example 4.6, f_c represents the contact force between the two gears. Summing the torques on each of the free-body diagrams gives

$$J_1\dot{\omega}_1 + B_1\omega_1 + K_1\theta_1 + r_1 f_c = \tau_{a_1}(t) \tag{38a}$$

$$J_2\dot{\omega}_2 + B_2\omega_2 + K_2\theta_2 - r_2 f_c = \tau_{a_2}(t) \tag{38b}$$

By the geometry of the gears,

$$\theta_1 = N\theta_2$$

$$\omega_1 = N\omega_2 \tag{39}$$

where $N = r_2/r_1$.

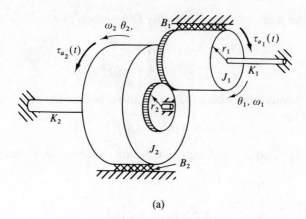

(a)

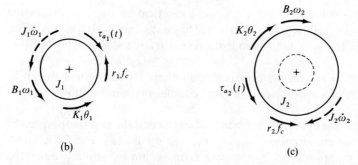

(b)   (c)

Figure 4.20 (a) System for Example 4.7. (b), (c) Free-body diagrams.

Selecting θ_2 and ω_2 as the state variables, we can write $\dot{\theta}_2 = \omega_2$ as the first state-variable equation and combine (38) and (39) to obtain the required equation for $\dot{\omega}_2$ in terms of θ_2, ω_2, $\tau_{a_1}(t)$, and $\tau_{a_2}(t)$. We first solve (38b) for f_c and substitute that expression into (38a). Then, substituting (39) into the result gives

$$(J_2 + N^2 J_1)\dot{\omega}_2 + (B_2 + N^2 B_1)\omega_2 + (K_2 + N^2 K_1)\theta_2 - N\tau_{a_1}(t) - \tau_{a_2}(t) = 0$$

$$(40)$$

At this point, it is convenient to define the parameters

$$J_{2_{eq}} = J_2 + N^2 J_1$$

$$B_{2_{eq}} = B_2 + N^2 B_1$$

$$K_{2_{eq}} = K_2 + N^2 K_1$$

$$(41)$$

which can be viewed as the combined moment of inertia, damping co-efficient, and spring constant, respectively, when the combined system is described in terms of the variables θ_2 and ω_2. For example, it is common to say that $N^2 J_1$ is the equivalent inertia of disk 1 when reflected to shaft 2. Similarly, $N^2 B_1$ and $N^2 K_1$ are the reflected viscous-friction coefficient and spring constant, respectively. Hence, the parameters $J_{2_{eq}}$, $B_{2_{eq}}$, and $K_{2_{eq}}$ defined in (41) are the sums of the parameters associated with shaft 2 and the corresponding parameters reflected from shaft 1.

With the new notation, we can rewrite (40) as

$$J_{2_{eq}} \dot{\omega}_2 + B_{2_{eq}} \omega_2 + K_{2_{eq}} \theta_2 - N\tau_{a_1}(t) - \tau_{a_2}(t) = 0 \tag{42}$$

and the state-variable equations are

$$\dot{\theta}_2 = \omega_2$$
$$\dot{\omega}_2 = \frac{1}{J_{2_{eq}}} [-K_{2_{eq}} \theta_2 - B_{2_{eq}} \omega_2 + N\tau_{a_1}(t) + \tau_{a_2}(t)] \tag{43}$$

Note that the driving torque $\tau_{a_1}(t)$ applied to shaft 1 has the value $N\tau_{a_1}(t)$ when reflected to shaft 2.

If we wanted the system model in terms of θ_1 and ω_1, straightforward substitutions would lead to the equations

$$\dot{\theta}_1 = \omega_1$$
$$\dot{\omega}_1 = \frac{1}{J_{1_{eq}}} \left[-K_{1_{eq}} \theta_1 - B_{1_{eq}} \omega_1 + \tau_{a_1}(t) + \frac{1}{N} \tau_{a_2}(t) \right]$$

where the combined parameters with the elements associated with shaft 2 reflected to shaft 1 are

$$J_{1_{eq}} = J_1 + \frac{1}{N^2} J_2$$

$$B_{1_{eq}} = B_1 + \frac{1}{N^2} B_2$$

$$K_{1_{eq}} = K_1 + \frac{1}{N^2} K_2$$

EXAMPLE 4.8   With the experience we have gained in deriving the mathe-matical models for separate translational and rotational mechanical systems, it is a straightforward matter to treat systems that combine both types of elements. To illustrate the procedure, derive the state-variable equations for the mechanical system shown in Figure 4.21, which uses a rack and a pinion

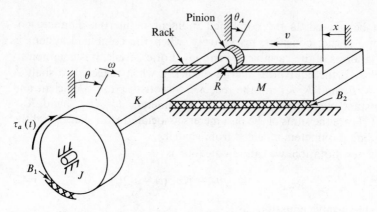

Figure 4.21 System for Example 4.8 with rack and pinion gear.

gear to convert rotational motion to translational motion. The moment of inertia J represents the rotor of a motor on which an applied torque $\tau_a(t)$ is exerted. The rotor is connected by a flexible shaft to a pinion gear of radius R that meshes with the linear rack. The rack is rigidly attached to the mass M, which might represent the bed of a milling machine.

Solution   The free-body diagrams for the moment of inertia J, the pinion gear, and the mass M are shown in Figure 4.22. The contact force between the rack and pinion is denoted by f_c. Forces and torques that will not appear in the equations of interest (e.g., the vertical force on the mass and the bearing forces on the rotor and pinion gear) have been omitted. Summing the torques in Figure 4.22(a) and Figure 4.22(b) and the forces in Figure 4.22(c) yields the three equations

$$J\dot{\omega} + B_1\omega + K(\theta - \theta_A) - \tau_a(t) = 0 \tag{44a}$$

$$Rf_c - K(\theta - \theta_A) = 0 \tag{44b}$$

$$M\dot{v} + B_2 v - f_c = 0 \tag{44c}$$

In addition, the geometric relationship

$$R\theta_A = x \tag{45}$$

must hold because of the contact between the rack and the pinion gear.

We are faced with a basic choice concerning the state variables. The fact that there are three energy-storing elements corresponding to the parameters J, M, and K suggests that the three variables ω, v, and $\Delta\theta = \theta - \theta_A$ would

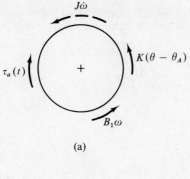

(a)

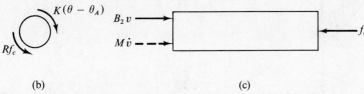

(b)   (c)

Figure 4.22 Free-body diagrams for Example 4.8. (a) Rotor. (b) Pinion gear. (c) Mass.

be appropriate. However, such a choice contains no information about x. Since the position of the mass is likely to be of interest in a practical situation, we select the four state variables θ, ω, x, and v. Using (45) to eliminate θ_A in (44a) gives

$$J\dot{\omega} + B_1\omega + K\theta - \frac{K}{R}x - \tau_a(t) = 0$$

and using (45) and (44b) to eliminate f_c in (44c) results in

$$M\dot{v} + B_2 v + \frac{K}{R^2}x - \frac{K}{R}\theta = 0$$

Thus the desired state-variable equations are

$$\dot{\theta} = \omega$$

$$\dot{\omega} = \frac{1}{J}\left[-K\theta - B_1\omega + \frac{K}{R}x + \tau_a(t)\right]$$

$$\dot{x} = v$$

$$\dot{v} = \frac{1}{M}\left(\frac{K}{R}\theta - \frac{K}{R^2}x - B_2 v\right)$$

(46)

EXAMPLE 4.9   As a final example of combined translational and rotational mechanical systems, consider the system shown in Figure 4.23. The mass

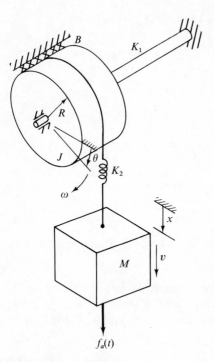

Figure 4.23 System for Example 4.9 with translational and rotational elements.

and spring are connected to the disk by a flexible cable. Actually, the spring might be used to represent the stretching of the cable. The mass M is subjected to the external force $f_a(t)$. Find the state-variable equations for the system, treating $f_a(t)$ and the weight of the mass as inputs.

Solution   Let θ and x be measured from references corresponding to the position where the shaft K_1 is not twisted and the spring K_2 is not stretched. The free-body diagrams for the disk and the mass are shown in Figure 4.24, where f_2 denotes the force exerted by the spring. Since the downward displacement of the top end of the spring is $R\theta$,

$$f_2 = K_2(x - R\theta) \tag{47}$$

Because of the four energy-storing elements corresponding to the parameters K_1, J, K_2, and M, we select θ, ω, x, and v as the state variables.

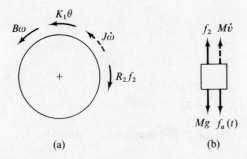

Figure 4.24 Free-body diagrams for Example 4.9. (a) Disk. (b) Mass.

From the free-body diagrams and with (47), we can write

$$J\dot{\omega} + B\omega + K_1\theta - RK_2(x - R\theta) = 0 \qquad (48a)$$

$$M\dot{v} + K_2(x - R\theta) = f_a(t) + Mg \qquad (48b)$$

Note that the reaction force $f_2 = K_2(x - R\theta)$ of the cable on the mass is not the same as the total external force $f_a(t) + Mg$ on the mass. As indicated by (48b), the difference is the inertial force $M\dot{v}$. Only if the mass were negligible would the external force be transmitted directly through the spring. From (48) and the identities $\dot{\theta} = \omega$ and $\dot{x} = v$, we can write the state-variable equations

$$\dot{\theta} = \omega$$

$$\dot{\omega} = \frac{1}{J}[-(K_1 + K_2 R^2)\theta - B\omega + K_2 Rx]$$

$$\dot{x} = v \qquad (49)$$

$$\dot{v} = \frac{1}{M}[K_2 R\theta - K_2 x + f_a(t) + Mg]$$

In order to emphasize the effect of the weight Mg, suppose that $f_a(t) = 0$ and that the mass and disk are not moving. Let θ_0 denote the constant angular displacement of the disk and x_0 the constant displacement of the mass under these conditions. Then (48) becomes

$$K_1\theta_0 = RK_2(x_0 - R\theta_0)$$

$$K_2(x_0 - R\theta_0) = Mg \qquad (50)$$

from which

$$\theta_0 = \frac{RMg}{K_1}$$

$$x_0 = \frac{Mg}{K_2} + \frac{R^2 Mg}{K_1}$$

These expressions represent the constant displacements caused by the gravitational force Mg.

Now reconsider the case where $f_a(t)$ is nonzero and where the system is in motion. Let

$$\theta = \theta_0 + \phi$$

$$x = x_0 + z$$

$$(51)$$

so that ϕ and z represent the additional angular and vertical displacements caused by the input $f_a(t)$. Note that $\omega = \dot{\theta} = \dot{\phi}$ and $v = \dot{x} = \dot{z}$. Substituting (51) into (48) gives

$$J\dot{\omega} + B\omega + K_1(\theta_0 + \phi) - RK_2(x_0 + z - R\theta_0 - R\phi) = 0$$

$$M\dot{v} + K_2(x_0 + z - R\theta_0 - R\phi) = f_a(t) + Mg$$

Using (50) to cancel those terms involving θ_0, x_0, and Mg, we are left with

$$J\dot{\omega} + B\omega + K_1\phi - RK_2(z - R\phi) = 0$$

$$M\dot{v} + K_2(z - R\phi) = f_a(t)$$

so the corresponding state-variable equations are

$$\dot{\phi} = \omega$$

$$\dot{\omega} = \frac{1}{J}[-(K_1 + K_2 R^2)\phi - B\omega + K_2 Rz]$$

$$\dot{z} = v$$

$$(52)$$

$$\dot{v} = \frac{1}{M}[K_2 R\phi - K_2 z + f_a(t)]$$

We see that (49) and (52) have the same form, except that in the latter case the term Mg is missing and θ and x have been replaced by ϕ and z. As long as the stiffness elements are linear, we can ignore the gravitational force

Mg if we measure all displacements from the equilibrium positions corresponding to no inputs except gravity. This agrees with the conclusion
reached in Example 2.3.

PROBLEMS

4.1 Find the value of K_{eq} in Figure P4.1(b) such that the relationship between $\tau_a(t)$ and θ is the same as for Figure P4.1(a).

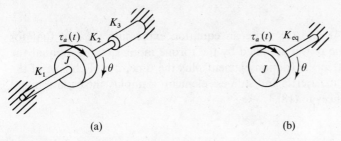

(a)   (b)

Figure P4.1

4.2 The left side of the fluid drive element denoted by B in Figure P4.2 moves with the angular velocity $\omega_a(t)$. Find the input-output differential equation relating ω_2 and $\omega_a(t)$.

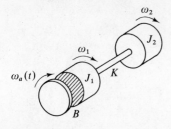

Figure P4.2

4.3 Choose a set of state variables for the system shown in Figure P4.3, assuming that the values of the angular displacements θ_1, θ_2, θ_3, and θ_4 with respect to a fixed reference are not of interest.

    a. Write the state-variable equations describing the system.

b. Find the input-output differential equation relating ω_4 and $\tau_a(t)$.

c. Repeat part a if the torque input $\tau_a(t)$ is replaced by the angular velocity input $\omega_1(t)$.

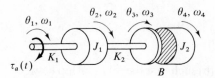

Figure P4.3

4.4 Using (1), (4), and (8), write an equation expressing the fact that for Figure 4.11(a) the power supplied by the torque input $\tau_a(t)$ must equal the power dissipated in the friction element plus the time rate of change of the energy stored in the inertia and stiffness elements. Simplify the equation and show that it reduces to (18).

4.5 A torque input $\tau_a(t)$ is applied to the lower gear shown in Figure P4.5.

a. Draw the free-body diagram for each gear and write a pair of simultaneous differential equations describing the system, where the contact force f_c is included.

b. Derive a single input-output differential equation relating θ_1 and $\tau_a(t)$.

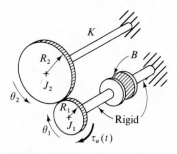

Figure P4.5

4.6 The input for the drive system shown in Figure P4.6 is the applied torque $\tau_a(t)$, while a load attached to the moment of inertia J_2 produces the load torque $\tau_L = A|\omega_2|\omega_2$.

a. Taking as state variables $\Delta\theta_1 = \theta_1 - \theta_a$, $\Delta\theta_2 = \theta_2 - \theta_b$, ω_1, ω_2, and ω_a, write the state-variable equations.

b. Write an algebraic output equation for the contact force on gear a, with the positive sense upward.

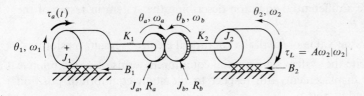

Figure P4.6

4.7 Use the p-operator technique to derive (35).

4.8 Starting with (46), find the input-output differential equation for the system shown in Figure 4.21, taking $\tau_a(t)$ as the input and x as the output.

4.9 Starting with (49), find the input-output differential equation for the system shown in Figure 4.23, with $f_a(t)$ and the gravitational force as inputs and θ as the output.

4.10 Solve part a of Problem 2.20 when the pulley has moment of inertia J and bearing friction B_2.

4.11 In the mechanical system shown in Figure P4.11, the input is the force $f_a(t)$.

    a. Write the system model as a pair of differential equations involving only the variables x, θ, $f_a(t)$, and their derivatives. When $\theta = x = 0$, the springs are undeflected.

    b. Find a single input-output differential equation relating x and $f_a(t)$.

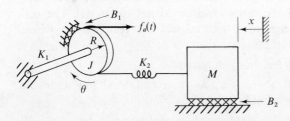

Figure P4.11

4.12 A mass and translational spring are suspended by cables wrapped around two sections of a drum as shown in Figure P4.12. The cables are assumed not to stretch, and the moment of inertia of the drum is J. The viscous-friction coefficient between the moment of inertia and a fixed surface is denoted by B. The spring is neither stretched nor compressed when $\theta = 0$.

a. Write a differential equation describing the system in terms of the variable θ.

b. For what value of θ will the rotational element remain motionless?

c. Rewrite the system's differential equation in terms of ϕ, the relative angular displacement with respect to the static-equilibrium position you found in part b.

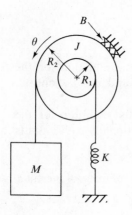

Figure P4.12

4.13 In the system shown in Figure P4.13, the two masses are equal and the cable wrapped around the drum is assumed not to stretch. The displacements x_1 and x_2 are measured with respect to a rest position where K is neither stretched nor compressed.

a. Draw the necessary free-body diagrams.

b. Choose v_1, v_2, and $z = x_1 + x_2$ as state variables, and write the system model in state-variable form.

c. Repeat part b when the state variables are chosen to be x_1, v_1, x_2, and v_2.

d. Explain why θ, v_1, and v_2 do not consitute a satisfactory set of state variables.

e. Draw a translational system capable of only horizontal motion that is equivalent to Figure P4.13, i.e., one that is described by the same set of equations. Label all element values and applied forces in terms of those in the original system.

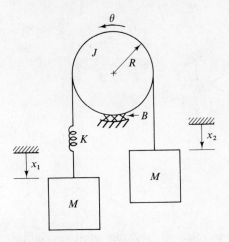

Figure P4.13

4.14 The input to the system in Figure P4.14 is the applied torque $\tau_a(t)$. The springs K_1 and K_2 are undeflected when $\theta_1 = \theta_2 = x = 0$.

a. Draw free-body diagrams for the mass and the two moments of inertia.

b. Write differential equations describing the system in terms of θ_1, θ_2, x, and $\tau_a(t)$.

c. Select a set of state variables and write the system model in state-variable form.

d. Find the constant value of $\tau_a(t)$ for which the system will reach an equilibrium position with the mass remaining motionless. Determine the corresponding deflections of the springs K_1 and K_2.

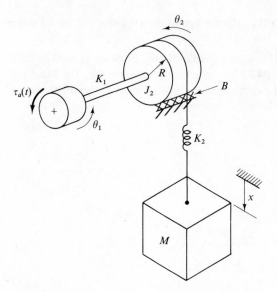

Figure P4.14

4.15 The mass M in Figure P4.15 is released from an arbitrary initial position. Using θ_1, ω_1, θ_2, and ω_2 as state variables, write the state-variable equations. The reference positions for θ_1 and θ_2 correspond to undeflected conditions for the springs K_1 and K_2.

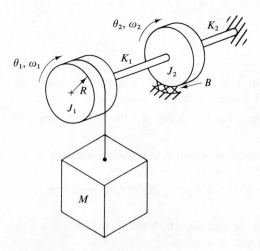

Figure P4.15

4.16 In Figure P4.16, $\tau_a(t)$ is an applied torque and points C and D are attached to the rim of the disk. Assume that the angular displacement of the disk is small. The reference positions for θ_1, θ_2, and x_3 correspond to undeflected conditions for the springs K_1, K_2, and K_3.

    a. Draw free-body diagrams for the mass M and the moment of inertia J.

    b. Write a pair of coupled differential equations describing the system in terms of θ_2, x_3, and $\tau_a(t)$.

    c. Write the system model in state-variable form.

    d. Write an algebraic output equation for θ_1.

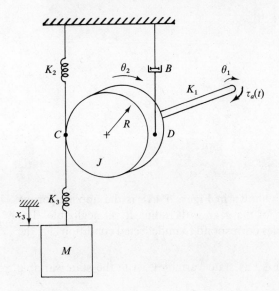

Figure P4.16

4.17 In the system shown in Figure P4.17, the translational elements B, K_2, and M are suspended from the rim of the disk, which is connected to the wall by a flexible shaft. The input is the displacement $x(t)$, where $x = \theta = 0$ corresponds to undeflected conditions for the springs.

    a. Choose an appropriate set of state variables and write the state-variable equations.

    b. Write the input-output differential equation when the output is θ.

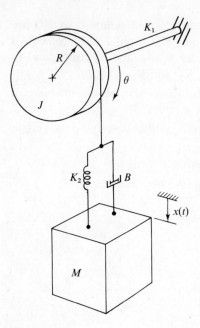

Figure P4.17

4.18 The input to the system shown in Figure P4.18 is the applied torque $\tau_a(t)$. The moment of inertia of the gear with radius R_2 is negligible. The reference values of the variables correspond to undeflected conditions for the springs.

    a. Taking x, v, θ_1, and ω_1 as state variables, write the state-variable equations.

    b. Write an algebraic output equation for θ_3.

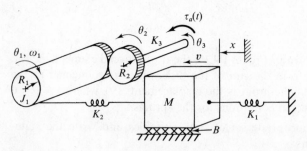

Figure P4.18

c. Write an algebraic output equation for the contact force on the gear with radius R_2. Take the positive sense upward.

d. Write the state-variable equations if the input $\tau_a(t)$ is replaced by the angular displacement $\theta_3(t)$, again taking as state variables x, v, θ_1, and ω_1.

4.19 The element law for the shaft connecting the drum labeled J_2 in Figure P4.19 to the wall is $\tau(\theta_2) = 2|\theta_2|\theta_2$. When $\theta_1 = \theta_2 = 0$, the shaft is undeflected. The mass M is suspended by a cable that does not stretch.

a. Define a suitable set of state variables and write the system model in state-variable form.

b. Draw an equivalent rotational system in the form of Figure 4.11(a) by labeling the parameters and torques in the diagram in terms of the parameters and torques in Figure P4.19.

c. Find expressions for θ_{1_0} and θ_{2_0}, the constant angular displacements when $\tau_a(t) = 0$ and the system is in static equilibrium.

d. Write the equations in part a with all angular displacements replaced by relative angular displacements from the static-equilibrium condition you found in part c.

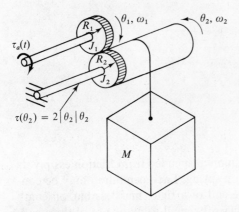

Figure P4.19

4.20 The mass M_1 shown in Figure P4.20 is at the end of an ideal lever that has two identical springs connected to its midpoint. When $\theta = x = 0$, the springs are neither stretched nor compressed. Assume that $|\theta|$ is small, so that $\sin \theta \simeq \theta$ and $\cos \theta \simeq 1$.

a. Draw free-body diagrams for the lever and for the mass M_2. For the lever, sum the torques about the pivot, where the inertial torque corresponding to M_1 is the inertial force multiplied by L. Sum the forces acting on M_2.

b. Select a set of state variables and write the system model in state-variable form.

c. Find the value of θ that corresponds to the static-equilibrium position where $f_a(t) = 0$ and the masses are stationary.

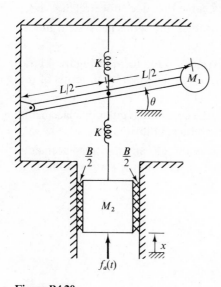

Figure P4.20

4.21 Figure P4.21 shows two pendulums suspended from frictionless pivots and connected at their midpoints by a spring. Each pendulum may be considered to be a point mass M at the end of a rigid, massless bar of length L. Assume that $|\theta_1|$ and $|\theta_2|$ are sufficiently small to allow use of the small-angle approximations $\sin \theta \simeq \theta$ and $\cos \theta \simeq 1$. The spring is unstretched when $\theta_1 = \theta_2$.

a. Draw a free-body diagram for each pendulum.

b. Define a set of state variables and write the state-variable equations.

c. Write an algebraic output equation for the spring force. Consider the force to be positive when the spring is in tension.

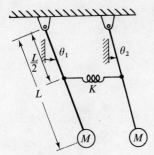

Figure P4.21

CHAPTER 5
DEVELOPING A FIXED LINEAR MODEL

In all but one of the examples used in earlier chapters, the stiffness and friction elements were linear. In practice, however, such elements are inherently nonlinear and may be considered linear over only a limited range of operating conditions. Also, a system model may have one or more time-varying coefficients.

When confronted with a mathematical model that contains either non-linearities or time-varying coefficients, the analyst has essentially three choices: (1) to attempt to solve the differential equations directly, (2) to derive a fixed linear model that can be analyzed, or (3) to obtain computer solutions of the response for specific numerical cases. The first alternative is possible only in specialized cases and will not be pursued. We shall present the linearization approach in this chapter and the computer approach in Chapter 7.

In this chapter, we give a method for linearizing an element law that depends on a single variable and then incorporating it into the system model. We also consider linearizing nonlinearities that depend on two or more variables. The chapter concludes by showing how to apply the same technique to developing fixed linear models for time-varying systems.

5.1 LINEARIZATION OF AN ELEMENT LAW

The object of linearization is to derive a linear model whose response will agree closely with that of the nonlinear model. Although the responses of the linear and nonlinear models will not agree exactly and under some conditions may differ significantly, there will generally be a set of inputs and initial conditions for which the agreement will be satisfactory. In this section, we consider the linearization of a single element law that is a nonlinear function of a single variable. We can express such an element law as a function $f(x)$, where x would generally be a state variable. If x represents the total length of a nonlinear spring and $f(x)$ the force on the spring, the function $f(x)$ might appear as shown in Figure 5.1(a), where x_0 denotes the free or unstretched length.

We shall carry out the linearization of the element law with respect to an *operating point*, which is a specific point on the nonlinear characteristic denoted by $\bar{x}$ and $\bar{f}$. A sample operating point is shown in Figure 5.1(b). We discuss the procedure for determining the operating point in the following section; for now, we shall assume that the values of $\bar{x}$ and $\bar{f}$ are known.

We can write $x(t)$ as the sum of a constant portion, which is its value at the operating point, and a time-varying portion $\hat{x}(t)$ such that

$$x(t) = \bar{x} + \hat{x}(t) \tag{1}$$

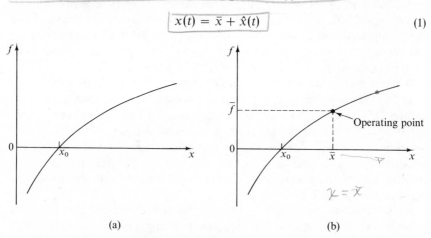

(a)   (b)

Figure 5.1 (a) A nonlinear spring characteristic. (b) Nonlinear spring characteristic with operating point.

The constant term $\bar{x}$ is called the *nominal value* of x, and the time-varying term $\hat{x}(t)$ is the *incremental variable* corresponding to x. Likewise, we can write $f(t)$ as the sum of its nominal value $\bar{f}$ and the incremental variable $\hat{f}(t)$:

$$f(t) = \bar{f} + \hat{f}(t) \tag{2}$$

where the dependence of $\hat{f}$ on time is shown explicitly. Because $\bar{x}$ and $\bar{f}$ always denote a point that lies on the curve for the nonlinear element law,

$$\bar{f} = f(\bar{x}) \tag{3}$$

Having defined the necessary terms, we shall develop two methods of linearizing the element law to relate the incremental variables $\hat{x}$ and $\hat{f}$ to one another. The first uses a graphical approach, while the second is based on a Taylor-series expansion.

GRAPHICAL APPROACH

Figure 5.2 shows the nonlinear element law $f(x)$ with the tangent to the curve at the operating point appearing as the straight line. For the moment,

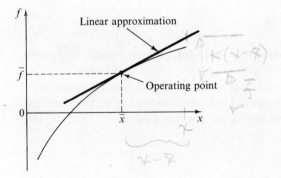

Figure 5.2 Nonlinear spring characteristic with linear approximation.

we note that the tangent line will be a good approximation to the nonlinear curve provided that the independent variable x does not deviate greatly from its nominal value $\bar{x}$ and the curvature of the curve $f(x)$ is small in the vicinity of the operating point. The slope of the tangent line is

$$k = \left. \frac{df}{dx} \right|_{x = \bar{x}}$$

or, written more concisely,

$$k = \left. \frac{df}{dx} \right|_{\bar{x}} \tag{4}$$

where the subscript $\bar{x}$ after the vertical line indicates that the derivative must be evaluated at $x = \bar{x}$. The tangent passes through the operating point that has the coordinates $(\bar{x}, \bar{f})$ and is described by the equation

$$f = \bar{f} + k(x - \bar{x})$$

which can be written as

$$f - \bar{f} = k(x - \bar{x}) \tag{5}$$

Noting from (1) and (2) that the incremental variables are

$$\hat{x} = x - \bar{x} \tag{6a}$$

$$\hat{f} = f - \bar{f} \tag{6b}$$

we see that (5) reduces to

$$\hat{f} = k\hat{x} \tag{7}$$

where k is given by (4).

We can represent (7) in graphical form by redrawing the nonlinear function $f(x)$ with a coordinate system whose axes are $\hat{x}$ and $\hat{f}$ and whose origin is located at the operating point, as depicted in Figure 5.3. The incremental variables $\hat{x}$ and $\hat{f}$ are linearly related with the constant of proportionality k being the slope of the tangent line at the operating point.

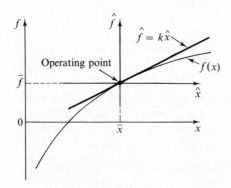

Figure 5.3 Nonlinear spring characteristic with incremental-variable coordinates.

Obviously, the accuracy of the linear approximation depends on the curvature of $f(x)$ in the vicinity of the operating point $(\bar{x}, \bar{f})$ and on the extent to which x deviates from $\bar{x}$ as the system responds to its excitations. We expect that the linearized model should be a good approximation for those values of x for which the straight line closely approximates the original curve. If the nonlinear spring is part of a larger system, we would probably not know in advance the range of values we would encounter for x. The problem is further complicated by the fact that usually we are primarily interested in how well certain responses of the overall system are approximated by calculations made on the linearized model. For example, we can expect some of the elements in a particular system to play a more critical role than the other elements in determining the response of interest. Generally, the only way to assess with certainty the quality of the approximations is to compare computer solutions for the nonlinear model and the linearized model.

SERIES-EXPANSION APPROACH

As an alternative to the geometric arguments just presented, we can derive the linearized approximation in (7) by expressing $f(x)$ in terms of its Taylor-series expansion (see Purcell, Section 23.7) about the operating point $(\bar{x}, \bar{f})$. This expansion is

$$f(x) = f(\bar{x}) + \frac{df}{dx}\bigg|_{\bar{x}} (x - \bar{x}) + \frac{1}{2!} \frac{d^2f}{dx^2}\bigg|_{\bar{x}} (x - \bar{x})^2 + \cdots$$

where the subscript $\bar{x}$ after the vertical line indicates that the associated derivative is evaluated at $x = \bar{x}$. We can find the first two terms of this expansion provided that f and its first derivative exist for $x = \bar{x}$. Since we seek a linear approximation to the actual curve, we shall neglect subsequent terms, which are higher-order in $x - \bar{x}$. The justification for truncating the series after the first two terms is that if x is sufficiently close to $\bar{x}$, then the higher-order terms are negligible compared to the constant and linear terms. Hence, we write

$$f(x) \simeq f(\bar{x}) + \frac{df}{dx}\bigg|_{\bar{x}} (x - \bar{x}) \tag{8}$$

Since $\bar{f} = f(\bar{x})$ and $k = [df/dx]|_{\bar{x}}$, this equation reduces to (7).

The accuracy of the linearized approximation in (8) depends on the extent to which the higher-order terms we have omitted from the Taylor-series expansion are truly negligible. This depends on the magnitude of $x - \bar{x}$, which is the incremental independent variable, and on the values of the higher-order derivatives of $f(x)$ at the operating point. Before considering the linearization of the complete model, we shall consider a numerical example illustrating the linearization of a nonlinear element law.

EXAMPLE 5.1   A nonlinear translational spring obeys the force-displacement relationship $f(x) = |x|x$ where x is the elongation of the spring from its unstretched length. Determine the linearized element law in numerical form for each of the operating points corresponding to the nominal spring elongations $\bar{x}_1 = -1$, $\bar{x}_2 = 0$, $\bar{x}_3 = 1$, and $\bar{x}_4 = 2$.

Solution   We can rewrite $f(x)$ as

$$f(x) = \begin{cases} -x^2 & \text{for } x < 0 \\ x^2 & \text{for } x \geq 0 \end{cases} \tag{9}$$

It follows that the slope of the tangent at the operating point is

$$\left. \frac{df}{dx} \right|_{\bar{x}} = \begin{cases} -2\bar{x} & \text{for } \bar{x} < 0 \\ 2\bar{x} & \text{for } \bar{x} \geq 0 \end{cases}$$

$$= 2|\bar{x}| \quad \text{for all } \bar{x} \tag{10}$$

Thus the linear approximation to the spring characteristic is

$$\hat{f} = 2|\bar{x}|\hat{x}$$

where the coefficient $k = 2|\bar{x}|$ may be thought of as an effective spring constant whose numerical value depends on the nominal value of the spring's elongation $\bar{x}$. Substituting the four specified values of $\bar{x}$ into (9) and (10) gives the values of $\bar{f}$ and k that appear in Table 5.1. Figure 5.4 shows the four linear approximations superimposed on the nonlinear spring characteristic. Note that the value of the effective spring constant k is strongly dependent on the location of the operating point. In fact, k vanishes for $\bar{x} = 0$, which implies that the spring would not appear in the linearized model of a system having $\bar{x} = 0$ as its operating point. It is also interesting to note that k is the same for $\bar{x} = -1$ and $\bar{x} = +1$, although the values of $\bar{f}$ differ.

Concerning the accuracy of the approximation, one might say that for deviations in x of 0.25 from the operating point, the approximation seems to be quite good; for deviations exceeding 1.0 it would be poor. It is difficult to make a definitive statement, however, without knowing the system in which the element is to appear.

Table 5.1 Nominal elongations, nominal forces, and effective spring constants

i	$\bar{x}_i$	$\bar{f}_i$	k_i
1	-1	-1	2
2	0	0	0
3	1	1	2
4	2	4	4

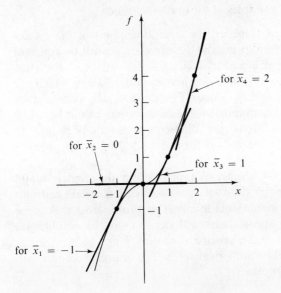

Figure 5.4 Nonlinear spring characteristic and linear approximations for four values of $\bar{x}$.

5.2 LINEARIZATION OF THE MODEL

We shall now consider the process of incorporating one or more linearized element laws into a system model. Starting with a given nonlinear model, we need to do the following:

1. Determine the operating point of the model by writing and solving the appropriate nonlinear algebraic equations and selecting the desired solution if more than one exists.

2. Rewrite all linear terms in the mathematical model as the sum of their nominal and incremental variables, noting that the derivatives of constant terms are zero.

3. Replace all nonlinear terms by the first two terms of their Taylor-series expansions, i.e., the constant and linear terms.

4. Using the algebraic equation(s) defining the operating point, cancel the constant terms in the differential equations, leaving only linear terms involving incremental variables.

5. Determine the initial conditions of all incremental variables in terms of the initial conditions of the variables of the nonlinear model.

For all situations we shall consider, the operating point of the system will be a condition of equilibrium in which each variable will be constant and equal to its nominal value and in which all derivatives of the state variables will be zero. Inputs will take on their nominal values, which are typically selected to be their average values. For example, if a system input is $u(t) = A + B \sin \omega t$, then the nominal value of the input would be taken as $\bar{u} = A$. Under these conditions, the differential equations reduce to algebraic equations that can be solved for the operating point, using a computer if necessary.

Upon completion of step 4, the terms remaining in the model should involve only incremental variables and they should all be linear with constant coefficients. In general, the coefficients involved in those terms that came from the expansion of nonlinear terms will depend on the equilibrium conditions. Hence, we must find a specific operating point before we can express the linearized model in numerical form. The entire procedure will be illustrated by several examples.

EXAMPLE 5.2 Derive a linearized model for the translational mechanical system shown in Figure 5.5(a), where the nonlinear spring characteristic $f_K(x)$ is given in Figure 5.5(b) and where the average value of the applied force $f_a(t)$ is zero.

Solution First we derive the nonlinear model by drawing the free-body diagram shown in Figure 5.5(c) and summing forces, getting

$$M\ddot{x} + B\dot{x} + f_K(x) = f_a(t) \tag{11}$$

To find the operating point, we replace $f_a(t)$ by its average value $\bar{f}_a$ and x by $\bar{x}$, getting

$$M\ddot{\bar{x}} + B\dot{\bar{x}} + f_K(\bar{x}) = \bar{f}_a$$

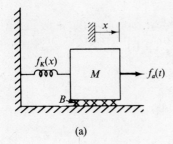

(a)

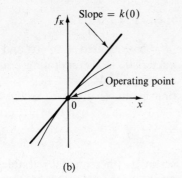

(b)

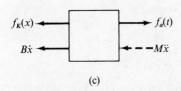

(c)

Figure 5.5 (a) Nonlinear system for Example 5.2. (b) Nonlinear spring characteristic. (c) Free-body diagram.

Noting that $\dot{\bar{x}} = \ddot{\bar{x}} = 0$ because $\bar{x}$ is a constant and that $\bar{f}_a$ was specified to be zero, we see that

$$\bar{f}_K = f_K(\bar{x}) = 0$$

Thus the operating point is at $\bar{x} = 0, \bar{f}_K = 0$, which corresponds to the origin of the spring characteristic in Figure 5.5(b).

The next step is to rewrite the linear terms in (11) in terms of the incremental variables $\hat{x} = x - \bar{x}$ and $\hat{f}_a(t) = f_a(t) - \bar{f}_a$, to obtain

$$M(\ddot{\bar{x}} + \ddot{\hat{x}}) + B(\dot{\bar{x}} + \dot{\hat{x}}) + f_K(x) = \bar{f}_a + \hat{f}_a(t)$$

Since $\dot{\bar{x}} = \ddot{\bar{x}} = 0$, we can rewrite the equation as

$$M\ddot{\hat{x}} + B\dot{\hat{x}} + f_K(x) = \bar{f}_a + \hat{f}_a(t) \tag{12}$$

Expanding the spring force $f_K(x)$ about $\bar{x} = 0$ gives

$$f_K(x) = f_K(0) + \left.\frac{df_K}{dx}\right|_{x=0} \hat{x} + \cdots$$

Substituting the first two terms into (12) yields the approximate equation

$$M\ddot{\hat{x}} + B\dot{\hat{x}} + f_K(0) + k(0)\,\hat{x} = \bar{f}_a + \hat{f}_a(t)$$

The constant $k(0)$ denotes the derivative df_K/dx evaluated at $x = 0$ and is the slope of the tangent to the spring characteristic at the operating point, as indicated in Figure 5.5(b).

Since the spring force at the operating point is $f_K(0) = \bar{f}_a = 0$, the linearized model is

$$M\ddot{\hat{x}} + B\dot{\hat{x}} + k(0)\,\hat{x} = \hat{f}_a(t) \tag{13}$$

which is a fixed linear differential equation in the incremental variable $\hat{x}$ with the incremental input $\hat{f}_a(t)$. The coefficients are the constants M, B, and $k(0)$. To solve (13), we must know the initial values $\hat{x}(0)$ and $\dot{\hat{x}}(0)$. We find them from the initial values $x(0)$ and $\dot{x}(0)$, according to

$$\hat{x}(0) = x(0) - \bar{x}$$

$$\dot{\hat{x}}(0) = \dot{x}(0) - \dot{\bar{x}}$$

In this example, $\bar{x} = \dot{\bar{x}} = 0$, so $\hat{x}(0) = x(0)$ and $\dot{\hat{x}}(0) = \dot{x}(0)$. Once we have solved the linearized model, we find the approximate solution of the non-linear model by adding the nominal value $\bar{x}$ to the incremental solution $\hat{x}(t)$. Remember that the sum of the terms, $\bar{x} + \hat{x}$, is only an approximation to the actual solution of the nonlinear model.

Should we want to put the linearized model given by (13) into state-variable form, we need only define the incremental velocity $\hat{v} = v - \bar{v}$, where $\bar{v} = 0$. Then $\hat{v} = \dot{\hat{x}}$, and we can write the pair of first-order equations

$$\dot{\hat{x}} = \hat{v}$$

$$\dot{\hat{v}} = \frac{1}{M}[-k(0)\,\hat{x} - B\hat{v} + \hat{f}_a(t)]$$

Since $\bar{x} = \bar{v} = 0$, the appropriate initial conditions are $\hat{x}(0) = x(0)$ and $\hat{v}(0) = \dot{x}(0)$.

EXAMPLE 5.3   Repeat Example 5.2 with the applied force $f_a(t)$ having a nonzero average value of $\bar{f}_a$ for positive time, as shown in Figure 5.6(a).

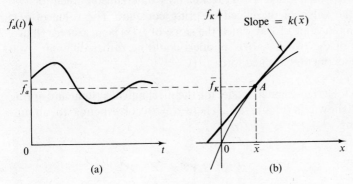

(a)   (b)

Figure 5.6 (a) Applied force for Example 5.3. (b) Nonlinear spring characteristic with new operating point.

Solution   The form of the nonlinear model given by (11) is unaffected by the value of $\bar{f}_a$. However, a new operating point will exist that is defined by the equation

$$f_K(\bar{x}) = \bar{f}_K = \bar{f}_a \tag{14}$$

For the spring characteristic shown in Figure 5.6(b) and the value of $\bar{f}_a$ shown in Figure 5.6(a), the operating point is point A. In graphical terms, the straight line $\bar{f}_a$ in Figure 5.6(a) is projected horizontally onto the curve of the spring characteristic in Figure 5.6(b), intersecting it at the operating point, the coordinates of which are $x = \bar{x}$, $f_K = \bar{f}_K$.

Upon substituting $x = \bar{x} + \hat{x}$ and $f_a(t) = \bar{f}_a + \hat{f}_a(t)$ into (11) and using $\dot{x} = \ddot{x} = 0$, we obtain

$$M\ddot{\hat{x}} + B\dot{\hat{x}} + f_K(x) = \bar{f}_a + \hat{f}_a(t) \tag{15}$$

which is identical to (12). The first two terms in the Taylor series for the spring force are

$$f_K(x) = f_K(\bar{x}) + \frac{df_K}{dx}\bigg|_{\bar{x}} \hat{x} \tag{16}$$

where $f_K(\bar{x}) = \bar{f}_K$ and where $\bar{x}$ must satisfy (14). Substituting (16) for $f_K(x)$ into (15) and invoking (14) yield the desired linear model, namely

$$M\ddot{\hat{x}} + B\dot{\hat{x}} + k(\bar{x})\hat{x} = \hat{f}_a(t) \tag{17}$$

where $k(\bar{x}) = [df_K/dx]|_{\bar{x}}$ and is the slope of the straight line in Figure 5.6(b). Note that the form of the model given by (17) with $\bar{f}_a \neq 0$ is the same as that with $\bar{f}_a = 0$, which is given by (13). The only difference between the two equations is the value of the effective spring constant. The value of $k(\bar{x})$ will depend on the value of $\bar{x}$ at which the slope of $f_K(x)$ is measured. Hence, the responses of the two linearized models could be rather different, even for the same incremental applied force $\hat{f}_a(t)$.

EXAMPLE 5.4   Derive a linear model for the mechanical system and spring characteristic shown in Figure 5.7, where $x = 0$ corresponds to an unstretched spring.

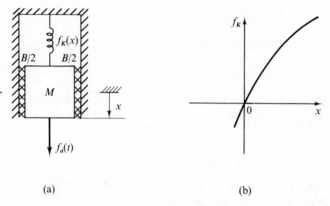

(a)   (b)

Figure 5.7 (a) Mechanical system for Example 5.4. (b) Nonlinear spring characteristic.

Solution   We obtain the nonlinear model of the system by drawing the free-body diagram shown in Figure 5.8(a) and setting the sum of the vertical forces equal to zero. Since the mass is constrained to move vertically, we must include its weight Mg in the free-body diagram. The resulting non-linear model is

$$M\ddot{x} + B\dot{x} + f_K(x) = f_a(t) + Mg$$

Note that its form is similar to that given by (11), the nonlinear model for the two preceding examples. By setting $x = \bar{x}$ and $f_a(t) = \bar{f}_a$ and by noting that $\dot{x} = \ddot{x} = 0$, we find the algebraic equation for the operating point to be

$$f_K(\bar{x}) = \bar{f}_a + Mg$$

which is the same as (14) except for inclusion of the force Mg on the right side.

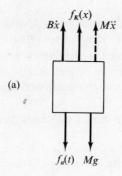

(a)

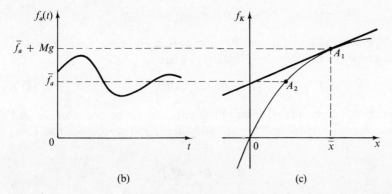

(b)   (c)

Figure 5.8 (a) Free-body diagram for Example 5.4. (b) Input. (c) Nonlinear spring characteristic with two operating points.

The nonlinear spring characteristic in Figure 5.7(b) is repeated in Figure 5.8(c). Figure 5.8(b) shows $f_a(t)$ and $\bar{f}_a$ as in Example 5.3, but it also shows the total nominal force $\bar{f}_a + Mg$ that must be projected to the characteristic curve in Figure 5.8(c) to establish the operating point A_1. The point A_2, which is obtained by projecting the force $\bar{f}_a$ from Figure 5.8(b), would be the operating point if the motion of the mass were horizontal, as it is in Example 5.3. For applied forces having the same average values, the two systems will have different linearized spring characteristics if the curve of $f_K(x)$ does not have the same slope at points A_1 and A_2. If the spring were linear, however, the slope of the characteristic in Figure 5.8(c) would be constant and the presence of the weight would have no influence on the effective spring constant, as was observed in Example 2.3. With the provision that $k(\bar{x})$ is the slope of $f_K(x)$ measured at the point A_1 rather than at point A_2, the resulting linearized model is again given by (17).

EXAMPLE 5.5   A high-speed vehicle of mass M moves along a horizontal track and is subject to a linear retarding force of Bv caused by viscous friction associated with the bearings and a nonlinear retarding force of $D|v|v$ caused by air drag. Obtain a linear model that is valid when the driving force $f_a(t)$ undergoes variations about a positive nominal value $\bar{f}_a$.

Solution   We readily find the nonlinear differential equation governing the vehicle's velocity to be

$$M\dot{v} + Bv + D|v|v = f_a(t) \tag{18}$$

Setting $v = \bar{v}$ and $f_a(t) = \bar{f}_a$ and noting that $\dot{\bar{v}} = 0$, we have

$$B\bar{v} + D|\bar{v}|\bar{v} = \bar{f}_a$$

for the operating-point equation. Because $\bar{f}_a$ is positive, we know that $\bar{v}$ is positive and we can replace $D|\bar{v}|\bar{v}$ by $D(\bar{v})^2$. Then

$$D(\bar{v})^2 + B\bar{v} - \bar{f}_a = 0 \tag{19}$$

By inspection, we see that (19) will have two real roots, one positive and the other negative. However, we are only interested in the positive root, which is

$$\bar{v} = \frac{-B + \sqrt{B^2 + 4\bar{f}_a D}}{2D} \tag{20}$$

The negative root was introduced when we replaced $|\bar{v}|\bar{v}$ by $(\bar{v})^2$ and is not a root of the actual operating-point equation.

Provided that v always remains positive, we can replace $D|v|v$ in (18) by Dv^2. Then, using $v = \bar{v} + \hat{v}$ and $f_a(t) = \bar{f}_a + \hat{f}_a(t)$ and noting that $\dot{\bar{v}} = 0$, we can rewrite (18) as

$$M\dot{\hat{v}} + B(\bar{v} + \hat{v}) + Dv^2 = \bar{f}_a + \hat{f}_a(t) \tag{21}$$

To linearize the term v^2, we replace it by the constant and linear terms in its Taylor series, namely

$$(\bar{v})^2 + \frac{d}{dv}(v^2)\bigg|_{\bar{v}} (v - \bar{v}) = (\bar{v})^2 + 2\bar{v}\hat{v} \tag{22}$$

Substituting (22) for v^2 into (21) and regrouping, we have

$$M\dot{\hat{v}} + (B + 2D\bar{v})\hat{v} + B\bar{v} + D(\bar{v})^2 = \bar{f}_a + \hat{f}_a(t)$$

The constant terms cancel because of (19), the operating-point equation. Thus the desired linearized model, which holds for $\bar{f}_a > 0$ and $v > 0$, is

$$M\dot{\hat{v}} + b\hat{v} = \hat{f}_a(t)$$

where b denotes the effective damping coefficient

$$b = B + 2D\bar{v}$$

with $\bar{v}$ given by (20) as a function of the average driving force $\bar{f}_a$.

It is worthwhile to observe that in this particular case the Taylor series of the nonlinearity has only three terms, and we could have obtained it without differentiation by writing

$$v^2 = (\bar{v}+\hat{v})^2 = (\bar{v})^2 + 2\bar{v}\hat{v} + (\hat{v})^2$$

In this example, we can see that $(\hat{v})^2$ is the error introduced by replacing v^2 by $(\bar{v})^2 + 2\bar{v}\hat{v}$. Provided that $\bar{v} \gg |\hat{v}|$, the error will be small compared to the two terms that are retained.

EXAMPLE 5.6   A nonlinear system obeys the state-variable equations

$$\dot{x} = y \tag{23a}$$

$$\dot{y} = -|x|x - 2x - 2y^3 - 3 + 0.2\cos t \tag{23b}$$

Find the operating point and develop the linearized model in numerical form.

Solution   The operating point, described by $\bar{x}$ and $\bar{y}$, must satisfy the conditions $\dot{x} = \dot{y} = 0$ with the incremental portion of the input set to zero. Hence, the operating-point equations reduce to

$$\bar{y} = 0$$

$$|\bar{x}|\bar{x} + 2\bar{x} + 3 = 0 \tag{24}$$

which have the solution $\bar{x} = -1$, $\bar{y} = 0$. We replace the two nonlinear elements in (23b) by the first two terms in their respective Taylor-series expansions. For $|x|x$, we write

$$|\bar{x}|\bar{x} + 2|\bar{x}|\hat{x}$$

while we replace y^3 by

$$(\bar{y})^3 + 3(\bar{y})^2\hat{y}$$

By substituting these approximations into (23) and using $x = \bar{x} + \hat{x}$ and $y = \bar{y} + \hat{y}$, we obtain

$$\dot{\hat{x}} = \bar{\dot{y}} + \hat{y}$$

$$\dot{\hat{y}} = -(|\bar{x}|\bar{x} + 2|\bar{x}|\hat{x}) - 2(\bar{x} + \hat{x}) - 2[(\bar{y})^3 + 3(\bar{y})^2\hat{y}] - 3 + 0.2\cos t \qquad (25)$$

With $\bar{x} = -1$ and $\bar{y} = 0$, (25) reduces to

$$\dot{\hat{x}} = \hat{y}$$

$$\dot{\hat{y}} = -4\hat{x} + 0.2\cos t \qquad (26)$$

which is the linearized model in state-variable form. Comparing (26) with (25), we note that (1) all the constant terms have been cancelled, (2) the coefficient of $\hat{y}$ in the second equation is zero since $\bar{y} = 0$, and (3) the coefficient of $\hat{x}$ reflects the combined effects of the linear and nonlinear terms.

EXAMPLE 5.7   Consider the pendulum sketched in Figure 5.9(a) and derive linear models that will be valid for small variations in θ about any equilibrium conditions that can exist. The pendulum may be considered to be a point mass M attached to a massless bar of length L, which has a frictionless pivot at its other end. A torque $\tau_a(t)$ with an average value of zero is applied by a shaft connected to its pivoted end. The pivot is such that the pendulum is free to undergo any number of revolutions.

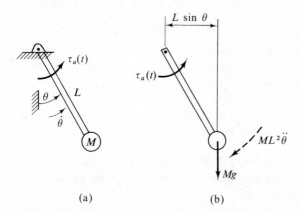

(a)   (b)

Figure 5.9 (a) Pendulum for Example 5.7. (b) Partial free-body diagram.

Solution   The partial free-body diagram shown in Figure 5.9(b) includes all terms that exert a torque about the pivot, namely the applied and inertial torques and the weight. Because of the assumption of a point mass, the moment of inertia about the pivot is $J = ML^2$. Note that the support force at the pivot and the inertial force resulting from the centripetal acceleration $L(\dot\theta)^2$ along the bar have been omitted from the free-body diagram, because neither results in a torque about the pivot. Applying D'Alembert's law yields the nonlinear differential equation

$$ML^2\ddot\theta + MgL \sin\theta - \tau_a(t) = 0 \qquad (27)$$

where the nonlinearity results from the presence of the factor $\sin\theta$.

To obtain a linearized version of (27), we first solve for any equilibrium conditions and then expand $\sin\theta$ in its Taylor series about any such equilibrium values of θ. Noting that $\bar\tau_a = 0$ and setting $\ddot\theta = 0$, we find from (27) that an equilibrium condition must satisfy

$$\sin\bar\theta = 0 \qquad (28)$$

Equation (28) has an infinite set of solutions, namely $\bar\theta = 0$, $\pm\pi$ rad, $\pm 2\pi$ rad, ... It is obvious from Figure 5.9(a) that we can restrict our consideration to the two values $\bar\theta_1 = 0$ rad (where the mass M is directly below the pivot) and $\bar\theta_2 = \pi$ rad (where M is directly above the pivot), since all other solutions to (28) differ from one of these cases by an integer multiple of 2π rad.

The Taylor-series expansion for $\sin\theta$ is

$$\sin\theta = \sin\bar\theta + \left.\frac{d}{d\theta}\sin\theta\right|_{\bar\theta}(\theta - \bar\theta) + \cdots$$
$$= \sin\bar\theta + (\cos\bar\theta)\hat\theta + \cdots \qquad (29)$$

For the operating point $\bar\theta_1 = 0$, the first two terms in (29) reduce to $\hat\theta$. Substituting this expression into (27) and noting that $\ddot\theta = \ddot{\hat\theta}$ and $\tau_a(t) = \hat\tau_a(t)$, we obtain the linearized model

$$ML^2\ddot{\hat\theta} + MgL\hat\theta = \hat\tau_a(t)$$

The quality of the linearization is reasonably good for $|\theta| \leq 0.5$ rad; this is indicated by the fact that when $\theta = 0.5$ rad, the deviation of θ from $\sin\theta$ is only 4.2%.

For the operating point $\bar\theta_2 = \pi$ rad, when the pendulum is inverted, the first two terms in (29) reduce to $-\hat\theta$. Then the linearized model becomes

$$ML^2\ddot{\hat\theta} - MgL\hat\theta = \hat\tau_a(t)$$

5.3 MULTIVARIABLE NONLINEAR ELEMENTS

In Section 5.1, we showed that when an element law is a nonlinear function of a single variable, we can develop a linearized version by using the constant and linear terms in its Taylor-series expansion. In modeling some devices, such as motors and transistor circuits, one encounters element laws that are functions of two or possibly more variables. The analytical technique introduced in Section 5.1 can be extended to such situations by using the Taylor-series expansion for functions of more than one variable (see Wylie). Because the method is easily generalized to more than two variables (although the notation is somewhat cumbersome), we shall restrict our attention to element laws that are functions of only two variables.

Assume that a nonlinear element law involves the function $f(x, y)$. The Taylor-series expansion about the point $x = \bar{x}$, $y = \bar{y}$ is

$$f(x, y) = f(\bar{x}, \bar{y}) + \frac{\partial f}{\partial x}\bigg|_{\bar{x}, \bar{y}} (x - \bar{x}) + \frac{\partial f}{\partial y}\bigg|_{\bar{x}, \bar{y}} (y - \bar{y}) + \cdots \tag{30}$$

where we have not explicitly shown the second- and higher-order terms involving $(x - \bar{x})^2$, $(x - \bar{x})(y - \bar{y})$, $(y - \bar{y})^2$, etc. because we intend to neglect them. The symbol $[\partial f/\partial x]|_{\bar{x}, \bar{y}}$ denotes the partial derivative of f with respect to x, evaluated at the operating point, i.e., for $x = \bar{x}$ and $y = \bar{y}$. The three terms shown in (30) are the constant and linear terms, which are used as the approximation to the original model. In contrast to the single-variable case, there are two linear terms—one associated with each of the variables x and y. Each of these is the product of a coefficient evaluated at the operating point and an incremental variable. We shall substitute the approximate expression

$$f(\bar{x}, \bar{y}) + \frac{\partial f}{\partial x}\bigg|_{\bar{x}, \bar{y}} \hat{x} + \frac{\partial f}{\partial y}\bigg|_{\bar{x}, \bar{y}} \hat{y} \tag{31}$$

for $f(x, y)$ into the system model. As a result, the constant term $f(\bar{x}, \bar{y})$ will be canceled because of the algebraic equation describing the operating point, leaving the two linear terms involving $\hat{x}$ and $\hat{y}$ in the linearized model. Should a nonlinear element law be a function of more than two variables, the expansion will still contain only one constant term, but it will include a linear term for each of the variables. A mathematical example will illustrate this technique, which we shall also use to develop linearized models for a transistor in Section 10.2 and for a motor in Section 11.3.

EXAMPLE 5.8 Develop the linearized model for the nonlinear system described by the pair of equations

$$\dot{x} = y - 2 \tag{32a}$$

$$\dot{y} = (2 - x)y^3 + 8 \tag{32b}$$

Give expressions for the initial values of the incremental variables in terms of the initial values of x and y.

Solution   Since the operating point must be a condition of equilibrium, it is described by the pair of equations

$$\bar{y} - 2 = 0$$

$$(2 - \bar{x})(\bar{y})^3 + 8 = 0$$

which have the solution $\bar{x} = 3$ and $\bar{y} = 2$. With the nonlinear term in (32b) denoted as

$$f(x, y) = (2 - x)y^3 \tag{33}$$

the required partial derivatives are

$$\frac{\partial f}{\partial x} = -y^3$$

$$\frac{\partial f}{\partial y} = 3(2 - x)y^2 \tag{34}$$

Evaluating these expressions for $\bar{x} = 3$ and $\bar{y} = 2$ and substituting the results into (31) give

$$-8 - 8\hat{x} - 12\hat{y} \tag{35}$$

as the approximate expression for $f(x, y)$. Writing $y = 2 + \hat{y}$ in (32a), substituting (35) for $(2 - x)y^3$ in (32b), and noting that $\dot{x} = \dot{\hat{x}}$ and $\dot{y} = \dot{\hat{y}}$ give the linearized model

$$\dot{\hat{x}} = \hat{y}$$

$$\dot{\hat{y}} = -8\hat{x} - 12\hat{y}$$

The initial values of the incremental variables are

$$\hat{x}(0) = x(0) - \bar{x} = x(0) - 3$$
$$\hat{y}(0) = y(0) - \bar{y} = y(0) - 2$$

5.4 LINEAR TIME-VARYING ELEMENTS

Linear time-varying systems obey the principle of superposition but are generally difficult or impossible to solve analytically. However, it is often possible to convert a linear time-varying model to a fixed linear model by using incremental variables and making approximations similar to those used in the linearization process described in the preceding sections.

We shall see that a time-varying parameter multiplying a variable will lead to two terms in the incremental model: (1) a constant coefficient multiplying an incremental system variable and (2) an incremental input proportional to the time-varying part of the parameter. First we shall examine a single time-varying element law, and then we shall present an example illustrating application of the procedure to a system model.

The general form of a linear time-varying element law is

$$f(x, t) = a(t)x \tag{36}$$

where x is the system variable and $a(t)$ is the time-varying coefficient. We can write the variable x as $x = \bar{x} + \hat{x}$ where the nominal portion $\bar{x}$ is defined in terms of an equilibrium condition corresponding to a nominal value of the coefficient $\bar{a}$. Likewise, we can write the time-varying coefficient as $a(t) = \bar{a} + \hat{a}(t)$. Generally $\bar{a}$ is taken to be the average value of $a(t)$. Expressing $a(t)$ and x in terms of their nominal and incremental components, we can rewrite (36) as

$$f(x, t) = [\bar{a} + \hat{a}(t)][\bar{x} + \hat{x}]$$
$$= \bar{a}\bar{x} + \bar{a}\hat{x} + \bar{x}\hat{a}(t) + \hat{a}(t)\hat{x} \tag{37}$$

The first term, $\bar{a}\bar{x}$, is a constant and will be canceled when the other variables in the model are replaced by the sum of their nominal and incremental components. The term $\bar{a}\hat{x}$ is linear in the incremental variable $\hat{x}$ and has a constant coefficient. The term $\bar{x}\hat{a}(t)$ appears as an input to the model since its time variation is that of the coefficient $a(t)$ rather than the system variable x. If $|\hat{a}(t)| \ll |\bar{a}|$, then the term $\hat{a}(t)\hat{x}$, which is the product of two incremental variables, is negligible compared to the second term in

(37). If $|\hat{x}| \ll |\bar{x}|$, then $\hat{a}(t)\hat{x}$ is negligible compared to the third term in this equation. In either of these two cases, the approximate element law becomes

$$f(x, t) \simeq \bar{a}\bar{x} + \bar{a}\hat{x} + \bar{x}\hat{a}(t) \tag{38}$$

which can be substituted into the system model.

Equation (38) can also be derived by using the Taylor-series expansion of $f = ax$ and retaining only the constant and linear terms. Adapting (31), we have for the approximate expression for $f(x, t)$

$$f(\bar{a}, \bar{x}) + \left.\frac{\partial f}{\partial x}\right|_{\bar{a}, \bar{x}} \hat{x} + \left.\frac{\partial f}{\partial a}\right|_{\bar{a}, \bar{x}} \hat{a}$$

where $\partial f/\partial x = a$ and $\partial f/\partial a = x$. Substituting the appropriate expressions for $f(\bar{a}, \bar{x})$ and the two partial derivatives evaluated at the operating point results in (38).

EXAMPLE 5.9   Develop a fixed linear model for the system described by the equation

$$\ddot{x} + (1 + \alpha \sin \omega t)x = u(t) \tag{39}$$

Solution   Because of the coefficient $(1 + \alpha \sin \omega t)$ that multiplies x, (39) is a time-varying, though linear, equation. Substituting $x = \bar{x} + \hat{x}$ and $u(t) = \bar{u} + \hat{u}$, where $\bar{x}$ is as yet undetermined and $\bar{u}$ is the nominal value of the input, into (39) gives

$$\ddot{\bar{x}} + (1 + \alpha \sin \omega t)(\bar{x} + \hat{x}) = \bar{u} + \hat{u}(t)$$

We next expand the product term to obtain

$$\ddot{\hat{x}} + \bar{x} + \hat{x} + \bar{x}\alpha \sin \omega t + \alpha(\sin \omega t)\hat{x} = \bar{u} + \hat{u}(t) \tag{40}$$

If we select $\bar{x}$ such that $\bar{x} = \bar{u}$ and drop the term $\alpha(\sin \omega t)\hat{x}$ (which is the product of incremental variables), (40) becomes

$$\ddot{\hat{x}} + \hat{x} = \hat{u}(t) - \bar{x}\alpha \sin \omega t \tag{41}$$

Comparing this result to (39), the original time-varying version, we see that in (41) the coefficient of $\hat{x}$ is the constant portion of the time-varying coefficient in (39). The other main difference is that the sinusoidal variation in (39) appears as a second input in the fixed version. If the parameter α is zero, (41) reduces to (39) except that the former is expressed in terms of incremental variables. Since $\bar{x} = \bar{u}$, however, the two versions are indeed equivalent.

PROBLEMS

In Problems 5.1 through 5.4, use a Taylor-series expansion to derive the linearized model for the element law and operating point(s) specified. In each case, show the linearized characteristic on a sketch of the nonlinear element law.

5.1 $f(x) = 0.5x^3$ where $\bar{x} = -2, 0,$ and 2.

5.2 $f(\theta) = A \sin \theta$ where $\bar{\theta} = 0$ and π rad.

5.3 $f(y) = 1/y$ where $y > 0$ and $\bar{y} = 0.5$.

5.4 $f(z) = \begin{cases} \sqrt{z} & \text{for } z \geq 0 \\ -\sqrt{|z|} & \text{for } z < 0 \end{cases}$ where $\bar{z} = -2, 0,$ and 2.

5.5 A nonlinear spring characteristic $f_K(x)$ is shown in Figure P5.5, where

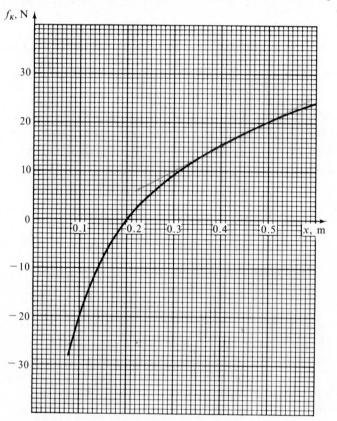

Figure P5.5

x denotes the total length. For each of the operating points specified, determine graphically the force exerted by the spring at the operating point and evaluate graphically the linearized spring constant.

    a. $\bar{x} = 0.1$ m

    b. $\bar{x} = 0.2$ m

    c. $\bar{x} = 0.3$ m

    d. $\bar{x} = 0.4$ m

5.6 The nonlinear mechanical system shown in Figure P5.6 has $M = 1.5$ kg, $B = 0.5$ N·s/m and the spring characteristic $f_K(x)$ plotted in Figure P5.5. The gravitational constant is 9.807 m/s². The variable x denotes the total length of the spring.

    a. Write the nonlinear differential equation obeyed by the system in numerical form.

    b. Solve for the operating point $\bar{x}$.

    c. Derive the linearized differential equation that is valid in the vicinity of the operating point.

    d. Give the range of $\hat{x}$ for which the linearized spring force is within 25% of the nonlinear spring force.

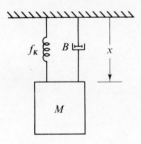

Figure P5.6

5.7 a. Repeat Problem 5.6 for the system shown in Figure P5.7, where the mass is attached to the wall by a series combination of a linear spring and a nonlinear spring. The parameter values are $M = 4.0$ kg, $B = 0.3$ N·s/m, and $K = 25$ N/m. The applied force is $f_a(t) = 10 + 2 \sin 3t$, and the nonlinear spring characteristic $f_K(x)$ is plotted in Figure P5.5. The position $z = 0$ corresponds to $f_a(t) = 0$ with both springs undeflected.

    b. Repeat part a for the case in which both springs are nonlinear with the characteristic $f_K(x)$ plotted in Figure P5.5.

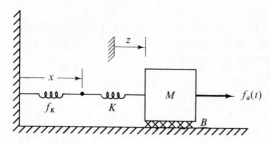

Figure P5.7

5.8 The mechanical system shown in Figure P5.8 has a linear spring with spring constant K and a nonlinear viscous damper that exerts the force $f_B(v) = v + v^3$.

a. Write the nonlinear differential equation that describes the system if it is displaced from its equilibrium position.

b. Define appropriate incremental variables and derive a linear differential equation that describes the motion about the equilibrium point.

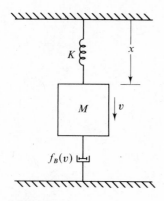

Figure P5.8

5.9 In developing the linearized model for a high-speed vehicle in Example 5.5, we assumed that $\bar{f}_a$, the nominal value of the driving force, was positive. Repeat the development for $\bar{f}_a < 0$, and find an expression for b that is valid for both cases.

5.10 A mechanical system containing a nonlinear spring obeys the differential equation

$$\ddot{x} + 2\dot{x} + f(x) = A + B \sin 3t$$

where

$$f(x) = \begin{cases} 4\sqrt{x} & \text{for } x \geq 0 \\ -4\sqrt{|x|} & \text{for } x < 0 \end{cases}$$

a. Derive the linearized model in numerical form that is valid for $A = 8$.

b. Find the operating point $\bar{x}$ and the linearized spring constant for (1) $A = 4$ and (2) $A = -4$.

c. When $A = 4$, $x(0) = 1.5$, and $\dot{x}(0) = 0.5$, find $\hat{x}(0)$ and $\dot{\hat{x}}(0)$.

5.11 A nonlinear system obeys the equation

$$\ddot{x} + 2\dot{x} + \dot{x}^3 + \frac{4}{x} = A + B \cos t$$

a. Solve for the operating-point conditions on $\bar{x}$ and $\dot{\bar{x}}$. What restriction must be placed on the value of A?

b. Derive the linearized model, evaluating any coefficients in terms of A or numbers.

5.12 A system is described by the nonlinear equation

$$\ddot{y} + 2\dot{y}^3 + 2y + |y|y = A + B \cos t$$

a. For $A = -3$, find the operating point and derive the linearized model, expressing all coefficients in numerical form.

b. Repeat part a for $A = 15$.

5.13 The model of a nonlinear system is described by the equation

$$\ddot{x} + \dot{x} + 3x\sqrt{|x|} = A + B \sin \omega t$$

The equilibrium position is known to be $\bar{x} = 4$. Determine the linearized incremental model, and evaluate $\hat{x}(0)$ and $\dot{\hat{x}}(0)$ when $x(0) = 5$ and $\dot{x}(0) = 0$.

5.14 A nonlinear system obeys the equation

$$\dot{x} + 0.5x^2 = 2 + A \sin t$$

a. Sketch the nonlinear term $0.5x^2$ and indicate all possible operating points.

b. For each operating point you found in part b, derive the linearized model for the system.

5.15 A nonlinear system obeys the differential equation

$$\dot{y} + 4\sqrt{y} = 8 + B \cos 2t$$

and has the initial condition $y(0) = 5$.

a. Derive the linearized incremental model corresponding to the specified input.

b. Sketch the nonlinear element law for $y \geq 0$, and indicate the operating point and the linear approximation.

c. Evaluate the appropriate initial condition for the incremental variable.

5.16 A system obeys the differential equation

$$\ddot{x} + 3|\dot{x}|\dot{x} + 4x^3 = A + B \sin 2t$$

and has the initial conditions $x(0) = 2$ and $\dot{x}(0) = 1$.

a. Derive the linearized model for $A = 4$ and find the initial values of $\hat{x}$ and $\dot{\hat{x}}$.

b. Repeat part a for $A = 32$.

5.17 The disk shown in Figure P5.17 is supported by a nonlinear torsional spring and is subject to both linear and nonlinear frictional torques. The applied torque is $\tau_a(t) = 8 + \hat{\tau}_a(t)$.

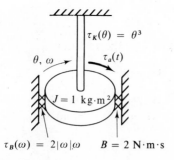

$$\tau_K(\theta) = \theta^3$$

$$\tau_a(t)$$

$$\theta, \omega$$

$$J = 1 \text{ kg·m}^2$$

$$\tau_B(\omega) = 2|\omega|\omega \qquad B = 2 \text{ N·m·s}$$

Figure P5.17

a. Find the operating point $\bar{\theta}$.

b. Derive the linearized input-output equation in terms of $\hat{\theta}(t) = \theta - \bar{\theta}$.

c. Find the initial values of the incremental variables $\hat{\theta}$ and $\dot{\hat{\theta}}$ if $\theta(0) = 0.5$ rad and $\dot{\theta}(0) = -0.5$ rad/s.

5.18 The mass-spring system shown in Figure P5.18 has a linear and a nonlinear spring and is subjected to the applied force $f_a(t)$.

a. Show that the nonlinear model is $2\ddot{x} + 3x + |x|x = f_a(t)$.

b. Solve for the operating point $\bar{x}$ when $\bar{f}_a = 10$ N.

c. Derive the linearized model when $f_a(t) = 10 + \hat{f}_a(t)$.

d. Find the initial values $\hat{x}(0)$ and $\dot{\hat{x}}(0)$ if $x(0) = 3$ m and $\dot{x}(0) = 1$ m/s.

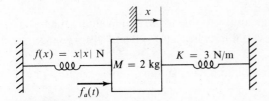

Figure P5.18

5.19 The rotating cylinder shown in Figure P5.19 has damping vanes and a linear frictional torque such that the motion is described by the non-linear equation

$$\dot{\omega} + 2|\omega|\omega + 2\omega = \tau_a(t)$$

where the applied torque is $\tau_a(t) = 12 + \hat{\tau}_a(t)$. Find the operating point and derive a linearized model in terms of the incremental angular velocity $\hat{\omega}$.

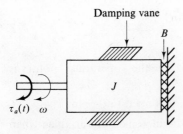

Figure P5.19

5.20 Derive a linearized input-output equation for the system described in Problem 2.18. Assume that $\bar{f}_a = 0$.

5.21 Derive a linearized model for the system described in Problem 2.19 in the form of a single differential equation involving only the variable $\hat{x}_2$.

5.22 Derive linearized state-variable equations for the system described in Problem 4.19.

5.23 A nonlinear system obeys the state-variable equations

$$\dot{x} = -x + y$$

$$\dot{y} = \frac{1}{y} + 4 + B \sin t$$

Find the linearized equations in state-variable form, and evaluate the initial conditions on the incremental variables when $x(0) = 0$ and $y(0) = -0.5$.

5.24 A second-order nonlinear system having state variables x and y obeys the equations

$$\dot{x} = -2x + y^3$$

$$\dot{y} = x + 4 + \cos t$$

a. Find the operating-point values $\bar{x}$ and $\bar{y}$.

b. Find the linearized state-variable equations in numerical form.

c. Find the linearized model as an input-output equation relating $\hat{x}$ and its derivatives to the incremental input.

5.25 A nonlinear system with state variables x and y and input $u(t)$ obeys the equations

$$\dot{x} = -2|x|x - y + u(t) - 6$$

$$\dot{y} = x - y - 6$$

a. Verify that when $u(t) = 2 + B \cos 2t$, the operating point is $\bar{x} = 0.7808$, $\bar{y} = -5.2192$.

b. Evaluate the linearized model about this operating point.

c. Evaluate the initial conditions for the incremental variables when $x(0) = 1$ and $y(0) = -6$.

5.26 Use (31) to linearize the following functions of the two variables x and y:

    a. $f(x, y) = x/y$

    b. $f(x, y) = \sqrt{x^2 + y^2}$

    c. $f(x, y) = xy$

5.27 Derive the linearized model and the operating point equations for a system described by

$$\dot{x} = -x + y$$
$$\dot{y} = xy + u(t)$$

where x and y are the state variables and $u(t)$ is the input. The non-linear system has two possible operating points. State any restrictions that must be placed on $\bar{u}$.

5.28 A linear time-varying system obeys the differential equation

$$\ddot{y} + 4\dot{y} + (10 + \cos 4t) y = u(t)$$

Assuming that the input $u(t)$ can be written as $u(t) = \bar{u} + \hat{u}(t)$, develop a fixed linear model for the system.

CHAPTER 6
SIMULATION DIAGRAMS

A simulation diagram is an interconnection of symbols representing certain basic mathematical operations, such as summation and integration, in such a way that the overall diagram obeys the system's mathematical model. In the diagram, the lines interconnecting the blocks represent the variables describing the system behavior, e.g., the input and state variables. Inspecting a simulation diagram of a system may provide new insight into the system's structure and behavior beyond that available from the differential equations themselves. As suggested by their name, such diagrams are useful in preparing a computer simulation (especially on an analog computer), although there are some differences between simulation diagrams and the patching diagrams that describe the interconnection of computer components.

We begin by defining the symbols for the basic mathematical operations that we will use. Then we consider the construction of the complete diagram, starting with the model in state-variable form and proceeding to the input-output form. The chapter concludes with an illustration of the nonstandard form that typically results from the coupled second-order equations formulated directly from free-body diagrams.

147

6.1 DIAGRAM BLOCKS

The operations we generally use in simulation diagrams are those found on analog computers, namely summation, gain, integration, and multiplication. Other operations, such as division and exponentiation, may be defined as needed. In this section, we describe the blocks representing these basic operations. Inputs and functions of time that may appear as coefficients in the differential equations are presumed to be available for use in the simulation diagram and hence need not be generated.

SUMMER

The addition and subtraction of variables is represented by a *summer*, or *summing junction*. A summer is represented by a circle having any number of arrows directed toward it (denoting inputs) and a single arrow directed away from it (denoting the output). Next to each entering arrowhead is a plus or minus symbol indicating the sign associated with the variable the particular arrow represents. The output variable, appearing at the one arrow leaving the circle, is defined to be the sum of all of the incoming variables, with the associated signs taken into account. A summer having the three inputs x_1, x_2, and x_3 appears in Figure 6.1.

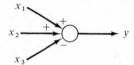

Figure 6.1 Summer, $y = x_1 + x_2 - x_3$.

GAIN

The multiplication of a single variable by a constant is represented by a *gain* block. We place no restriction on the value of the gain, which may be positive or negative. It may be an algebraic function of other constants and/or system parameters. Several self-explanatory examples are shown in Figure 6.2.

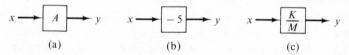

Figure 6.2 Gains. (a) $y = Ax$. (b) $y = -5x$. (c) $y = (K/M)x$.

INTEGRATOR

Dynamic elements are elements that obey either an integral or a differential equation. The only dynamic element we use in simulation diagrams is the *integrator*. If its input is denoted by $x(t)$ and its output by $y(t)$, the integrator obeys the relationship

$$y(t) = y(t_0) + \int_{t_0}^{t} x(\lambda)\, d\lambda \tag{1}$$

where the initial time t_0 is usually taken as zero and where λ is a dummy variable of integration. We can also write (1) in its differential form as

$$\dot{y}(t) = x(t) \tag{2}$$

We represent the integrator by a block containing an integral sign, with one input arrow and one output arrow, as shown in Figure 6.3(a). Generally, the initial condition $y(t_0)$ is not explicitly shown. However, it can be included, as shown in Figure 6.3(b).

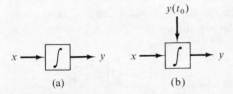

(a) (b)

Figure 6.3 Integrators. (a) Without initial condition shown. (b) With initial condition shown.

One is sometimes tempted to include in simulation diagrams a differentiator whose output y is the derivative of its input x. However, since differentiation does not normally occur in physical devices and since its simulation by computers is very sensitive to errors, we should avoid the use of a differentiator. In a later section of this chapter, we present techniques for constructing simulation diagrams without resorting to differentiators for those systems whose models involve one or more derivatives of the input. In fact, such a situation arose in Example 3.6, where we defined a new state variable that was a linear combination of the original state variable and the input in order to get the equations into state-variable form.

MULTIPLIER

A *multiplier* is represented by a block that is labeled MULT and has two inputs and one output. The output variable is the product of the two input variables; thus it is a nonlinear block, unlike the summer, gain, and integrator. Using multipliers in simulation diagrams, we may implement a variety of nonlinear characteristics and time-varying coefficients. Several examples are shown in Figure 6.4. In Figure 6.4(c), the time-varying coefficient $a(t)$ is assumed to be available as an input to the simulation diagram, as are the inputs to the system.

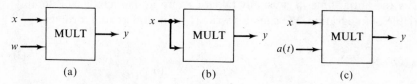

Figure 6.4 Multipliers. (a) $y = wx$. (b) $y = x^2$. (c) $y = a(t)x$.

OTHER NONLINEARITIES

Nonlinear operations other than multiplication can be represented by a rectangular box that shows the appropriate symbol (if there is one) or a sketch of the function. Three examples are shown in Figure 6.5.

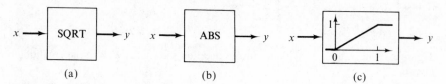

Figure 6.5 Nonlinearities.

(a) $y = \sqrt{x}$. (b) $y = |x|$. (c) $y = \begin{cases} 0 & x < 0 \\ x & 0 \le x \le 1 \\ 1 & x > 1 \end{cases}$

6.2 DIAGRAMS FOR STATE-VARIABLE MODELS

Having defined the necessary building blocks, we next develop techniques for constructing the simulation diagram when the system model is available in state-variable form. We shall demonstrate that because the state-variable model consists of a set of simultaneous first-order differential equations, it is particularly well suited for representation by a simulation diagram. We shall describe the procedure in detail and illustrate it first for a first-order system and then for systems of arbitrary order.

FIRST-ORDER SYSTEMS

The general form of the state-variable model for an nth-order system appears in (3.1) and (3.2). For a first-order system having a single input and a single output, the equations reduce to

$$\dot{q} = f(q, u, t) \tag{3a}$$

$$y = g(q, u, t) \tag{3b}$$

where q is the state variable, u is the input, and y is the output. The presence of t in the arguments of $f(q, u, t)$ and $g(q, u, t)$ allows for coefficients that may vary with time. The input u is presumed to be available for use in forming $\dot{q}$ and y, as are any time-varying coefficients. The derivative $\dot{q}$ can be formed by combining gains, summers, and any required nonlinear blocks, provided that the state variable q is available. However, if we make $\dot{q}$ the input to an integrator, the output of the integrator will be the state variable q, which now becomes available for use in generating $\dot{q}$. The resulting structure is circular in that $\dot{q}$ leads to q via the integrator, while q, along with u, is used to obtain $\dot{q}$ by simulating the function f. Having completed this portion of the diagram, we can ·form the output y by simulating the function g whose inputs are q and u, both of which are available.

To be more specific, we shall consider the important case of a fixed linear system that can be described by

$$\dot{q} = aq + bu \tag{4a}$$

$$y = cq + du \tag{4b}$$

where a, b, c, and d are constants. The steps in the process are as follows, and the simulation diagram as it appears after each step is shown in Figure 6.6.

1. Draw an integrator block with input $\dot{q}$ and output q.

2. Draw a summer to the left of the integrator having $\dot{q}$ as its output and bu and aq as its inputs. Label both input arrows with plus signs.

3. Draw a block with gain a having q as its input, and connect its output to the arrow labeled aq entering the summer.

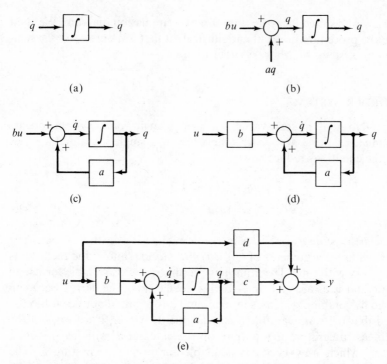

Figure 6.6 Simulation diagram for (4). (a), (b). (c) Partially complete. (d) Complete for (4a). (e) Complete.

4. Draw a block with gain b having u as its input, and connect its output to the arrow labeled bu entering the summer. This step completes the simulation of the state-variable equation (4a).

5. Append blocks with gains of c and d to the right side of the diagram and make their inputs be q and u, respectively. Then draw a summer with inputs cq and du and output y. The diagram is now complete.

If the system had been time-varying rather than fixed, the simulation diagram would be similar to Figure 6.6 except that we would diagram any time-varying coefficients by using a multiplier and showing the coefficient as an input to the multiplier, as shown in Figure 6.4(c).

It is interesting to note that Figure 6.6 clearly illustrates the feedback nature of the system, i.e., the fact that the rate of change of the output is dependent on the output itself, in addition to the input. The simulation diagrams of most dynamic systems have such feedback paths from the outputs of one or more integrators back to the integrator input by way of gain blocks and summing junctions.

EXAMPLE 6.1 Draw a simulation diagram for the high-speed vehicle discussed in Example 5.5 and described by the equation

$$M\dot{v} + Bv + D|v|v = f_a(t)$$

where $f_a(t)$ is the driving force and v is the velocity of the vehicle.

Solution The model for this vehicle is a first-order nonlinear differential equation with the velocity v as both the state variable and the output. The input is the applied force $f_a(t)$. In state-variable form, the model is

$$\dot{v} = \frac{1}{M}[-Bv - D|v|v + f_a(t)] \tag{5}$$

We initiate the simulation diagram by drawing an integrator with input $\dot{v}$ and output v. The coefficient $1/M$ appears as a gain whose output is $\dot{v}$ and whose input corresponds to the three terms inside the brackets in (5). The bracketed term is the output of a summer with the three inputs Bv, $D|v|v$, and $f_a(t)$, each having the appropriate sign at the corresponding arrowhead. The term $|v|v$ is formed by using a nonlinear block for the absolute value and a multiplier. The completed diagram is shown in Figure 6.7.

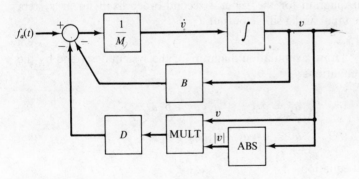

Figure 6.7 Simulation diagram for high-speed vehicle.

SECOND- AND HIGHER-ORDER SYSTEMS

An nth-order system having m inputs and p outputs is represented in state-variable form by a set of n first-order differential equations, such as (3.1), each having the form

$$\dot{q}_i = f_i(q_1, q_2, \ldots, q_n, u_1, \ldots, u_m, t) \qquad i = 1, 2, \ldots, n \tag{6}$$

As in (3.2), the outputs are given by algebraic equations of the form

$$y_j = g_j(q_1, q_2, ..., q_n, u_1, ..., u_m, t) \qquad j = 1, 2, ..., p \tag{7}$$

Generalizing the method we have just outlined for a single state-variable equation, we draw n integrators. We label the input and output of each integrator $\dot{q}_i$ and q_i, respectively, where $i = 1, 2, ..., n$. Then we construct the derivative of each state variable according to (6) from the n state variables that appear as integrator outputs and the m inputs, using gains, summers, and nonlinear blocks as required. Finally, we form the p outputs according to (7) from the state variables and inputs, using nondynamic elements.

In the case of a fixed linear model, (6) and (7) can be written as

$$\dot{q}_i = a_{i1} q_1 + a_{i2} q_2 + \cdots + a_{in} q_n + b_{i1} u_1 + \cdots + b_{im} u_m \qquad i = 1, 2, ..., n \tag{8a}$$

$$y_j = c_{j1} q_1 + c_{j2} q_2 + \cdots + c_{jn} q_n + d_{j1} u_1 + \cdots + d_{jm} u_m \qquad j = 1, 2, ..., p \tag{8b}$$

In the following example, we shall illustrate this procedure by constructing the simulation diagram for the general second-order fixed linear system having a single input and a single output.

EXAMPLE 6.2   Draw a simulation diagram for the system described by the state-variable equations

$$\dot{q}_1 = a_{11} q_1 + a_{12} q_2 + b_1 u \tag{9a}$$

$$\dot{q}_2 = a_{21} q_1 + a_{22} q_2 + b_2 u \tag{9b}$$

and the output equation

$$y = c_1 q_1 + c_2 q_2 + du \tag{10}$$

Solution   The system has the two state variables q_1 and q_2, so we need two integrators. Both $\dot{q}_1$ and $\dot{q}_2$ are the sums of three linear terms having constant coefficients. Thus each of the integrator inputs is the output of a summer which itself has three inputs as defined by (9a) and (9b). The output y is also formed by using a summer with three inputs according to (10). The resulting diagram is shown in Figure 6.8.

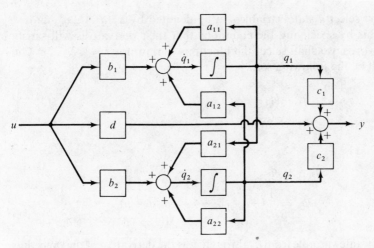

Figure 6.8 Simulation diagram for general fixed linear second-order system with a single input and single output in state-variable form.

6.3   FIXED LINEAR INPUT-OUTPUT MODELS

The input-output form of the model of an nth-order fixed linear system with input u and output y is the single differential equation

$$a_n y^{(n)} + a_{n-1} y^{(n-1)} + \cdots + a_0 y = b_m u^{(m)} + \cdots + b_0 u \qquad (11)$$

where $y^{(k)}$ denotes $d^k y/dt^k$, and where in practice $m \leq n$. Because (11) involves only the input u and the output y, we are free to define a set of n state variables and to select an output equation, provided only that the input-output relationship is unchanged. Specifically, we select the state variables such that the resulting simulation diagram has a simple structure and is easily drawn. Remember, however, that the choice of state variables and output equation satisfying (11) is not unique. (Desoer presents several forms that a simulation diagram derived from an input-output differential equation may take and discusses their properties in Section III-6.)

We shall develop the method in two stages. First we consider a system model that is of arbitrary order but has no input derivatives, and then we extend the method to include input derivatives.

MODELS WITHOUT INPUT DERIVATIVES

If input derivatives are excluded, (11) becomes

$$a_n y^{(n)} + a_{n-1} y^{(n-1)} + \cdots + a_0 y = b_0 u \qquad (12)$$

We must select n state variables, to be denoted by $q_1, q_2, ..., q_n$, and we shall do this by specifying the equations that their derivatives will satisfy. In other words, we shall select the functions f_i in (6) for $i = 1, 2, ..., n$. Our choice will be the following set:

$$\dot{q}_1 = q_2$$
$$\dot{q}_2 = q_3$$
$$\vdots \tag{13}$$
$$\dot{q}_{n-1} = q_n$$
$$\dot{q}_n = \frac{1}{a_n}(-a_{n-1}q_n - \cdots - a_0 q_1 + u)$$

When each state variable from q_2 through q_n is the derivative of the preceding one, the state variables are known as *phase variables* and the resulting simulation diagram will have a very simple structure.

Using (13), we can now derive the differential equation obeyed by q_1. From the first $n-1$ equations of (13), we have

$$q_2 = \dot{q}_1$$
$$q_3 = \ddot{q}_1$$
$$\vdots \tag{14}$$
$$q_n = q_1^{(n-1)}$$

Likewise,

$$\dot{q}_n = q_1^{(n)} \tag{15}$$

Substituting (14) and (15) into the last equation in (13) and rearranging, we get

$$a_n q_1^{(n)} + a_{n-1} q_1^{(n-1)} + \cdots + a_0 q_1 = u \tag{16}$$

which is identical to (12) except that it is in terms of the state variable q_1 rather than the output y and that the coefficient b_0 is not present on the right side.

To complete the new form of the system model, we take the output equation to be

$$y = b_0 q_1 \tag{17}$$

Now we verify that the combination of (13), the state-variable equations, and (17), the output equation, is equivalent to (12), the original input-output equation. We show that the left side of (12) is equal to $b_0 u$ when (16) and (17) are used. Because of (17), $y^{(k)} = b_0 q_1^{(k)}$ for $k = 0, 1, \ldots, n$. Thus

$$a_n y^{(n)} + a_{n-1} y^{(n-1)} + \cdots + a_0 y = a_n b_0 q_1^{(n)} + a_{n-1} b_0 q_1^{(n-1)} + \cdots + a_0 b_0 q_1$$
$$= b_0 [a_n q_1^{(n)} + a_{n-1} q_1^{(n-1)} + \cdots + a_0 q_1]$$
$$= b_0 u$$

Now we can draw the simulation diagram using the method presented in Section 6.2. However, because each state variable from q_2 through q_n is the derivative of the preceding one, it is more natural to arrange the n integrators in series. Then the input to the integrator at the far left is $\dot{q}_n$ and the output of the integrator at the far right is q_1, resulting in the partial diagram shown in Figure 6.9(a). We form the variable $\dot{q}_n$ by using a gain of $1/a_n$ and a summer

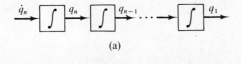

(a)

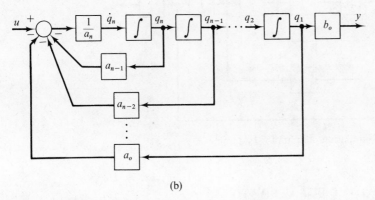

(b)

Figure 6.9 Simulation diagram of nth-order fixed linear system in input-output form. (a) Partial diagram. (b) Complete diagram.

with $n+1$ inputs consisting of the system input u and the n state variables with their appropriate gains. Adding the single gain of b_0 at the right implements the output equation and yields the system output y. The completed diagram is shown in Figure 6.9(b).

EXAMPLE 6.3   Draw the simulation diagram for the system described by

$$2\ddot{y} + 3\ddot{y} + 4\dot{y} + y = 3u$$

Solution   The system is third-order and hence has three state variables. We define the state variables q_1, q_2, and q_3 to obey the equations

$$\dot{q}_1 = q_2$$
$$\dot{q}_2 = q_3 \tag{18}$$
$$\dot{q}_3 = \tfrac{1}{2}(-3q_3 - 4q_2 - q_1 + u)$$

and select the output equation to be

$$y = 3q_1 \tag{19}$$

The simulation diagram corresponding to (18) and (19) is shown in Figure 6.10. The reader should verify that (18) and (19) reduce to the original input-output equation.

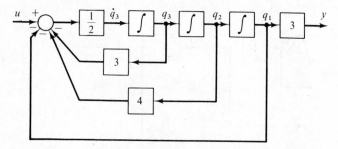

Figure 6.10 Simulation diagram for Example 6.3.

MODELS WITH INPUT DERIVATIVES

The general form of the input-output equation when input derivatives are present is given by (11). Although we could derive a simulation diagram for arbitrary n and for $m \le n$, the notation and proofs become cumbersome. Instead, we shall consider the two cases $m < n$ and $m = n$ for a general second-order system. The generalization to higher-order systems will then become clear. The reader should develop the results for the corresponding third-order cases in Problems 6.14 and 6.16.

m less than n   For a second-order system $n = 2$, and the most general case for $m < n$ is

$$a_2 \ddot{y} + a_1 \dot{y} + a_0 y = b_1 \dot{u} + b_0 u \tag{20}$$

To construct the simulation diagram, we select the state variables as if the term $b_1 \dot{u}$ did not exist in (20). Referring to the general form given by (13) with $n = 2$, we take

$$\dot{q}_1 = q_2 \tag{21a}$$

$$\dot{q}_2 = \frac{1}{a_2}(-a_1 q_2 - a_0 q_1 + u) \tag{21b}$$

For the output equation, we select

$$y = b_1 q_2 + b_0 q_1 \tag{22}$$

which reduces to (17) if $b_1 = 0$.

To verify that (21) and (22) reduce to (20), we first replace q_2 by $\dot{q}_1$ and $\dot{q}_2$ by $\ddot{q}_1$ in (21b) and (22), obtaining

$$a_2 \ddot{q}_1 + a_1 \dot{q}_1 + a_0 q_1 = u$$
$$y = b_1 \dot{q}_1 + b_0 q_1 \tag{23}$$

Substituting these expressions into the left side of (20) gives

$$a_2 \ddot{y} + a_1 \dot{y} + a_0 y = a_2(b_1 \ddot{q}_1 + b_0 \dot{q}_1) + a_1(b_1 \ddot{q}_1 + b_0 \dot{q}_1) + a_0(b_1 \dot{q}_1 + b_0 q_1)$$
$$= b_1(a_2 \ddot{q}_1 + a_1 \ddot{q}_1 + a_0 \dot{q}_1) + b_0(a_2 \ddot{q}_1 + a_1 \dot{q}_1 + a_0 q_1)$$
$$= b_1 \dot{u} + b_0 u$$

as required. The simulation diagram corresponding to (21) and (22) is shown in Figure 6.11. We observe that the state variables q_1 and q_2 are generated in exactly the same way as in Figure 6.9 for $n = 2$. The only effect of the $b_1 \dot{u}$ term on the simulation diagram is to include q_2 in the output equation.

Having followed this development of the simulation diagram for a second-order system with $m < n$, the reader should be able to generalize (21) and (22) for a system of arbitrary order. The state-variable equations that result from such a derivation are precisely those given in (13), and the output equation is

$$y = b_m q_{m+1} + b_{m-1} q_m + \cdots + b_0 q_1 \tag{24}$$

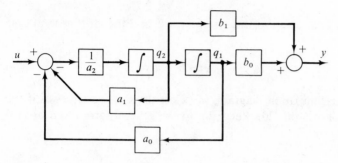

Figure 6.11 Simulation diagram for input-output model with $n = 2$ and $m = 1$.

The simulation diagram will be the one shown in Figure 6.9(b), except that the output variable y is the output of a summer that has $m+1$ arrows directed into it, with each arrow corresponding to one of the terms on the right side of (24). We shall give several examples after we discuss the case for $m = n$.

m equal to n   When the input-output form for a second-order system contains the second derivative of the input, the differential equation is (11) with $n = m = 2$, namely

$$a_2 \ddot{y} + a_1 \dot{y} + a_0 y = b_2 \ddot{u} + b_1 \dot{u} + b_0 u \qquad (25)$$

Just as we do when $m < n$, we define the state variables q_1 and q_2 to obey (21). The reader should verify that if the output variable y is taken as

$$y = b_2 \dot{q}_2 + b_1 q_2 + b_0 q_1 \qquad (26)$$

the combination of (21) and (26) is equivalent to (25).

The simulation diagram corresponding to this set of state-variable and output equations is shown in Figure 6.12. Note, however, that (26) does not agree with the form of the output equation specified by (7), because y is given as a function of the derivative of one of the state variables, in addition to the state variables themselves. In spite of this feature, (26) is a useful form of the output equation for many purposes, particularly because each of the six gains appearing in the simulation diagram is a coefficient of the input-output differential equation. Thus we can draw the diagram without performing any calculations to evaluate the gains.

It is not a difficult matter to modify (26) so that the output equation does not involve any state-variable derivatives. Using (21b) to replace $\dot{q}_2$ in (26), we obtain

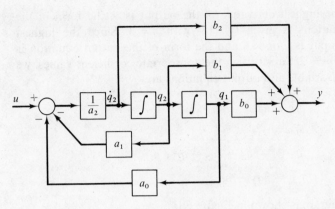

Figure 6.12 Simulation diagram for input-output model with $n = m = 2$ using $\dot{q}_2$.

$$y = \frac{b_2}{a_2}(-a_1 q_2 - a_0 q_1 + u) + b_1 q_2 + b_0 q_1$$

$$= \left(b_1 - \frac{a_1 b_2}{a_2}\right) q_2 + \left(b_0 - \frac{a_0 b_2}{a_2}\right) q_1 + \frac{b_2}{a_2} u \qquad (27)$$

which gives y in the form specified by (7). The simulation diagram corresponding to (21) and (27) is shown in Figure 6.13.

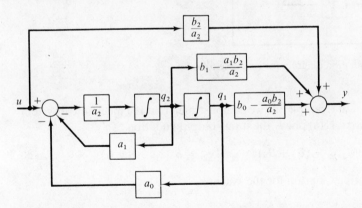

Figure 6.13 Simulation diagram for $n = m = 2$ with the output dependent on only the state variables and the input.

EXAMPLE 6.4 Draw the simulation diagram for the third-order system described by

$$\dddot{y} + 5\ddot{y} + 2\dot{y} + y = 3\ddot{u} - 4u$$

Solution Since the highest derivative of the output is $\ddot{y}$, the form of the state-variable equations is given by (13) with $n = 3$. Since the highest derivative of the input is $\ddot{u}$, $m < n$ and the form of the output equation is given by (24) with $m = 2$. Substituting the appropriate coefficient values, we find that the state-variable and output equations are

$$\begin{cases} \dot{q}_1 = q_2 \\ \dot{q}_2 = q_3 \\ \dot{q}_3 = -5q_3 - 2q_2 - q_1 + u \\ y = 3q_3 - 4q_1 \end{cases}$$

The simulation diagram is shown in Figure 6.14.

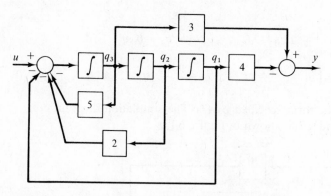

Figure 6.14 Simulation diagram for Example 6.4.

EXAMPLE 6.5 In Example 3.6, we demonstrated that the mechanical system shown in Figure 6.15(a) obeys the input-output equation

$$M_2 \ddot{x}_2 + (B_2 + B_3)\dot{x}_2 + Kx_2 = B_3 \dot{x}_1 + Kx_1(t) \tag{28}$$

Draw a simulation diagram for the system.

Solution We define the state variables q_1 and q_2 as in (21), with $a_2 = M_2$, $a_1 = B_2 + B_3$, and $a_0 = K$. Then

$$\dot{q}_1 = q_2$$

$$\dot{q}_2 = \frac{1}{M_2}[-(B_2 + B_3)q_2 - Kq_1 + x_1(t)]$$

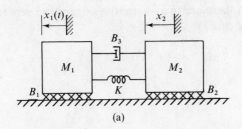

(a)

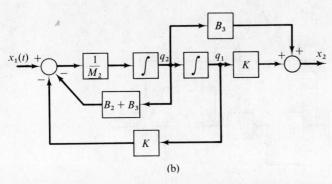

(b)

Figure 6.15 (a) Mechanical system for Example 6.5. (b) Simulation diagram.

where $x_1(t)$ is the system input since it denotes a specified displacement. The output equation for the displacement x_2 follows from (22) with $b_1 = B_3$ and $b_0 = K$. It is

$$x_2 = B_3 q_2 + K q_1$$

The simulation diagram is shown in Figure 6.15(b).

EXAMPLE 6.6   Draw a simulation diagram for the input-output equation

$$\tfrac{1}{2}\dddot{y} + 2\ddot{y} + y = 2\dddot{u} - 3\dot{u} + u \tag{29}$$

Solution   Since $n = m = 3$, we define the state variables q_1, q_2, and q_3 to obey

$$\dot{q}_1 = q_2 \tag{30a}$$

$$\dot{q}_2 = q_3 \tag{30b}$$

$$\dot{q}_3 = 2(-2q_3 - q_1 + u) \tag{30c}$$

For the output equation there are two alternatives. First, the generalization of (26) to the third-order case is

$$y = b_3 \dot{q}_3 + b_2 q_3 + b_1 q_2 + b_0 q_1$$

which, for this example, becomes

$$y = 2\dot{q}_3 - 3q_2 + q_1 \tag{31}$$

In Figure 6.16(a), the simulation diagram corresponding to (30) and (31), one of the terms being used to form the output is $\dot{q}_3$.

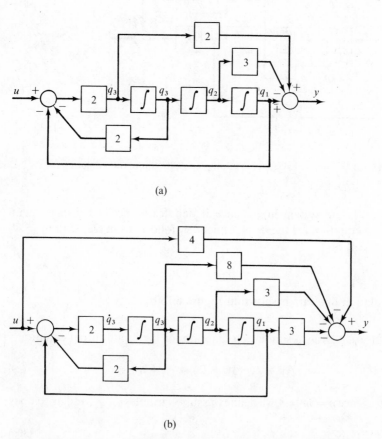

(a)

(b)

Figure 6.16 Simulation diagrams for the third-order system described by (29). (a) Using $\dot{q}_3$ in the output equation. (b) Using only state variables and the input in the output equation.

Alternatively, we can use (30c) to eliminate $\dot{q}_3$ in (31), giving

$$
\begin{aligned}
y &= 2[2(-2q_3 - q_1 + u)] - 3q_2 + q_1 \\
&= -8q_3 - 3q_2 - 3q_1 + 4u
\end{aligned}
\tag{32}
$$

which is a function of only the state variables and the input. The simulation diagram implementing (30) and (32) is shown in Figure 6.16(b).

6.4 MODELS IN NONSTANDARD FORM

The free-body diagram for a single mass or moment of inertia in a mechanical system will generally result in a second-order differential equation that contains one or more variables not directly associated with the motion of that element. If the system consists of several such elements, each having its own free-body diagram, then the overall model will probably appear first as a set of coupled second-order equations.

In such situations, one may wish to draw a simulation diagram directly from the nonstandard form of the model, which is in neither state-variable nor input-output form. Because the variables associated with a single mass are the position, velocity, and acceleration, the corresponding portion of the simulation diagram generally consists of a chain of two integrators with the variables being fed back to a summer at the input of the first integrator. Hence, the complete diagram is an interconnection of subdiagrams that contain two-integrator chains. These ideas are illustrated in the following example.

EXAMPLE 6.7   Draw a simulation diagram for the mechanical system shown in Figure 6.17 and modeled in Example 4.8.

Solution   In the solution of Example 4.8, we drew free-body diagrams for the rotor, the pinion gear, and the mass. These diagrams lead to the three equations

$$
J\ddot{\theta} + B_1\dot{\theta} + K(\theta - \theta_A) = \tau_a(t)
\tag{33a}
$$

$$
M\ddot{x} + B_2\dot{x} = f_c
\tag{33b}
$$

$$
Rf_c = K(\theta - \theta_A)
\tag{33c}
$$

which are identical to (4.44) except for a minor rearrangement of the terms and the use of the identities $\dot{\theta} = \omega$ and $\dot{x} = v$. In addition to (33), we have

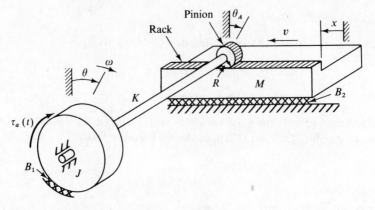

Figure 6.17 Mechanical system for Example 6.7.

the geometric relationship $\qquad R\theta_A = x$ (34)

We can draw the simulation diagram for the system by using the two differential equations (33a) and (33b) to construct a pair of two-integrator chains, one yielding θ and the other x. Then we use the two algebraic equations (33c) and (34) to obtain θ_A and f_c in terms of the variables θ and x. The completed diagram is shown in Figure 6.18(a). Note that the two parallel two-integrator chains are coupled through the term $K(\theta - \theta_A)$, which represents the torque transmitted by the shaft connecting the pinion gear to the rotational inertia.

We can also draw a simulation diagram to correspond to the state-variable form of the model given by (4.46), which is repeated here:

$$\dot{\theta} = \omega$$

$$\dot{\omega} = \frac{1}{J}\left[-K\theta - B_1\omega + \frac{K}{R}x + \tau_a(t)\right]$$

$$\dot{x} = v$$

$$\dot{v} = \frac{1}{M}\left(\frac{K}{R}\theta - \frac{K}{R^2}x - B_2 v\right)$$

For comparative purposes, this diagram is shown in Figure 6.18(b), and the reader should reconstruct it from the state-variable equations. As drawn, the diagram has the two-integrator chains arranged in series, rather than parallel. Although the contact force f_c is not explicitly shown in this version, it can be generated from θ and x by using gains and a summer. In comparing the two diagrams, remember that in order to draw the state-variable version we had to use the two algebraic equations (33c) and (34) to eliminate f_c and θ_A. In contrast, the diagram in Figure 6.18(a) can be drawn directly from the free-body diagram and (34).

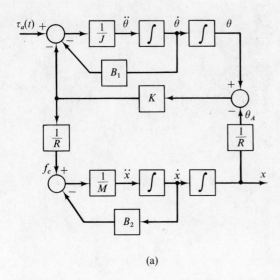

(a)

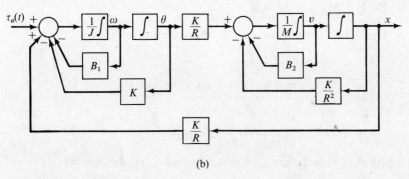

(b)

Figure 6.18 Simulation diagrams for the mechanical system of Example 6.7. (a) Nonstandard form. (b) State-variable form.

PROBLEMS

In Problems 6.1 through 6.4, draw simulation diagrams for each of the system models.

6.1 a. $2\ddot{y} + 3\dot{y} + 6y = 4u$

    b. $2\ddot{y} + 3\dot{y} + 6y = -2\dot{u} + 3u$

    c. $6\dddot{y} + 2\ddot{y} + 3\dot{y} + y = 5u$

    d. $\ddot{y} + 3\dot{y}^2 + 2y = u$

6.2 a. $\ddot{z} + 4\dot{z} + 2z = u$

b. $\ddot{z} + 4\dot{z} + 2z = 3\dot{u} + 2u$

c. $\dot{x} = 4x + 6y + 2u$
$\dot{y} = -2x - 3y$

d. $\ddot{y} + 2|\dot{y}|\dot{y} + y^3 = 5u$

6.3 a. $\ddot{y} + 5\dot{y} + 6y = 12u$

b. $\ddot{y} + 5\dot{y} + 6y = 2\ddot{u} + \dot{u} + 2u$

c. $\dot{x}_1 = 3x_1 + 5x_2 + 3$
$\dot{x}_2 = 4x_1 + 6x_2$

d. $\dot{v}_1 = 3v_1 + 2|v_2|v_2 + 3\cos 2t$
$\dot{v}_2 = v_1 + v_2$

6.4 a. $\dot{y} + 3y = 2u$

b. $\dot{y} + 3y = 2\dot{u} + u$

c. $\dddot{y} + 2\ddot{y} + 3\dot{y} + 4y = 2\ddot{u} - \dot{u} + 4u$

d. $\ddot{z} + 4\dot{z} + 3z = \ddot{u} + 2u$

6.5 a. Draw a simulation diagram for the system described by

$$\dot{\theta} = \omega$$

$$\dot{\omega} = -8\theta - 4\omega + 2x$$

$$\dot{x} = v$$

$$\dot{v} = 6\theta - 3x + u(t)$$

where θ, ω, x, and v are state variables and $u(t)$ is the input.

b. Verify that when θ is the output, the input-output differential equation is

$$\frac{d^4\theta}{dt^4} + 4\frac{d^3\theta}{dt^3} + 11\frac{d^2\theta}{dt^2} + 12\frac{d\theta}{dt} + 120\theta = 2u(t)$$

and draw the corresponding simulation diagram. Compare it with your answer to part a.

6.6 Draw a simulation diagram for the rotational mechanical system described by the following two equations, where $\theta_a(t)$ is the input.

$$2\ddot{\theta}_1 + \dot{\theta}_1 + 3(\dot{\theta}_1 - \dot{\theta}_2) + \theta_1 = \dot{\theta}_a + \theta_a(t)$$

$$2\ddot{\theta}_2 + \dot{\theta}_2 = 3(\dot{\theta}_1 - \dot{\theta}_2) + \theta_2$$

6.7 In Examples 2.2, 3.2, and 3.10, we derived three different forms for the model of the translational mechanical system shown in Figure 2.14(a). Draw a simulation diagram corresponding to each of these three forms and comment on their similarities and differences.

6.8 Draw simulation diagrams corresponding to the two versions for the model of the translational system given by (3.7) and (3.21), which were derived in Examples 3.4 and 3.9, respectively.

6.9 Draw a simulation diagram for the system modeled in Example 3.6.

6.10 Draw a simulation diagram for the system modeled in Example 3.7. Include the reaction force f_r as the output.

6.11 A rotational mechanical system is modeled in Example 4.2. Draw simulation diagrams for the forms of the model given in (4.21) and (4.23).

6.12 Draw simulation diagrams for the system models given by (4.34) and (4.35) in Example 4.5. *Hint:* For (4.35) use $z = \dot{\omega}_2 + (1/J_2)\tau_L(t)$.

6.13 Draw simulation diagrams for the system models given by (4.49) and (4.52) in Example 4.9. Comment on the difference between the two versions.

6.14 Generalize (21) and (22) to simulate (11) with $n = 3$ and $m = 2$. Verify that your result does indeed satisfy the specified input-output equation.

6.15 Verify that (21) and (26) satisfy (25).

6.16 Generalize (21) and (26) to simulate (11) with $n = 3$ and $m = 3$. Verify that your result satisfies the specified input-output equation.

6.17 Draw simulation diagrams for the equations presented in the statements of the following problems.

    a. Problem 3.21.

    b. Problem 3.22.

In Problems 6.18 through 6.23, draw simulation diagrams for the system described in the problem indicated.

6.18 Parts a and b of Problem 2.9.

6.19 Problem 2.17.

6.20 Problem 3.26.

6.21 Problem 3.27.

6.22 Parts a, b, and c of Problem 4.18.

6.23 Problem 4.21.

CHAPTER 7
NUMERICAL SOLUTIONS

It is not usually possible to find a closed-form solution for nonlinear or time-varying dynamic systems. Although we can find the responses of fixed linear systems analytically for systems of any order, it is not practical to do so if the order exceeds three or four. When an analytical solution is impossible or impractical, one can use a digital computer to obtain an approximation to the actual response at discrete points in time. This process of *numerical solution* for the response is often referred to as *digital simulation*, and it requires the numerical integration of the differential equations comprising the system model.

The methods for obtaining numerical solutions require that we start with the system model in state-variable form, so we shall assume that the state-variable and output equations are available in the standard form of (3.1) and (3.2). We shall also need the initial values of the state variables. We first present the simplest possible method of solution, which is known as Euler's method. Then we discuss the general form that any computer program for the numerical solution must have and illustrate it with examples using FORTRAN and two computer languages developed specifically for solving models of dynamic systems. The chapter concludes with a description of several numerical methods other than Euler's method and some comments on their relative merits in terms of computation time and accuracy.

7.1 EULER'S METHOD

Because it is the simplest, Euler's method is a logical introduction to the many methods for the numerical solution of dynamic models. In spite of its simplicity, the method can be of considerable value in solving practical problems, although one or more of the methods in Section 7.4 are generally employed. In order to avoid undue complexities in notation, we shall limit our initial discussion to first-order systems and then generalize the results to second- and higher-order systems.

FIRST-ORDER SYSTEMS

Consider a first-order system that has the single state variable $q(t)$ and the single input $u(t)$. Assume that the state variable is known at some initial time t_0 and that the input is known for all $t \geq t_0$. Hence, we can compute numerical values for $q(t_0)$ and for $u(t)$ at any $t \geq t_0$.

The state-variable equation has the form

$$\dot{q} = f(q, u, t) \tag{1}$$

where the function f is an algebraic function of the variables $q(t)$, $u(t)$, and, for time-varying systems, t itself. If the system is fixed and linear, (1) has the form $\dot{q} = aq + bu$ where a and b are constants.

For a typical system, the analytical solution to (1) for a specific input and for a specific initial value of the state variable is indicated by the curve labeled $q(t)$ in Figure 7.1. The results of a possible numerical solution (obviously not a very accurate one) are indicated by the dots that occur at $t = t_0, t_0 + T, t_0 + 2T, \ldots$ and have the values $\tilde{q}(t_0 + kT)$ where k takes on the integer values $0, 1, 2, \ldots$. The symbol $\tilde{q}$ emphasizes that, except for some trivial cases, the numerical solution $\tilde{q}$ is different from the analytical solution q. Furthermore, use of the integer k in the argument of $\tilde{q}$ indicates the discrete rather than continuous nature of the numerical solution.

The parameter T is the amount of time between the discrete numerical-solution points. It is known as the *step size*. For our discussion of Euler's method, we shall assume that T is constant and that it has been specified.

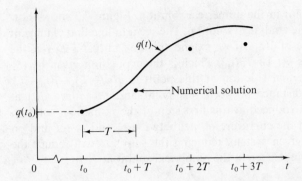

Figure 7.1 Comparison of numerical and analytical solutions for a typical system.

We select the initial point $\tilde{q}(t_0)$ in the numerical solution to be the specified initial value of the state variable, giving

$$\tilde{q}(t_0) = q(t_0) \tag{2}$$

Next, we must devise an algorithm by which the computer can evaluate the next point in the numerical solution, $\tilde{q}(t_0 + T)$. To assist in developing such an algorithm, we expand $q(t)$ in its Taylor series about the point $t = t_0$, getting

$$q(t) = q(t_0) + \dot{q}(t_0)(t - t_0) + \frac{1}{2!}\ddot{q}(t_0)(t - t_0)^2 + \cdots \tag{3}$$

When $t = t_0 + T$, (3) reduces to

$$q(t_0 + T) = q(t_0) + T\dot{q}(t_0) + \frac{T^2}{2!}\ddot{q}(t_0) + \cdots \tag{4}$$

which would give the exact value of $q(t_0 + T)$ if we could evaluate all the derivatives of $q(t)$ at $t = t_0$ and sum the resulting infinite series. Unfortunately, neither of these conditions is satisfied; we must settle for an approximate result.

We presumably know the initial value $q(t_0)$ in (4), and we can calculate $\dot{q}(t_0)$ from (1). Since the computer cannot carry out differentiation in analytical form, we cannot evaluate the remaining terms on the right side of (4) exactly. However, if we take the step size T sufficiently small, we can make the sum of those terms involving $\ddot{q}$ and higher-order derivatives arbitrarily small—so small they can be neglected. From (1) and (2), we can write the resulting approximation as

$$\tilde{q}(t_0 + T) = \tilde{q}(t_0) + Tf[\tilde{q}(t_0), u(t_0), t_0] \tag{5}$$

which is the second point in the numerical solution. Figure 7.2 shows how $\tilde{q}(t_0 + T)$ is related to the analytical solution. The straight line that is tangent to $q(t)$ at t_0 has a slope of $\dot{q}(t_0)$. If we extend this line until $t = t_0 + T$, the value of the ordinate is $q(t_0) + T\dot{q}(t_0)$, which is the expression given by (5). The error in the numerical solution at this point is $q(t_0 + T) - \tilde{q}(t_0 + T)$, which is the vertical distance between this point and the curve $q(t)$ at $t = t_0 + T$. The error introduced at this first step of the numerical solution is clearly dependent on the curvature of $q(t)$ between t_0 and $t_0 + T$ and on the step size T. For a given system, reducing the step size will reduce the error introduced by a single step in the numerical solution.

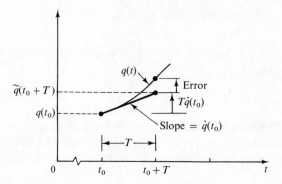

Figure 7.2 Geometric interpretation of Euler's method.

Having calculated the approximate response at $t = t_0 + T$, we calculate the next point in the sequence by expanding the approximate solution $\tilde{q}$ about $t = t_0 + T$ and again retaining only the first two terms in the expansion. The result is

$$\tilde{q}(t_0 + 2T) = \tilde{q}(t_0 + T) + Tf[\tilde{q}(t_0 + T), u(t_0 + T), t_0 + T] \qquad (6)$$

Note that all the terms on the right side of (6) are defined at $t = t_0 + T$ and thus can be calculated only after the first step is complete. At this point, it should be clear that the general algorithm for going from the $(k-1)$st step to the kth step is

$$\tilde{q}[t_0 + kT] = \tilde{q}[t_0 + (k-1)T]$$
$$+ Tf\{\tilde{q}[t_0 + (k-1)T], u[t_0 + (k-1)T], t_0 + (k-1)T\} \quad (7)$$

The algorithm represented by (7) is known as *Euler's method*, and it is the simplest of a wide variety of methods available for the numerical solution of ordinary differential equations.

We can construct an alternative development of (7) by using heuristic reasoning rather than a Taylor-series expansion. The value of the derivative $\dot{q}(t)$ at time t_0 is defined as

$$\dot{q}(t_0) = \lim_{T \to 0} \frac{q(t_0 + T) - q(t_0)}{T}$$

Hence, if T is sufficiently small but nonzero, we can use (1) to write

$$\frac{q(t_0 + T) - q(t_0)}{T} \simeq f[q(t_0), u(t_0), t_0]$$

which can then be rearranged as

$$q(t_0 + T) \simeq q(t_0) + Tf[q(t_0), u(t_0), t_0] \tag{8}$$

If we use the symbol $\tilde{q}$ in place of q in (8), we can replace the approximation by an equality, resulting in (5). If we apply the same reasoning at the general time $t = t_0 + kT$, then Euler's method as given by (7) results.

We can simplify the notation used to express Euler's method by taking the initial time as $t_0 = 0$ and by not showing the step size T in the arguments of the variables. Hence, we can write the state variable $\tilde{q}(t_0 + kT)$ as $\tilde{q}^{(k)}$, the input $u(t_0 + kT)$ as $u^{(k)}$, and time $t_0 + kT$ as k. With this notation, Euler's method for a first-order system leads to the following set of equations:

$$\begin{aligned}
\tilde{q}^{(0)} &= q(0) \\
\tilde{q}^{(1)} &= \tilde{q}^{(0)} + Tf[\tilde{q}^{(0)}, u^{(0)}, 0] \\
\tilde{q}^{(2)} &= \tilde{q}^{(1)} + Tf[\tilde{q}^{(1)}, u^{(1)}, 1] \\
&\vdots \\
\tilde{q}^{(k)} &= \tilde{q}^{(k-1)} + Tf[\tilde{q}^{(k-1)}, u^{(k-1)}, k-1]
\end{aligned} \tag{9}$$

If the output variable y is given by the algebraic equation

$$y(t) = g[q, u, t]$$

where the function g is an algebraic function of the state variable, the input, and time t, we can evaluate the numerical approximation to the output by using

$$\tilde{y}^{(k)} = g[\tilde{q}^{(k)}, u^{(k)}, k]$$

after computing the corresponding state variable.

EXAMPLE 7.1 Use Euler's method with $T = 2$ to obtain the first six points in a numerical solution of the equation

$$\dot{q} + 0.2q = u(t)$$

where $q(0) = 1$ and $u(t) = 1$ for $t \geq 0$.

Solution Since $T = 2$ and $t_0 = 0$, $\tilde{q}^{(k)} = \tilde{q}(2k)$. The function $f(q, u, t)$ is $f = -0.2q + u(t)$ where $u(t) = 1$. With $\tilde{q}^{(0)} = 1$, evaluation of (9) gives

$$\tilde{q}(0) = \tilde{q}^{(0)} = 1$$

$$\tilde{q}(2) = \tilde{q}^{(1)} = 1 + 2[(-0.2 \times 1) + 1] = 2.60$$

$$\tilde{q}(4) = \tilde{q}^{(2)} = 2.60 + 2[(-0.2 \times 2.60) + 1] = 3.56$$

$$\tilde{q}(6) = \tilde{q}^{(3)} = 3.56 + 2[(-0.2 \times 3.56) + 1] = 4.136$$

$$\tilde{q}(8) = \tilde{q}^{(4)} = 4.136 + 2[(-0.2 \times 4.136) + 1] = 4.482$$

$$\tilde{q}(10) = \tilde{q}^{(5)} = 4.482 + 2[(-0.2 \times 4.482) + 1] = 4.689$$

The analytical solution to the differential equation and its initial condition is $q(t) = 5 - 4\epsilon^{-t/5}$, which approaches 5.0 as time approaches infinity.* The first six points in the numerical solution, along with the corresponding points in the exact solution and the error, appear in Table 7.1. The two solutions are plotted in Figure 7.3. It is apparent that for the step size used, the numerical solution is neither very close to nor very far from the exact solution. If greater accuracy is required, two alternatives are available. We may either repeat Euler's method with a smaller step size or use a different integration method that gives greater accuracy for the same step size.

Table 7.1 Numerical solution for Example 7.1

t	k	Numerical solution, $\tilde{q}(kT)$	Exact solution, $q(t)$	Error, $q(t) - \tilde{q}(kT)$
0	0	1.000	1.000	0
2	1	2.600	2.319	−0.281
4	2	3.560	3.203	−0.357
6	3	4.136	3.795	−0.341
8	4	4.482	4.192	−0.290
10	5	4.689	4.459	−0.230

* The symbol ϵ denotes the base of the system of natural logarithms and has the value 2.71828 ....

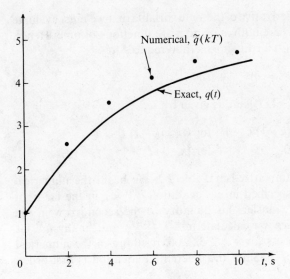

Figure 7.3 Comparison of analytical and numerical solutions for Example 7.1.

EXAMPLE 7.2 Use Euler's method to solve numerically the first-order linear time-varying equation

$$\dot{q} = a(t)q \tag{10}$$

for the coefficient

$$a(t) = \begin{cases} t - 1 & \text{for } 0 \le t < 1 \\ 0 & \text{for } t \ge 1 \end{cases} \tag{11}$$

sketched in Figure 7.4. The initial condition is $q(0) = 1$. Use a step size of $T = 0.25$, and continue the solution until $t = 1.5$.

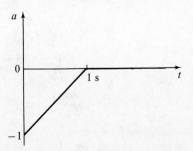

Figure 7.4 Time-varying coefficient for Example 7.2.

Solution From (10), the derivative of the state variable that we must evaluate at each step is $f = a(t)q$ where both $a(t)$ and q are functions of time. Hence, the general expression in (9) for Euler's method reduces to

$$\tilde{q}^{(k)} = \tilde{q}^{(k-1)} + Ta^{(k-1)}\tilde{q}^{(k-1)} \tag{12}$$

where $a^{(k-1)}$ denotes $a[(k-1)T]$ and is given by

$$a^{(k-1)} = \begin{cases} (k-1)T - 1 & \text{for } 0 \le (k-1)T < 1 \\ 0 & \text{for } (k-1)T \ge 1 \end{cases} \tag{13}$$

The calculations are summarized in Table 7.2. We begin the numerical solution by entering the specified initial condition $\tilde{q}^{(0)} = 1$ in the far right of the first row. Next we transfer this quantity to the second row in the column labeled $\tilde{q}^{(k-1)}$. Then we calculate $a^{(k-1)}$, $f^{(k-1)}$, and finally $\tilde{q}^{(k)}$ for $k = 1$. We repeat this process for $k = 2, 3, ..., 6$, resulting in the numerical solution plotted in Figure 7.5. Since the differential equation is first-order,

Table 7.2 Numerical solution for Example 7.2

t	k	$\tilde{q}^{(k-1)}$	$a^{(k-1)}$	$f^{(k-1)} = a^{(k-1)}\tilde{q}^{(k-1)}$	$\tilde{q}^{(k)}$
0	0	—	—	—	1.0
0.25	1	1.0	−1.0	$(-1.0)(1.0) = -1.0$	$1.0 + 0.25(-1.0) = 0.75$
0.50	2	0.75	−0.75	$(-0.75)(0.75) = -0.5625$	$0.75 + 0.25(-0.5625) = 0.6094$
0.75	3	0.6094	−0.50	$(-0.50)(0.6094) = -0.3047$	$0.6094 + 0.25(-0.3047) = 0.5332$
1.00	4	0.5332	−0.25	$(-0.25)(0.5332) = -0.1333$	$0.5332 + 0.25(-0.1333) = 0.4999$
1.25	5	0.4999	0	$(0)(0.4999) = 0$	$0.4999 + 0.25(0) = 0.4999$
1.50	6	0.4999	0	$(0)(0.4999) = 0$	$0.4999 + 0.25(0) = 0.4999$

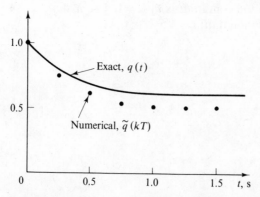

Figure 7.5 Comparison of analytical and numerical solutions for Example 7.2.

we can solve it analytically, obtaining

$$q(t) = \begin{cases} \epsilon^{(0.5t^2 - t)} & \text{for } 0 \leq t < 1 \\ \epsilon^{-0.5} & \text{for } t \geq 1 \end{cases}$$

which is shown in Figure 7.5 for comparison. As the reader can verify, we can improve the accuracy of the numerical solution by using a step size smaller than 0.25.

SECOND- AND HIGHER-ORDER SYSTEMS

If a system has more than one state variable, there will be an equation like (1) for each state variable, and the numerical solutions of these equations must proceed in parallel. In other words, regardless of the integration algorithm we use, we must compute each of the state variables through the $(k-1)$st step before going to the kth step. Conceptually, the numerical solution of such models is no more difficult than the numerical solution of first-order models.

For a system with two state variables and a single input, we can write the state-variable equations as

$$\begin{aligned} \dot{q}_1 &= f_1(q_1, q_2, u, t) \\ \dot{q}_2 &= f_2(q_1, q_2, u, t) \end{aligned} \tag{14}$$

where f_1 and f_2 are two different algebraic functions of the state variables, the input, and t. By expanding both $q_1(t)$ and $q_2(t)$ in their respective Taylor series and repeating the steps that led to (9), we obtain the set of equations

$$\begin{aligned} \tilde{q}_1^{(0)} &= q_1(0) \\ \tilde{q}_2^{(0)} &= q_2(0) \end{aligned} \tag{15}$$

$$\begin{aligned} \tilde{q}_1^{(1)} &= \tilde{q}_1^{(0)} + Tf_1[\tilde{q}_1^{(0)}, \tilde{q}_2^{(0)}, u^{(0)}, 0] \\ \tilde{q}_2^{(1)} &= \tilde{q}_2^{(0)} + Tf_2[\tilde{q}_1^{(0)}, \tilde{q}_2^{(0)}, u^{(0)}, 0] \end{aligned} \tag{16}$$

$$\begin{aligned} \tilde{q}_1^{(2)} &= \tilde{q}_1^{(1)} + Tf_1[\tilde{q}_1^{(1)}, \tilde{q}_2^{(1)}, u^{(1)}, 1] \\ \tilde{q}_2^{(2)} &= \tilde{q}_2^{(1)} + Tf_2[\tilde{q}_1^{(1)}, \tilde{q}_2^{(1)}, u^{(1)}, 1] \end{aligned} \tag{17}$$

$$\vdots$$

$$\begin{aligned} \tilde{q}_1^{(k)} &= \tilde{q}_1^{(k-1)} + Tf_1[\tilde{q}_1^{(k-1)}, \tilde{q}_2^{(k-1)}, u^{(k-1)}, k-1] \\ \tilde{q}_2^{(k)} &= \tilde{q}_2^{(k-1)} + Tf_2[\tilde{q}_1^{(k-1)}, \tilde{q}_2^{(k-1)}, u^{(k-1)}, k-1] \end{aligned} \tag{18}$$

There are two important features to note about (15) through (18). First, we must calculate both $\tilde{q}_1^{(1)}$ and $\tilde{q}_2^{(1)}$ before we calculate $\tilde{q}_1^{(2)}$ and $\tilde{q}_2^{(2)}$, since the right sides of (17) involve both the state variables computed for $k = 1$. Second, the state-variable formulation provides a natural form for the numerical solution of a second- or higher-order system, because it consists of simultaneous equations for the derivatives of each of the state variables. This is true regardless of the integration method we use. In fact, virtually all such algorithms are programmed for the solution of simultaneous first-order differential equations. This characteristic is an important reason for introducing the state-variable formulation early in this book.

If the output of the second-order system described by (14) is given by

$$y(t) = g(q_1, q_2, u, t)$$

we can evaluate the numerical approximation to the output at each time point after we have calculated the state variable. Specifically, for the kth time point,

$$\tilde{y}^{(k)} = g[q_1^{(k)}, \tilde{q}_2^{(k)}, u^{(k)}, k]$$

For an nth-order system having m inputs and p outputs and described by the state-variable and output equations

$$\dot{q}_i = f_i(q_1, q_2, ..., q_n, u_1, u_2, ..., u_m, t) \qquad \text{for } i = 1, 2, ..., n$$

$$y_j = g_j(q_1, q_2, ..., q_n, u_1, u_2, ..., u_m, t) \qquad \text{for } j = 1, 2, ..., p$$

applying Euler's method yields the corresponding equations

$$\tilde{q}_i^{(k)} = \tilde{q}_i^{(k-1)} + Tf_i[\tilde{q}_1^{(k-1)}, \tilde{q}_2^{(k-1)}, ..., \tilde{q}_n^{(k-1)}, u_1^{(k-1)},$$

$$u_2^{(k-1)}, ..., u_m^{(k-1)}, k-1] \qquad \text{for } i = 1, 2, ..., n$$

$$\tilde{y}_j^{(k)} = g_j[\tilde{q}_1^{(k)}, \tilde{q}_2^{(k)}, ..., \tilde{q}_n^{(k)}, u_1^{(k)}, ..., u_m^{(k)}, k] \qquad \text{for } j = 1, 2, ..., p$$

$$(19)$$

Although the notation becomes rather involved in the general case, the structure of the numerical solution and the basic concepts are exactly the same as those we encountered in the first- and second-order cases.*

---

* With the notation of Chapter 17, we can write (19) as

$$\tilde{\mathbf{q}}^{(k)} = \tilde{\mathbf{q}}^{(k-1)} + T\mathbf{f}[\tilde{\mathbf{q}}^{(k-1)}, \mathbf{u}^{(k-1)}, k-1]$$

$$\tilde{\mathbf{y}}^{(k)} = \mathbf{g}[\tilde{\mathbf{q}}^{(k)}, \mathbf{u}^{(k)}, k]$$

where $\tilde{\mathbf{q}}, \mathbf{u}, \tilde{\mathbf{y}}, \mathbf{f}$, and $\mathbf{g}$ are vectors.

EXAMPLE 7.3   In Example 3.1, the state-variable equations for the mass-spring-dashpot system shown in Figure 7.6 were found to be

$$\dot{x} = v$$

$$\dot{v} = \frac{1}{M}[-Kx - Bv + f_a(t)]$$

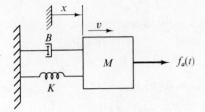

Figure 7.6 Mechanical system for Example 7.3.

Use Euler's method to calculate the response of the system to the input $f_a(t) = \sin \omega t$, using as parameter values $M = 1$ kg, $B = 2$ N·s/m, $K = 4$ N/m, and $\omega = 0.5$ rad/s and the initial conditions $x(0) = 1$ m, $v(0) = 0$. Use a step size of $T = 0.2$ s and compute the numerical approximation to the displacement x over the interval $0 \le t \le 1.0$ s.

Solution   In numerical form, the state-variable equations are

$$\dot{x} = v$$

$$\dot{v} = -4x - 2v + \sin 0.5t$$

where $x(0) = 1$ m and $v(0) = 0$. Retaining x and v as the state variables rather than using q_1 and q_2, we have for the functions f_1 and f_2 in (14)

$$f_1(x, v, u, t) = v$$

$$f_2(x, v, u, t) = -4x - 2v + \sin 0.5t$$

The results of using (15) through (18) are as follows:

$t = 0$ s:
$$\tilde{x}(0) = 1.0 \text{ m}$$
$$\tilde{v}(0) \doteq 0 \text{ m/s}$$

$t = 0.2$ s:
$$\tilde{x}(0.2) = 1.0 + 0.2(0) = 1.0 \text{ m}$$
$$\tilde{v}(0.2) = 0 + 0.2[-4(1.0) - 2(0) + \sin 0] = -0.80 \text{ m/s}$$

$t = 0.4$ s:
$$\tilde{x}(0.4) = 1.0 + 0.2(-0.80) = 0.840 \text{ m}$$
$$\tilde{v}(0.4) = -0.80 + 0.2[-4(1.0) - 2(-0.80) + \sin 0.10]$$
$$= -1.260 \text{ m/s}$$

$t = 0.6$ s:
$$\tilde{x}(0.6) = 0.840 + 0.2(-1.260) = 0.588 \text{ m}$$
$$\tilde{v}(0.6) = -1.260 + 0.2[-4(0.84) - 2(-1.26) + \sin 0.20]$$
$$= -1.388 \text{ m/s}$$

$t = 0.8$ s:
$$\tilde{x}(0.8) = 0.588 + 0.2(-1.388) = 0.310 \text{ m}$$
$$\tilde{v}(0.8) = -1.388 + 0.2[-4(0.588) - 2(-1.388) + \sin 0.30]$$
$$= -1.244 \text{ m/s}$$

$t = 1.0$ s:
$$\tilde{x}(1.0) = 0.310 + 0.2(-1.244) = 0.061 \text{ m}$$
$$\tilde{v}(1.0) = -1.244 + 0.2[-4(0.310) - 2(-1.244) + \sin 0.40]$$
$$= -0.917 \text{ m/s}$$

7.2 COMPUTER IMPLEMENTATION

By this point, it should be apparent that calculating numerical solutions by hand can be tedious and is susceptible to error. To attain the potential benefits of numerical solutions, one must program a digital computer, or perhaps a programmable calculator, to carry out the calculations and present the results in an appropriate fashion. Following a brief discussion of the methods for implementing such calculations on a digital computer, we shall present computed results for the system discussed in Example 7.3.

In addition to means for the repetitive solution of equations such as (18), any digital computer program used for implementing such calculations must include means for initializing and terminating the calculations, for specifying the appropriate parameter values, and for generating appropriate output in the form of tables and graphs. The necessary operations and the sequence in which they must be performed are summarized in Figure 7.7.

Although it is possible to write a program in a general language such as FORTRAN or BASIC to implement the simulation for a reasonably specific model, it is usually more efficient in terms of the user's time to employ one of the many simulation languages that are available. Two such languages are DYNAMO (DYNAmic MOdels) and CSMP (Continuous System Modeling Program). In nearly all cases, the model must be in state-variable form—that is, it must be a set of simultaneous first-order differential

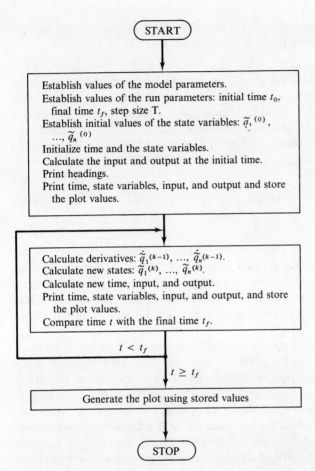

Figure 7.7 Operations required for the numerical solution of system models.

equations—before programming begins. To illustrate the required procedure, we shall consider examples in which FORTRAN, DYNAMO, and
CSMP are used to simulate the response of a simple translational
mechanical system.

EXAMPLE 7.4   Write a FORTRAN program to obtain the response of the
mechanical system discussed in Example 7.3 and shown in Figure 7.6, for
which the state-variable equations are

$$\dot{x} = v$$

$$\dot{v} = -4x - 2v + \sin 0.5t$$

The initial conditions are $x(0) = 1$ m and $v(0) = 0$, and the simulation should cover the interval $0 \leq t \leq 17.5$ s with a step size of $T = 0.03125$ s.

Solution A FORTRAN program that will implement the numerical solution of the model with the parameter values and initial conditions to be read from data cards is given in Figure 7.8(a). Except that a plot has not been included and the results are not printed at every time point, the program follows the outline presented in Figure 7.7. The tabulated numerical output generated by the program is shown in Figure 7.8(b).

EXAMPLE 7.5 Repeat Example 7.4, using DYNAMO as the computer language.

Solution DYNAMO is a computer language* developed especially for solving models of dynamic systems. It uses Euler's method and has instructions for generating tabulated output and plots with a minimum of statements by the user. For the system under consideration, the set of DYNAMO statements in Figure 7.9(a) results in the tabulated and plotted output shown in Figure 7.9(b) and Figure 7.9(c).

The model parameters and initial-condition values are specified on the C cards (for *c*onstant), and the state variables are initialized on the N cards (for *i*nitialization). The parameters for the run are entered on the SPEC card (for *spec*ification). The two state variables are computed according to Euler's method by the L cards (for *l*evel, which is analogous to state), and the derivatives by the R cards (for *r*ate, which is analogous to derivative). The input $f_a(t)$ is calculated by the A card (for *a*uxiliary, which is any algebraic expression). The RUN card at the end signifies the end of the model and causes processing to begin. If we want additional runs using different parameter values or different initial conditions, we can change any of the numerical values entered on the C cards or the SPEC card without having to recompile the program.

Comparing the FORTRAN and DYNAMO programs shown in Figure 7.8(a) and Figure 7.9(a), respectively, reveals that using DYNAMO involves fewer statements, in spite of the fact that the FORTRAN program has no plotting capability as written. Of considerable advantage is the fact that the DYNAMO user does not have to include format statements for the input and output or statements generating the time variable and testing for completion of the run. Comparison of Figure 7.8(b) and Figure 7.9(b) indicates that both programs resulted in the same numerical solution for x and v, as they must since both use the same step size and algorithm. Any apparent

---

* Instructions for using DYNAMO are given in Pugh.

```
C   EXAMPLE 7-4: FORTRAN SOLUTION OF
C       MASS-SPRING-DASHPOT SYSTEM
C   INITIALIZATION
        REAL M,K
        READ(5,5)M,B,K,OMEGA
        READ(5,10)TZERO,TFIN,TSTEP,TPRT
        READ(5,15)XZERO,VZERO
        TIME=TZERO
        TPNXT=TZERO+TPRT
        X=XZERO
        V=VZERO
        FA=SIN(OMEGA*TIME)
        WRITE(6,20)
        WRITE(6,25)TIME,X,V,U
        DEL=0.5*TSTEP
C   START SOLUTION
    100 CONTINUE
        XDOT=V
        VDOT=(-K*X-B*V+FA)/M
        X=X+TSTEP*XDOT
        V=V+TSTEP*VDOT
        TIME=TIME+TSTEP
        FA=SIN(OMEGA*TIME)
        IF(ABS(TIME-TPNXT).GT.DEL)GO TO 100
        TPNXT=TPNXT+TPRT
        WRITE(6,25)TIME,X,V,FA
        IF(ABS(TIME-TFIN).GT.DEL)GO TO 100
        STOP
      5 FORMAT(4F10.4)
     10 FORMAT(4F10.4)
     15 FORMAT(2F10.4)
     20 FORMAT(1H1,8X,'TIME',14X,'POSITION',10X,
      C   'VELOCITY',10X,'APPLIED FORCE')
     25 FORMAT(1X,4F18.6)
        END
```

TIME	POSITION	VELOCITY	APPLIED FORCE
0.0	1.000000	0.0	0.0
0.500000	0.671093	-1.068062	0.247404
1.000000	0.176726	-0.753453	0.479425
1.500000	-0.045631	-0.110700	0.681639
2.000000	0.000885	0.255636	0.841471
2.500000	0.145466	0.279847	0.948985
3.000000	0.255669	0.142081	0.997495
3.500000	0.293037	0.007782	0.983986
4.000000	0.277773	-0.062465	0.909297
4.500000	0.240285	-0.083579	0.778073
5.000000	0.197121	-0.089049	0.598472
5.500000	0.150753	-0.098121	0.381661
6.000000	0.098490	-0.112138	0.141120
6.500000	0.039349	-0.124251	-0.108195
7.000000	-0.024145	-0.128355	-0.350783
7.500000	-0.087327	-0.122332	-0.571561
8.000000	-0.145273	-0.107149	-0.756803
8.500000	-0.193884	-0.084919	-0.894989
9.000000	-0.230133	-0.057703	-0.977530
9.500000	-0.251950	-0.027272	-0.999293
10.000000	-0.258124	0.004677	-0.958924
10.500000	-0.248313	0.036340	-0.858934
11.000000	-0.223104	0.065816	-0.705540
11.500000	-0.184031	0.091251	-0.508279
12.000000	-0.133506	0.111028	-0.279415
12.500000	-0.074669	0.123891	-0.033179
13.000000	-0.011186	0.129040	0.215120
13.500000	0.052992	0.126159	0.450044
14.000000	0.113873	0.115435	0.656987
14.500000	0.167672	0.097535	0.823081
15.000000	0.211047	0.073573	0.938000
15.500000	0.241300	0.045038	0.994599
16.000000	0.256550	0.013701	0.989358
16.500000	0.255850	-0.018488	0.922604
17.000000	0.239241	-0.049528	0.798487
17.500000	0.207758	-0.077488	0.624724

Figure 7.8 (a) FORTRAN program for Example 7.4. (b) Computer output.

```
*        EXAMPLE 7-5: DYNAMO SOLUTION OF
NOTE        MASS-SPRING-DASHPOT SYSTEM
NOTE
C  M=1/B=2/K=4/OMEGA=0.5
C  XZRO=1/VZRO=0
N  X=XZRO
N  V=VZRO
L  X.K=X.J+DT*XDOT.JK
L  V.K=V.J+DT*VDOT.JK
R  XDOT.KL=V.K
R  VDOT.KL=(-K*X.K-B*V.K+FA.K)/M
A  FA.K=SIN(OMEGA*TIME.K)
PRINT X,V,FA
PLOT   X=X
SPEC   DT=.03125/LENGTH=17.5/PRTPER=.5/PLTPER=.25
RUN
```

TIME E+00	X E+00	V E+00	FA E+00
.0	1.0000	.0000	.00000
.5	.5711	-1.0681	.24740
1.	.1767	-.7535	.47943
1.5	.0456	-.1107	.68164
2.	-.0009	.2556	.84147
2.5	.1455	.2798	.94898
3.	.2557	.1421	.99749
3.5	.2930	.0078	.98399
4.	.2778	-.0625	.90930
4.5	.2403	-.0836	.77807
5.	.1971	-.0890	.59847
5.5	.1508	-.0981	.38166
6.	.0985	-.1121	.14112
6.5	.0393	-.1243	-.10820
7.	-.0241	-.1284	-.35078
7.5	-.0873	-.1223	-.57156
8.	-.1453	-.1071	-.75680
8.5	-.1939	-.0849	-.89499
9.	-.2301	-.0577	-.97753
9.5	-.2519	-.0273	-.99929
10.	-.2581	.0047	-.95892
10.5	-.2483	.0363	-.85893
11.	-.2231	.0658	-.70554
11.5	-.1840	.0913	-.50828
12.	-.1335	.1110	-.27942
12.5	-.0747	.1239	-.03318
13.	-.0112	.1290	.21512
13.5	.0530	.1262	.45004
14.	.1139	.1154	.65699
14.5	.1677	.0975	.82308
15.	.2110	.0736	.93800
15.5	.2413	.0450	.99460
16.	.2566	.0137	.98936
16.5	.2558	-.0185	.92260
17.	.2392	-.0495	.79849
17.5	.2078	-.0775	.62472

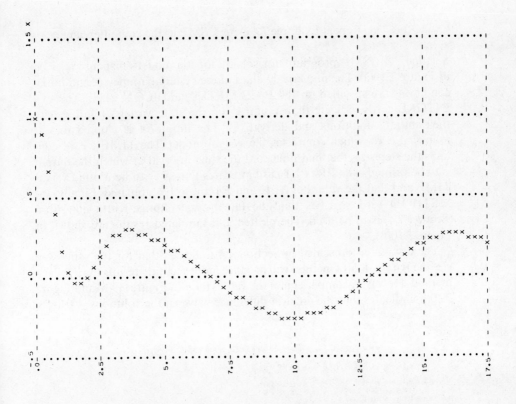

Figure 7.9 (a) DYNAMO program for Example 7.5. (b) Tabulated output. (c) Plotted output.

differences in the numerical results occur in the least significant digit and are caused by rounding-off the number for printing.

As indicated at the beginning of this section, there are a wide variety of other dynamic simulation languages with capabilities similar to or more extensive than those of DYNAMO. Although DYNAMO was used to obtain the numerical solutions in this book, other simulation languages can be used just as well.

EXAMPLE 7.6 Repeat Example 7.4 using CSMP as the computer language.

Solution A CSMP program* for solving for the model's response is given in Figure 7.10(a). The numerical values of the system parameters and initial conditions are specified on the PARAMETER and ICON cards. The two INTGRL cards indicate that X and V are the state variables and specify their initial conditions and derivatives. The input FA is entered on the following card, which completes the system model. The TIMER card contains the step size, the final time, and the time interval at which the output is to be printed. The PRTPLT card indicates that X is to be plotted versus TIME and that the values of X, V, and FA are to be tabulated. Finally, the METHOD card specifies that the fourth-order Runge-Kutta algorithm (Section 7.4) is to be used. The plotted and tabulated results are shown in Figure 7.10(b).

The discussion presented in Section 7.4 indicates that for the same step size the Runge-Kutta method will produce more accurate results than Euler's method. However, comparison of the values for x in Figure 7.9(b) and Figure 7.10(b) indicates that the greatest difference between the solutions is 0.0258,

```
        ****CONTINUOUS SYSTEM MODELING PROGRAM****
      ***PROBLEM INPUT STATEMENTS***
*    EXAMPLE 7-6:   CSMP SOLUTION OF
*        MASS-SPRING-DASHPOT SYSTEM
PARAMETER M=1.,B=2.,K=4.
PARAMETER OMEGA=0.5
INCON X0=1.,V0=0.
X=INTGRL(X0,V)
V=INTGRL(V0,(-K*X-B*V+FA)/M)
FA=SIN(OMEGA*TIME)
TIMER  DELT=0.03125,FINTIM=17.5,OUTDEL=0.5
PRTPLT X(X,V,FA)
METHOD  RKSFX
END
STOP
```

Figure 7.10 (a) CSMP program for Example 7.6.

* Instructions for using CSMP are given in IBM publication H20-0367-3.

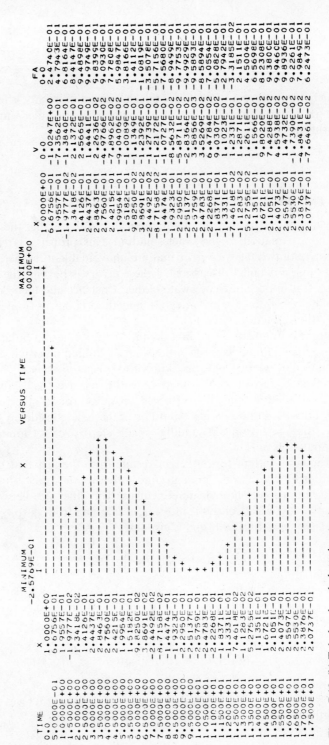

Figure 7.10 (b) Tabulated and plotted output.

189

which occurs at $t = 1.50$ s. Since this maximum difference is less than 3% of the maximum value of x, the results obtained with Euler's method should be sufficiently accurate for most purposes.

7.3 ERRORS

An inherent characteristic of a numerical solution is that the result is only an approximation to the exact solution. This property is caused by the presence of two types of error that are unavoidable in numerical solutions. Although a detailed discussion of these errors is beyond the scope of this book, it is important to be aware of their existence and their dependence on the choices of step size and integration method. (Consult Conte or Hamming for a thorough discussion.)

Because numbers stored in a digital computer have a finite word length, the results of numerical calculations have a limited precision. Errors caused by the finite word length are known as *round-off errors*. The cumulative effect of these errors tends to increase as the step size T is decreased, because more calculations must be performed. However, round-off error is generally not a problem in numerical solutions of the type considered here, because one tries to avoid an unduly small step size to limit expense in computation time.

The more serious form of error results from the fact that the computed solution $\tilde{q}(t_0 + T)$ would not agree precisely with the exact solution $q(t_0 + T)$ even if no round-off error were incurred. The difference between the exact and the numerical solutions introduced in going from t_0 to $t_0 + T$ is called the *local truncation error* and is given by

$$e(t_0) = q(t_0 + T) - \tilde{q}(t_0 + T) \tag{20}$$

To examine the local truncation error introduced by Euler's method, we use Taylor's remainder theorem (Purcell, Section 23.8) to write the exact solution at $t_0 + T$ as

$$q(t_0 + T) = q(t_0) + T\dot{q}(t_0) + \frac{T^2}{2}\ddot{q}(t_0 + \tau) \tag{21}$$

where τ must lie in the interval $0 \le \tau \le T$. Recall that (5) gives the numerical solution by Euler's method as

$$\tilde{q}(t_0 + T) = q(t_0) + T\dot{q}(t_0) \tag{22}$$

Substituting (21) and (22) into (20) shows the local truncation error to be

$$e(t_0) = \frac{T^2}{2} \ddot{q}(t_0 + \tau) \tag{23}$$

where $t_0 + \tau$ is some unknown value of t between the two ends of the time interval under consideration. In this case, the local truncation error is proportional to T^2 and is said to be of order T^2. Hence, if the step size T is halved, we would expect the local truncation error to be reduced to approximately one-fourth of its previous value, although twice as many steps would be needed in order to compute the solution.

Note that we have not given a numerical value for the error (if we could, we would know the exact solution), and we have said nothing about how the local truncation errors introduced at each step propagate as the solution proceeds. These are difficult questions to answer, but we can assess solution accuracy for our purposes by running the solution with several different step sizes and noting the differences in the results. When the changes in the computed solution caused by a further decrease in step size are small enough to be negligible to the user, we have found a satisfactory step size.

To demonstrate the effect of the step size on the truncation error for a first-order system using Euler's method, the equation $\dot{q} + 0.2q = 1$ with $q(0) = 1$ has been solved numerically for step sizes of $T = 0.5, 2, 10$, and 15. We solved this equation in Example 7.1 for $T = 2$ over the interval $0 \le t \le 10$. The results of these calculations are compared with the exact solution in Figure 7.11. Whereas the numerical solution with $T = 0.5$ agrees very closely with the exact solution, and the solution with $T = 2$ is reasonably accurate, there is no similarity at all between the exact and numerical solutions when $T = 10$ or 15.

The numerical result in Figure 7.11(d) is typical of what is known as *numerical instability*. A moderate to substantial amount of truncation error was introduced at each step, and these errors grew rapidly as the solution proceeded.

The numerical solution plotted in Figure 7.11(c) is interesting in that it neither converges to the correct steady-state value, as do those in Figure 7.11(a) and Figure 7.11(b), nor increases without limit, as does that in Figure 7.11(d). This step size is referred to as the *critical step size* (see Franks, Section 3-14) for this particular model and this particular integration method (Euler's method). If either the system model or the integration method is changed, then the critical step size will be different. In order to obtain meaningful numerical results, we must choose a step size that is sufficiently less than the critical value for the particular combination of model and integration method we are using.

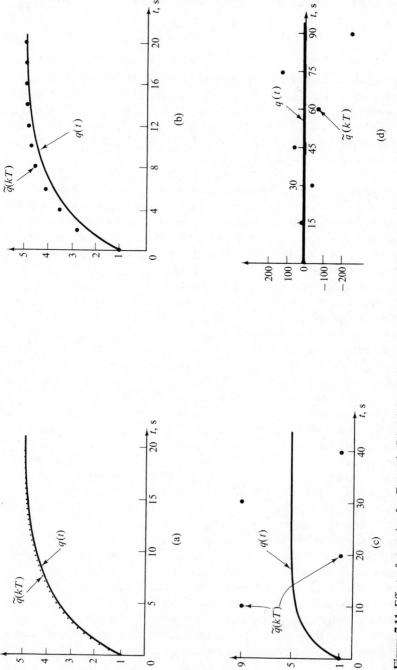

Figure 7.11 Effect of step size for Example 7.1. (a) $T = 0.5$. (b) $T = 2$. (c) $T = 10$. (d) $T = 15$.

7.4 OTHER INTEGRATION METHODS

Euler's method is only the simplest of a wide variety of integration methods that we can use in the numerical solution of dynamic models. Several widely used methods are described briefly in this chapter. Since a comprehensive treatment of these methods is beyond the scope and objectives of this book, we merely state some of the algorithms and comment briefly on their important characteristics (see Conte or Hamming for more complete discussions and further references).

For simplicity in notation, we shall consider the solution of only the single first-order equation $\dot{q} = f(q, u, t)$. Extension to multiple state-variable equations was discussed in Section 7.1 for Euler's method, and it is similar for other integration methods.

SECOND-ORDER RUNGE-KUTTA METHOD

The order of any Runge-Kutta method refers to the number of derivative evaluations needed for each step and is not related to the order of the system model. Rather than using only the derivative $\dot{\tilde{q}}^{(k-1)}$ in going from $\tilde{q}^{(k-1)}$ to $\tilde{q}^{(k)}$, as in Euler's method, the second-order Runge-Kutta method uses the average of two derivative evaluations. With $\tilde{q}^{(k-1)}$ and $u^{(k-1)}$ known, the calculation of $\dot{\tilde{q}}^{(k-1)}$ follows directly from the state-variable equation

$$\dot{\tilde{q}}^{(k-1)} = f[\tilde{q}^{(k-1)}, u^{(k-1)}, k-1]$$

We then use Euler's method to generate a preliminary estimate $\tilde{q}_E^{(k)}$ of the state-variable at the end of the interval, namely

$$\tilde{q}_E^{(k)} = \tilde{q}^{(k-1)} + Tf[\tilde{q}^{(k-1)}, u^{(k-1)}, k-1]$$

We use this estimated value of the state variable, along with the known input $u^{(k)}$, in the state-variable equation to form the derivative

$$\dot{\tilde{q}}_E^{(k)} = f[\tilde{q}_E^{(k)}, u^{(k)}, k]$$

Finally, we take the actual increment to be added to $\tilde{q}^{(k-1)}$ to be T times the average of the two derivatives $\dot{\tilde{q}}^{(k-1)}$ and $\dot{\tilde{q}}_E^{(k)}$, so

$$\tilde{q}^{(k)} = \tilde{q}^{(k-1)} + \frac{T}{2}\{f[\tilde{q}^{(k-1)}, u^{(k-1)}, k-1] + f[\tilde{q}_E^{(k)}, u^{(k)}, k]\}$$

Because we have averaged derivative evaluations at each end of the time interval, the local truncation error of the second-order Runge-Kutta method is of order T^3. Hence, halving the step size will reduce the truncation error to approximately one-eighth of its previous value. For comparable accuracy, one can use a larger step size with this method than with Euler's method, but at each step two derivative evaluations are required rather than one.

FOURTH-ORDER RUNGE-KUTTA METHOD

A fourth-order Runge-Kutta method, which requires four derivative evaluations at each step, is widely used. The form of the algorithm is

$$\tilde{q}^{(k)} = \tilde{q}^{(k-1)} + \frac{T}{6}[A_1 + 2A_2 + 2A_3 + A_4]$$

where A_1 is the derivative evaluated at $t = t_0 + (k-1)T$, A_2 and A_3 are two different derivative evaluations at the midpoint $t = t_0 + kT - T/2$, and A_4 is evaluated at $t = t_0 + kT$. Formulas for the parameters $A_1, ..., A_4$ are given in introductory textbooks on numerical methods. The local truncation error is of order T^5 and will be appreciably less than that for the second-order Runge-Kutta method or Euler's method for the same value of T.

The Runge-Kutta methods are referred to as *single-step methods* because to calculate $\tilde{q}^{(k)}$ we need to know $\tilde{q}^{(k-1)}$ but no previous values of the state variable. Since the calculations can proceed from a known initial state, such methods are referred to as *self-starting methods*.

MULTISTEP METHODS

A *multistep method* uses the values of several previously computed state variables such as $\tilde{q}^{(k-2)}, \tilde{q}^{(k-3)}, ...$ in order to evaluate $\tilde{q}^{(k)}$. Such methods have the disadvantage that they are not self-starting; for the initial step, only $\tilde{q}^{(0)}$ is known. However, if we start the solution by using a Runge-Kutta formula, we can continue it by a multistep method with fewer derivative evaluations for comparable accuracy.

A widely used algorithm of this type is the fourth-order Adams-Bashforth method given by

$$\tilde{q}^{(k)} = \tilde{q}^{(k-1)} + \frac{T}{24}[55f^{(k-1)} - 59f^{(k-2)} + 37f^{(k-3)} - 9f^{(k-4)}] \tag{24}$$

where $f^{(j)}$ denotes the derivative function $f[\tilde{q}^{(j)}, u^{(j)}, j]$. Although (24) appears to require four evaluations of the derivative function $f(q, u, t)$, those denoted by $f^{(k-2)}, f^{(k-3)}$, and $f^{(k-4)}$ were computed previously and can be saved in the computer memory. Hence we need evaluate only $f^{(k-1)}$ to go from the $(k-1)$st step to the kth step.

PREDICTOR-CORRECTOR METHODS

Predictor-corrector methods combine formulas in an iterative fashion to yield improved accuracy. One formula, the predictor, is used to compute an approximate value of the state variable at the kth step, denoted by $\tilde{q}_P^{(k)}$. Then another formula, the corrector, which has $\tilde{q}_P^{(k)}$ on its right side, is used to compute a refined or corrected value of the state variable, denoted by $q_C^{(k)}$.

We can use the difference between the predicted and corrected values, $\tilde{q}_C^{(k)} - \tilde{q}_P^{(k)}$, to indicate the accuracy of the result. If this difference is larger than desired, we repeat the correction process, with $\tilde{q}_C^{(k)}$ replacing $\tilde{q}_P^{(k)}$ on the right side of the corrector formula. Alternatively, an unsatisfactory difference can be taken to indicate that the step size T should be reduced. Hence, we can program an adjustable step-size feature as part of the integration algorithm. In fact, very small differences between predicted and corrected values on several successive steps can be taken as an indication that the step size should be increased, speeding up the calculations.

The Adams-Moulton predictor-corrector method uses the Adams-Bashforth formula given by (24) as the predictor and the fourth-order Adams-Moulton formula as the corrector. The resulting pair of equations is

$$\tilde{q}_P^{(k)} = \tilde{q}^{(k-1)} + \frac{T}{24}[55f^{(k-1)} - 59f^{(k-2)} + 37f^{(k-3)} - 9f^{(k-4)}]$$

$$\tilde{q}_C^{(k)} = \tilde{q}^{(k-1)} + \frac{T}{24}\{9f[\tilde{q}_P^{(k)}, u^{(k)}, k] + 19f^{(k-1)} - 5f^{(k-2)} + f^{(k-3)}\}$$

PROBLEMS

7.1 Use Euler's method to evaluate numerically the response of the system described by the differential equation $\dot{y} + 0.5y = 1$ with $y(0) = 0$ for the step sizes $T = 1, 2, 4,$ and 8. In each case, calculate the first five points. For example, for $T = 1$ find $y(0), y(1), ..., y(4)$. Indicate which of these step sizes appear to be satisfactory for this equation.

7.2 Repeat Problem 7.1 for the equation $4\dot{y} + y = 0$ with $y(0) = 10$.

7.3 Repeat Problem 7.1 for the equation $\dot{y} + 0.4y = t$ with $y(0) = 0$.

7.4 Repeat Problem 7.1 for the equation $\dot{y} + 0.8y = \epsilon^{-t}$ with $y(0) = 4.0$.

7.5 Use Euler's method with $T = 0.5$ to solve numerically the equation $\dot{y} + |y| y = 1$ for the initial condition $y(0) = 0$. Compute the solution through $t = 2.0$.

7.6 Repeat Problem 7.5 for the equation $\dot{y} + 0.2y^3 = t$ with $y(0) = 0$.

7.7 Repeat Problem 7.5 for the equation $\dot{y} + f(y) = 0$ where

$$f(y) = \begin{cases} \sqrt{y} & \text{for } y \geq 0 \\ -\sqrt{|y|} & \text{for } y < 0 \end{cases}$$

when

    (1) $y(0) = 10$

    (2) $y(0) = -5$

7.8 Use Euler's method with $T = 0.5$ to compute the numerical solution to the equation $\ddot{y} + 2\dot{y} + 4y = 0$ for the initial conditions $y(0) = 2$, $\dot{y}(0) = -1$ for the time points $t = 0$, 0.5, 1.0, and 1.5. Convert the equation to a pair of simultaneous first-order equations to which Euler's method can be applied.

7.9 a. Repeat Problem 7.8 for the differential equation $\ddot{y} + 2\dot{y} + 4y = 2t$ with zero initial conditions.

    b. Repeat part a using the step size $T = 0.25$. Comment on the effect of changing the step size.

7.10 Repeat Problem 7.8 for the equation $\ddot{y} + 0.5\dot{y}^3 + 2|y|y = 0$ for the initial conditions $y(0) = 1$, $\dot{y}(0) = 1$.

7.11 Using Euler's method with $T = 0.5$, solve for $y(0)$, $y(0.5)$, $y(1.0)$, and $y(1.5)$ for the system obeying the equation $\dddot{y} + 2\ddot{y} + 3\dot{y} + y = 0$ with the initial conditions $y(0) = -2$, $\dot{y}(0) = 1$, $\ddot{y}(0) = -1$. Before applying Euler's method, convert this third-order equation to state-variable form.

7.12 Repeat Problem 7.11 for the equation $\dddot{y} + 2\ddot{y} + 3\dot{y} + y = 2t^2$ with zero initial conditions.

7.13 Use Euler's method with step size $T = 0.5$ to calculate $y(0)$, $y(0.5)$, ..., $y(2)$ for the equation $2\ddot{y} + 3\dot{y} + y = 1 - t$ where $y(0) = 4$ and $\dot{y}(0) = -2$.

7.14 Using a programmable calculator or a digital computer, prepare a program to implement Euler's method for the equations and initial conditions specified in the following problems. In each case, run the solution for several step sizes and select a reasonable value for the step size, stating the basis for your choice. Run the solution long enough to establish the general behavior of the solution.

    a. Problem 7.1.

    b. Part a of Problem 7.7.

    c. Problem 7.8.

    d. Problem 7.12.

In Problems 7.15 through 7.17, use a digital computer to simulate the responses of both the nonlinear model given in the indicated problem and its linearized approximation. In each case, find an appropriate step size by trial and error. Comment on the accuracy of the linearized model.

7.15 Problem 5.13 with $\omega = 4$. Simulate the cases $B = 4$ and 100.

7.16 Problem 5.23. Simulate the cases $B = 1$ and 4.

7.17 Problem 5.25. Simulate the cases $B = 1$ and 10.

7.18 Use a computer with Euler's method to simulate the response of the third-order system that obeys the equation $\dddot{y} + 7\ddot{y} + 15\dot{y} + 25y = 50u(t)$ for each of the following conditions:

    (1) $u(t) = 1$ for $t \geq 0$ with $y(0) = \dot{y}(0) = \ddot{y}(0) = 0$
    (2) $u(t) = 0$ for $t \geq 0$ with $y(0) = 10$, $\dot{y}(0) = \ddot{y}(0) = 0$

    a. Run case (1) for $0 \leq t \leq 6$, using step sizes of $T = 0.01, 0.02, 0.1$, and 0.25. Comment on the effects of the step size on the accuracy of the solution, and indicate a reasonable value for T.

    b. Run case (2) for the interval $0 \leq t \leq 6$, using the appropriate step size you found in part a.

    c. Rerun case (1) with $T = 0.5$.

7.19 For this problem, use a programmable calculator or a computer.

    a. For the system, input, and initial conditions in Example 7.2, obtain a numerical solution for $0 \leq t \leq 1.5$ when the step size is 0.1. Compare your results with Table 7.2.

    b. Find a numerical solution for $0 \leq t \leq 1.5$ using the second-order Runge-Kutta method with $T = 0.25$. Compare your results with Table 7.2.

7.20 Using a digital computer, simulate the responses of the rotational mechanical system modeled in Example 4.2 when the parameter values are

$$J_1 = 2.0 \text{ kg} \cdot \text{m}^2 \qquad B_1 = 20 \text{ N} \cdot \text{m} \cdot \text{s/rad} \qquad K_1 = 500 \text{ N} \cdot \text{m/rad}$$
$$J_2 = 0.5 \text{ kg} \cdot \text{m}^2 \qquad B_2 = 10 \text{ N} \cdot \text{m} \cdot \text{s/rad} \qquad K_2 = 400 \text{ N} \cdot \text{m/rad}$$

Run the program for each of the following three cases. Except as indicated, the initial conditions and input are zero.

(1) $\theta_2(0) = 0.1$ rad

(2) $\tau_a(t) = 100$ N·m for all $t > 0$

(3) $\tau_a(t) = 200$ N·m for $0 < t \le 0.1$ s and zero otherwise

You should find an acceptable step size within the interval $0.001 \le T \le 0.05$ seconds.

CHAPTER 8
ANALYTICAL SOLUTION OF FIXED
LINEAR MODELS

The first several chapters of this book presented methods of formulating the mathematical model for a system, while Chapter 7 considered performing numerical solutions with the aid of a digital computer. Analytical solutions, which in general are feasible only for fixed linear models, are examined in this chapter.

Section 8.1 summarizes the basic procedures for solving differential equations that will be needed in later sections. For many readers, this will be a review. Throughout the chapter, we assume that we have fixed linear differential equations with real coefficients. Furthermore, the methods we describe are not sufficient for all possible inputs, although they are satisfactory for those commonly encountered. Proofs and detailed justifications are omitted, but some general references are included in Appendix D. In the rest of the chapter, we consider in detail the solution of first- and second-order models, examine some specific inputs that are useful in system analysis, and define a number of important terms. Some comments about systems of arbitrary order appear in the concluding section.

8.1 THE COMPLETE SOLUTION OF DIFFERENTIAL EQUATIONS

We assume that the system model has been put into input-output form with all other variables eliminated, as discussed in Section 3.2. The model for an nth-order system with input $u(t)$ and output $y(t)$ can be written, as in (3.19), as

$$a_n y^{(n)} + \cdots + a_2 \ddot{y} + a_1 \dot{y} + a_0 y = b_m u^{(m)} + \cdots + b_1 \dot{u} + b_0 u \qquad (1)$$

where $y^{(n)} = d^n y/dt^n$, etc. The collection of terms on the right side of (1), which involves the input and its derivatives, is represented by

$$F(t) = b_m u^{(m)} + \cdots + b_1 \dot{u} + b_0 u \qquad (2)$$

where $F(t)$ is called the *forcing function*. With this definition, we may rewrite (1) as

$$a_n y^{(n)} + \cdots + a_2 \ddot{y} + a_1 \dot{y} + a_0 y = F(t) \qquad (3)$$

The desired solution, $y(t)$ for $t \geq 0$, must satisfy the differential equation for $t \geq 0$ and also n specified initial conditions, which are usually $y(0)$, $\dot{y}(0), \ldots, y^{(n-1)}(0)$.

HOMOGENEOUS DIFFERENTIAL EQUATIONS

If $F(t)$ is replaced by zero, (3) reduces to the *homogeneous differential equation*

$$a_n y_H^{(n)} + \cdots + a_2 \ddot{y}_H + a_1 \dot{y}_H + a_0 y_H = 0 \qquad (4)$$

where the subscript H has been added to emphasize that $y_H(t)$ is the solution to the homogeneous equation. Assume that a solution of (4) has the form

$$y_H(t) = K e^{rt}$$

where we must determine the constant r so that (4) is satisfied. Since multiplying y_H by a constant just multiplies the entire left side of (4) by that constant, any nonzero value will be satisfactory for the constant K. Substituting the assumed solution into (4) gives

$$(a_n r^n + \cdots + a_2 r^2 + a_1 r + a_0) K e^{rt} = 0$$

Since $K e^{rt} \neq 0$ for a nontrivial solution, we must have

$$a_n r^n + \cdots + a_2 r^2 + a_1 r + a_0 = 0 \qquad (5)$$

which is called the *characteristic equation*. Note that the coefficients in this algebraic equation are identical to those on the left side of the differential equation, so we can write (5) by inspection. An nth-order algebraic equation has n roots, which we denote by $r_1, r_2, ..., r_n$. Not only are $K_1 \epsilon^{r_1 t}, K_2 \epsilon^{r_2 t}, ...,$ $K_n \epsilon^{r_n t}$ all solutions of (4) for arbitrary K_i, but

$$y_H(t) = K_1 \epsilon^{r_1 t} + K_2 \epsilon^{r_2 t} + \cdots + K_n \epsilon^{r_n t} \tag{6}$$

is a solution for any values of the arbitrary constants K_1 through K_n. If all the roots of the characteristic equation are different, then (6) is the most general solution.

If two or more of the characteristic roots are identical, i.e., if there is a repeated root of (5), then we must modify the form of (6) to contain n independent terms and to represent the most general solution. If $r_1 = r_2$ with the remaining roots being distinct, the solution is

$$y_H(t) = K_1 \epsilon^{r_1 t} + K_2 t \epsilon^{r_1 t} + K_3 \epsilon^{r_3 t} + \cdots + K_n \epsilon^{r_n t}$$

If $r_1 = r_2 = r_3$, the most general solution to the homogeneous differential equation is

$$y_H(t) = (K_1 + K_2 t + K_3 t^2) \epsilon^{r_1 t} + K_4 \epsilon^{r_4 t} + \cdots + K_n \epsilon^{r_n t}$$

If any of the roots of the characteristic equation are complex numbers, they must occur in complex conjugate pairs because the coefficients in the characteristic equation are real. Suppose that $r_1 = \alpha + j\beta$ and $r_2 = \alpha - j\beta$, where $j = \sqrt{-1}$,* and that the characteristic equation has no other roots. Then

$$y_H(t) = K_1 \epsilon^{(\alpha + j\beta)t} + K_2 \epsilon^{(\alpha - j\beta)t}$$

However, because the solution is a real function of time, we should rewrite this equation by using suitable trigonometric identities. Factoring out $\epsilon^{\alpha t}$ from the right side, then using the first two entries in Table 8.1 with $\theta = \beta t$, and finally collecting like terms, we have

$$y_H(t) = \epsilon^{\alpha t} [K_1 \epsilon^{j\beta t} + K_2 \epsilon^{-j\beta t}]$$
$$= \epsilon^{\alpha t} [K_1 (\cos \beta t + j \sin \beta t) + K_2 (\cos \beta t - j \sin \beta t)]$$
$$= \epsilon^{\alpha t} [(K_1 + K_2) \cos \beta t + j(K_1 - K_2) \sin \beta t]$$

* The symbol i is also used for $\sqrt{-1}$, but we reserve i for electrical current.

Table 8.1 Useful trigonometric identities

$$\epsilon^{j\theta} = \cos \theta + j \sin \theta$$

$$\epsilon^{-j\theta} = \cos \theta - j \sin \theta$$

$$\cos \theta = \tfrac{1}{2}(\epsilon^{j\theta} + \epsilon^{-j\theta})$$

$$\sin \theta = \frac{1}{2j}(\epsilon^{j\theta} - \epsilon^{-j\theta})$$

$$A \cos \theta + B \sin \theta = \sqrt{A^2 + B^2} \sin\left(\theta + \tan^{-1}\frac{A}{B}\right)$$

$$A \cos \theta + B \sin \theta = \sqrt{A^2 + B^2} \cos\left(\theta - \tan^{-1}\frac{B}{A}\right)$$

With $K_3 = K_1 + K_2$ and $K_4 = j(K_1 - K_2)$, the equation becomes

$$y_H(t) = \epsilon^{\alpha t}(K_3 \cos \beta t + K_4 \sin \beta t) \tag{7}$$

By using the last entry in Table 8.1, we may rewrite (7) in the form

$$y_H(t) = K\epsilon^{\alpha t} \cos(\beta t + \phi) \tag{8}$$

where the constants K and ϕ are functions of K_3 and K_4. In these expressions involving complex roots, K_1 and K_2 will be complex numbers, with K_2 the complex conjugate of K_1; but K_3, K_4, K, and ϕ will be real constants. Either of the two forms given by (7) and (8) should be used.

We treat repeated complex characteristic roots much like repeated real roots. For a fourth-order characteristic equation with $r_1 = r_3 = \alpha + j\beta$ and $r_2 = r_4 = \alpha - j\beta$, the most general solution to the homogeneous differential equation has the form

$$y_H(t) = K_A \epsilon^{\alpha t} \cos(\beta t + \phi_A) + K_B t \epsilon^{\alpha t} \cos(\beta t + \phi_B)$$

NONHOMOGENEOUS DIFFERENTIAL EQUATIONS

We now consider the nonhomogeneous differential equation, where the forcing function $F(t)$ in (3) is nonzero. For mathematical convenience, we express the solution as the sum of two parts, namely

$$y(t) = y_H(t) + y_P(t) \tag{9}$$

where $y_H(t)$ and $y_P(t)$ are known as the complementary and the particular solutions, respectively. The *complementary solution* $y_H(t)$ is the solution to the homogeneous differential equation in (4); examples are given by (6) through (8). The *particular solution* $y_P(t)$ must satisfy the entire original differential equation, so

$$a_n y_P^{(n)} + \cdots + a_2 \ddot{y}_P + a_1 \dot{y}_P + a_0 y_P = F(t) \qquad (10)$$

A general procedure for finding $y_P(t)$ is the variation-of-parameters method (see Boyce and DiPrima, Section 5-5). When the forcing function $F(t)$ has only a finite number of different derivatives, however, the method of undetermined coefficients is satisfactory. Some examples of functions possessing only a finite number of independent derivatives are given in the left-hand column of Table 8.2. Others include $\cosh \omega t$ and $t^2 \epsilon^{zt} \cos(\omega t + \phi)$. An example of a forcing function that does *not* fall into this category is $1/t$.

In using the *method of undetermined coefficients*, we normally assume that the form of $y_P(t)$ consists of terms similar to those in $F(t)$ and its derivatives, with each term multiplied by a constant to be determined. We find the values of these constants by substituting the assumed $y_P(t)$ into the differential equation and then equating corresponding coefficients. The particular solution to be assumed for some common forcing functions is shown in Table 8.2.

If $F(t)$ or one of its derivatives contains a term identical to a term in $y_H(t)$, the corresponding terms in the right column of Table 8.2 should be multiplied by t. Thus if $y_H(t) = K_1 \epsilon^{-t} + K_2 \epsilon^{-2t}$ and $F(t) = 3\epsilon^{-t}$, then $y_P(t) = At\epsilon^{-t}$ should be used. If a term in $F(t)$ corresponds to a double root of the characteristic equation, the normal form for $y_P(t)$ is multiplied by t^2. If, for example, $y_H(t) = K_1 \epsilon^{-t} + K_2 t\epsilon^{-t}$ and $F(t) = \epsilon^{-t}$, then $y_P(t) = At^2\epsilon^{-t}$.

In the *general solution* $y(t) = y_H(t) + y_P(t)$ for an nth-order nonhomogeneous differential equation, the complementary solution $y_H(t)$ will contain n arbitrary constants, which we earlier denoted by K_1 through K_n. The

Table 8.2 Usual form of the particular solution

$F(t)$	$y_P(t)$
α	A
$\alpha_1 t + \alpha_0$	$At + B$
ϵ^{zt}	$A\epsilon^{zt}$
$\cos \omega t$	$A \cos \omega t + B \sin \omega t$
$\sin \omega t$	$A \cos \omega t + B \sin \omega t$

original differential equation in (3) will be satisfied regardless of the values of these constants, so their evaluation requires n separately specified initial conditions: $y(0), \dot{y}(0), \ldots, y^{(n-1)}(0)$. Since these initial values are associated with the entire solution and not just with the complementary solution, we cannot evaluate the arbitrary constants until we have found both $y_H(t)$ and $y_P(t)$.

EXAMPLE 8.1   If $\dot{y} + 2y = F(t)$ and $y(0) = 2$, find $y(t)$ for $t \geq 0$ for each of the following forcing functions:

1. $F(t) = 10$
2. $F(t) = 10 \cos 2t$
3. $F(t) = 10 + 10 \cos 2t$

Solution   By inspection of the left side of the differential equation, we find that the characteristic equation is $r + 2 = 0$, which has a single root at $r = -2$. Thus the complementary solution is

$$y_H(t) = K\epsilon^{-2t}$$

To find the particular solution for case 1 when $F(t) = 10$, we assume that $y_P(t) = A$. Substituting these expressions into the original differential equation gives $0 + 2A = 10$, so $A = 5$ and for $t \geq 0$

$$y(t) = y_H(t) + y_P(t)$$
$$= K\epsilon^{-2t} + 5$$

We replace t by zero in the last equation and use $y(0) = 2$ to obtain

$$2 = K + 5$$

Thus $K = -3$, and the response for $t \geq 0$ to $F(t) = 10$ is

$$y(t) = -3\epsilon^{-2t} + 5$$

In case 2, the forcing function is $F(t) = 10 \cos 2t$, so we assume that $y_P(t) = A \cos 2t + B \sin 2t$. Substituting these expressions into $\dot{y}_P + 2y_P = F(t)$, we have

$$(-2A \sin 2t + 2B \cos 2t) + 2(A \cos 2t + B \sin 2t) = 10 \cos 2t$$

or, collecting like terms,

$$(2A + 2B - 10) \cos 2t + (-2A + 2B) \sin 2t = 0$$

Since this must be an identity for all $t \geq 0$, we require that $2A + 2B - 10 = 0$ and that $-2A + 2B = 0$. Solving these equations gives $A = B = \frac{5}{2}$, so

$$y(t) = K\epsilon^{-2t} + \tfrac{5}{2}(\cos 2t + \sin 2t)$$

Replacing t by zero and $y(t)$ by 2, we have $2 = K + \frac{5}{2}$. Thus $K = -\frac{1}{2}$, and for all $t \geq 0$ the solution is

$$y(t) = -\tfrac{1}{2}\epsilon^{-2t} + \tfrac{5}{2}(\cos 2t + \sin 2t)$$

Finally, for case 3, $F(t) = 10 + 10 \cos 2t$ and the particular solution is the sum of the particular solutions for the individual terms in $F(t)$. Thus

$$y(t) = K\epsilon^{-2t} + 5 + \tfrac{5}{2}(\cos 2t + \sin 2t)$$

For the right side to reduce to 2 when t is replaced by zero, we require that $K = -\frac{11}{2}$. Hence for $t \geq 0$

$$y(t) = -\tfrac{11}{2}\epsilon^{-2t} + 5 + \tfrac{5}{2}(\cos 2t + \sin 2t)$$

EXAMPLE 8.2   Find the solution to the differential equation $\ddot{y} + 2\dot{y} + 5y = t$ for $t \geq 0$ if $y(0) = 0$ and $\dot{y}(0) = 2$.

Solution   The characteristic equation is $r^2 + 2r + 5 = 0$, which by the quadratic formula has roots at

$$r = \frac{-2 \pm \sqrt{4 - 20}}{2} = -1 \pm j2$$

As in (7), we can write the complementary solution as

$$y_H(t) = \epsilon^{-t}[K_1 \cos 2t + K_2 \sin 2t]$$

The particular solution is assumed to have the form $y_P(t) = At + B$. Substituting this expression into the differential equation gives

$$0 + 2A + 5(At + B) = t$$

or

$$(5A - 1)t + (2A + 5B) = 0$$

which requires that $5A = 1$ and $2A + 5B = 0$. Thus $A = \frac{1}{5}$, $B = -\frac{2}{25}$, and the general solution for $t \geq 0$ is

$$y(t) = \epsilon^{-t}[K_1 \cos 2t + K_2 \sin 2t] + \tfrac{1}{5}t - \tfrac{2}{25}$$

The derivative of this general solution is

$$\dot{y}(t) = \epsilon^{-t}[(2K_2 - K_1)\cos 2t - (K_2 + 2K_1)\sin 2t] + \tfrac{1}{5}$$

At $t = 0$ and with the given initial conditions, the last two equations reduce to

$$0 = K_1 - \tfrac{2}{25}$$

$$2 = 2K_2 - K_1 + \tfrac{1}{5}$$

from which $K_1 = \tfrac{2}{25}$ and $K_2 = \tfrac{47}{50}$. For $t \geq 0$,

$$y(t) = \epsilon^{-t}[\tfrac{2}{25}\cos 2t + \tfrac{47}{50}\sin 2t] + \tfrac{1}{5}t - \tfrac{2}{25}$$

8.2 FIRST-ORDER SYSTEMS

In this section, we consider the solution of first-order systems, which usually consist of one energy-storing element and any number of dissipative elements. We also introduce some definitions that will be used throughout the rest of the book. Although they are illustrated in the context of first-order systems, we shall define the terms in a general way to make their extension to systems of arbitrary order obvious.

A first-order system is described by a single differential equation of the form

$$\dot{y} + \frac{1}{\tau}y = F(t) \tag{11}$$

where τ is a real nonzero constant. To find the response for $t \geq 0$, we must know the forcing function $F(t)$ for all positive values of time and the initial condition $y(0)$. Since the characteristic equation $r + (1/\tau) = 0$ has one root at $r = -1/\tau$, the complementary solution is

$$y_H(t) = K\epsilon^{-t/\tau} \tag{12}$$

STABILITY

Typical curves for the complementary solution are shown in Figure 8.1. If $y_H(t)$ decays to zero as t approaches infinity, the system is said to be *stable*. If, on the other hand, $y_H(t)$ increases without limit as t becomes large, the system is *unstable*. A first-order system is stable if $\tau > 0$ and unstable if

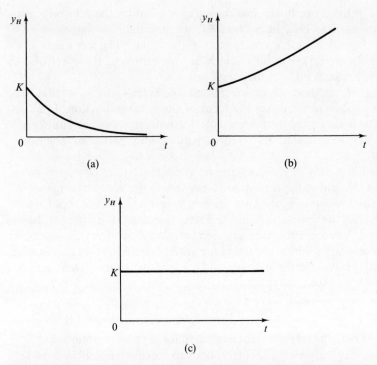

y_H

K

0 t

(a)

y_H

K

0 t

(b)

y_H

K

0 t

(c)

Figure 8.1 The complementary solution for a first-order system. (a) $\tau > 0$. (b) $\tau < 0$. (c) τ infinite.

$\tau < 0$. If the magnitude of τ approaches infinity, $y_H(t)$ becomes constant, as shown in Figure 8.1(c). Such a system is said to be *marginally stable*.

PARTS OF THE COMPLETE SOLUTION

A variety of names are used to identify the terms in the response of a fixed linear system, and these names can be easily confused. We shall explain the general terminology and relate the new terms to the complementary and particular solutions.

We have seen that we can find the form of the complementary solution by letting $F(t) = 0$ and solving for the roots of the characteristic equation. Thus the form of $y_H(t)$ depends not on the input but only on the system elements and their interconnections. The form of $y_H(t)$ represents the natural behavior of the system when the external input is removed and the system is excited by some initial stored energy. Thus $y_H(t)$ is also called the *free response*.

The size of the terms in the free response is given by the arbitrary constants, such as K in (12). In Section 8.1, we saw that the values of such constants depend on $y_P(t)$ and hence on the input. The constants also depend on the initial conditions, which in turn represent the effect of the history of the system prior to the initial instant.

Studying the method of undetermined coefficients, we saw that the particular solution depends on the form of the forcing function. For this reason, $y_P(t)$ is called the *forced response*. In summary, the free and forced responses are identical to the complementary and particular solutions, respectively.

An alternative way of dividing the total response into two parts is to regard it as the sum of the zero-state response and the zero-input response. The *zero-state response* $y_{zs}(t)$ is the complete response to the input when the initial values of the state variables are zero. It includes both the particular solution and the complementary solution, where the arbitrary constants in the complementary solution are found for initial conditions that correspond to zero initial values for the state variables. For a first-order system, we can write

$$y_{zs}(t) = y_P(t) + K_{zs}\, \epsilon^{-t/\tau}$$

where K_{zs} is the arbitrary constant evaluated for zero-state conditions.

The *zero-input response* $y_{zi}(t)$ corresponds to specified initial values of the state variables and a forcing function of zero. Hence, it consists of the complementary solution but not the particular solution, and for a first-order system we can denote it by

$$y_{zi}(t) = K_{zi}\, \epsilon^{-t/\tau}$$

where K_{zi} is the arbitrary constant evaluated for zero input. Note that although both y_{zs} and y_{zi} contain a term of the form $\epsilon^{-t/\tau}$, the multiplying constants in front of the terms are not the same.

A linear system with a specified input and specified initial values for its state variables has a response that we can find by superposition of the zero-state and zero-input responses. Thus

$$y(t) = y_{zs}(t) + y_{zi}(t)$$

Combining these expressions for a first-order system gives

$$y(t) = y_P(t) + (K_{zs} + K_{zi})\epsilon^{-t/\tau}$$

By comparing this equation to (9), we see that the two terms $K_{zs}\epsilon^{-t/\tau}$ and $K_{zi}\epsilon^{-t/\tau}$ combine to give the free response $y_H(t)$. We shall not pursue the concept of the zero-state and zero-input responses in this chapter, which is primarily concerned with input-output differential equations. These responses will be important, however, in Chapters 13 and 17.

It is often convenient to decompose the total response into transient and steady-state components. The *transient response* $y_t(t)$ consists of those terms that decay to zero as t approaches infinity. The *steady-state response* $y_{ss}(t)$ is the part of the solution that remains after the transient terms have disappeared. For a first-order system, $y_H(t) = K\epsilon^{-t/\tau}$ will be part of the transient response if $\tau > 0$, i.e., if the system is stable. If $y_P(t)$ has a form such as A, $A\cos\omega t + B\sin\omega t$, or $At + B$, then $y_P(t)$ will constitute the steady-state response. On the other hand, if $y_P(t) = At\epsilon^{-t} + B\epsilon^{-t}$, then $y_P(t)$ will be part of the transient response. For systems that are stable and for which the terms in y_P do not decay to zero, the transient and steady-state components are identical to the free and the forced responses, respectively.

THE COMPLETE RESPONSE TO A CONSTANT INPUT

If $F(t)$ has a constant value of A, then the forced response to the first-order differential equation

$$\dot{y} + \frac{1}{\tau}y = F(t)$$

is $y_P(t) = \tau A$, as discussed in Example 8.1. Assume that $\tau > 0$, so that the free response

$$y_H(t) = K\epsilon^{-t/\tau}$$

decays to zero. Then $y_P(t) = y_{ss}(t)$ and we can write the complete solution as

$$y(t) = y_{ss} + K\epsilon^{-t/\tau} \tag{13}$$

If $y(0)$ is known, we can evaluate this solution at $t = 0$, giving $y(0) = y_{ss} + K$ or $K = -[y_{ss} - y(0)]$. Thus for $t \geq 0$,

$$y(t) = y_{ss} - [y_{ss} - y(0)]\epsilon^{-t/\tau} \tag{14}$$

which is shown in Figure 8.2(a). Special cases of this result are shown in Figure 8.2(b) and Figure 8.2(c).

To give significance to the value of $y(\tau)$ shown in Figure 8.2(a), first note from (14) that, for $t = \tau$, we have

$$y(\tau) = y_{ss} - [y_{ss} - y(0)]\epsilon^{-1}$$

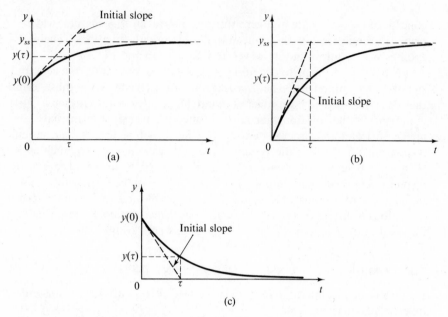

Figure 8.2 Response of a stable first-order system to a constant input. (a) y_{ss} and $y(0)$ both nonzero. (b) $y(0) = 0$. (c) $y_{ss} = 0$.

Since $\epsilon^{-1} = 0.3679$,

$$y(\tau) = y(0) + 0.6321[y_{ss} - y(0)]$$

Thus after τ seconds, the response to a constant input is approximately 63% of the way from the initial value to the steady-state value. Since $\epsilon^{-4} = 0.0183$, the response after 4τ seconds is approximately 98% of the way from the initial value to the steady-state value. Thus the system parameter τ, which is called the *time constant*, is a measure of the system's speed of response. Another interpretation of the time constant results from considering the initial slope of the response. Differentiating (14) gives

$$\dot{y}(t) = \frac{1}{\tau}[y_{ss} - y(0)]\epsilon^{-t/\tau}$$

and setting $t = 0$ gives

$$\tau\dot{y}(0) = y_{ss} - y(0)$$

Thus if the slope of the response curve were maintained at its initial value of $\dot{y}(0)$, it would take τ seconds, instead of an infinite time, for the response to reach its steady-state value.

EXAMPLE 8.3   Find the response of each of the first-order systems shown in Figure 8.3. The mass of the block shown in Figure 8.3(a) is assumed to be negligible. The systems are at rest with no stored energy at $t = 0$, and for $t > 0$ the respective inputs are $f_a(t) = A$ and $\omega_a(t) = A$.

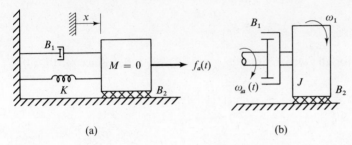

(a)   (b)

Figure 8.3 First-order mechanical systems for Example 8.3. (a) Translational. (b) Rotational.

Solution   The translational system shown in Figure 8.3(a) is described by the differential equation

$$(B_1 + B_2)\dot{x} + Kx = f_a(t)$$

or

$$\dot{x} + \frac{K}{B_1 + B_2}x = \frac{1}{B_1 + B_2}f_a(t)$$

Comparing this equation to (11), we see that the time constant is

$$\tau = \frac{B_1 + B_2}{K} \tag{15}$$

Since $f_a(t) = A$ for all positive values of time, $x_{ss} = A/K$. Since $x(0) = 0$, (14) gives as the complete response

$$x = \frac{A}{K}(1 - \epsilon^{-t/\tau}) \tag{16}$$

where τ is given by (15).

For the rotational system shown in Figure 8.3(b),

$$J\dot{\omega}_1 + (B_1 + B_2)\omega_1 = B_1\omega_a(t)$$

or

$$\dot{\omega}_1 + \frac{B_1 + B_2}{J}\omega_1 = \frac{B_1}{J}\omega_a(t)$$

The time constant is

$$\tau = \frac{J}{B_1 + B_2} \tag{17}$$

With $\omega_a(t) = A$ for all positive time, $(\omega_1)_{ss} = AB_1/(B_1 + B_2)$. Since $\omega_1(0) = 0$,

$$\omega_1 = \frac{AB_1}{B_1 + B_2}(1 - \epsilon^{-t/\tau}) \tag{18}$$

8.3 THE STEP FUNCTION AND IMPULSE

Two of the most important inputs we encounter in the analysis of dynamic systems are the unit step function and the unit impulse.

THE UNIT STEP FUNCTION

One frequently encounters inputs that are zero before some reference time and that have a nonzero constant value thereafter. To treat such inputs mathematically, we define the *unit step function*, which is denoted by $U(t)$. This function is defined to be zero for $t \le 0$ and unity for $t > 0$; it is shown in Figure 8.4(a).* If the step discontinuity occurs at some later time t_1, as shown in Figure 8.4(b), the function is defined by

$$U(t - t_1) = \begin{cases} 0 & \text{for } t \le t_1 \\ 1 & \text{for } t > t_1 \end{cases} \tag{19}$$

This notation is consistent with the fact that if any function $f(t)$ is plotted versus t, replacing every t in $f(t)$ by $t - t_1$ shifts the curve t_1 units to the

---

* Note that a capital U is used for the unit step function $U(t)$, in contrast to the lower-case letter in the symbol $u(t)$ for a general input. The value of the unit step function at time zero could be defined to be unity, or its value could be left undefined at this instant. Defining $U(0) = 0$ will be convenient in Chapter 12.

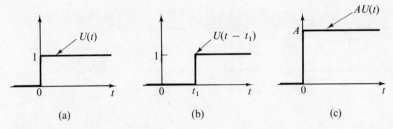

Figure 8.4 Step functions. (a) $U(t)$. (b) $U(t-t_1)$. (c) $AU(t)$.

right. If the height of the step is A rather than unity, we simply have A times the unit step function, as shown in Figure 8.4(c).

From the definition of $U(t-t_1)$, we note that

$$f(t)\,U(t-t_1) = \begin{cases} 0 & \text{for } t \le t_1 \\ f(t) & \text{for } t > t_1 \end{cases} \tag{20}$$

Thus the output of a system that was at rest for $t \le 0$ is often written in the form $f(t)\,U(t)$, where the multiplying factor $U(t)$ is used in place of the phrase "for $t > 0$."

We define the *unit step response* of a system, denoted by $y_U(t)$, as the output that occurs when the input is the unit step function $U(t)$ and when the system contains no initial stored energy, i.e., $y_U(t)$ is the zero-state response to the input $U(t)$. For the translational system shown in Figure 8.3(a),

$$y_U(t) = \frac{1}{K}[1 - \epsilon^{-Kt/(B_1 + B_2)}] \qquad \text{for } t \ge 0$$

while for the rotational system shown in Figure 8.3(b),

$$y_U(t) = \frac{B_1}{B_1 + B_2}[1 - \epsilon^{-(B_1 + B_2)t/J}] \qquad \text{for } t \ge 0$$

We can represent any function that consists of horizontal and vertical lines as the sum of step functions. Consider, for example, the pulse shown in Figure 8.5(a). Since this function is the sum of the two functions shown in Figure 8.5(b) and Figure 8.5(c),

$$f_1(t) = AU(t) - AU(t-t_1) \tag{21}$$

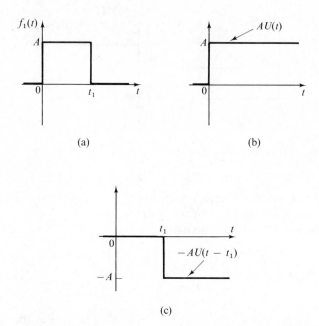

Figure 8.5 (a) Rectangular pulse. (b), (c) Formation of the rectangular pulse by the sum of two step functions.

Suppose that the pulse in (21) is the input to a linear first-order system described by the differential equation

$$\dot{y} + \frac{1}{\tau}y = u(t) \tag{22}$$

and for which $y(0) = 0$. From Section 8.2, we know that the response to $AU(t)$ is $A\tau(1-\epsilon^{-t/\tau})$ for $t \geq 0$, and this is also the response for $0 \leq t \leq t_1$ to the input in (21). For $t > t_1$, we may use superposition and sum the responses to the components $AU(t)$ and $-AU(t-t_1)$ to obtain

$$y(t) = A\tau[1 - \epsilon^{-t/\tau}] - A\tau[1 - \epsilon^{-(t-t_1)/\tau}]$$
$$= A\tau(-1 + \epsilon^{t_1/\tau})\epsilon^{-t/\tau}$$

Hence, we may write

$$y(t) = \begin{cases} A\tau(1-\epsilon^{-t/\tau}) & \text{for } 0 \leq t \leq t_1 & \text{(23a)} \\ A\tau(\epsilon^{t_1/\tau}-1)\epsilon^{-t/\tau} & \text{for } t > t_1 & \text{(23b)} \end{cases}$$

which is shown in Figure 8.6(a). As expected, the response for the first t_1 seconds is the same as in Figure 8.2(b), while for $t > t_1$ the output decays exponentially to zero with a time constant τ. The output is the superposition

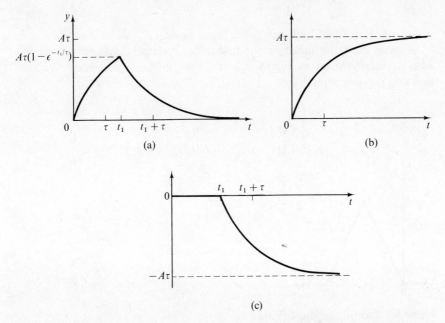

Figure 8.6 Responses of a first-order system to the inputs shown in Figure 8.5.

of the two functions shown in Figure 8.6(b) and Figure 8.6(c). It can also be written in the alternative form

$$y(t) = A\tau(1 - \epsilon^{-t/\tau})\,U(t) - A\tau(1 - \epsilon^{-(t-t_1)/\tau})\,U(t - t_1)$$

It is instructive to rewrite the pulse response in (23b) with $\epsilon^{t_1/\tau}$ replaced by its Taylor-series expansion

$$\epsilon^{t_1/\tau} = 1 + \frac{t_1}{\tau} + \frac{1}{2!}\left(\frac{t_1}{\tau}\right)^2 + \cdots$$

Then, for $t > t_1$

$$y(t) = A\tau\left[\frac{t_1}{\tau} + \frac{1}{2!}\left(\frac{t_1}{\tau}\right)^2 + \cdots\right]\epsilon^{-t/\tau}$$

If the pulse width is small compared to the time constant of the system, i.e., if $t_1 \ll \tau$, we can neglect all the terms inside the bracket except the first and write

$$y(t) \simeq At_1 \, \epsilon^{-t/\tau} \qquad \text{for } t > t_1 \tag{24}$$

where At_1 is the area underneath the input pulse.

Suppose the pulse input to our first-order system is changed to the triangular input shown in Figure 8.7 and that $y(0)$ is again zero. We can describe the new input by

$$f_2(t) = \begin{cases} 0 & \text{for } t \le 0 \text{ and } t > t_1 \\ 4At/t_1 & \text{for } 0 < t \le 0.5t_1 \\ 4A(1 - t/t_1) & \text{for } 0.5t_1 < t \le t_1 \end{cases}$$

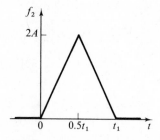

Figure 8.7 Triangular pulse with area At_1.

Note that the area under the input curve is still At_1. We can find the response for $t > t_1$ to this new input and again make the approximation $t_1 \ll \tau$. In Problem 8.13, the reader is asked to show that the result is

$$y(t) \simeq At_1 \, \epsilon^{-t/\tau} \qquad \text{for } t > t_1 \tag{25}$$

which is identical to (24).

As long as the width t_1 of any input pulse that is the input to a first-order system is small compared to the system's time constant, the response for $t > t_1$ depends on the area underneath the pulse but not on its shape. The responses in (24) and (25) to the inputs in Figure 8.5(a) and Figure 8.7 provide an example.

As background for another important property, note that the response of the system modeled by (22) to a step function of height At_1, with $y(0) = 0$, is At_1 times the unit-step response, namely

$$At_1 \, y_U(t) = At_1 \, \tau(1 - \epsilon^{-t/\tau}) \qquad \text{for } t \ge 0$$

Note that

$$\frac{d}{dt}[At_1 \, y_U(t)] = At_1 \, \epsilon^{-t/\tau} \tag{26}$$

for all positive values of t. Since the right side of (26) is identical to (24) and (25), we see that if $t_1 \ll \tau$, the response for $t > t_1$ to a pulse of area At_1 is the derivative of the response to a step function of height At_1.

THE UNIT IMPULSE

Since the response of a first-order system to a pulse of given area appears to be independent of the pulse shape as long as the pulse width t_1 is small compared to the time constant τ, it is reasonable to try to define an idealized pulse function whose width is small compared to the time constant of all first-order systems. However, in order to have $t_1 \ll \tau$ for all nonzero values of τ, t_1 must be infinitesimally small; and to have a nonzero pulse area, the height of the pulse must become infinitely large. Although such an idealized pulse creates conceptual and mathematical difficulties, let us reconsider the rectangular and the triangular input pulses, which are shown in Figure 8.8(a) and Figure 8.8(b), respectively.

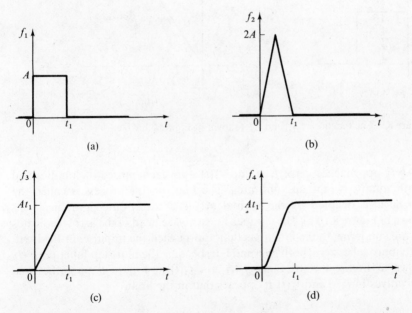

Figure 8.8 (a) Rectangular pulse with area At_1. (b) Triangular pulse with area At_1. (c) $f_3(t) = \int_0^t f_1(\lambda) \, d\lambda$. (d) $f_4(t) = \int_0^t f_2(\lambda) \, d\lambda$.

Note that $f_1(t)$ and $f_2(t)$ are the derivatives of $f_3(t)$ and $f_4(t)$, respectively. The initial part of $f_4(t)$ consists of two parabolas described by the equation

$$f_4(t) = \begin{cases} \left(\dfrac{2A}{t_1}\right)t^2 & \text{for } 0 < t \le 0.5t_1 \\[2ex] At_1 - \dfrac{2A}{t_1}(t-t_1)^2 & \text{for } 0.5t_1 < t \le t_1 \end{cases}$$

Let us specify $A = 1/t_1$ in Figure 8.8 so that the area underneath $f_1(t)$ and $f_2(t)$ is unity and so that the value of $f_3(t)$ and $f_4(t)$ for $t > t_1$ is also unity. If we continually decrease the value of t_1, then the heights of the pulses $f_1(t)$ and $f_2(t)$ increase in order to maintain unit areas, while $f_3(t)$ and $f_4(t)$ rise to their final values more rapidly. The dashed lines in Figure 8.9 show the changes in $f_1(t)$ and $f_3(t)$ when t_1 is halved.

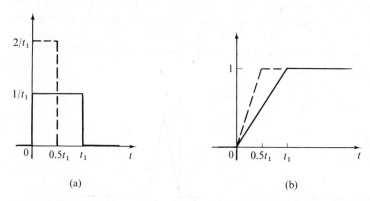

(a)   (b)

Figure 8.9 The functions $f_1(t)$ and $f_3(t)$ shown in Figure 8.8 when $A = 1/t_1$.

As t_1 approaches zero, $f_1(t)$ and $f_2(t)$ approach pulses of infinitesimal width, infinite height, and unit area. The limit of this process is called the *unit impulse*, denoted by the symbol $\delta(t)$. It is represented graphically as shown in Figure 8.10(a). The number 1 next to the head of the arrow indicates the area underneath the functions that approached the impulse. In the limit, as t_1 approaches zero, both $f_3(t)$ and $f_4(t)$ become the unit step function $U(t)$, shown in Figure 8.10(b). Since $f_1(t)$ and $f_2(t)$ in Figure 8.8 are the time derivatives of $f_3(t)$ and $f_4(t)$, it appears that in the limit,

$$\delta(t) = \frac{d}{dt}U(t) \qquad (27)$$

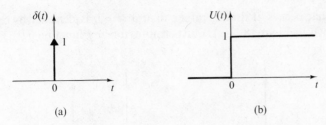

Figure 8.10 (a) The unit impulse. (b) The unit step function.

The arguments we used in the preceding paragraph are heuristic and not mathematically rigorous. Questions can be raised about (27) because differentiation is a limiting process and we introduced $\delta(t)$ as the result of another limiting process. Then, without mathematical justification, we interchanged the order of the two limiting processes. Furthermore, if we were to put our description of the unit impulse into equations, we might write

$$\delta(t) = 0 \qquad \text{for } t \neq 0 \tag{28a}$$

$$\int_{-\varepsilon}^{\varepsilon} \delta(t)\, dt = 1 \qquad \text{for } \varepsilon > 0 \tag{28b}$$

where (28b) suggests that the unit impulse has unit area. Although we shall not discuss the matter in detail here, it happens that (28) violates the axioms of real-function theory and is an impermissible way of defining a function. In fact, $\delta(t)$ is not a function in the usual sense, and we have purposely avoided calling it the unit impulse "function." However, results obtained in this section can be justified by rigorous mathematical arguments.*

We can define the unit impulse formally in terms of an integral and in a way that is consistent with distribution theory. For any function $f(t)$ that is continuous at $t = 0$, the unit impulse $\delta(t)$ must satisfy the integral expression

$$\int_{-a}^{b} f(t)\,\delta(t)\, dt = f(0) \qquad \text{for } a > 0,\, b > 0 \tag{29}$$

Note that (28b) is a special case of this equation, with $f(t) = 1$ for all values of t.

---

* The unit impulse is also called the *Dirac delta function* and is used in many areas of science and engineering. Both the unit impulse and ordinary functions can be regarded as special cases of generalized functions or distributions. For an introductory discussion of distribution theory, see Kuo, Appendix B, or Swisher, Section 2.4).

A unit impulse that occurs at time t_a rather than at $t = 0$ is denoted by $\delta(t - t_a)$ and is shown in Figure 8.11(a). Furthermore, for any function $f(t)$

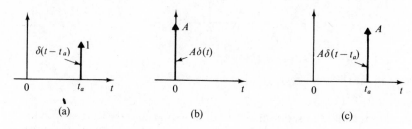

Figure 8.11 Impulses. (a) $\delta(t - t_a)$. (b) $A\delta(t)$. (c) $A\delta(t - t_a)$.

that is continuous at $t = t_a$, we can replace (29) by

$$\int_b^c f(t)\delta(t - t_a)\, dt = f(t_a) \qquad \text{for } b < t_a < c \tag{30}$$

which is referred to as the *sampling property* of the impulse. The product $A\delta(t)$, which is shown in Figure 8.11(b), is called an impulse of weight A. We can visualize it as the limit of a high, narrow pulse of area A. Equation (27) may be replaced by the more general relationship

$$A\delta(t - t_a) = \frac{d}{dt}[AU(t - t_a)] \tag{31}$$

i.e., differentiating a step function of height A occurring at t_a gives rise to an impulse of weight A at $t = t_a$ which is indicated in Figure 8.11(c).

The *unit impulse response* $h(t)$ is defined as the output that occurs when the input is $\delta(t)$ and when the system contains no stored energy before the impulse is applied. Thus $h(t)$ is the zero-state response to $\delta(t)$. The zero-state response for $t > t_1$ to a pulse of any shape is approximately equal to $h(t)$ times the area underneath the original pulse, as long as the pulse width t_1 is small compared to the time constant of the system.

Because a unit impulse applied to a system can cause an infinite power flow that instantaneously changes the energy storage in the system, the initial conditions we would need to calculate $h(t)$ directly are often hard to find. For this reason, the property described in the following paragraphs often proves useful.

Assume that we know the response to a certain input for a linear system that contains no initial stored energy. Then, if we substitute a new input that is the derivative of the original input, the new response will be the derivative of the old response. We can demonstrate the plausibility of this property by considering the general input-output differential equation

$$a_n y^{(n)} + \cdots + a_2 \ddot{y} + a_1 \dot{y} + a_0 y = b_m u^{(m)} + \cdots + b_1 \dot{u} + b_0 u \qquad (32)$$

Differentiating both sides of the equation gives

$$a_n y^{(n+1)} + \cdots + a_2 \dddot{y} + a_1 \ddot{y} + a_0 \dot{y} = b_m u^{(m+1)} + \cdots + b_1 \ddot{u} + b_0 \dot{u} \quad (33)$$

Since (33) is identical to (32) except that $\dot{u}$ and $\dot{y}$ have replaced u and y, respectively, the system's differential equation is still satisfied when the original input and output are replaced by their respective derivatives.

For a complete proof, we would also need to show that the appropriate initial conditions are satisfied when these replacements are made. Rather than doing this, however, we shall include a proof in Section 13.2.

Because $\delta(t)$ is the derivative of $U(t)$, the unit impulse response and the unit step response for a linear system are related by

$$h(t) = \frac{d}{dt} y_U(t) \qquad (34)$$

Thus once the unit step response $y_U(t)$ is known, we can obtain $h(t)$ from it by using (34).

EXAMPLE 8.4   For the translational system shown in Figure 8.12, find $y_U(t)$ and $h(t)$ when the output is the velocity $v(t)$. Then repeat the problem when the output is the acceleration $a(t)$.

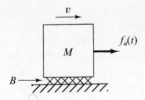

Figure 8.12 Translational system for Example 8.4.

Solution   The system is described by the differential equation

$$M\dot{v} + Bv = f_a(t)$$

or

$$\dot{v} + \frac{B}{M} v = \frac{1}{M} f_a(t)$$

Since the system's time constant is $\tau = M/B$ and $v_{ss} = 1/B$ if $f_a(t) = 1$ for all $t > 0$, the unit step response of the velocity is

$$(y_U)_v = \frac{1}{B}(1 - \epsilon^{-Bt/M}) \qquad \text{for } t > 0 \tag{35}$$

From (34), the unit impulse response of the velocity can be found by differentiating $y_U(t)$ and is

$$h_v(t) = \frac{1}{M}\epsilon^{-Bt/M} \qquad \text{for } t > 0 \tag{36}$$

These responses are shown in Figure 8.13.

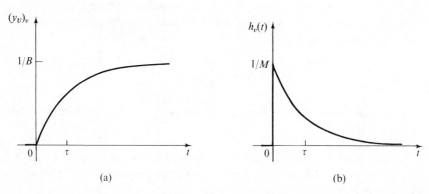

Figure 8.13 Responses for the system shown in Figure 8.12 when the output is the velocity. (a) Unit step response. (b) Unit impulse response.

Since the acceleration $a(t)$ of the mass is the derivative of its velocity, the unit step response of the acceleration is the derivative of the velocity step response in (35). But the derivative of the right side of (35) is $h_v(t)$, as given by (36). This can be written as

$$(y_U)_a = \frac{1}{M}[\epsilon^{-Bt/M}] U(t) \tag{37}$$

and is shown in Figure 8.14(a). When differentiating this expression to obtain the impulse response of the acceleration, we note that in addition to the term $-(B/M^2)\epsilon^{-Bt/M}$ that holds for $t > 0$, there is an impulse of weight $1/M$ at $t = 0$. We can see this by noting that the step response shown in Figure 8.14(a) has a discontinuity of height $1/M$ at $t = 0$. Upon differ-

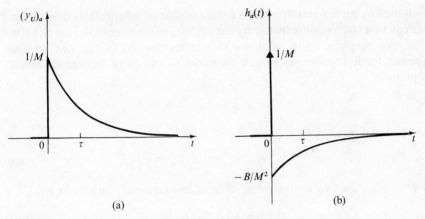

Figure 8.14 Responses for the system shown in Figure 8.12 when the output is the acceleration. (a) Unit step response. (b) Unit impulse response.

entiation, this discontinuity leads to the impulse $(1/M)\delta(t)$. In summary,

$$h_a(t) = \frac{1}{M}\delta(t) - \left[\frac{B}{M^2}\epsilon^{-Bt/M}\right]U(t) \tag{38}$$

which is shown in Figure 8.14(b).

Note in Figure 8.13(a) and Figure 8.14(a) that when $f_a(t)$ is a step function, the velocity of the mass does not change instantaneously at $t = 0$. However, there is a discontinuity in the acceleration curve. When $f_a(t)$ is an impulse, the velocity of the mass does undergo an instantaneous change. In the first case, there is no change in the kinetic energy of the mass at $t = 0$. In the latter case, however, the impulse causes an instantaneous increase in the energy of the mass.

8.4 SECOND-ORDER SYSTEMS

We can write the differential equation for a fixed linear second-order system as

$$\ddot{y} + a_1\dot{y} + a_0 y = F(t) \tag{39}$$

where, without loss of generality, the coefficient of $\ddot{y}$ has been made unity and where a_0 and a_1 are real constants. To find the response, as in Example 8.2, we need to know $F(t)$ for $t \geq 0$ and also the initial conditions $y(0)$ and $\dot{y}(0)$.

If the forcing function $F(t)$ has a finite number of independent derivatives, we can find the forced response by the method of undetermined coefficients. The free response $y_H(t)$ contains two terms instead of the one that is present for first-order systems. If the roots r_1 and r_2 of the characteristic equation

$$r^2 + a_1 r + a_0 = 0$$

are real and distinct, then

$$y_H(t) = K_1 \epsilon^{r_1 t} + K_2 \epsilon^{r_2 t} \tag{40}$$

For $r_1 = r_2$, the two roots are not distinct and we must replace (40) by

$$y_H(t) = K_1 \epsilon^{r_1 t} + K_2 t \epsilon^{r_1 t} \tag{41}$$

If the roots are complex, they must have the form $r_1 = \alpha + j\beta$ and $r_2 = \alpha - j\beta$, and we write the free response as

$$y_H(t) = \epsilon^{\alpha t} [K_1 \cos \beta t + K_2 \sin \beta t]$$

or, equivalently,

$$y_H(t) = K \epsilon^{\alpha t} \cos(\beta t + \phi) \tag{42}$$

Examples of $y_H(t)$ are shown in Figure 8.15 for the three cases represented by (40) through (42) when r_1, r_2, and α are negative numbers. The light lines in Figure 8.15(a) and Figure 8.15(b) indicate the two individual functions that are added to give $y_H(t)$, which is the heavy curve. In Figure 8.15(c), dashed lines labeled $\epsilon^{\alpha t}$ and $-\epsilon^{\alpha t}$ form the envelope of the damped oscillations. The values of the arbitrary constants K_1, K_2, K, and ϕ in (40) through (42) depend on $y_P(0)$ and the initial conditions $y(0)$ and $\dot{y}(0)$.

THE COMPLEX PLANE

When the roots of the characteristic equation are plotted in a complex plane, inspection of the plot reveals the nature of the system's free response. The root locations corresponding to the typical free responses shown in Figure 8.15 are indicated by the crosses in the respective parts of Figure 8.16.

Recall from Section 8.2 that a stable system is one for which the free response $y_H(t)$ decays to zero as t approaches infinity. All the examples in Figure 8.15 fall into this category. Examples of unstable systems, where $y_H(t)$ increases without bound, are shown in Figure 8.17. The characteristic-root locations appear directly under the corresponding sketches of the free

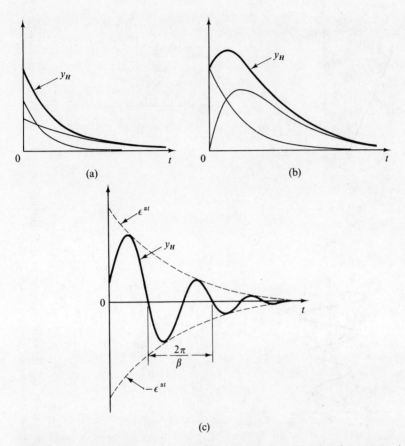

Figure 8.15 Typical curves for the free response of a second-order system. (a) Real distinct negative roots of the characteristic equation. (b) Identical negative roots. (c) Complex roots with $\alpha < 0$.

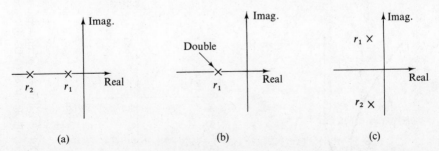

Figure 8.16 Roots of the characteristic equation corresponding to Figure 8.15. (a) Real distinct negative roots. (b) Identical negative roots. (c) Complex roots with $\alpha < 0$.

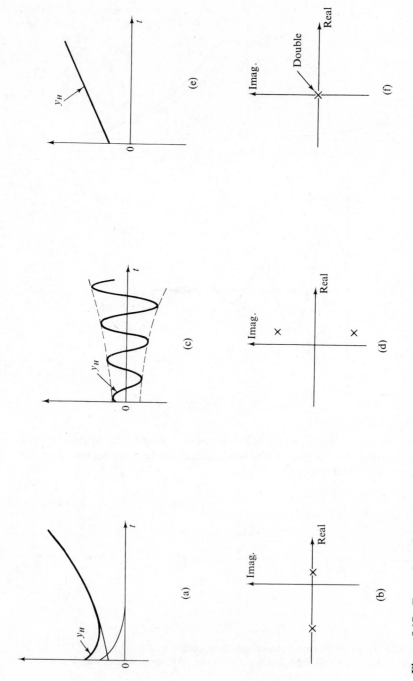

Figure 8.17 Examples of unstable second-order systems. (a), (b) One real root in the right half-plane. (c), (d) Complex roots in the right half-plane. (e), (f) Double root at the origin of the complex plane.

response. Figure 8.17(a) and Figure 8.17(b) correspond to a positive value of r_1 in (40), Figure 8.17(c) and Figure 8.17(d) correspond to a positive value of α in (42), and Figure 8.17(e) and Figure 8.17(f) correspond to $r_1 = 0$ in (41).

In addition to these stable and unstable classes of systems, it is possible (at least in an idealized case) for a linear system to be marginally stable and have a free response that neither decays to zero nor grows without bound. For second-order systems, such a response occurs when the characteristic equation has either a single root at $r = 0$ with the remaining root in the left half-plane or a pair of imaginary roots at $r = j\beta$ and $r = -j\beta$. The first case corresponds to the characteristic equation $r^2 + a_1 r = 0$ and to the free response

$$y_H(t) = K_1 + K_2 \, \epsilon^{-a_1 t}$$

where $a_1 > 0$. The second case corresponds to the characteristic equation $r^2 + \beta^2 = 0$ and to the free response

$$y_H(t) = K \cos(\beta t + \phi)$$

A system whose free response is a constant-amplitude sine or cosine function is often called a *simple harmonic oscillator*. Sample root locations and typical free-response curves for the two types of marginally stable second-order systems are shown in Figure 8.18.

We can summarize the discussion about stability in a way that is applicable to fixed linear systems of any order as follows. If all the roots of the characteristic equation are inside the left half-plane, the system is stable. If there are any roots inside the right half-plane or repeated roots on the imaginary axis, the system is unstable. If all the roots are inside the left half-plane except one or more distinct roots on the imaginary axis, the system is marginally stable.

DAMPING RATIO AND UNDAMPED NATURAL FREQUENCY

When the roots of the characteristic equation are complex, the parameter a_0 in (39) is positive, and it is useful to rewrite the differential equation in the standard form

$$\ddot{y} + 2\zeta\omega_n \dot{y} + \omega_n^2 y = F(t) \tag{43}$$

The parameter ω_n is called the *undamped natural frequency* and has units of radians per second. The parameter ζ is dimensionless and is known as the *damping ratio*. The characteristic equation is

$$r^2 + 2\zeta\omega_n r + \omega_n^2 = 0 \tag{44}$$

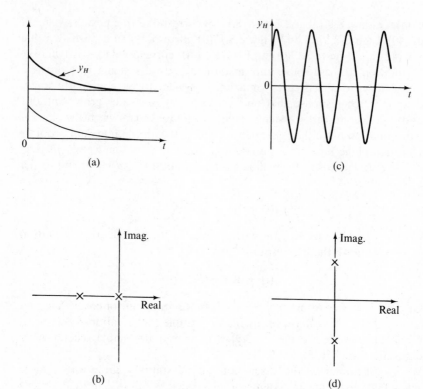

Figure 8.18 Examples of marginally stable second-order systems. (a), (b) One root at the origin of the complex plane. (c), (d) A pair of complex roots on the imaginary axis.

For $\zeta > 1$, the roots are distinct negative numbers, and $y_H(t)$ consists of two decaying exponentials. For $\zeta = 1$, there is a repeated root at $r = -\omega_n$, and $y_H(t)$ consists of terms having the form $\epsilon^{-\omega_n t}$ and $t\epsilon^{-\omega_n t}$. For $0 \leq \zeta < 1$, the roots are complex and are

$$r_1 = -\zeta\omega_n + j\omega_n\sqrt{1-\zeta^2}$$
$$r_2 = -\zeta\omega_n - j\omega_n\sqrt{1-\zeta^2}$$

Then by (42), the free response is

$$y_H(t) = K\epsilon^{-\zeta\omega_n t}\cos(\omega_n\sqrt{1-\zeta^2}\,t + \phi) \tag{45}$$

For $\zeta < 0$, the system is unstable.

The advantage of introducing the parameters ζ and ω_n becomes appparent when the characteristic roots are complex and are plotted in the complex plane, as indicated in Figure 8.19(a). Their distances from the origin are the same and are denoted by d. The distance d is the square root of the sum of the squares of the real and imaginary parts of the root. For the upper root,

$$d = \sqrt{(-\zeta\omega_n)^2 + \omega_n^2(1-\zeta^2)} = \omega_n$$

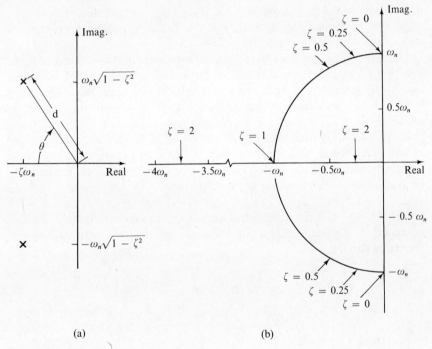

(a)   (b)

Figure 8.19 Characteristic-root locations in terms of ζ and ω_n. (a) General complex roots. (b) Locations for constant ω_n and varying ζ.

Hence, when $0 \le \zeta < 1$, the complex characteristic roots lie on a circle of radius ω_n centered at the origin. It is easy to show that their locations on the circle depend only on the damping ratio ζ. Specifically, the angle θ between the negative real axis and the line from the origin to r_1 in Figure 8.19(a) satisfies the relationship $\cos\theta = \zeta\omega_n/\omega_n = \zeta$. Thus

$$\theta = \cos^{-1}\zeta$$

The geometric relationships between ζ, ω_n, and the roots of (44) are summarized in Figure 8.19(b) for $\zeta \geq 0$.

It is instructive to observe the effect of the damping ratio ζ on the responses of a system described by the equation

$$\ddot{y} + 2\zeta\omega_n\dot{y} + \omega_n^2 y = \omega_n^2 u(t) \tag{46}$$

The unit step response is shown in Figure 8.20 for several values of ζ, with ω_n held constant. For $\zeta > 0$, the steady-state response is unity. Note that the value of ζ determines to what extent, if any, the response will overshoot its steady-state value. The overshoot is 100% for $\zeta = 0$ and decreases to zero when the damping ratio is unity. For $\zeta > 1$, the response approaches its steady-state value monotonically.

The case $\zeta = 1$ is the boundary between responses that oscillate and those that do not, and the system is said to have *critical damping*. Second-order systems for which $\zeta > 1$ have more than critical damping, those for which $0 < \zeta < 1$ are less than critically damped. When $\zeta = 0$ the system is *undamped*.

Figure 8.21 shows the unit impulse response of the general second-order system described by (46) for the values of ζ used in Figure 8.20. Note that the steady-state response is zero for $\zeta > 0$ and that the value of the damping ratio establishes the character of the response.

EXAMPLE 8.5   Consider the ideal pendulum shown in Figure 8.22 and discussed in Example 5.7. Find the form of the free response of the linearized model.

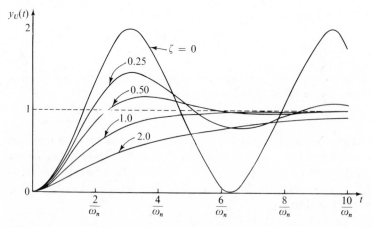

Figure 8.20 The unit step response for a second-order system described by (46).

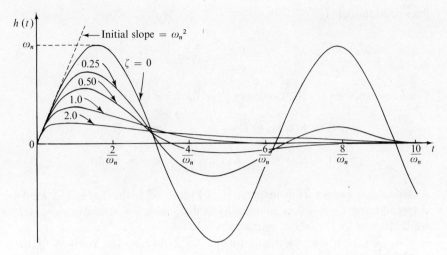

Figure 8.21 The unit impulse response for a second-order system described by (46).

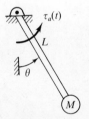

Figure 8.22 Ideal pendulum for Example 8.5.

Solution From the original nonlinear input-output equation

$$ML^2\ddot{\theta} + MgL \sin \theta = \tau_a(t)$$

the linearized model was shown in Example 5.7 to be

$$\ddot{\hat{\theta}} + \frac{g}{L}\hat{\theta} = \frac{1}{ML^2}\hat{\tau}_a(t)$$

when the operating point is $\bar{\theta} = 0$, i.e., with the mass M directly below the pivot. When $\bar{\theta} = \pi$ rad, i.e., with the mass M directly above the pivot, the linearized model is

$$\ddot{\hat{\theta}} - \frac{g}{L}\hat{\theta} = \frac{1}{ML^2}\hat{\tau}_a(t)$$

The free responses for the two cases have the form

$$\hat{\theta}_H(t) = K\cos(\sqrt{g/L}\ t + \phi) \qquad \text{for } \bar{\theta} = 0$$

$$\hat{\theta}_H(t) = K_1\,\epsilon^{\sqrt{g/L}\ t} + K_2\,\epsilon^{-\sqrt{g/L}\ t} \qquad \text{for } \bar{\theta} = \pi \text{ rad}$$

As we would expect from inspection of Figure 8.22, the linearized model corresponding to $\bar{\theta} = 0$ is marginally stable, with $\zeta = 0$ and $\omega_n = \sqrt{g/L}$, while the one for $\bar{\theta} = \pi$ rad is unstable.

If we add a linear frictional torque $B\dot{\theta}$ retarding the motion of the pendulum, the input-output equation becomes

$$ML^2\ddot{\theta} + B\dot{\theta} + MgL\sin\theta = \tau_a(t)$$

For $\bar{\theta} = 0$, the linearized model becomes

$$\ddot{\hat{\theta}} + \frac{B}{ML^2}\dot{\hat{\theta}} + \frac{g}{L}\hat{\theta} = \frac{1}{ML^2}\hat{\tau}_a(t) \qquad (47)$$

which is stable and which has

$$\omega_n = \sqrt{\frac{g}{L}}$$

$$\zeta = \frac{B}{2Mg^{1/2}L^{3/2}}$$

As expected, increasing B increases the damping ratio ζ.

For $\bar{\theta} = \pi$ rad, the linearized model is given by

$$\ddot{\hat{\theta}} + \frac{B}{ML^2}\dot{\hat{\theta}} - \frac{g}{L}\hat{\theta} = \frac{1}{ML^2}\hat{\tau}_a(t) \qquad (48)$$

and is again unstable, since one root of the characteristic equation is in the right half-plane.

8.5 SYSTEMS OF ORDER THREE AND HIGHER

Except for situations involving repeated characteristic roots, the free response of third- and higher-order systems is composed of those functions that comprise the free response of first- and second-order systems. The basis for this property is the fact that all the characteristic roots must either be real or occur in complex-conjugate pairs. For example, the third-order differential equation

$$\dddot{y} + a_2 \ddot{y} + a_1 \dot{y} + a_0 y = F(t)$$

has as its characteristic equation

$$r^3 + a_2 r^2 + a_1 r + a_0 = 0 \tag{49}$$

If two of its roots are complex, we may write this equation in factored form as

$$(r - \alpha - j\beta)(r - \alpha + j\beta)(r - \gamma) = 0$$

where the roots are $r_1 = \alpha + j\beta$, $r_2 = \alpha - j\beta$, and $r_3 = \gamma$. The general form of the free response is

$$y_H(t) = K_1 \epsilon^{r_1 t} + K_2 \epsilon^{r_2 t} + K_3 \epsilon^{r_3 t}$$

which, by using (42) for the terms corresponding to the complex roots, we can write as

$$y_H(t) = K_4 \epsilon^{\alpha t} \cos(\beta t + \phi) + K_3 \epsilon^{\gamma t}$$

Figure 8.23 shows a typical set of characteristic roots in the complex plane and a sample free response for such a third-order system. Figure 8.23(b) shows the individual response curves, with the response associated with roots r_1 and r_2 designated as y_{12}. Note that the term $y_3 = K_3 \epsilon^{\gamma t}$ decays more slowly than the envelope of y_{12}, because $|\alpha| > |\gamma|$ in Figure 8.23(a). In graphical terms, r_3 is further to the right in the complex plane than r_1 and r_2. Figure 8.23(c) shows the free response, which is the sum of the two individual responses.

We find the forced response $y_P(t)$ of a third- or higher-order system just as we found it for first- and second-order systems. Although it requires no new concepts or techniques, the solution is more cumbersome than for the lower-order cases.

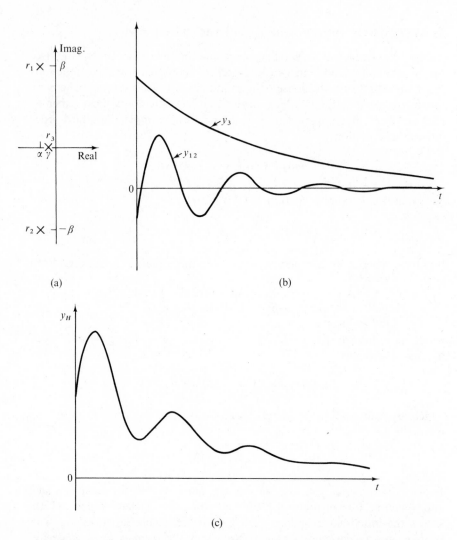

Figure 8.23   (a) Characteristic roots for a third-order system with $\alpha = -2$, $\beta = 10$, and $\gamma = -1$. (b) Components of the free response. (c) The free response.

PROBLEMS

8.1 a. Find and sketch the response of the system described by the equation $\dot{y} + 4y = u(t)$ when the input is $u(t) = 3$ for $t \geq 0$ and when $y(0) = 2$.

b. Repeat part a when $u(t) = t$ for $t \geq 0$ and $y(0) = 0$.

8.2 a. Find the response of the system described by

$$\ddot{y} + 3\dot{y} + 2y = \dot{u} + 2u(t)$$

when $u(t) = t$ for $t \geq 0$, $y(0) = 0$, and $\dot{y}(0) = 1$.

b. Repeat part a when $u(t) = te^{-t}$ for $t \geq 0$ and $y(0) = \dot{y}(0) = 0$.

8.3 a. Prove that (6) is a solution of (4) for the case $n = 2$.

b. Repeat part a for $n = 3$.

8.4 Prove that $y_H(t) = (K_1 + K_2 t)e^{r_1 t}$ is a solution of (4) when $n = 2$ and when r_1 is a double root of the characteristic equation.

8.5 Consider the differential equation $\ddot{y} + b\dot{y} + cy = F(t)$.

a. Find the complete response for $t \geq 0$ when $b = 7$, $c = 6$, $F(t) = 3$, $y(0) = 0$, and $\dot{y}(0) = -2$.

b. Repeat part a when the initial conditions are changed to $y(0) = 2$, $\dot{y}(0) = 0$.

c. Show that the free response increases without limit if either b or c is negative.

8.6 For the linear first-order differential equation given by (11), sketch $y(t)$ vs t for $t \geq 0$ when $\tau = 3$, $F(t) = -4$, and $y(0) = 6$. Find the value of time for which $y(t) = 0$.

8.7 Find the response $y(t)$ for $t \geq 0$ for a linear first-order system described by $\dot{y} + 0.5y = F(t)$ for each of the following conditions.

a. $F(t) = \sin t + \cos t$ and $y(0) = 0$.

b. $F(t) = e^{-t/2}$ and $y(0) = 1$.

8.8 The response of a system described by $\dot{y} + (1/\tau)y = A$ is given by (14).

a. Find the number of time constants it takes for the response to be 90% of the way from the initial value to the steady-state value.

b. Find the number of time constants it takes for the response to be 99.9% of the way from the initial value to the steady-state value.

8.9 A velocity input $v_a(t)$ is applied to point A in the mechanical system shown in Figure P8.9.

a. Write the system's differential equation in terms of the velocity v_1.

b. What is the time constant τ for the system?

c. Sketch the response when $v_a(t) = 0$ for $t \geq 0$ and $v_1(0) = 10$.

d. Repeat parts a, b, and c when the velocity input is replaced by a force $f_a(t)$ applied at point A, with the positive sense to the right. Explain why the expression for τ differs from the answer to part b.

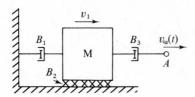

Figure P8.9

8.10 Consider a first-order system with input $u(t)$ and output y that is described by $\dot{y} + 0.5y = u(t)$.

    a. Find and sketch the unit step response $y_U(t)$.

    b. Find and sketch the response to $u(t) = 2U(t)$ with $y(0) = -1$.

    c. Find and sketch the response to $u(t) = U(t) - U(t-2)$ with $y(0) = 0$.

    d. Find and sketch the unit impulse response $h(t)$.

8.11 The unit ramp function $f_r(t)$ is defined to be zero for $t \le 0$ and t for $t > 0$, as shown in Figure P8.11(a). A unit ramp function delayed by a units of time is shown in Figure P8.11(b).

    a. Find the response of the first-order system described by (22) when $u(t) = f_r(t)$ and $y(0) = 0$.

    b. Repeat part a when $u(t) = f_r(t-a)$ and $y(a) = 0$.

    c. Note that $U(t) = d/dt[f_r(t)]$. Verify that $y_U(t) = \tau[1 - \epsilon^{-t/\tau}]U(t)$ is the derivative of the unit ramp response.

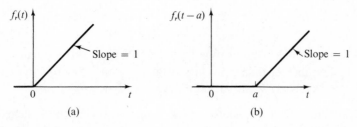

Figure P8.11

8.12 Any waveform that consists of straight lines can be represented as the sum of step functions and ramp functions.

a. Verify graphically that the triangular pulse shown in Figure 8.7 can be written as

$$f_2(t) = (4A/t_1)[f_r(t) - 2f_r(t - t_1/2) + f_r(t - t_1)]$$

where the ramp function $f_r(t)$ is shown in Figure P8.11(a).

b. Sketch the function

$$f_r(t) - f_r(t-1) - 2U(t-1) + f_r(t-2) - f_r(t-3).$$

8.13 a. Find the response of the first-order system described by (22) when $u(t)$ is the function $f_2(t)$ shown in Figure 8.7 and when $y(0) = 0$.

b. Show that for $t > t_1$, the solution reduces to

$$y(t) = \frac{4A\tau^2}{t_1}(1 - \epsilon^{t_1/2\tau})^2 \epsilon^{-t/\tau}$$

c. Derive (25) by using the expansion $\epsilon^x = 1 + x + x^2/2! + \cdots$, with x replaced by $t_1/2\tau$ and with $t_1/\tau \ll 1$.

8.14 a. Find the response of the first-order system described by (22) when $u(t)$ is the function $f(t)$ shown in Figure P8.14 and when $y(0) = 0$.

b. Show that for $t > t_1$, the solution reduces to

$$y(t) = \frac{2A\tau^2}{t_1}\left(1 - \epsilon^{t_1/\tau} + \frac{t_1}{\tau}\epsilon^{t_1/\tau}\right)\epsilon^{-t/\tau}$$

c. By using the Taylor-series expansion $\epsilon^x = 1 + x + x^2/2! + \cdots$, with x replaced by t_1/τ and with $t_1/\tau \ll 1$, show that the solution reduces to (25).

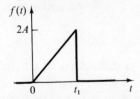

Figure P8.14

8.15 For each of the following differential equations, find and sketch the unit step response $y_U(t)$ and the unit impulse response $h(t)$.

a. $\ddot{y} + 5\dot{y} + 4y = u(t)$

b. $\ddot{y} + 2\dot{y} + 2y = u(t)$

8.16 a. For the system described by the equation $\ddot{y} + 4\dot{y} + 25y = 50u(t)$, find the damping ratio ζ and the undamped natural frequency ω_n.

b. Find and sketch the unit step response $y_U(t)$ and the unit impulse response $h(t)$.

8.17 a. For the system described by the equation $4\ddot{y} + 3\dot{y} + 9y = 4u(t)$, find the damping ratio ζ and the undamped natural frequency ω_n.

b. Sketch the approximate response to the initial conditions $y(0) = -1$ and $\dot{y}(0) = 0$, taking $u(t) = 0$ but without finding the analytical solution.

c. Find the analytical solution for the conditions in part b.

8.18 Repeat Problem 8.17 for a system described by $\ddot{y} + 8\dot{y} + 36y = 4u(t)$.

8.19 A second-order system is described by $\ddot{y} + 2\dot{y} + 4y = 4u(t)$.

a. Find and sketch the response when $u(t) = U(t)$, $y(0) = 0$, and $\dot{y}(0) = 1$.

b. Find the maximum value of $y(t)$. Determine the overshoot $y_{max} - y_{ss}$.

8.20 a. For the system shown in Figure 3.9(a) and modeled by (3.10), find expressions for the damping ratio ζ and the undamped natural frequency ω_n.

b. Find the values of B_2 for which the system will be stable if $M = 2$ kg, $B_3 = 3$ N·s/m, and $K = 4$ N/m. Find the values of B_2 for which the free response will decay to zero without oscillations.

8.21 The unit step response of a certain linear system is

$$y_U(t) = [2 + 2\epsilon^{-t}\cos(2t + \pi/4)]\,U(t)$$

a. Find numerical values for the damping ratio ζ and the undamped natural frequency ω_n.

b. Find the unit impulse response $h(t)$. Sketch $y_U(t)$ and $h(t)$.

8.22 A certain third-order system is described by the differential equation $\dddot{y} + 2\ddot{y} + \dot{y} + 2y = 4$. Find $y(t)$ when $y(0) = 0$, $\dot{y}(0) = 5$, and $\ddot{y}(0) = -3$. The roots of the system's characteristic equation are $r_1 = -2$, $r_2 = j$, and $r_3 = -j$.

8.23 a. Find the form of the free response for the system shown in Figure P2.16 when $M_1 = 1$ kg, $M_2 = 2$ kg, and $K = 2$ N/m for each spring. (*Hint:* To factor a characteristic equation of the form

$$a_4 r^4 + a_2 r^2 + a_0 = 0$$

first use the quadratic formula to find the two values of r^2 that satisfy the equation. Then find the four values of r.)

b. Determine if the system is stable, marginally stable, or unstable. Explain your answer in both mathematical and physical terms.

8.24 A linear fourth-order system is described by the input-output equation

$$a_4 \, y^{(iv)} + a_3 \, y^{(iii)} + a_2 \, \ddot{y} + a_1 \, \dot{y} + a_0 \, y = F(t)$$

Write the form of the complete response when $F(t) = A$ for each of the following sets of characteristic-root locations. Do not evaluate the multiplying constants in front of the individual terms. Identify the transient and steady-state response.

a. $0, -2, -2, -5$

b. $-1, -2, \pm j3$

c. $-2, -2, -2 \pm j3$

d. $-2 \pm j3, -2 \pm j3$

8.25 Repeat Problem 8.24 when $F(t) = \cos 3t$.

CHAPTER 9 —————————————————————————
LINEAR ELECTRICAL CIRCUITS

Except at quite high frequencies, electrical circuits can usually be considered an interconnection of lumped elements. In such cases, which include a large and very important portion of the applications of electrical phenomena, we can model a circuit using ordinary differential equations and apply the solution techniques discussed in this book.

In this chapter, we shall consider fixed linear circuits using the same approach we used for mechanical systems. We shall introduce the element and interconnection laws and then combine them to form procedures for finding the model of a circuit. After developing a general modeling technique, we shall present several specialized results applicable to the important cases of resistive and first-order circuits. We conclude the chapter with a discussion of circuit models in state-variable form. Nonlinear circuits and time-varying linear circuits are considered in the following chapter.

9.1 VARIABLES

The variables most commonly used to describe the behavior of electrical circuits are

e, voltage in volts (V)

i, current in amperes (A)

The related variables

q, charge in coulombs (C)

ϕ, flux in webers (Wb)

Λ, flux linkage in weber-turns

may be used on occasion. Current is the time derivative of charge, so i and q are related by the expressions

$$i = \frac{dq}{dt} \tag{1}$$

and

$$q(t) = q(t_0) + \int_{t_0}^{t} i(\lambda)\, d\lambda \tag{2}$$

Flux and flux linkage are related by the number of turns N in a coil of wire, such that if all the turns are linked by all the flux, then $\Lambda = N\phi$.

We represent the current into and out of a circuit element by arrows drawn on the circuit diagram as shown in Figure 9.1. The arrows point in the direction in which positive charge, i.e., positive ions, flows when the current has a positive value. Equivalently, a positive current can also correspond to electrons (which have a negative charge) flowing in the opposite direction.

Because a net charge cannot exist within any circuit element, the current entering one end of a two-terminal element must leave the other end. Hence,

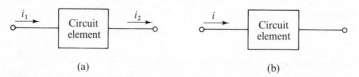

(a)   (b)

Figure 9.1 Conventions for denoting current. (a) Acceptable. (b) Preferred.

$i_1 = i_2$ in Figure 9.1(a) at all times, so only one current arrow need be shown, as in Figure 9.1(b).

The voltage at a point in a circuit is a measure of the difference between the electrical potential of that point and the potential of an arbitrarily established reference point called the *ground node*, or *ground* for short. The ground associated with a circuit is denoted by the symbol shown in the lower part of Figure 9.2(a). Any point in the circuit that has the same potential as the ground has a voltage of zero, by definition. The voltage e_1 shown in Figure 9.2(a) is positive if the point with which it is associated is at a higher potential than the ground and negative if the potential of the point with which it is associated is lower than that of the ground.

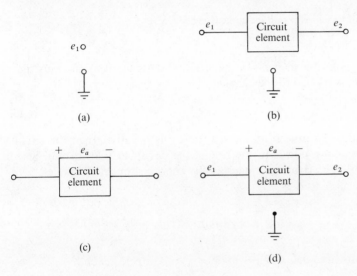

Figure 9.2 Conventions for denoting voltages.

We can define the voltages of the two terminals of a circuit element individually with respect to ground by writing appropriate symbols next to the terminals, as shown in Figure 9.2(b). We define the voltage between the terminals of an element by placing a symbol next to the element and plus and minus signs on either side of the element or at the terminals, as shown in Figure 9.2(c). When the element voltage e_a is positive, the terminal marked with the plus sign is at a higher potential than the other terminal.

In Figure 9.2(d), e_1 and e_2 denote the terminal voltages with respect to ground, and e_a is the voltage across the element with its positive sense indicated by the plus and minus signs. These three voltages are related by the equation

$$e_a = e_1 - e_2$$

Interchanging the plus and minus signs will reverse the sign of the voltage e_a in any equation in which it appears.

When we define the positive senses of the current and voltage associated with a circuit element as shown in Figure 9.3, such that a positive current

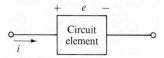

Figure 9.3 Positive senses of voltage and current for (3).

is assumed to enter the element at the terminal designated by the plus sign, then the power supplied to the element is

$$p = ei \qquad (3)$$

which has units of watts. If at some instant p is negative, then the circuit element is supplying power to the rest of the circuit at that instant. Since power is the time derivative of energy, the energy supplied to the element over the interval t_0 to t_1 is

$$\int_{t_0}^{t_1} p(t)\, dt$$

which has units of joules, where 1 joule = 1 volt-ampere-second.

9.2 ELEMENT LAWS

The elements in the electrical circuits that we shall consider are resistors, capacitors, inductors, and sources. The first three of these are referred to as *passive elements* because, while they can store or dissipate energy that is present in the circuit, they cannot introduce additional energy. They are analogous to the dashpot, mass, and spring for mechanical systems. In contrast, sources are *active elements* that can introduce energy into the circuit and that serve as the inputs. They are analogous to the force or displacement inputs for mechanical systems.

RESISTOR

A *resistor* is an element for which there is an algebraic relationship between the voltage across its terminals and the current through it—that is, an element that can be described by a curve of e vs i. A linear resistor is one for which the voltage and current are directly proportional to each other—that is, one described by *Ohm's law*

$$e = Ri \tag{4}$$

or

$$i = \frac{1}{R}e \tag{5}$$

where R is the *resistance* in ohms (Ω). A resistor and its current and voltage are denoted as shown in Figure 9.4. If we reversed either the current arrow or the voltage polarity (but not both) in the figure, we would introduce a minus sign into (4) and (5). The resistance of a body of length ℓ and constant cross-sectional area A made of a material with resistivity ρ is $R = \rho\ell/A$.

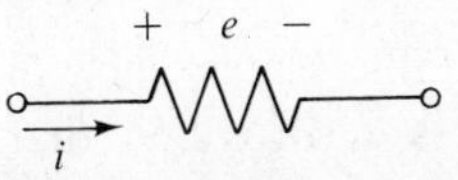

Figure 9.4 A resistor and its variables.

Analogous to the frictional element of mechanical systems, a resistor dissipates any energy supplied to it by converting it into heat. We can write the power ei dissipated by a linear resistor as

$$p = Ri^2 = \frac{1}{R}e^2$$

CAPACITOR

A *capacitor* is an element that obeys an algebraic relationship between the voltage and the charge, where the charge is the integral of the current. We use the symbol shown in Figure 9.5 to represent a capacitor. For a linear

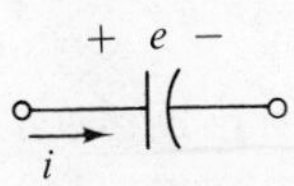

Figure 9.5 A capacitor and its variables.

capacitor, the charge and voltage are related by

$$q = Ce \tag{6}$$

where C is the *capacitance* in farads (F). For a fixed linear capacitor, the capacitance is a constant. If (6) is differentiated and $\dot{q}$ replaced by i, the element law for a fixed linear capacitor becomes

$$i = C\frac{de}{dt} \tag{7}$$

To express the voltage across the terminals of the capacitor in terms of the current, we solve (7) for de/dt and then integrate, getting

$$e(t) = e(t_0) + \frac{1}{C}\int_{t_0}^{t} i(\lambda)\, d\lambda \tag{8}$$

where $e(t_0)$ is the voltage corresponding to the initial charge, and where the integral is the charge delivered to the capacitor between the times t_0 and t.

One form of a capacitor consists of two parallel metallic plates, each of area A, separated by a dielectric material of thickness d. Provided that fringing of the electric field is negligible, the capacitance of this element is $C = \varepsilon A/d$, where ε is the permittivity of the dielectric material. The values of practical capacitances are typically expressed in microfarads (μF), where $1\ \mu\text{F} = 10^{-6}$ F. However, for numerical convenience we shall use farads in our examples.

The energy supplied to a capacitor is stored in its electrical field and can affect the response of the circuit at future times. For a fixed linear capacitor, the stored energy is

$$w = \tfrac{1}{2}Ce^2$$

Because the energy stored is a function of the voltage across its terminals, the initial voltage $e(t_0)$ of a capacitor is one of the conditions we need to find the complete response of a circuit for $t \geq t_0$.

INDUCTOR

An *inductor* is an element for which there is an algebraic relationship between the voltage across its terminals and the derivative of the flux linkage. The symbol for an inductor and the convention for defining its current and voltage are shown in Figure 9.6. For a linear inductor,

$$e = \frac{d}{dt}(Li)$$

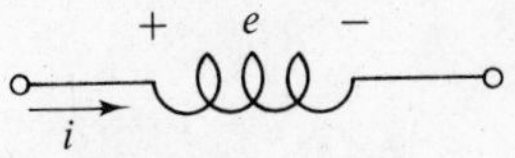

Figure 9.6 An inductor and its variables.

where L is the *inductance* and has units of henries (H). For a fixed linear inductor, L is constant and we can write the element law as

$$e = L\frac{di}{dt} \tag{9}$$

We can find an expression for the current through the inductor by using (9) to integrate di/dt, giving

$$i(t) = i(t_0) + \frac{1}{L}\int_{t_0}^{t} e(\lambda)\, d\lambda \tag{10}$$

where $i(t_0)$ is the initial current through the inductor.

For a linear inductor made by winding N turns of wire around a toroidal core of a material having a constant permeability μ, cross-sectional area A, and mean circumference ℓ, the inductance is $L = \mu N^2 A/\ell$. Typical values of inductance are usually less than 1 henry and are often expressed in millihenries (mH).

The energy supplied to an inductor is stored in its magnetic field, and for a fixed linear inductor this energy is given by

$$w = \tfrac{1}{2}Li^2$$

To find the complete response of a circuit for $t \geq t_0$, we need to know the initial current $i(t_0)$ for each inductor.

SOURCES

The inputs for electrical circuit models are provided by ideal voltage and current sources. A *voltage source* is any device that causes a specified voltage to exist between two points in a circuit, regardless of the current that may flow. A *current source* causes a specified current to flow through the branch containing the source, regardless of the voltage that may be required. The symbols used to represent general voltage and current sources are shown in Figure 9.7(a) and Figure 9.7(b). We often represent physical sources by the combination of an ideal source and a resistor.

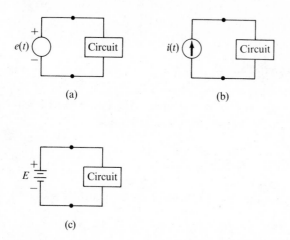

(a)   (b)

(c)

Figure 9.7 Sources. (a) Voltage. (b) Current. (c) Constant voltage source.

A voltage source that has a constant value for all time is often represented as shown in Figure 9.7(c). The symbol E denotes the value of the voltage, and the terminal connected to the longer line is the positive terminal. A battery is often represented in this fashion.

OPEN AND SHORT CIRCUITS

An *open circuit* is any element through which current cannot flow. For example, a switch in the open position provides an open circuit, as shown in Figure 9.8(a). Likewise, we can consider a current source that has a value of $i(t) = 0$ over a nonzero time interval an open circuit and redraw it as shown in Figure 9.8(b).

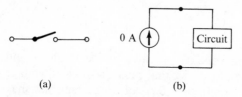

(a)   (b)

Figure 9.8 Examples of open circuits. (a) Open switch. (b) Zero current source.

A *short circuit* is any element across which there is no voltage. A switch in the closed position, as shown in Figure 9.9(a), is an example of a short circuit. Another example is a voltage source with $e(t) = 0$, as indicated in Figure 9.9(b).

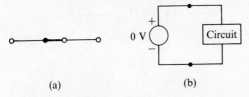

(a)   (b)

Figure 9.9 Examples of short circuits. (a) Closed switch. (b) Zero voltage source.

9.3 INTERCONNECTION LAWS

Two interconnection laws are used in conjunction with the appropriate element laws in modeling electrical circuits. These laws are known as Kirchhoff's voltage law and Kirchhoff's current law.

KIRCHHOFF'S VOLTAGE LAW

If a closed path, i.e., a loop, is traced through any part of a circuit, the algebraic sum of the voltages across the elements comprising the loop must equal zero. This property is known as *Kirchhoff's voltage law*. It may be written as

$$\sum_j e_j = 0 \quad \text{around a loop} \tag{11}$$

where e_j denotes the voltage across the jth element in the loop.

It follows that summing the voltages across individual elements in any two different paths from one point to another will give the same result. For instance, in the portion of a circuit sketched in Figure 9.10(a), summing the voltages around the loop, going in a counterclockwise direction, and taking into account the polarities indicated on the diagram give

$$e_1 + e_2 - e_3 - e_4 = 0$$

Reversing the direction in which the loop is traversed yields

$$e_4 + e_3 - e_2 - e_1 = 0$$

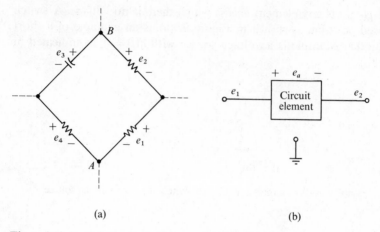

(a)   (b)

Figure 9.10 Partial circuits to illustrate Kirchhoff's voltage law.

Likewise, going from point A to point B by each of the two paths shown gives

$$e_1 + e_2 = e_4 + e_3$$

which is, of course, equivalent to both of the foregoing loop equations. In fact, we invoked (11) for the circuit element shown in Figure 9.2(d), which is repeated in Figure 9.10(b), when we stated that $e_a = e_1 - e_2$, since it follows from the voltage law that $e_2 + e_a - e_1 = 0$.

KIRCHHOFF'S CURRENT LAW

When the terminals of two or more circuit elements are connected together, the common junction is referred to as a *node*. All the joined terminals are at the same voltage and can be considered part of the node. Because it is not possible to accumulate any net charge at a node, the algebraic sum of the currents at any node must be zero at all times. This property is known as *Kirchhoff's current law*. It may be written as

$$\sum_j i_j = 0 \quad \text{at any node} \tag{12}$$

where the summation is over the currents through all the elements joined to the node.

In applying (12), we must take into account the directions of the current arrows. We shall use a plus sign in (12) for a current arrow directed away from the node being considered and a minus sign for a current arrow toward the node. This is consistent with the fact that the current i entering a node is equivalent to the current $-i$ leaving the node.* For the partial circuit shown in Figure 9.11, applying (12) at the node to which the three elements are connected gives $i_1 + i_2 + i_3 = 0$.

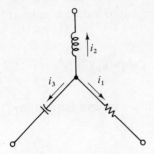

Figure 9.11 Partial circuit to illustrate Kirchhoff's current law.

It is a common practice to write the current-law equation directly in terms of the element values and the voltages of the nodes. Consider, for example, the circuit segment shown in Figure 9.12, where e_A, e_B, e_D, and e_F represent the voltages of the nodes with respect to ground. By Kirchhoff's voltage law, the voltage across the resistor is $e_A - e_B$; and by the element law, the current through the resistor is $i_1 = (e_A - e_B)/R$. Similarly, the current

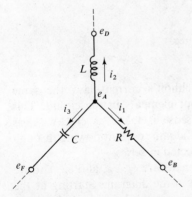

Figure 9.12 Partial circuit to illustrate Kirchhoff's current law written in terms of node voltages.

* Instead of interpreting the left side of (12) as the algebraic sum of the currents leaving the node, it would also be correct to use the algebraic sum of the currents entering the node.

through the inductor is

$$i_2 = i_2(0) + \frac{1}{L} \int_0^t (e_A - e_D) \, d\lambda$$

and that through the capacitor is $i_3 = C(\dot{e}_A - \dot{e}_F)$. Thus we can write the current law in terms of the node voltages and the initial inductor current as

$$\frac{1}{R}(e_A - e_B) + i_2(0) + \frac{1}{L} \int_0^t (e_A - e_D) \, d\lambda + C(\dot{e}_A - \dot{e}_F) = 0$$

In the following two examples, Kirchhoff's voltage law and current law are used to derive a circuit model.

EXAMPLE 9.1 Derive the model for the circuit shown in Figure 9.13.

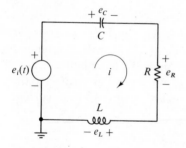

Figure 9.13 Series *RLC* circuit with a voltage source.

Solution By a trivial application of Kirchhoff's current law, the same current must flow through each of the four elements in the circuit. This current is denoted by *i* and its positive sense is taken as clockwise, as indicated in Figure 9.13. Because they have the same current flowing through them, the four elements are said to be connected in *series*.

The voltages across the three passive elements are e_L, e_R, and e_C, and we have assigned them the polarities indicated in the diagram. Starting at the ground node and proceeding counterclockwise around the single loop, we have by Kirchhoff's voltage law

$$e_L + e_R + e_C - e_i(t) = 0 \tag{13}$$

The element laws (4), (8), and (9) give expressions for e_R, e_C, and e_L:

SUB INTO
VOLTAGE LAW
& Differentiate
to get 2ND
DEGREE D.E.

$$e_R = Ri$$

$$e_C = e_C(0) + \frac{1}{C}\int_0^t i(\lambda)\, d\lambda \tag{14}$$

$$e_L = L\frac{di}{dt}$$

where the initial time has been taken as $t_0 = 0$ in (8). Substituting (14) into (13) and rearranging gives the circuit model as the integral-differential equation

$$L\frac{di}{dt} + Ri + \frac{1}{C}\int_0^t i(\lambda)\, d\lambda = e_i(t) - e_C(0) \tag{15}$$

To eliminate the constant term and the integral, we differentiate (15) term by term, giving

$$L\frac{d^2i}{dt^2} + R\frac{di}{dt} + \frac{1}{C}i = \dot{e}_i$$

which is a second-order differential equation for the current i with the derivative of the applied voltage acting as the forcing function.

EXAMPLE 9.2   Obtain the input-output differential equation relating the input $i_i(t)$ to the output e_o for the circuit shown in Figure 9.14.

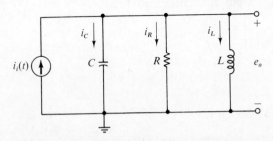

Figure 9.14 Parallel *RLC* circuit with a current source.

Solution Each of the four circuit elements in Figure 9.14 has one terminal connected to the ground node and the other terminal connected to a single node. By a trivial application of Kirchhoff's voltage law, we see that the voltage across each element is e_o. Hence, we say that the elements are connected in *parallel*.

Because the circuit has a single node whose voltage is unknown, we shall apply Kirchhoff's current law at that node in order to obtain the circuit model. We could also apply the current law at the ground node, but we would obtain no new information. The currents through the three passive elements are i_C, i_R, and i_L. As indicated by the arrows in Figure 9.14, each of these currents is considered positive when it flows from the upper node to the ground node.

Applying Kirchhoff's current law by summing the currents leaving the upper node, we write

$$i_C + i_R + i_L - i_i(t) = 0 \tag{16}$$

From the element laws given by (5), (7), and (10), we have

$$i_R = \frac{1}{R} e_o$$

$$i_C = C \dot{e}_o \tag{17}$$

$$i_L = i_L(0) + \frac{1}{L} \int_0^t e_o(\lambda) \, d\lambda$$

where the initial time has been taken as $t_0 = 0$.

Substituting (17) into (16) and rearranging the result give the model as

$$C\dot{e}_o + \frac{1}{R} e_o + \frac{1}{L} \int_0^t e_o(\lambda) \, d\lambda = i_i(t) - i_L(0) \tag{18}$$

Differentiating (18) term by term eliminates the constant term and the integral, resulting in the input-output differential equation

$$C\ddot{e}_o + \frac{1}{R} \dot{e}_o + \frac{1}{L} e_o = \frac{di_i}{dt}$$

9.4 OBTAINING THE CIRCUIT MODEL

Two general procedures for developing models of electrical circuits are the node-equation method and the loop-equation method. Example 9.1 was actually a simple illustration of the loop-equation method, and Example 9.2

used the node-equation method. In the *loop-equation method*, a rather trivial application of the current law allows us to express the current through every element in terms of one or more loop currents. We then write an appropriate set of simultaneous equations by using the voltage law and the element laws. In the *node-equation method*, we use Kirchhoff's voltage law in a trivial way to express the voltage across every element in terms of node voltages. Then we write a set of simultaneous equations by using Kirchhoff's current law and the element laws.

We shall emphasize the node-equation method, partly because in some circuits the loop-equation method requires us to use fictitious loop currents that do not correspond to measurable currents through individual elements. Furthermore, the node-equation method is well suited to handling the current sources that appear in models of transistor circuits. (References that cover both methods in detail are listed in Appendix D.)

When we use the node-equation method, we start by labeling the voltage of each node with respect to the ground node. If a voltage source is connected between a particular node and ground, the voltage of that node is the known source voltage. Where they are needed, we introduce symbols to define the voltages of the other nodes with respect to ground. Once we have done this, we can express the voltage across each passive element in terms of the node voltages by a trivial application of Kirchhoff's voltage law, as illustrated by the discussion of Figure 9.2(d) and Figure 9.12. We write a current-law equation for each of the nodes whose voltage is unknown, using the element laws to express the currents through the passive elements in terms of the node voltages. We need only rearrange the resulting set of equations into input-output form to complete the model.

EXAMPLE 9.3   Derive the input-output equation for the circuit shown in Figure 9.15(a), using the node-equation method. The input and output voltages are $e_i(t)$ and e_o, respectively.

Solution   The first step is to define all unknown node voltages and redraw the circuit diagram with all node voltages shown, as in Figure 9.15(b). We use the heavy lines to emphasize that the ground node extends across the bottom of the entire circuit and that the node whose voltage is e_o extends from L to R_3. We show the source voltage $e_i(t)$ at the upper left node and denote the voltage of the remaining node with respect to ground by e_A. Since e_A and e_o are unknown node voltages, we shall write a current-law equation at each of these nodes, using the appropriate element laws.

To assist in writing the equations, we draw separate sketches for each node, as shown in Figure 9.16 (analogous to the free-body diagrams drawn for mechanical systems). For each element, the voltage across its terminals is

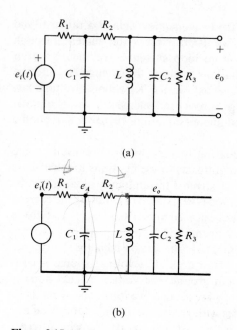

(a)

(b)

Figure 9.15 (a) Circuit for Example 9.3. (b) Circuit with node voltages shown.

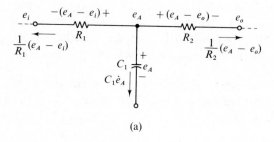

(a)

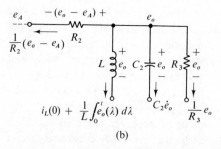

(b)

Figure 9.16 Nodes with currents expressed by element laws for Example 9.3.

shown in terms of the node voltages, with the plus sign placed at the node in question. Then we use the appropriate element law to write an expression for the current leaving the node.

We can apply Kirchhoff's current law to each of the nodes shown in Figure 9.16 by setting the algebraic sum of the currents leaving each node equal to zero. The result is the pair of equations

$$\frac{e_A - e_i}{R_1} + C_1 \dot{e}_A + \frac{e_A - e_o}{R_2} = 0 \tag{19a}$$

$$\frac{e_o - e_A}{R_2} + i_L(0) + \frac{1}{L} \int_0^t e_o(\lambda)\, d\lambda + C_2 \dot{e}_o + \frac{e_o}{R_3} = 0 \tag{19b}$$

With a bit of experience, the reader should be able to write these equations directly from the circuit diagram, without drawing the sketches shown in Figure 9.16. It is worthwhile to note that the current through R_2 is labeled $(e_A - e_o)/R_2$ in the sketch for node A and $(e_o - e_A)/R_2$ in Figure 9.16(b). However, the reference arrows for this current are also reversed on the two parts of the figure, so there is no inconsistency in the expressions.

We can now differentiate (19b) to eliminate the constant term and the integral. Doing this and rearranging terms give the circuit model as the following pair of coupled differential equations for the node voltages e_A and e_o:

$$C_1 \dot{e}_A + \left(\frac{1}{R_1} + \frac{1}{R_2}\right) e_A - \frac{1}{R_2} e_o = \frac{1}{R_1} e_i(t)$$

$$C_2 \ddot{e}_o + \left(\frac{1}{R_2} + \frac{1}{R_3}\right) \dot{e}_o + \frac{1}{L} e_o - \frac{1}{R_2} \dot{e}_A = 0 \tag{20}$$

By combining the equations in (20) to eliminate e_A, we obtain the input-output differential equation relating $e_i(t)$ and e_o. To simplify the calculations, assume that the passive elements have the numerical values $R_1 = R_2 = 2\ \Omega$, $R_3 = 4\ \Omega$, $C_1 = 1$ F, $C_2 = 4$ F, and $L = \frac{1}{2}$ H. With these parameter values, (20) becomes

$$\dot{e}_A + e_A - \tfrac{1}{2} e_o = \tfrac{1}{2} e_i(t)$$

$$4\ddot{e}_o + \tfrac{3}{4}\dot{e}_o + 2e_o - \tfrac{1}{2}\dot{e}_A = 0$$

By using the p-operator method or by using a combination of substitution and differentiation, we can show that the circuit obeys the third-order equation

$$16\dddot{e}_o + 19\ddot{e}_o + 10\dot{e}_o + 8e_o = \dot{e}_i \tag{21}$$

for the prescribed element values. In order to solve (21) for e_o for all $t \geq 0$, we would need to know the three initial conditions $e_o(0)$, $\dot{e}_o(0)$, and $\ddot{e}_o(0)$, the last of which may be difficult to obtain.

It is interesting to note that if $e_i(t)$ is a step function, the steady-state solution to (21) will be $(e_o)_{ss} = 0$. Thus there will be no voltage across the output terminals as time approaches infinity with the input voltage held constant. We can reach this conclusion in another way by noting that the inductor L is connected directly across the output terminals and that its element law as given by (9) is $e_o = L \, di_L/dt$. In the steady state, $di_L/dt = 0$ so $(e_o)_{ss} = 0$. In this steady-state condition, the inductor is said to behave as a short circuit since there is no voltage across its terminals.

EXAMPLE 9.4   Write the input-output differential equation for the circuit shown in Figure 9.17(a) where the inputs are the two voltage sources $e_1(t)$ and $e_2(t)$ and the output is the voltage e_o.

Solution   In Figure 9.17(b), the circuit has been redrawn with the bottom node defined to be the ground and with the other node voltages shown. Two of the nodes are labeled $e_1(t)$ and e_o to correspond to the left voltage

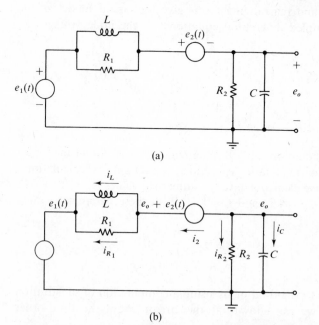

(a)

(b)

Figure 9.17 Circuit for Example 9.4. (a) As specified by the example statement. (b) With currents and node voltages defined.

source and the output voltage, respectively. In labeling the node to the right of L and R_1, we do not introduce a new symbol (such as e_A) but take advantage of the fact that $e_2(t)$ is a source voltage. By Kirchhoff's voltage law, this remaining node voltage is $e_o + e_2(t)$, as shown on the diagram. This approach simplifies the solution whenever there is a voltage source not connected to ground by avoiding the introduction of unnecessary variables.

The directions of the current reference arrows included in Figure 9.17(b) are arbitrary, but our equations must be consistent with the directions selected. Applying Kirchhoff's current law to the upper right node, we have

$$i_C + i_{R_2} + i_2 = 0 \tag{22}$$

Although we can express i_C and i_{R_2} in terms of the node voltage e_o by the element laws, the current i_2 through the voltage source cannot be directly related to $e_2(t)$. However, by applying the current law to the node labeled $e_o + e_2(t)$, we see that

$$i_2 = i_{R_1} + i_L$$

which, when inserted into (22), gives

$$i_C + i_{R_2} + i_{R_1} + i_L = 0 \tag{23}$$

Using the element laws in the forms given by (5), (7), and (10), we write

$$i_{R_1} = \frac{1}{R_1}[e_o + e_2(t) - e_1(t)] \tag{24a}$$

$$i_{R_2} = \frac{1}{R_2}e_o \tag{24b}$$

$$i_C = C\dot{e}_o \tag{24c}$$

$$i_L = i_L(0) + \frac{1}{L}\int_0^t (e_o + e_2 - e_1)\, d\lambda \tag{24d}$$

Substituting (24) into (23) and rearranging terms give the integral-differential equation

$$Ce_o + \left(\frac{1}{R_1} + \frac{1}{R_2}\right)e_o + \frac{1}{L}\int_0^t e_o\, d\lambda$$
$$= \frac{1}{R_1}[e_1(t) - e_2(t)] + \frac{1}{L}\int_0^t (e_1 - e_2)\, d\lambda - i_L(0)$$

Upon differentiation, this expression becomes the desired form of the model, namely

$$C\ddot{e}_o + \left(\frac{1}{R_1} + \frac{1}{R_2}\right)\dot{e}_o + \frac{1}{L}e_o = \frac{1}{R_1}(\dot{e}_1 - \dot{e}_2) + \frac{1}{L}[e_1(t) - e_2(t)] \qquad (25)$$

which is a second-order differential equation. To solve it, we must know the two initial conditions $e_o(0)$ and $\dot{e}_o(0)$, in addition to the source voltages $e_1(t)$ and $e_2(t)$.

If along with $e_1(0)$ and $e_2(0)$, we know $e_C(0)$ and $i_L(0)$, which account for the initial stored energy, we can calculate the values of $e_o(0)$ and $\dot{e}_o(0)$. First, we set $e_o(0) = e_C(0)$ since the output voltage is measured across the capacitor. To find $\dot{e}_o(0)$, we note from (24c) that $\dot{e}_o(0) = (1/C)i_C(0)$. Setting t equal to zero in (23) gives

$$i_C(0) = -[i_{R_1}(0) + i_{R_2}(0) + i_L(0)] \qquad (26)$$

Using (24a) and (24b) to find $i_{R_1}(0)$ and $i_{R_2}(0)$, we finally get

$$\dot{e}_o(0) = \frac{1}{C}\left\{-\left(\frac{1}{R_1} + \frac{1}{R_2}\right)e_C(0) - i_L(0) + \frac{1}{R_1}[e_1(0) - e_2(0)]\right\}$$

9.5 RESISTIVE CIRCUITS

Electrical circuits frequently contain combinations of two or more resistors that are connected to the remainder of the circuit by a single pair of terminals, as shown in Figure 9.18(a). In such situations, it is possible to replace the entire combination of resistors by a single equivalent resistor R_{eq}, as shown in Figure 9.18(b). Provided that R_{eq} is selected such that $e = R_{eq}i$ is satisfied,

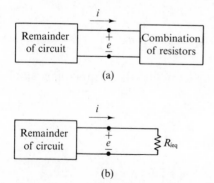

Figure 9.18 Replacement of a combination of resistors by an equivalent resistance.

the response of the remainder of the circuit will be identical in both cases. Once we have found R_{eq}, it will be easier to analyze the complete circuit because there will be fewer nodes and thus fewer equations.

There are many useful circuits that contain only resistors and sources, with no energy-storing elements. Such circuits are known as *resistive* circuits and are modeled by algebraic rather then differential equations. We shall develop rules for finding voltages and currents in such circuits and for calculating equivalent resistances.

RESISTORS IN SERIES

Two resistors are in series if a single terminal of each resistor is connected to a single terminal of the other with no other element connected to the common node, as shown in Figure 9.19(a). Obviously, two resistors in series must have the same current flowing through them.

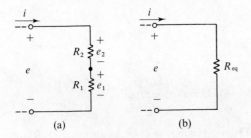

(a)   (b)

Figure 9.19 (a) Two resistors in series. (b) Equivalent resistance.

It follows from Ohm's law that $e_1 = R_1 i$ and $e_2 = R_2 i$ and from Kirchhoff's voltage law that $e = e_1 + e_2$. Thus

$$e = (R_1 + R_2)i \tag{27}$$

Since $e = R_{eq} i$, we know from (27) that the equivalent resistance shown in Figure 9.19(b) for the series combination shown in Figure 9.19(a) is

$$R_{eq} = R_1 + R_2 \tag{28}$$

Using (27) with the expressions for e_1 and e_2, we see that

$$e_1 = \left(\frac{R_1}{R_1 + R_2} \right) e \tag{29a}$$

$$e_2 = \left(\frac{R_2}{R_1 + R_2} \right) e \tag{29b}$$

which is known as the *voltage-divider rule*. From (29), the ratio of the individual resistor voltages is

$$\frac{e_1}{e_2} = \frac{R_1}{R_2} \tag{30}$$

RESISTORS IN PARALLEL

Two resistors are in parallel if each terminal of one resistor is connected to a separate terminal of the other resistor, as shown in Figure 9.20(a). It is apparent that two resistors in parallel must have the same voltage across their terminals.

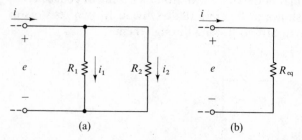

(a)   (b)

Figure 9.20 (a) Two resistors in parallel. (b) Equivalent resistance.

From Ohm's law, the individual currents are $i_1 = (1/R_1)e$ and $i_2 = (1/R_2)e$. From Kirchhoff's current law, $i = i_1 + i_2$. Thus

$$i = \left(\frac{1}{R_1} + \frac{1}{R_2}\right)e \tag{31}$$

For the equivalent resistance shown in Figure 9.20(b), we have $i = (1/R_{eq})e$. From (31), we see that for the parallel combination in Figure 9.20(a),

$$\frac{1}{R_{eq}} = \frac{1}{R_1} + \frac{1}{R_2}$$

or

$$R_{eq} = \frac{R_1 R_2}{R_1 + R_2} \tag{32}$$

To relate i_1 to the total current, we can write $i_1 = (1/R_1)e = (R_{eq}/R_1)i$. Doing this for i_2 and then using (32) to express R_{eq} in terms of R_1 and

R_2, we obtain

$$i_1 = \left(\frac{R_2}{R_1 + R_2}\right)i \tag{33a}$$

$$i_2 = \left(\frac{R_1}{R_1 + R_2}\right)i \tag{33b}$$

which is known as the *current-divider rule*. From (33), the ratio of the individual resistor currents is $i_1/i_2 = R_2/R_1$.

Calculating equivalent resistances and solving for the currents and voltages in most types of resistive networks can be simplified by using these rules for series and parallel combinations, as demonstrated in the following example.

EXAMPLE 9.5   The resistive circuit shown in Figure 9.21(a) consists of a voltage source connected to a combination of seven resistors. The output is the voltage e_o. Find the equivalent resistance R_{eq} of the seven-resistor combination and evaluate e_o.

Solution   To obtain R_{eq}, we use (28) and (32) repeatedly to combine series or parallel combinations of resistors into single equivalent resistors. Starting with the original circuit in Figure 9.21(a), we replace the 6-Ω and 3-Ω resistors that are in parallel by a single 2-Ω resistor. We also combine the 10-Ω and 2-Ω resistors in series into a 12-Ω resistor, giving the intermediate circuit diagram shown in Figure 9.21(b). Note that the output voltage e_o does not appear on this diagram. Next we replace the parallel combination of the 4-Ω and 12-Ω resistors by a 3-Ω resistor, giving Figure 9.21(c). Combining the series combination of 2 Ω and 3 Ω gives two parallel 5-Ω branches, which are replaced by a single resistor of $\frac{5}{2}$ Ω, resulting in Figure 9.21(d). Thus the equivalent resistance connected across the voltage source is

$$R_{eq} = \tfrac{1}{2} + \tfrac{5}{2} = 3\,\Omega$$

To find the output voltage e_o, we make repeated use of the voltage-divider rule given by (29a) to obtain, in turn, e_A, e_B, and finally e_o. From Figure 9.21(d),

$$e_A = \left(\frac{5/2}{1/2 + 5/2}\right)e_i(t) = \frac{5}{6}e_i(t)$$

and from Figure 9.21(c),

$$e_B = \left(\frac{3}{2+3}\right)e_A = \frac{1}{2}e_i(t)$$

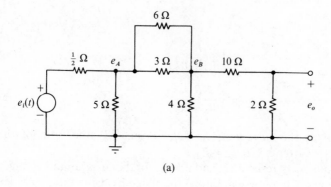

(a)

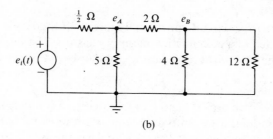

(b)

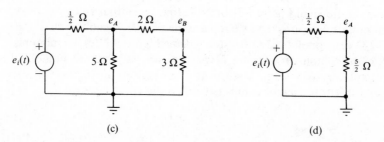

(c)   (d)

Figure 9.21 Circuits for Example 9.5. (a) Original circuit. (b), (c), (d) Equivalent circuits.

Then, from the original circuit diagram,

$$e_o = \left(\frac{2}{2+10}\right)e_B = \frac{1}{12}e_i(t)$$

Although the rules for combining series and parallel resistors will often simplify the process of modeling a circuit, there are circuits in which the resistances do not occur in series or parallel combinations. In such situations, we can find an equivalent resistance by writing and solving the appropriate node equations, which will be strictly algebraic when only resistors and sources are involved. The reader is asked to do this in Problem 9.5.

9.6 FIRST-ORDER CIRCUITS WITH CONSTANT INPUTS

A circuit consisting of a single energy-storing element (i.e., a capacitor or an inductor) and any number of resistors and sources can be described by a first-order differential equation. If the sources are constant for $t > 0$, the response of any voltage or current in the circuit depends only on the circuit's time constant, the value of the response immediately after $t = 0$ (denoted by $t = 0+$), and its steady-state value as t approaches infinity. The basis for this assertion is the discussion of the response of first-order systems in Section 8.2. It was shown in (8.14) that for an input that is constant for $t > 0$, the response must have the form

$$y(t) = y_{ss} + [y(0) - y_{ss}] \epsilon^{-t/\tau} \tag{34}$$

for $t > 0$. We observe that the only parameters involved in (34) are y_{ss}, $y(0)$, and τ.

The steady-state response y_{ss} and the time constant τ can be determined from inspection of the circuit's differential equation. The quantity $y(0)$ will depend on the energy contained in the circuit at $t = 0$ and on the input. We shall illustrate methods for finding these three quantities with an example and then develop short-cut methods for finding each of them.

We must be careful when the variable in question can undergo an instantaneous change at $t = 0$ caused by the opening or closing of a switch. Recall that the energy stored in an inductor or capacitor cannot change instantaneously unless there is an infinitely large power flow. Hence, neither the voltage across a capacitor nor the current through an inductor can change instantaneously. In other words, capacitor voltages and inductor currents must be continuous functions at all times, as long as all voltages and currents are finite.

Other variables can change instantaneously, however, and thus can have discontinuities. Included in this category are resistor voltages and currents, capacitor currents, and inductor voltages. When we must consider functions that have a discontinuity at $t = 0$, it is customary to use the notations $y(0-)$ and $y(0+)$ for the limiting values of $y(t)$ as t approaches zero through negative and positive values, respectively. The function $y(t)$ shown in Figure 9.22 has a discontinuity at $t = 0$, where it jumps instantaneously from $y(0-)$

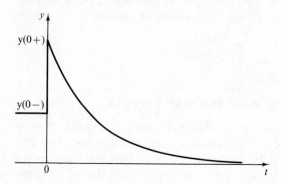

Figure 9.22 Typical response of a first-order circuit.

to $y(0+)$, and is typical of the responses of first-order circuits. To account for the possibility of discontinuous behavior, we modify (34) slightly to read

$$y(t) = y_{ss} + [y(0+) - y_{ss}] \epsilon^{-t/\tau} \tag{35}$$

This change is consistent with the fact that we must use $y(0+)$, the value of $y(t)$ immediately following the discontinuity, in computing the response for $t > 0$.

EXAMPLE 9.6 Evaluate and sketch the unit step response of the circuit shown in Figure 9.23(a).

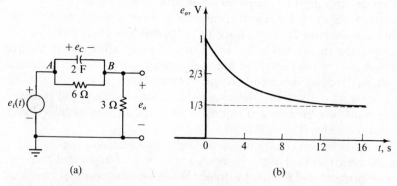

(a)

(b)

Figure 9.23 First-order RC circuit for Example 9.6. (a) Circuit diagram. (b) Unit step response.

Solution Setting the sum of the currents leaving node B equal to zero gives

$$2(\dot{e}_o - \dot{e}_i) + \tfrac{1}{6}[e_o - e_i(t)] + \tfrac{1}{3}e_o = 0 \tag{36}$$

Since $e_i(t) = U(t)$, $e_i = 1$ and $\dot{e}_i = 0$ for all $t > 0$. Thus (36) reduces to

$$\dot{e}_o + \tfrac{1}{4}e_o = \tfrac{1}{12} \qquad \text{for } t > 0 \tag{37}$$

Comparing (37) with (8.11), we see that the circuit has a time constant of 4 s and a steady-state response of $(e_o)_{ss} = (\tfrac{1}{12})4 = \tfrac{1}{3}$ V. Thus the unit step response is

$$e_o(t) = \tfrac{1}{3} + K\epsilon^{-t/4} \qquad \text{for } t > 0 \tag{38}$$

where the constant K is given by

$$K = e_o(0+) - (e_o)_{ss} \tag{39}$$

Because e_o is the voltage across a resistor, it can change instantaneously when $e_i(t)$ changes from 0 to 1 V at $t = 0$. However, the voltage across the capacitor must be continuous at $t = 0$ because it is related to the energy stored in the capacitor, which cannot change instantaneously. If the capacitor voltage is denoted by e_C with the polarity shown in Figure 9.23(a), we can write Kirchhoff's voltage law for a loop consisting of the voltage source, the capacitor, and the 3-Ω resistor, giving

$$e_i(t) - e_C(t) - e_o(t) = 0 \tag{40}$$

We can apply (40) at $t = 0+$ to find $e_o(0+)$. From the definition of the unit step function, we note that $e_i(0+) = 1$. Because the capacitor has no energy at $t = 0-$ and its energy is unchanged at $t = 0+$, it follows that $e_C(0+) = e_C(0-) = 0$. Thus at $t = 0+$, (40) gives

$$e_o(0+) = 1 - 0 = 1 \text{ V}$$

Substituting $e_o(0+) = 1$ V and $(e_o)_{ss} = \tfrac{1}{3}$ V into (39), we find that $K = \tfrac{2}{3}$ V, which we substitute into (38) to give the complete expression for the step response, namely

$$e_o(t) = \tfrac{1}{3} + \tfrac{2}{3}\epsilon^{-t/4} \qquad \text{for } t > 0 \tag{41}$$

which is sketched in Figure 9.23(b).

SHORT-CUT SOLUTION METHOD

From (35) and from Example 9.6, it is apparent that for first-order circuits with constant inputs, we need only three pieces of information to write the complete response: the time constant τ, the initial output $y(0+)$, and the steady-state output y_{ss}. For this class of circuits, it is possible to calculate these three quantities without finding and solving the input-output differential equation by using a short-cut solution method. After outlining the short-cut method, we shall illustrate it by repeating Example 9.6.

Briefly, we derive three equivalent circuits in a way that readily yields a single response parameter with a minimum of calculation. In each case, we can make simplifications for the source and the energy-storing element because we are seeking limited information.

First, in evaluating the time constant, we can make all the sources zero. This simplification is possible because τ occurs in only the free response $y_H(t)$, which is a solution of the differential equation when the inputs are zero. Thus we can redraw the circuit with the sources set equal to zero, i.e., with voltage sources replaced by short circuits and current sources replaced by open circuits. Then we can determine the equivalent resistance R_{eq} connected across the terminals of the single energy-storing element by the methods of Section 9.5. With these simplifications, we can calculate the time constant according to either $\tau = R_{eq} C$ or $\tau = L/R_{eq}$, depending on the type of energy-storing element in the circuit. Because C and R_{eq} have units of ampere-seconds per volt and volts per ampere, respectively, the time constant $R_{eq} C$ has units of seconds. Similarly, L has units of volt-seconds per ampere, and the ratio L/R_{eq} has units of seconds.

Next we find the value of $y(0+)$ by constructing a different circuit, with the capacitor or inductor replaced by an equivalent element that is valid only at $t = 0+$. If an inductor had zero current for $t < 0$ and the source voltage is finite, it must still have zero current at $t = 0+$ and thus can be replaced by an open circuit. If a capacitor had zero voltage at $t = 0-$ and the source supplies only a finite current, then the capacitor must still have zero voltage at $t = 0+$ and can be replaced by a short circuit. More generally, a capacitor having a nonzero voltage at $t = 0-$ can be replaced by a constant voltage source in the model that is valid for $t = 0+$. Similarly, an inductor with a current flowing at $t = 0-$ can be replaced by a current source as far as its effect on $y(0+)$ is concerned.

We can devise a third equivalent circuit that is valid only for the steady-state condition by noting that for a constant input, the steady-state values of all currents and voltages will be constants. Under these conditions, $di_L/dt = 0$ and $de_C/dt = 0$, and from the respective element laws we see that the voltage across any inductor and the current through any capacitor must be zero. Hence, in the steady-state equivalent circuit, an inductor becomes a short circuit and a capacitor becomes an open circuit.

EXAMPLE 9.7 Repeat Example 9.6 using the short-cut solution method just described.

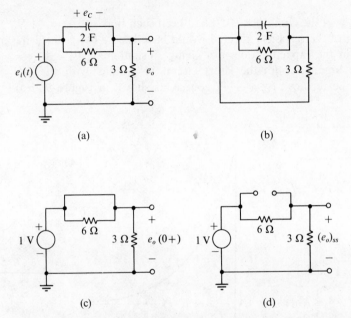

Figure 9.24 Circuits for Example 9.7 (a) Original. (b) For calculation of τ. (c) For calculation of $e_o(0+)$. (d) For calculation of $(e_o)_{ss}$.

Solution Figure 9.24 shows the original circuit and the three equivalent circuits used to find τ, $e_o(0+)$, and $(e_o)_{ss}$. In Figure 9.24(b), the voltage source has been replaced by a short circuit, putting the two resistors in parallel. The equivalent resistance is $R_{eq} = (6)(3)/(6+3) = 2\ \Omega$. Thus the circuit's time constant is $\tau = (2\ \Omega)(2\ F) = 4$ s.

The circuit shown in Figure 9.24(c) results from treating the capacitor as a short circuit, since $e_c(0-) = 0$, and setting the voltage source equal to 1 V. Because of the short circuit in parallel with the 6-Ω resistor, the source voltage appears across the 3-Ω resistor, giving $e_o(0+) = 1$ V.

In the steady-state circuit shown in Figure 9.24(d) the capacitor is replaced by an open circuit since no current can flow through it as t approaches infinity. Then the series combination of the 3-Ω and 6-Ω resistors is across the 1-V source, and applying the voltage-divider law indicates that $(e_o)_{ss} = \frac{1}{3}$ V. Substituting $\tau = 4$ s, $e_o(0+) = 1$ V, and $(e_o)_{ss} = \frac{1}{3}$ V into the general solution given by (35) results in the expression obtained in (41), namely

$$e_o(t) = \tfrac{1}{3} + \tfrac{2}{3}\epsilon^{-t/4} \qquad \text{for } t > 0 \tag{42}$$

If the behavior of the circuit for $t < 0$ had been such that $e_c(0-) = -1$ V, the only aspect of the response affected would have been $e_c(0+)$. Since the energy stored in the capacitor does not change instantaneously, $e_c(0+) = -1$ V, and in Figure 9.24(c) the short circuit in parallel with the 6-Ω resistor would be replaced by a -1-V source, as shown in Figure 9.25(a).

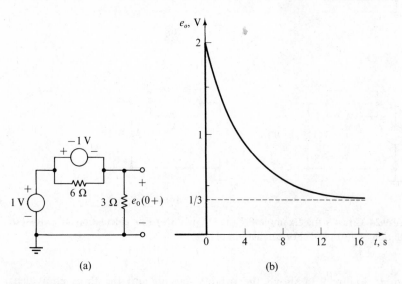

(a)   (b)

Figure 9.25 (a) Figure 9.24(c) modified to account for an initial capacitor voltage. (b) Response.

Since the combined sources result in 2 V across the 3-Ω resistor, it follows that $e_o(0+) = 2$ V. The complete response then becomes

$$e_o(t) = \tfrac{1}{3} + \tfrac{5}{3}\epsilon^{-t/4} \tag{43}$$

As indicated by comparing (42) and (43), the net effect of the initial stored energy is to change the size of the transient response. It has no effect on the steady-state value.

9.7 STATE-VARIABLE MODELS

To obtain the model of a circuit in state-variable form, we define an appropriate set of state variables and then derive an equation for the derivative of each state variable in terms of only the state variables and inputs. The choice of state variables is not unique, but they are normally related to the energy in each of the circuit's energy-storing elements. We generally select the capacitor voltages and inductor currents as the state variables. For fixed linear circuits, exceptions will occur only if there are capacitor voltages or inductor currents that are not independent of one another. This situation will be illustrated in Example 9.11.

For each capacitor or inductor, we want to express $\dot{e}_c$ or di_L/dt as an algebraic function of state variables and inputs. We do this by writing the capacitor and inductor element laws in their derivative forms as

$$\dot{e}_C = \frac{1}{C} i_C$$

$$\frac{di_L}{dt} = \frac{1}{L} e_L$$

and then obtaining expressions for i_C and e_L in terms of the state variables and inputs. To find these expressions, we use the resistor element laws and Kirchhoff's voltage and current laws.

All the techniques we have discussed can still be used. The only difference is that we want to retain the variables e_C and i_L wherever they appear in our equations and to express other variables in terms of them. For example, in applying Kirchhoff's current law to node A in the partial circuit shown in Figure 9.26, we would write

$$\frac{1}{R_1}(e_A - e_B) + \frac{1}{R_2}(e_A - e_D) + i_L = 0$$

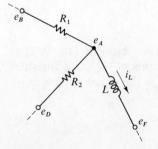

Figure 9.26 Partial circuit to illustrate Kirchhoff's current law in terms of state variables.

instead of

$$\frac{1}{R_1}(e_A - e_B) + \frac{1}{R_2}(e_A - e_D) + i_L(0) + \frac{1}{L}\int_0^t (e_A - e_F)\,d\lambda = 0$$

To solve the state-variable equations, we must know the inputs and the initial values of the state variables. However, if all the currents and voltages remain finite, e_C and i_L cannot change instantaneously. Thus when the capacitor voltages and inductor currents are taken as the state variables, it is not usually necessary to distinguish between the values of the state variables at $t = 0-$ and $t = 0+$.

EXAMPLE 9.8   Write the state-variable equations for the circuit shown in Figure 9.27, for which we found the input-output equation in Example 9.2.

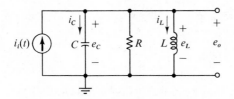

Figure 9.27 Parallel RLC circuit for Example 9.8.

Solution   Because the circuit contains an inductor and a capacitor, we select i_L and e_C as the state variables, with the positive senses indicated on the circuit diagram. Starting with the inductor element law in the form $di_L/dt = (1/L)e_L$, we note that e_L is the same as the capacitor voltage e_C because the two elements are in parallel. Thus we obtain the first state-variable equation by replacing e_L by e_C, getting

$$\frac{di_L}{dt} = \frac{1}{L}e_C \tag{44}$$

For the second equation, we write the capacitor element law as $\dot{e}_c = (1/C)i_C$. To express the capacitor current i_C in terms of the state variables and the input, we apply Kirchhoff's current law at the upper node, getting

$$i_C + \frac{1}{R}e_C + i_L - i_i(t) = 0 \tag{45}$$

where we have written the resistor current in terms of the state variable e_C.

Solving (45) for i_C gives

$$i_C = -i_L - \frac{1}{R}e_C + i_i(t) \tag{46}$$

the right side of which is written entirely in terms of the state variables and input. Substituting (46) into the capacitor element law, we find the second state-variable equation to be

$$\dot{e}_C = \frac{1}{C}\left[-i_L - \frac{1}{R}e_C + i_i(t)\right] \tag{47}$$

Finally, we find the voltage e_o from the algebraic output equation

$$e_o = e_C \tag{48}$$

EXAMPLE 9.9   Derive the state-variable equations for the circuit shown in Figure 9.28, which we studied in Example 9.3.

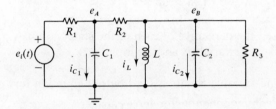

Figure 9.28 Circuit for Example 9.9.

Solution   We choose as state variables the inductor current i_L and the capacitor voltages e_A and e_B. For the inductor, we write

$$\frac{di_L}{dt} = \frac{1}{L}e_B$$

which requires no modification because e_B is a state variable.

We obtain the two other state-variable equations by writing the element law for each capacitor in the form

$$\dot{e}_A = \left(\frac{1}{C_1}\right)i_{C_1} \quad \text{and} \quad \dot{e}_B = \left(\frac{1}{C_2}\right)i_{C_2}$$

The current i_{C_1} will appear in a Kirchhoff current equation for node A, namely

$$i_{C_1} = \frac{1}{R_1}[e_i(t) - e_A] - \frac{1}{R_2}(e_A - e_B) \tag{49}$$

For i_{C_2} we consider node B, getting

$$i_{C_2} = \frac{1}{R_2}(e_A - e_B) - i_L - \frac{1}{R_3}e_B \tag{50}$$

Substituting (49) and (50) into the respective element-law equations gives the final two state-variable equations. The complete set of three equations is

$$\frac{di_L}{dt} = \frac{1}{L}e_B$$

$$\dot{e}_A = \frac{1}{C_1}\left[-\left(\frac{1}{R_1} + \frac{1}{R_2}\right)e_A + \frac{1}{R_2}e_B + \frac{1}{R_1}e_i(t)\right]$$

$$\dot{e}_B = \frac{1}{C_2}\left[-i_L + \frac{1}{R_2}e_A - \left(\frac{1}{R_2} + \frac{1}{R_3}\right)e_B\right]$$

As required, we have expressed the derivative of each of the state variables as an algebraic function of the state variables and the input $e_i(t)$.

EXAMPLE 9.10   Find the state-variable equations and output equation for the circuit shown in Figure 9.29.

Solution   We make the usual choice of i_L and e_C as state variables and start by writing

$$\frac{di_L}{dt} = \frac{1}{L}e_L = 2e_L$$

$$\dot{e}_C = \frac{1}{C}i_C = \frac{1}{2}i_C \tag{51}$$

Node B has the voltage e_L, and since we want to retain e_C in our equations, we use Kirchhoff's voltage law to write the voltage of node A as $e_L + e_C$ rather than denoting it by the symbol e_A. By applying Kirchhoff's current law to nodes B and A, we have

$$2e_L + i_L - i_C = 0$$

$$e_L + \dot{e}_C - e_i(t) + 2(e_L + e_C) + i_C = 0 \tag{52}$$

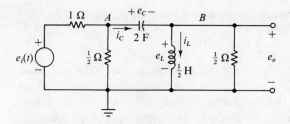

Figure 9.29 Circuit for Example 9.10.

We can solve (52) simultaneously for e_L and i_C in terms of the state variables i_L and e_C and the input $e_i(t)$. Doing this gives the algebraic equations

$$e_L = \tfrac{1}{5}[-i_L - 3e_C + e_i(t)] \tag{53a}$$

$$i_C = \tfrac{1}{5}[3i_L - 6e_C + 2e_i(t)] \tag{53b}$$

which, when substituted into (51), give the state-variable equations

$$\frac{di_L}{dt} = \tfrac{2}{5}[-i_L - 3e_C + e_i(t)]$$

$$\dot{e}_C = \tfrac{1}{10}[3i_L - 6e_C + 2e_i(t)] \tag{54}$$

We find the output equation by noting that $e_o = e_L$ and rewriting (53a) as

$$e_o = \tfrac{1}{5}[-i_L - 3e_C + e_i(t)]$$

EXAMPLE 9.11 Find the state-variable model for the circuit shown in Figure 9.30.

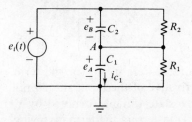

Figure 9.30 Circuit for Example 9.11.

Solution The fact that the circuit shown in Figure 9.30 has two capacitors suggests that we use the two voltages e_A and e_B as state variables. However, by applying Kirchhoff's voltage law to the left loop, we see that

$$e_A + e_B - e_i(t) = 0 \tag{55}$$

Equation (55) is an algebraic relationship between the proposed state variables and the input, which is not allowed. In other words, the capacitor voltages e_A and e_B are not independent and thus cannot both be chosen as state variables. In this example, we have a loop composed of only capacitors and voltage sources. An analogous situation occurs when a circuit has a node to which only inductors and current sources are connected. In such a case, one of the inductor currents is not independent and may not be selected as a state variable.

For the circuit shown in Figure 9.30, we select e_A, the voltage across C_1, as the state variable. Hence, we need an equation for $\dot{e}_A$ in terms of e_A and $e_i(t)$. First we write the element law for C_1 as

$$\dot{e}_A = \frac{1}{C_1} i_{C_1} \tag{56}$$

where the positive sense of i_{C_1} is downward. Then we apply Kirchhoff's current law at node A, obtaining

$$C_2(\dot{e}_A - \dot{e}_i) + \frac{1}{R_2}[e_A - e_i(t)] + \frac{1}{R_1}e_A + i_{C_1} = 0 \tag{57}$$

Solving (57) for i_{C_1} and substituting the result into (56), we find

$$\dot{e}_A = \frac{1}{C_1}\left[-C_2\,\dot{e}_A - \left(\frac{1}{R_1} + \frac{1}{R_2}\right)e_A + C_2\,\dot{e}_i + \frac{1}{R_2}e_i(t) \right]$$

which can be rearranged to yield

$$\dot{e}_A = \left(\frac{1}{C_1 + C_2}\right)\left[-\left(\frac{1}{R_1} + \frac{1}{R_2}\right)e_A + C_2\,\dot{e}_i + \frac{1}{R_2}e_i(t) \right] \tag{58}$$

Equation (58) would be in state-variable form were it not for the term involving $\dot{e}_i$ on the right side. Since the derivative of the input should not appear in the final equation, we define a new state variable, denoted by x, using the method developed in Section 6.3. Transferring the term involving $\dot{e}_i$ to the left side of (58), we have

$$\frac{d}{dt}\left[e_A - \left(\frac{C_2}{C_1 + C_2}\right)e_i(t) \right] = \left(\frac{1}{C_1 + C_2}\right)\left[-\left(\frac{1}{R_1} + \frac{1}{R_2}\right)e_A + \frac{1}{R_2}e_i(t) \right] \tag{59}$$

We define the bracketed term on the left to be the new state variable

$$x = e_A - \left(\frac{C_2}{C_1 + C_2}\right) e_i(t) \tag{60}$$

Then e_A is given by the output equation

$$e_A = x + \left(\frac{C_2}{C_1 + C_2}\right) e_i(t) \tag{61}$$

and, when we substitute (60) and (61) into (59), the state-variable equation becomes

$$\dot{x} = \left(\frac{1}{C_1 + C_2}\right)\left\{-\left(\frac{1}{R_1} + \frac{1}{R_2}\right)x + \left[\frac{1}{R_2} - \left(\frac{1}{R_1} + \frac{1}{R_2}\right)\left(\frac{C_2}{C_1 + C_2}\right)\right]e_i(t)\right\} \tag{62}$$

Notice that the circuit is first-order and can be modeled by a single state-variable equation. However, we did find that the state variable had to be a linear combination of the voltage across C_1 and the input. We can obtain the capacitor voltage e_A from the algebraic output equation (61) after solving the state-variable equation. Finally, we combine (55) and (61) to give the output equation for e_B, namely

$$e_B = \left(\frac{C_1}{C_1 + C_2}\right) e_i(t) - x$$

PROBLEMS

9.1 Use the node-equation method to find the input-output differential equation relating e_o and $e_i(t)$ for the circuit shown in Figure P9.1.

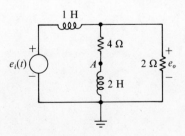

Figure P9.1

9.2 a. Find the input-output differential equation for the circuit shown in Figure P9.2(a).

b. Repeat part a for the circuit shown in Figure P9.2(b).

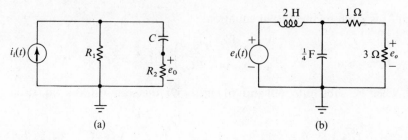

(a)   (b)

Figure P9.2

9.3 For the circuit shown in Figure P9.3, use the rules for series and parallel combinations of resistors to find i_o and the equivalent resistance connected across the source.

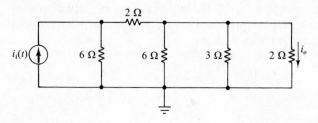

Figure P9.3

9.4 For the circuit shown in Figure P9.4, use the rules for series and parallel resistors to find e_o and the equivalent resistance connected across the source.

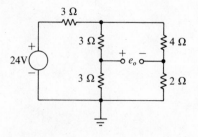

Figure P9.4

9.5 a. Explain why the rules for series and parallel resistors cannot be used for the circuit shown in Figure P9.5.

b. Use the node-equation method to find the voltages of nodes A and B with respect to the ground node.

c. Find the currents i_o and i_1 and the equivalent resistance connected across the source.

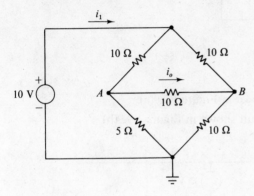

Figure P9.5

9.6 Use the node-equation method to find e_o for the circuit shown in Figure P9.6.

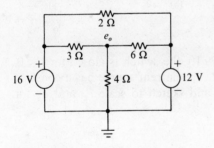

Figure P9.6

9.7 Repeat Problem 9.6 when the 12-V source is replaced by a current source of 5 A with the positive sense upward.

9.8 a. For the circuit shown in Figure P9.8, use the rules for series and parallel resistors to find e_o when $e_i = 24$ V and $i_a = 0$.

b. Repeat part a when $i_a = 5$ A and $e_i = 0$.

c. Use the node-equation method to find e_o when $e_i = 24$ V and $i_a = 5$ A. Compare the result to your answers to parts a and b.

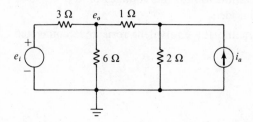

Figure P9.8

9.9 a. Find e_o for the circuit shown in Figure P9.9(a).

b. Repeat part a for the circuit shown in Figure P9.9(b).

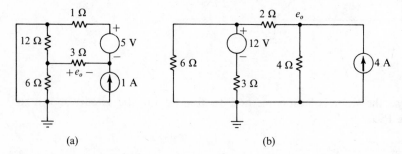

(a)   (b)

Figure P9.9

9.10 For the circuit shown in Figure P9.10, the switch is closed for $t \leq 0$, open for $0 < t \leq 2$, and closed for $t > 2$. The current source has a constant value of 3 A for all values of t. Find and sketch to scale e_C and e_A for $-2 \leq t \leq 8$.

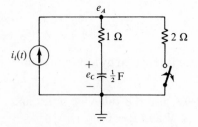

Figure P9.10

9.11 a. For the circuit shown in Figure P9.11, use the node-equation method to find the input-output differential equation.

b. Solve the equation you found in part a for the unit step response and sketch the result.

c. Use the short-cut method to find and sketch the unit step response.

d. Find and sketch the unit impulse response.

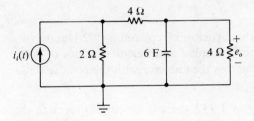

Figure P9.11

9.12 a. For the circuit shown in Figure P9.12, there is no current through the inductor for $t < 0$. Find and sketch e_o vs t if the switch closes at $t = 0$.

b. Repeat part a if the switch has been closed for $t < 0$ and opens at $t = 0$. Assume that steady-state conditions exist at $t = 0-$. Also find and sketch e_s, the voltage across the switch.

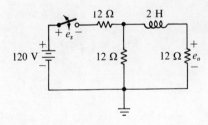

Figure P9.12

9.13 The current source $i_i(t)$ shown in Figure P9.13 is 2 A for all values of t. The switch is open for all $t < 0$ and closed for all $t > 0$.

a. Find and sketch e_C for both positive and negative values of time.

b. Find and sketch i_o for both positive and negative values of time.

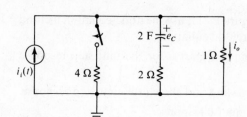

Figure P9.13

9.14 Repeat Problem 9.13 when the capacitor is replaced by a 2-H inductor. Find and sketch e_L and i_o for both positive and negative values of time, where e_L is defined as the voltage across the inductor with the positive sense upward.

9.15 For the circuit shown in Figure P9.15, there is no current through the inductor for $t < 0$. The switch closes at $t = 0$.

   a. Use the short-cut method described in Section 9.6 to find $e_o(0+)$, $(e_o)_{ss}$, and τ.

   b. Use the results of part a to write an expression for and sketch e_o vs t.

   c. Check your answer to part b by writing the input-output differential equation and then solving it.

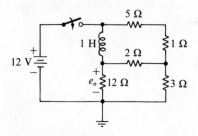

Figure P9.15

9.16 Repeat Problem 9.15 when the switch has been closed for $t < 0$ and then opens at $t = 0$. Assume that steady-state conditions exist at $t = 0-$.

9.17 Steady-state conditions exist at $t = 0-$ for the circuit shown in Figure P9.17.

   a. If the switch is open for all $t < 0$ and closed for all $t > 0$, find and sketch e_C vs t.

   b. Repeat part a when the switch is closed for all $t < 0$ and open for all $t > 0$.

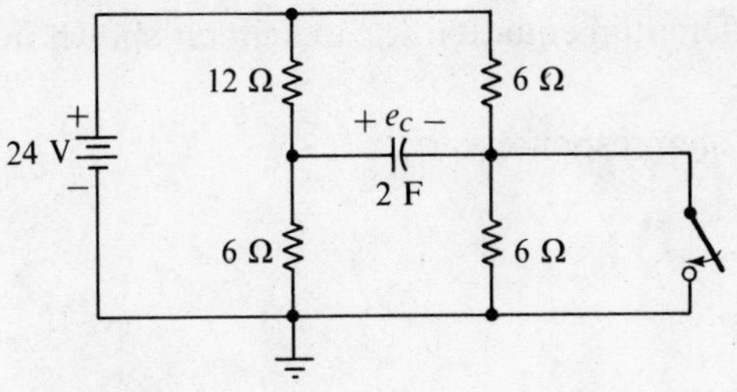

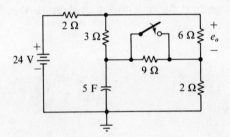

Figure P9.17

9.18 For the circuit shown in Figure P9.18, steady-state conditions exist at $t = 0-$ with the switch open. The switch closes at $t = 0$.

    a. Find $e_o(0-)$, $e_o(0+)$, $(e_o)_{ss}$, and the time constant of the circuit with the switch closed.

    b. Find and sketch e_o vs t.

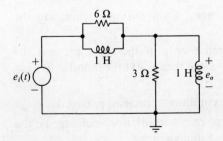

Figure P9.18

9.19 Repeat Problem 9.18 when the capacitor is replaced by a 5-H inductor.

9.20 a. Find the input-output differential equation relating e_o and $e_i(t)$ for the circuit shown in Figure P9.20.

    b. Find the unit step response $y_U(t)$ and the unit impulse response $h(t)$.

Figure P9.20

9.21 a. Find the input-output differential equation for the circuit shown in Figure P9.21.

b. Find and sketch the unit step response $y_U(t)$.

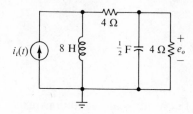

Figure P9.21

9.22 a. Find the input-output differential equation relating e_o to $e_i(t)$ and $i_a(t)$ for the circuit shown in Figure P9.22.

b. Write the free response in general form without evaluating the coefficients.

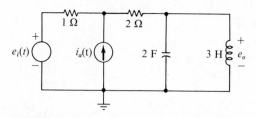

Figure P9.22

9.23 a. For the circuit shown in Figure P9.23, apply the node-equation method to nodes A and B to obtain a pair of coupled differential equations in the variables e_A, e_o, and $e_i(t)$.

b. Find the input-output differential equation relating e_o and $e_i(t)$.

c. Write the free response in general form without evaluating the coefficients.

d. Show that if there is no stored energy in the capacitor and the inductor at $t = 0-$ and if $e_i(t) = U(t)$, then $e_o(0+) = 0$ and $\dot{e}_o(0+) = 2$ V/s.

e. Find and sketch e_o vs t for the conditions specified in part d.

f. Explain why the form of the free response will not be changed if the capacitor and left-hand resistor are interchanged.

g. Find the steady-state response to $e_i(t) = U(t)$ when the capacitor and left-hand resistor are interchanged.

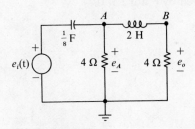

Figure P9.23

9.24 a. For the circuit shown in Figure P9.24, use the node-equation method to find the input-output differential equation relating e_o and $e_i(t)$.

b. Determine the damping ratio ζ and the undamped natural frequency ω_n.

c. Find $e_C(0-)$, $i_L(0-)$, $e_o(0-)$, $e_C(0+)$, $i_L(0+)$, $e_o(0+)$, and $\dot{e}_o(0+)$ when $e_i(t)$ is 4 V for all $t < 0$ and 8 V for all $t > 0$.

d. For the input in part c, use the input-output differential equation to find the steady-state value of e_o for large positive values of t. Check your answer by replacing the capacitor and the inductor by open and short circuits, respectively.

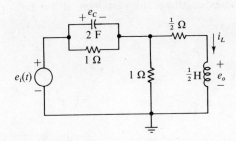

Figure P9.24

9.25 After the circuit shown in Figure P9.25 has reached steady-state conditions with the switches closed, both switches open at $t = 0$, removing the top two resistors from the circuit.

a. Write the state-variable equations for $t > 0$. Also write the algebraic output equation for e_o.

b. Find the value of each state variable at $t = 0+$, and also calculate $e_o(0+)$.

c. Find the differential equation obeyed by e_o for $t > 0$.

d. Repeat part c when the output is e_C.

e. Find and sketch e_C vs t.

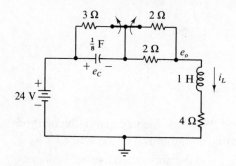

Figure P9.25

9.26 a. Write a set of state-variable equations describing the circuit shown in Figure P9.26. Define the variables and show their positive senses on the diagram.

b. Write an algebraic output equation for i_o, which is the current through the 6-V source.

c. If $i_i(t) = 2U(t)$, find the values of the state variables at $t = 0+$. Assume that $R_1 = R_2 = 2\,\Omega$.

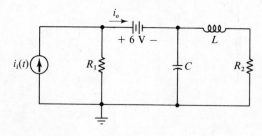

Figure P9.26

9.27 a. For the circuit shown in Figure P9.1, write a set of state-variable equations and an algebraic output equation for e_o.

b. Repeat part a for the circuit shown in Figure P9.2(b).

9.28 a. For the circuit shown in Figure P9.22, write a set of state-variable equations and an algebraic output equation for e_o.

b. Repeat part a for the circuit shown in Figure P9.24.

9.29 a. Find a set of state-variable equations for the circuit shown in Figure P9.29. Write an algebraic output equation for e_o.

b. If $e_i(t)$ is 6 V for all $t < 0$ and 2 V for all $t > 0$, find the values of each state variable at $t = 0+$ and calculate $e_o(0+)$.

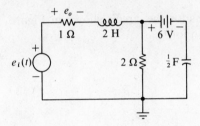

Figure P9.29

9.30 a. For the circuit shown in Figure P9.30(a), find a set of state-variable equations and write an algebraic output equation for e_o.

b. Repeat part a for the circuit shown in Figure P9.30(b).

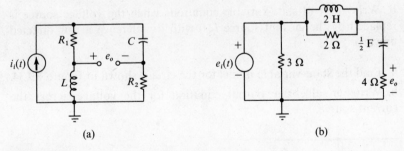

(a)   (b)

Figure P9.30

9.31 a. For the circuit shown in Figure P9.31(a), find a set of state-variable equations and write an algebraic output equation for i_o. Define the variables and show their positive senses on the diagram.

b. Repeat part a for the circuit shown in Figure P9.31(b).

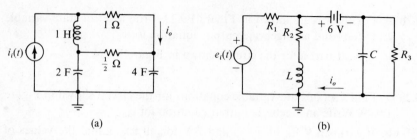

(a)   (b)

Figure P9.31

9.32 Find a set of state-variable equations for the circuit shown in Figure P9.32. Write the algebraic output equation for i_o.

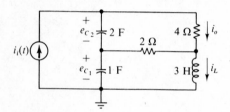

Figure P9.32

9.33 a. Write the state-variable equation for the circuit shown in Figure 9.30 when the initial choice of the state variable is e_B rather than e_A.

b. Write a set of state-variable equations when the voltage source is replaced by the current source $i_i(t)$ with its reference arrow directed upward.

9.34 a. Find the state-variable model for the circuit shown in Figure P9.34.

b. Write an algebraic output equation for the voltage across the current source.

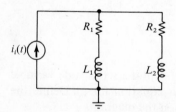

Figure P9.34

CHAPTER 10
NONLINEAR ELECTRICAL CIRCUITS

We have defined resistors, capacitors, and inductors, respectively, as elements for which there is an algebraic relationship between the voltage and current, voltage and charge, and current and flux linkage. If the two variables involved in the algebraic relationship are directly proportional to one another, then the element is linear, as was the case in all the examples in Chapter 9.

We discussed several examples of nonlinear mechanical systems in earlier chapters. We saw that the basic approaches for formulating the system model are similar for linear and nonlinear systems. Two methods for solving the model are to approximate it by a linearized model in the vicinity of the operating point (as explained in Chapter 5) or to obtain a digital-computer solution of the original model (as explained in Chapter 7). These same procedures apply to nonlinear electrical circuits.

The only nonlinear elements we consider in Section 10.1 are two-terminal resistors. The techniques for setting up the nonlinear state-variable equations or input-output equations are similar to those we developed in Chapter 9. The general procedure for obtaining a linearized model is illustrated by several examples of increasing complexity. In two cases, we compare the solution of the linearized model with a computer solution of the original nonlinear model.

Diodes are the most common type of nonlinear resistor. Two examples illustrate a graphical method useful in some simple semiconductor diode circuits. An idealization of the diode yields a current-voltage characteristic that consists of horizontal and vertical lines. In such cases, it is not normally possible to find a single linearized model that is valid for the desired range of inputs. The analysis of such circuits is illustrated by several examples.

In Section 10.2 we consider the transistor, which is described by algebraic relationships between its voltages and currents but which has three external terminals. The technique we developed in Section 5.3 for linearizing functions of more than one variable is used here to derive a linearized incremental model. Finally, in Section 10.3, we examine circuits with time-varying or nonlinear capacitors and inductors.

10.1 CIRCUITS WITH NONLINEAR RESISTORS

We first consider three examples of increasing complexity for circuits containing a fairly general type of nonlinearity. As in Chapter 5, we express the variables as the sum of a constant portion and a time-varying portion. For example, we write a voltage e_o as

$$e_o = \bar{e}_o + \hat{e}_o$$

where the constant term $\bar{e}_o$ is the nominal value, corresponding to a particular operating point, and where $\hat{e}_o$ is the incremental time-varying component. In the circuit diagrams, we indicate the fact that a resistor is nonlinear by drawing a curved line through its symbol.

EXAMPLE 10.1 The circuit shown in Figure 10.1(a) contains a nonlinear resistor obeying the element law $i_o = 2e_o^3$. Write the differential equation relating e_0 and $e_i(t)$. If $e_i(t) = 18 + A \cos \omega t$, find the operating point and derive the linearized input-output differential equation. Also determine the time constant of the linearized model.

Solution The right-hand resistor is nonlinear because the current i_o is not directly proportional to the voltage e_o. Summing the currents leaving the node at the upper right gives

$$\tfrac{1}{2}\dot{e}_o + e_o - e_i(t) + 2e_o^3 = 0$$

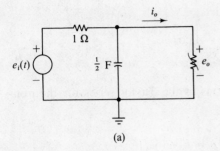

(a)

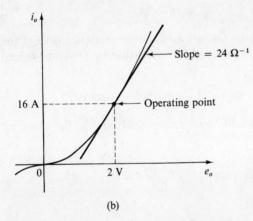

(b)

Figure 10.1 (a) Circuit for Example 10.1. (b) Characteristic curve for the nonlinear resistor.

Thus the input-output equation is

$$\tfrac{1}{2}\dot{e}_o + 2e_o^3 + e_0 = e_i(t) \tag{1}$$

To determine the operating point, we replace $e_i(t)$ by $\bar{e}_i = 18$ V, e_o by $\bar{e}_o$, and $\dot{e}_o$ by zero to obtain

$$2\bar{e}_o^3 + \bar{e}_o = 18$$

The only real value of $\bar{e}_o$ that satisfies this algebraic equation is

$$\bar{e}_o = 2 \text{ V} \tag{2}$$

From the nonlinear element law, we see that $i_o = 16$ A, which gives the operating point shown in Figure 10.1(b).

To develop a linearized model, we let

$$e_i(t) = 18 + A \cos \omega t$$

$$e_o = 2 + \hat{e}_o \qquad (3)$$

As in (5.8), we write the first two terms in the Taylor series for the non-linear term $2e_o^3$, which are

$$\bar{i}_o + \frac{di_o}{de_o}\bigg|_{\bar{e}_o} (e_o - \bar{e}_o) = \bar{i}_o + (6\bar{e}_o^2)\hat{e}_o$$

$$= 16 + 24\hat{e}_o \qquad (4)$$

This approximation describes the tangent to the characteristic curve at the operating point, as shown in Figure 10.1(b). Substituting (3) and (4) into (1) gives

$$\tfrac{1}{2}(\dot{\bar{e}}_o + \dot{\hat{e}}_o) + (16 + 24\hat{e}_o) + (2 + \hat{e}_o) = 18 + A \cos \omega t$$

Since $\dot{\bar{e}}_o = 0$ and since the constant terms cancel (as is always the case), we have for the linearized model

$$\tfrac{1}{2}\dot{\hat{e}}_o + 25\hat{e}_o = A \cos \omega t$$

or

$$\dot{\hat{e}}_o + 50\hat{e}_o = 2A \cos \omega t$$

By inspecting this equation, we see that the time constant of the linearized model is $\tau = \tfrac{1}{50}$ s.

EXAMPLE 10.2   Write the state-variable equation for the circuit shown in Figure 10.2(a), which contains a nonlinear resistor obeying the element law $i_2 = \tfrac{1}{8}e_C^3$. Find the operating point when $e_i(t) = 2 + \hat{e}_i(t)$, and derive the linearized state-variable equations in terms of the incremental variables. Plot and compare i_L vs t for the nonlinear and linearized models when $\hat{e}_i(t) = [A \sin t] U(t)$ for (1) $A = 0.1$ V, (2) $A = 1.0$ V, and (3) $A = 10.0$ V.

Solution   We choose e_C and i_L as state variables and note that

$$e_L = e_i(t) - e_C \qquad (5)$$

Applying Kirchhoff's current law to the upper right node gives

$$i_L + e_L - \tfrac{1}{8}e_C^3 - 2\dot{e}_C = 0$$

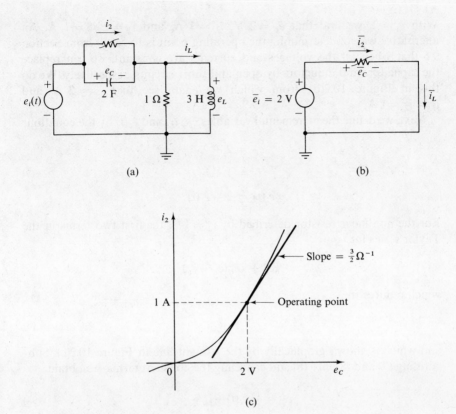

Figure 10.2 (a) Circuit for Example 10.2. (b) Circuit used for calculation of the operating point. (c) Characteristic curve for the nonlinear resistor.

Inserting (5) into this equation and into the element law $di_L/dt = \frac{1}{3}e_L$ gives

$$\dot{e}_C = \frac{1}{2}[e_i(t) + i_L - \frac{1}{8}e_C^3 - e_C]$$

$$\frac{di_L}{dt} = \frac{1}{3}[e_i(t) - e_C]$$

(6)

which consitute the nonlinear state-variable equations.

At the operating point, the derivatives of the state variables are zero and (6) reduces to the algebraic equations

$$\bar{e}_i + \bar{i}_L - \frac{1}{8}\bar{e}_C^3 - \bar{e}_C = 0$$

$$\bar{e}_i - \bar{e}_C = 0$$

With $\bar{e}_i = 2$, we find that $\bar{e}_C = 2$ V, $i_L = 1$ A, and $i_2 = 2^3/8 = 1$ A. An alternative way of determining the operating point is to recall from Section 9.6 that when all the voltages and currents are constant, we can replace the capacitors and inductors by open and short circuits, respectively. We do this in Figure 10.2(b), from which we again see that $\bar{e}_C = 2$ V and $\bar{i}_L = \bar{i}_2 = 1$ A.

Next, we define the incremental variables $\hat{e}_C$, $\hat{i}_L$, and $\hat{e}_i(t)$ by the equations

$$
\begin{aligned}
e_C &= 2 + \hat{e}_C \\
i_L &= 1 + \hat{i}_L \\
e_i(t) &= 2 + \hat{e}_i(t)
\end{aligned}
\tag{7}
$$

For the nonlinear resistor described by $i_2 = \tfrac{1}{8}e_C^3$, the first two terms in the Taylor series for i_2 are

$$\tfrac{1}{8}\bar{e}_C^3 + \tfrac{3}{8}\bar{e}_C^2(e_C - \bar{e}_C)$$

which reduces to

$$1 + \tfrac{3}{2}\hat{e}_C \tag{8}$$

and which is shown graphically by the straight line in Figure 10.2(c). Substituting (7) and (8) into (6) and canceling the constant terms, we obtain

$$\dot{\hat{e}}_C = \tfrac{1}{2}[\hat{e}_i(t) + \hat{i}_L - \tfrac{3}{2}\hat{e}_C] \tag{9a}$$

$$\frac{d\hat{i}_L}{dt} = \tfrac{1}{3}[\hat{e}_i(t) - \hat{e}_C] \tag{9b}$$

as the linearized state-variable equations.

If we want an input-output equation relating $\hat{i}_L$ to $\hat{e}_i(t)$, we can rearrange (9b) to get an expression for $\hat{e}_C$, which we then substitute into (9a). The result is

$$6\frac{d^2\hat{i}_L}{dt^2} + 7.5\frac{d\hat{i}_L}{dt} + \hat{i}_L = 2\dot{\hat{e}}_i + 1.5\hat{e}_i(t) \tag{10}$$

The characteristic equation corresponding to (10) is

$$6r^2 + 7.5r + 1 = 0$$

and its roots are $r_1 = -1.0982$ and $r_2 = -0.1518$. Thus the free response of the linearized model will consist of two exponentially decaying terms having time constants

$$\tau_1 = \frac{1}{1.0982} = 0.9106 \text{ s}$$

$$\tau_2 = \frac{1}{0.1518} = 6.589 \text{ s}$$

When $\hat{e}_i(t) = A \sin t$, the forced response will have the form $B \sin(t + \phi)$. The form of the complete response of the linearized model will be

$$\hat{i}_L = K_1 \epsilon^{-1.0982t} + K_2 \epsilon^{-0.1518t} + B \sin(t + \phi) \tag{11}$$

The results of a computer solution of the nonlinear model given by (6) with $e_i(t) = 2 + [A \sin t] U(t)$ are shown in Figure 10.3 for three different values of the amplitude A. Plotted on the same axes are curves obtained by calculating $\hat{i}_L$ from the linearized model in (9) with $\hat{e}_i(t) = [A \sin t] U(t)$ and then forming the quantity $\hat{i}_L + i_L$. Because $e_i(t) = \bar{e}_i$ for all $t < 0$, we used the initial conditions

$$e_C(0) = \bar{e}_C = 2 \text{ V}, \qquad \hat{e}_C(0) = 0$$

$$i_L(0) = i_L = 1 \text{ A}, \qquad \hat{i}_L(0) = 0$$

Note that the responses of the nonlinear and linearized models are almost identical when $A = 0.1$ V, are in close agreement when $A = 1.0$ V, but differ significantly when $A = 10.0$ V. As (11) indicates, the steady-state response of the linearized model is always a sinusoidal oscillation about the operating point. For large values of A, however, the steady-state response of the nonlinear model is not symmetrical about the operating point.

We have plotted the three sets of curves with different vertical scales and different origins in order to get a good comparison of the responses of the nonlinear and linearized models for each of the three values of A. If we repeat the example when $\bar{e}_i = 0$ and $\hat{e}_i(t) = [A \sin t] U(t)$, then in the steady state we would expect the response of the nonlinear model to be symmetrical about the operating point. When we carry out a computer run for this case, we obtain not only the expected symmetry but also good agreement of the nonlinear and linearized responses for all three values of A.

EXAMPLE 10.3 The voltage source $e_i(t)$ for the circuit shown in Figure 10.4(a) is

$$e_i(t) = \begin{cases} -2 & \text{for } t \le 0 \\ 2 + A \cos 4t & \text{for } t > 0 \end{cases}$$

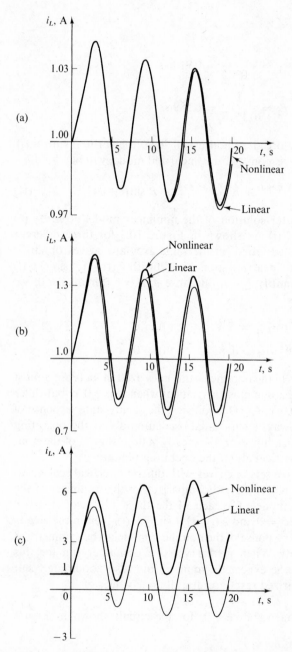

Figure 10.3 Results of computer simulation for Example 10.2 with $e_i(t) = 2 + [A \cos t] U(t)$. (a) $A = 0.1$ V. (b) $A = 1.0$ V. (c) $A = 10.0$ V.

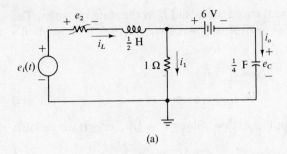

(a)

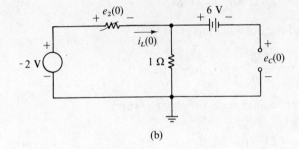

(b)

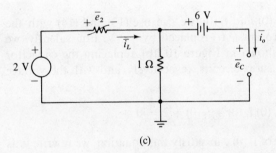

(c)

Figure 10.4 (a) Circuit for Example 10.3. (b) Circuit for the calculation of the initial conditions. (c) Circuit for the calculation of the operating point.

The element law for the nonlinear resistor is $e_2 = 2|i_L|i_L$.

a. Find a set of state-variable equations and an algebraic output equation for i_o. Give the initial conditions we need to solve the equations.

b. Find a linearized set of state-variable equations about the operating point corresponding to $\bar{e}_i = 2$ V. Write the necessary initial conditions and an algebraic output equation for i_o.

c. Simulate the responses of the nonlinear and linearized models and plot i_o vs time when $A = 1$ V, 5 V, and 20 V.

Solution We make our usual choice of e_C and i_L as the state variables. The voltage at the top of the 1-Ω resistor is $e_C + 6$, so $i_1 = e_C + 6$. By Kirchhoff's current law,

$$
\begin{aligned}
i_o &= i_L - i_1 \\
&= i_L - e_C - 6
\end{aligned} \tag{12}
$$

which is the desired output equation. Since $i_o = \frac{1}{4}\dot{e}_C$, one state-variable equation is

$$
\dot{e}_C = 4(i_L - e_C - 6) \tag{13}
$$

Summing voltages around the outside loop gives

$$
e_C + 6 + \frac{1}{2}\frac{di_L}{dt} + 2|i_L|i_L - e_i(t) = 0
$$

from which we have for the second state-variable equation

$$
\frac{di_L}{dt} = 2[e_i(t) - 2|i_L|i_L - e_C - 6] \tag{14}
$$

To find the values of $e_C(0)$ and $i_L(0)$, we can use (13) and (14) with the derivatives replaced by zero and $e_i(t)$ replaced by -2 V. Equivalently, we can redraw the circuit as shown in Figure 10.4(b), replacing the capacitor and inductor by open and short circuits, respectively, and with the voltage source -2 V. Then

$$
2|i_L(0)|i_L(0) + i_L(0) + 2 = 0
$$

Since only a negative value of $i_L(0)$ can satisfy this equation, we rewrite it as

$$
-2[i_L(0)]^2 + i_L(0) + 2 = 0 \tag{15}
$$

which has roots at -0.7808 and 1.2808. The latter root results from re-writing the equation in quadratic form and is extraneous, so

$$
\begin{aligned}
i_L(0) &= -0.7808 \text{ A} \\
e_C(0) &= -0.7808 - 6 = -6.7808 \text{ V}
\end{aligned} \tag{16}
$$

We can find the operating point corresponding to $\bar{e}_i = 2$ V from (13) and (14) or from the simplified circuit shown in Figure 10.4(c). We see that

$$
2|i_L|i_L + i_L - 2 = 0
$$

This time only a positive value of i_L can satisfy the equation, so we rewrite it as

$$2i_L^2 + i_L - 2 = 0 \tag{17}$$

Since (17) has the same form as (15) except that $i_L(0)$ is replaced by $-i_L$, we know that the desired root is at 0.7808. Thus

$$i_L = 0.7808 \text{ A}$$
$$\bar{e}_C = 0.7808 - 6 = -5.2192 \text{ V} \tag{18}$$
$$i_o = 0$$

We now introduce the incremental variables

$$\hat{e}_i(t) = e_i(t) - 2$$
$$\hat{\imath}_L = i_L - 0.7808$$
$$\hat{e}_C = e_C + 5.2192 \tag{19}$$
$$\hat{\imath}_o = i_o$$

The first two terms of the Taylor series for e_2, the voltage across the non-linear resistor, are

$$2|i_L|i_L + \left.\frac{de_2}{di_L}\right|_{i_L} (i_L - i_L)$$

Since i_L is positive, $|i_L| = i_L$ and the approximate expression for e_2 becomes

$$2i_L^2 + (4i_L)\hat{\imath}_L$$

which reduces to

$$1.2193 + 3.1232\hat{\imath}_L \tag{20}$$

Substituting (19) and (20) into (12), (13), and (14) and canceling the constant terms give the two state-variable equations

$$\dot{\hat{e}}_C = 4(\hat{\imath}_L - \hat{e}_C)$$
$$\frac{d\hat{\imath}_L}{dt} = 2[\hat{e}_i(t) - 3.1232\hat{\imath}_L - \hat{e}_C] \tag{21}$$

and the output equation

$$\hat{\imath}_o = \hat{\imath}_L - \hat{e}_C \tag{22}$$

The required initial conditions are

$$\hat{\imath}_L(0) = i_L(0) - 0.7808 = -1.5616 \text{ A}$$
$$\hat{e}_C(0) = e_C(0) + 5.2192 = -1.5616 \text{ V}$$

(23)

The results of a computer solution of the nonlinear model when $e_i(t) = 2 + A\cos 4t$ for $t > 0$ and for the initial conditions in (16) are shown in Figure 10.5. A solution of the linearized model when $\hat{e}_i(t) = A\cos 4t$ and for comparable initial conditions is shown on the same set of axes. Because $i_o = 0$, we can compare i_o from the nonlinear model to $\hat{\imath}_o$ from the linearized model directly. The responses agree closely for $A = 1$ V. Note, however, that when $A = 5$ V and 20 V, the output waveshape for the nonlinear model begins to look triangular in the steady state. For the linearized model, the steady-state response to a sinusoidal input must remain sinusoidal regardless of the size of the input.

Reflecting on the results of this example and the preceding one, we see that even in the steady state, a nonlinearity can alter the average value of the output and also its waveshape. Its effect may be different for different operating points, and it is difficult to predict in advance the range of incremental inputs for which a linearized model will give a satisfactory approximation. Because it is difficult to establish general guidelines for approximating nonlinear systems, we must usually rely heavily on computer simulations.

DIODES

Diodes are two-terminal resistive elements for which the current-voltage relationship depends strongly on the sign of the voltage. The symbol for a diode is shown in Figure 10.6(a).

For a limited range of positive voltages and for negative voltages whose magnitude is not sufficient to cause electrical breakdown, a semiconductor diode at room temperature is described by the equation

$$i = I_o[\epsilon^{39e} - 1]$$

(24)

which is illustrated in Figure 10.6(b). The constant I_o depends on the dimensions and materials of the device. For negative values of the voltage e, the current i is negative but with a magnitude less than I_o. If the voltage becomes positive, however, the current increases very rapidly with an increase in voltage. In fact, (24) is useful only for negative voltages and for positive voltages less than some fraction of a volt.

Diodes have many important applications, often in circuits that do not contain inductors or capacitors. The next example develops a graphical

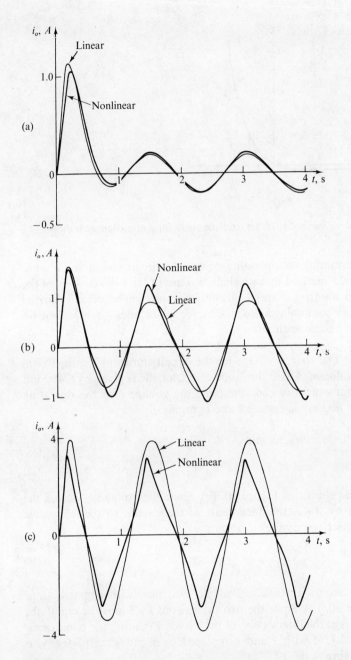

Figure 10.5 Results of computer simulation for Example 10.3 with $e_i(t) = 2 + A \cos 4t$ for $t > 0$. (a) $A = 1$ V. (b) $A = 5$ V. (c) $A = 20$ V.

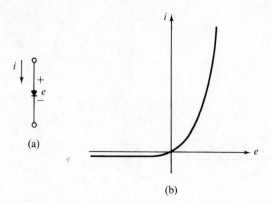

Figure 10.6 (a) Diode symbol. (b) Characteristic curve for a semiconductor diode.

method for determining the operating point for a circuit containing a semiconductor diode, a method that we shall also use in later discussions of the transistor. Then another example illustrates construction of a linearized model for a circuit containing a diode. Chapter 18 includes a comprehensive example of a tunnel diode circuit.

EXAMPLE 10.4   The source voltage for the circuit shown in Figure 10.7(a) has a constant value of A, and the diode characteristic is given by (24). Find an equation from which we can determine the voltage $\bar{e}$ at the operating point, and show how to determine $\bar{e}$ and $\bar{i}$ graphically.

Solution   The diode characteristic

$$i = I_o[\epsilon^{39e} - 1] \tag{25}$$

is the curved line shown in Figure 10.7(b). Since the voltage e_o across the resistor is given by $A - e$, the linear part of the circuit, to the left of the dashed line, is described by the equation

$$i = \frac{1}{R}(A - e) \tag{26}$$

This describes the straight line, called the *load line*, with negative slope shown in Figure 10.7(b). Since the two expressions for i must be equal, the operating point is at the intersection of the curves. Equating the two expressions in (25) and (26) with i and e replaced by $\bar{i}$ and $\bar{e}$, respectively, is equivalent to saying that

$$I_o[\epsilon^{39\bar{e}} - 1] + \frac{1}{R}(\bar{e} - A) = 0 \tag{27}$$

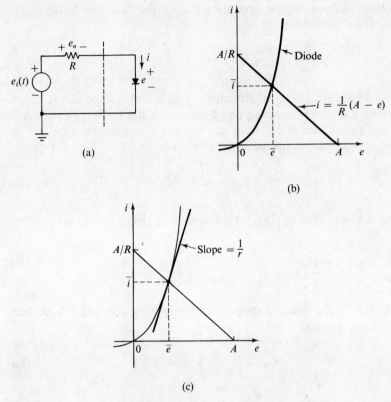

Figure 10.7 (a) Circuit for Example 10.4. (b) Determining the operating point. (c) Determining the diode resistance for the linearized model.

which can be regarded as the Kirchhoff current-law equation at the node comprising the upper right corner of Figure 10.7(a). Because of the transcendental term $\epsilon^{39\bar{e}}$, we cannot solve this equation analytically and must use either the graphical method or some numerical root-finding procedure.

EXAMPLE 10.5   Let $e_i(t) = A + \hat{e}_i(t)$ for the circuit shown in Figure 10.7(a). Find the linearized equation relating $\hat{e}$ to $\hat{e}_i(t)$. Also find the equation relating $\hat{e}_o$ to $\hat{e}_i(t)$.

Solution   Let $\bar{e}$ and $\bar{i}$ denote the diode voltage and current corresponding to $\bar{e}_i = A$, as found by solving (27) and as indicated in Figure 10.7(b). Inserting $e_i(t) = A + \hat{e}_i(t)$ and $e = \bar{e} + \hat{e}$ into (26) gives

$$i = \frac{1}{R}[A + \hat{e}_i(t) - (\bar{e} + \hat{e})] \qquad (28)$$

If the diode characteristic is approximated by the first two terms in its Taylor series, (25) becomes

$$i = i + \frac{di}{de}\bigg|_{\bar{e}} (e - \bar{e})$$

As shown in Figure 10.7(c), the quantity $di/de|_{\bar{e}}$ is the slope of the diode curve evaluated at the operating point. Because it has the units of ohms^{-1} and relates the incremental current and voltage, it is given the symbol $1/r$. Then the approximation for the current becomes

$$i = i + \frac{1}{r}\hat{e} \tag{29}$$

Equating the right sides of (28) and (29) and using the fact that $i = (A - \bar{e})/R$ give

$$\frac{1}{R}[A + \hat{e}_i(t) - \bar{e} - \hat{e}] = \frac{1}{R}(A - \bar{e}) + \frac{1}{r}\hat{e}$$

which reduces to the desired expression relating the incremental diode and input voltages, namely

$$\hat{e} = \left(\frac{r}{R+r}\right)\hat{e}_i(t) \tag{30}$$

The incremental input and output voltages are related by

$$\hat{e}_o = \hat{e}_i(t) - \hat{e} = \left(\frac{R}{R+r}\right)\hat{e}_i(t) \tag{31}$$

As expected, the last two equations indicate that if the diode is replaced by the linear resistance r, then the incremental voltages are given by the voltage-divider rule. Note that the value of the diode resistance depends on the operating point.

If we modified the circuit shown in Figure 10.7(a) by including capacitors and inductors, we could find the operating point by first replacing the capacitors and inductors by open and short circuits, respectively. We could analyze the resulting resistive circuit using the graphical method of Example 10.4. The diode resistance r is still the reciprocal of the slope of its characteristic curve at the operating point. When calculating the incremental voltages and currents, we replace the diode by the resistance r, suppress the constant value of the source, and return the inductors and capacitors to the circuit, resulting in a linear RLC circuit whose mathematical model can be found by any of the techniques discussed in Chapter 9.

IDEAL DIODES

The technique of linearizing the diode characteristic about a single operating point is not very helpful in the majority of applications. Recall that the semiconductor diode characteristic given by (24) and shown in Figure 10.6(b) has a slope that is extremely large in the first quadrant compared to that in the third quadrant. When $e = -0.1$ V and -0.5 V, the currents given by (24) are $-0.98I_o$ and $-1.00I_o$, respectively. In contrast, the currents corresponding to $e = 0.1$ V and 0.2 V are approximately $50I_o$ and $2440I_o$, respectively. If the operation of the circuit involves both positive and negative diode voltages, then there is no single straight line that can satisfactorily approximate the device's behavior over its range of operation. In fact, the differences in slope are so great that the actual characteristic shown in Figure 10.8(a) is often approximated by the idealized characteristic shown in Figure 10.8(b).

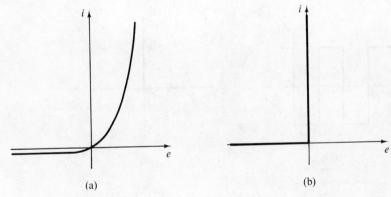

(a)   (b)

Figure 10.8 Characteristic curves for diodes. (a) Semiconductor diode. (b) Ideal diode.

The idealized diode characteristic represents a device that acts like a switch whose position depends on the conditions existing in the circuit. It behaves like an open circuit (infinite resistance) when the voltage is negative and when no current can flow, and like a short circuit (zero resistance) when the current is positive and the voltage is zero. Hence, the operation of a network containing a diode depends critically on which half of the diode characteristic is in effect at a given moment, i.e., whether the switch is open or closed. When the segment of the diode characteristic being used changes, we can expect a sudden change in the circuit's output. Several commonly used diode circuits are modeled and analyzed in the following examples.

EXAMPLE 10.6   Assume that the diode in Figure 10.9(a) has the idealized characteristic given in Figure 10.8(b). For the input voltage shown in Figure 10.9(b), find and sketch the voltages e_o and e_R.

Solution   When the voltage $e_i(t)$ is positive, the diode acts like a short circuit and e_o is zero. When $e_i(t)$ is negative, the diode is an open circuit, no current flows, and $e_o = e_i(t)$. The waveshape of the output voltage is shown in

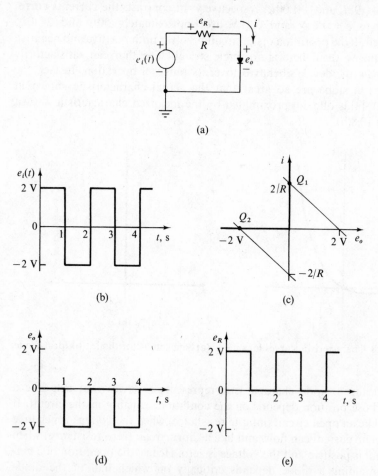

Figure 10.9 (a) Circuit for Example 10.6. (b) Input. (c) Operating points for $\bar{e}_i = 2$ V and $\bar{e}_i = -2$ V. (d) Diode voltage. (e) Resistor voltage.

Figure 10.9(d). Since the configuration of the circuit is similar to the one shown in Figure 10.7(a), we can also obtain the solution by the method shown in Figure 10.7(b), but with A first equal to 2 V and then to -2 V. In Figure 10.9(c), which shows the idealized diode curve, Q_1 and Q_2 are the operating points for $\bar{e}_i = 2$ V and $\bar{e}_i = -2$ V, respectively. The voltage across the resistor R is given by $e_R = e_i(t) - e_o$ and is shown in Figure 10.9(e). Figure 10.9(a) is an example of a *rectifier circuit* that removes either the positive or the negative portions of the incoming signal, depending on where the output is taken.

EXAMPLE 10.7   Sketch the output voltage e_o of the circuit shown in Figure 10.10(a) for the input voltages $e_1(t)$ and $e_2(t)$ shown in Figure 10.10(b) and Figure 10.10(c), respectively.

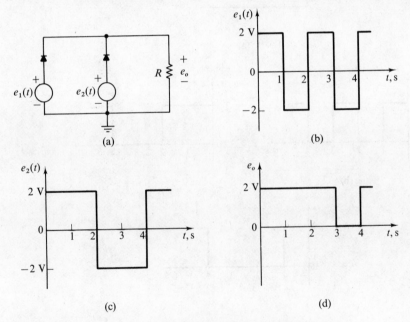

Figure 10.10 (a) Circuit for Example 10.7. (b), (c) Inputs. (d) Output.

Solution   If either $e_1(t)$ or $e_2(t)$ or both are positive, then at least one of the diodes will act like a short circuit, and the output voltage e_o will be the same as the positive source voltage. If neither $e_1(t)$ nor $e_2(t)$ is positive, both diodes will act like open circuits and $e_o = 0$. For the input waveshapes shown in Figure 10.10(b) and Figure 10.10(c), the output voltage is shown in

Figure 10.10(d). The circuit is an example of an "OR" gate, because the output voltage is positive if either $e_1(t)$ *or* $e_2(t)$ or both are positive. Another way to characterize the circuit's behavior is to say that e_0 is the largest of $e_1(t)$, $e_2(t)$, and zero.

EXAMPLE 10.8 Sketch the output voltage e_o of the circuit shown in Figure 10.11(a) for the input voltages $e_1(t)$ and $e_2(t)$ shown in Figure 10.11(b) and Figure 10.11(c), respectively.

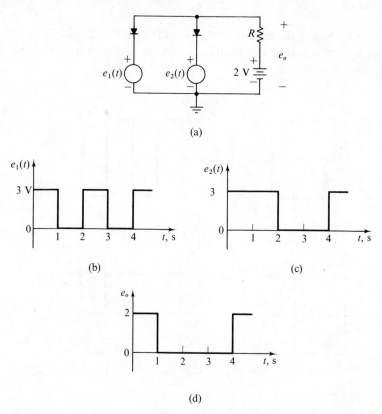

(a)

(b)

(c)

(d)

Figure 10.11 (a) Circuit for Example 10.8. (b), (c) Inputs. (d) Output.

Solution If either $e_1(t)$ or $e_2(t)$ is zero, the corresponding diode will act like a short circuit and $e_o = 0$. If both $e_1(t)$ and $e_2(t)$ are 2 V or more, then both diodes act like open circuits and $e_o = 2$ V. The output waveform is shown in Figure 10.11(d), and the circuit is an example of an "AND" gate, since $e_o = 2$ V only if $e_1(t)$ *and* $e_2(t)$ are at least 2 V.

In general, the output voltage shown in Figure 10.11(a) is the smallest of $e_1(t)$, $e_2(t)$, and 2 V. If the orientation of both diodes were reversed, e_o would be the largest of $e_1(t)$, $e_2(t)$, and 2 V.

EXAMPLE 10.9   A *Zener diode* has a characteristic curve similar to that for the semiconductor diode shown in Figure 10.6(b), except that when the voltage becomes more negative than a critical value, the magnitude of the negative current increases very rapidly. An idealized representation of a typical Zener diode is shown in Figure 10.12(a). The circuit shown in

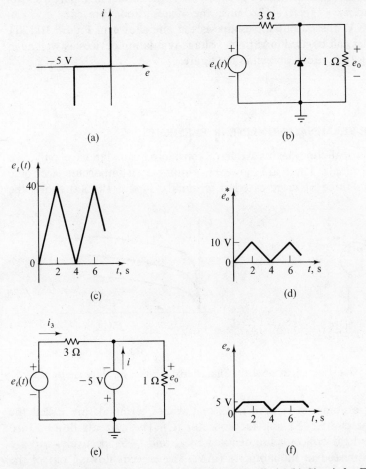

(a)

(b)

(c)

(d)

(e)

(f)

Figure 10.12 (a) Characteristic curve for the Zener diode. (b) Circuit for Example 10.9. (c) Input. (d) Output without the Zener diode. (e) Equivalent circuit for $e_i(t) > 5$ V. (f) Output with the Zener diode.

Figure 10.12(b) contains the Zener diode and has the input voltage shown in Figure 10.12(c). Sketch the output voltage e_o.

Solution If the Zener diode were removed from the circuit, the output voltage would be the variable e_o^* plotted in Figure 10.12(d). Because of the diode orientation, a positive value of e_o corresponds to a negative diode voltage. Thus when e_o reaches 5 V, the diode operates on the left vertical line of the characteristic curve and a large current can flow. Enough additional current passes through the 3-Ω resistor to keep e_o at 5 V, even though $e_i(t)$ increases further. For $e_i(t) > 5$ V, the model in Figure 10.12(e) applies and $i_3 = \frac{1}{3}[e_i(t) - 5]$. Since the Zener diode prevents e_o from exceeding 5 V, the output waveshape is the one shown in Figure 10.12(f). Zener diodes can be used in various voltage-regulating devices as well as in the clipping-type circuit shown in the figure.

10.2 THREE-TERMINAL NONLINEAR ELEMENTS

All the electrical components we have considered thus far have only two external terminals. There are, however, a number of important electronic devices with three or more external terminals. Figure 10.13(a) shows the

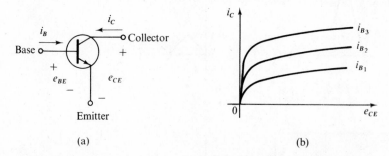

Figure 10.13 Transistor. (a) Symbol. (b) Characteristic curves for three values of i_B.

symbol for a common type of *transistor*, whose terminals are called the base (B), the emitter (E), and the collector (C). The currents flowing into the base and the collector are denoted by i_B and i_C, respectively, and are typically expressed in milliamperes (mA). The current flowing out of the emitter is not shown on the diagram, because it is the sum of i_B and i_C and is thus not independent of the other currents. Two of the three different voltages that exist between pairs of terminals are shown. The voltage between the base and the collector equals the difference between e_{BE} and e_{CE} and is hence not independent of the other voltages.

As might be expected, we can express any two of the four currents and voltages shown in Figure 10.13(a) in terms of the other two. One common representation is to express i_C and e_{BE} in terms of i_B and e_{CE}. We can indicate this by the pair of algebraic equations

$$i_C = f_1(i_B, e_{CE}) \tag{32a}$$

$$e_{BE} = f_2(i_B, e_{CE}) \tag{32b}$$

where f_1 and f_2 are nonlinear functions of the two arguments. A typical graphical representation of (32a) over part of its operating range is shown in Figure 10.13(b). Curves for (32b) are less common since e_{BE} is typically less than some fraction of a volt when e_{CE} is positive. In our examples, we shall assume that $e_{BE} = 0$ in such a case. For the general analysis of a circuit, we must combine the nonlinear relationships for the transistor with the element and interconnection laws for the external network and then solve the model by some convenient numerical or graphical method.

We first consider circuits that contain only transistors, sources, and resistive elements. A graphical analysis similar to that used in Example 10.4 is convenient, as illustrated in the following example. Later in this section, we show how to replace the transistor by a linearized model. This is especially valuable when the circuit also contains capacitors and inductors.

EXAMPLE 10.10   For the transistor shown in Figure 10.14(a), assume that $e_{BE} = 0$ for positive values of e_{CE} and use the collector characteristics shown in Figure 10.14(b). Find the values of i_C when the source current $i_i(t)$ is 0, 0.5 mA, and -0.5 mA. Then sketch the waveform for i_C when $i_i(t)$ is the square wave shown in Figure 10.14(c).

Solution   When $i_i(t) = 0$, i_B is the current through R_1, namely

$$i_B = \frac{1}{R_1} E_1 = 1 \text{ mA}$$

The part of the circuit to the right of the transistor may be represented on the collector characteristics by a load line, which we can determine in a manner similar to that used in Example 10.4. Specifically, if i_C is in milliamperes and e_{CE} in volts,

$$i_C = \frac{1}{R_2}(E_2 - e_{CE}) = 200 - 10e_{CE}$$

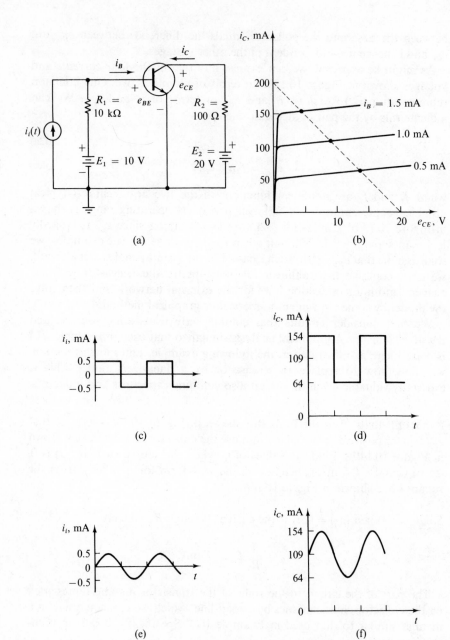

(a)

(b)

(c)

(d)

(e)

(f)

Figure 10.14 (a) Circuit for Example 10.10. (b) Collector characteristics. (c), (d) Square-wave input and the corresponding output. (e), (f) Sinusoidal input and the corresponding output.

which describes the dashed line in Figure 10.14(b). Since the collector current must satisfy this equation and must also correspond to $i_B = 1$ mA, the solution must be given by the intersection of the dashed line and the curve for $i_B = 1$ mA. At this intersection, $i_C = 109$ mA and $e_{CE} = 9$ V.

In general,

$$i_B = \frac{1}{R_1} E_1 + i_i(t)$$

When $i_i(t) = 0.5$ mA, then $i_B = 1.5$ mA, and the solution is given by the intersection of the dashed line with the curve for $i_B = 1.5$ mA, namely

$$i_C = 154 \text{ mA} \quad \text{and} \quad e_{CE} = 4.5 \text{ V}$$

Similarly, when $i_i(t) = -0.5$ mA, then $i_B = 0.5$ mA and the solution, given by the intersection of the dashed line with the curve for $i_B = 0.5$ mA, is

$$i_C = 64 \text{ mA} \quad \text{and} \quad e_{CE} = 13.5 \text{ V}$$

With these results, we see that for the given square-wave input, the output current is the waveform shown in Figure 10.14(d). If we subtract the constant value of 109 mA, corresponding to zero input, from the i_C curve, the result is a square wave that fluctuates between 45 mA and −45 mA. Note that, except for the addition of a constant component, the input signal has been amplified by a factor of 90.

For a sinusoidal input, i_B and i_C would take on a continuous range of values instead of only two values. By extending the previous analysis to the input shown in Figure 10.14(e), we find that the collector current has the waveform shown in Figure 10.14(f).

This example illustrates the use of the transistor as an amplifier. If the i_C vs e_{CE} characteristic curves are a family of equally spaced, parallel straight lines in the region of operation, then the amplification will be the same for all parts of the input signal. In practice, there is a sufficient region where the characteristic curves do approximately satisfy this requirement. The values of E_1 and E_2 are chosen such as to put the operating point near the middle of this linear region. If the curves are not equally spaced straight lines, the amplification will not be constant (e.g., the positive and negative parts of the input will not be amplified equally), and the output waveshape will be a distorted version of the input waveshape.

CIRCUITS WITH CAPACITORS OR INDUCTORS

If the elements connected to a transistor consist of only resistors and sources, then all the equations are algebraic. If there are capacitors or inductors, we can determine an operating point corresponding to the constant component of the input by replacing the capacitors or inductors by open or short circuits, respectively (again resulting in algebraic equations).

If a constant component in the output of an amplifier is undesirable, we can often remove it by adding an additional RC branch. As a preliminary step, consider the steady-state response of the simple circuit shown in Figure 10.15. If $e_i(t)$ is a constant, then in the steady state the capacitor

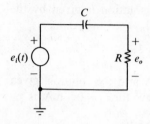

Figure 10.15 RC circuit for which $e_o \simeq e_i(t)$ at high frequencies.

acts like an open circuit, as discussed in Chapter 9. In Problem 10.17, the reader is asked to show that the steady-state response to the input

$$e_i(t) = \sin \omega t \tag{33}$$

is

$$e_o = \frac{\omega RC}{\sqrt{1+(\omega RC)^2}} \sin\left[\omega t + \tan^{-1}\left(\frac{1}{\omega RC}\right) \right] \tag{34}$$

Note that if $\omega RC \gg 1$, $e_o \simeq e_i(t)$. Thus for a rapidly changing signal (that is, for a large value of ω) and a sufficiently large capacitance, we can replace the capacitor by a short circuit.

EXAMPLE 10.11   The circuit shown in Figure 10.16(a) is identical to that shown in Figure 10.14(a) except for the addition of an RC branch at the right. Assume that the transistor characteristics are the same as for Example 10.10, and sketch the steady-state output current i_o for the input current shown in Figure 10.16(b).

Solution   For the constant component of the signals, the capacitor acts like an open circuit. Specifically, when $i_i(t) = 0$, $E_1 = 10$ V, and $E_2 = 20$ V, the collector current i_C is 109 mA, as we found in Example 10.10.

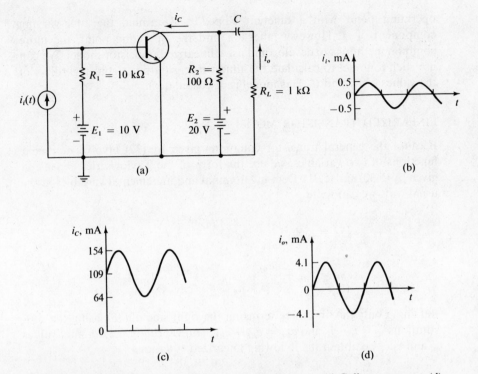

Figure 10.16 (a) Circuit for Example 10.11. (b) Input current. (c) Collector current. (d) Output current.

For the time-varying components of the signals, assume that C is sufficiently large that the capacitor acts like a short circuit. Because $R_L \gg R_2$, the transistor's behavior is not affected significantly by the addition of the RC branch, and the peak-to-peak swing in i_C is still about 90 mA. When we are considering only the time-varying components, we can replace the capacitor and the voltage source E_2 by short circuits. Then, by the current-divider rule, i_o will be about $\frac{1}{11}$ times the time-varying component of i_C, as shown in Figure 10.16(d).

Note that if the value of R_L were not large compared to R_2, the time-varying component of i_C would be affected significantly. If we regarded the capacitor as a short circuit and replaced E_2 by zero, the effective resistance in the collector circuit would be $R_2 R_L/(R_2 + R_L)$.

When R_L is not large compared to R_2, one method is to use the load line shown in Figure 10.14(b) to determine the operating point graphically, just as before. Then we could use a new load line, passing through the operating point with a different slope, to determine the time-varying component of i_C. However, after we find the operating point, the more common method is to develop and use a linearized transistor model. We can use such a model to calculate the time-varying output signal regardless of the values of R_L and C. This model is discussed next.

LINEARIZED TRANSISTOR MODEL

Because the general transistor equations given by (32) involve nonlinear functions of two variables, we use the form of the Taylor-series expansion given by (5.30) and (5.31). Defining nominal and incremental variables in the usual way, we can write

$$
\begin{aligned}
i_C &= \bar{i}_C + \left.\frac{\partial i_C}{\partial i_B}\right|_{\bar{i}_B, \bar{e}_{CE}} \hat{i}_B + \left.\frac{\partial i_C}{\partial e_{CE}}\right|_{\bar{i}_B, \bar{e}_{CE}} \hat{e}_{CE} + \cdots \\
e_{BE} &= \bar{e}_{BE} + \left.\frac{\partial e_{BE}}{\partial i_B}\right|_{\bar{i}_B, \bar{e}_{CE}} \hat{i}_B + \left.\frac{\partial e_{BE}}{\partial e_{CE}}\right|_{\bar{i}_B, \bar{e}_{CE}} \hat{e}_{CE} + \cdots
\end{aligned}
\tag{35}
$$

Retaining only the first three terms on the right side of (35), using the substitutions $i_C = \bar{i}_C + \hat{i}_C$ and $e_{BE} = \bar{e}_{BE} + \hat{e}_{BE}$, and canceling the nominal values $\bar{i}_C$ and $\bar{e}_{BE}$, we obtain the following linearized equations:

$$
\begin{aligned}
\hat{i}_C &= \left.\frac{\partial i_C}{\partial i_B}\right|_{\bar{i}_B, \bar{e}_{CE}} \hat{i}_B + \left.\frac{\partial i_C}{\partial e_{CE}}\right|_{\bar{i}_B, \bar{e}_{CE}} \hat{e}_{CE} \\
\hat{e}_{BE} &= \left.\frac{\partial e_{BE}}{\partial i_B}\right|_{\bar{i}_B, \bar{e}_{CE}} \hat{i}_B + \left.\frac{\partial e_{BE}}{\partial e_{CE}}\right|_{\bar{i}_B, \bar{e}_{CE}} \hat{e}_{CE}
\end{aligned}
\tag{36}
$$

where the partial derivatives are evaluated at $i_B = \bar{i}_B$ and $e_{CE} = \bar{e}_{CE}$ and hence are constants. Because of the units associated with these coefficients, we often use special symbols for them, allowing us to rewrite (36) as

$$
\hat{i}_C = \beta \hat{i}_B + \frac{1}{r_o} \hat{e}_{CE}
\tag{37a}
$$

$$
\hat{e}_{BE} = r_n \hat{i}_B + \mu \hat{e}_{CE}
\tag{37b}
$$

where

$$
\begin{aligned}
\beta &= \left.\frac{\partial i_C}{\partial i_B}\right|_{\bar{i}_B, \bar{e}_{CE}} & \frac{1}{r_o} &= \left.\frac{\partial i_C}{\partial e_{CE}}\right|_{\bar{i}_B, \bar{e}_{CE}} \\
r_n &= \left.\frac{\partial e_{BE}}{\partial i_B}\right|_{\bar{i}_B, \bar{e}_{CE}} & \mu &= \left.\frac{\partial e_{BE}}{\partial e_{CE}}\right|_{\bar{i}_B, \bar{e}_{CE}}
\end{aligned}
\tag{38}
$$

The quantities β and μ are dimensionless, while r_n and r_o have units of ohms. Figure 10.17 illustrates the determination of β and r_o from the collector characteristics. The curves are assumed to be a family of equally spaced, parallel straight lines near the operating point Q. From (38),

$$\beta = \frac{B - C}{2A} \qquad (39)$$

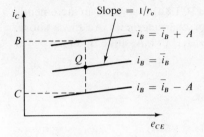

Figure 10.17 Graphical evaluation of β and r_o.

and $1/r_o$ is the slope of the curve for $i_B = \bar{i}_B$. For the curves shown in Figure 10.14(b) with $\bar{i}_B = 1.0$ mA, this graphical method gives $\beta = 100$ and $r_o = 1$ kΩ.

It is possible to construct a circuit diagram relating the incremental voltages and currents by using (37). The equation for $\hat{e}_{BE}$ is the sum of two terms, which suggest two elements in series since the total voltage across a series connection is the sum of the individual voltages. The $r_n i_B$ term suggests the voltage across a resistance r_n, through which the current $\hat{i}_B$ flows. The $\mu \hat{e}_{CE}$ term represents a voltage caused by the voltage $\hat{e}_{CE}$ that appears elsewhere in the circuit. We represent such a term by a *controlled source*, a source that cannot be independently specified but that is controlled by some other voltage or current. Corresponding to (37b), therefore, we have the left side of Figure 10.18.

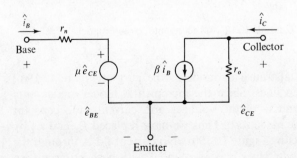

Figure 10.18 Incremental transistor model.

In (37a), the fact that $\hat{i}_C$ is the sum of two terms suggests the presence of two elements in parallel on the collector side of the incremental model. The term $\hat{e}_{CE}/r_o$ describes the current through a resistance r_o, while the term $\beta \hat{i}_B$ is represented by a current-controlled current source. The complete model is shown in Figure 10.18.

EXAMPLE 10.12   For the circuit analyzed in Example 10.10 and shown in Figure 10.14(a), which is repeated in Figure 10.19(a), construct an incremental model and use it to find $\hat{i}_C$. The collector characteristics are those shown in Figure 10.14(b), for which $\beta = 100$ and $r_o = 1$ kΩ. Assume that $r_n = \mu = 0$.

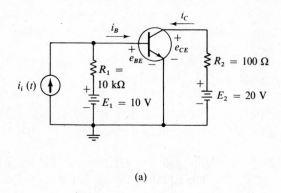

(a)

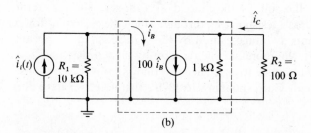

(b)

Figure 10.19 (a) Circuit for Example 10.12. (b) Incremental circuit model.

Solution   The incremental transistor model is the part of Figure 10.19(b) that is within the dashed lines. Since the incremental model relates only the time-varying components of the voltages and currents, the constant voltage sources should be made zero. Thus we have replaced E_1 and E_2 by short circuits in constructing Figure 10.19(b). No current flows through the 10-kΩ resistor in the model since it is in parallel with a short circuit, so

$\hat{\imath}_B = \hat{\imath}_i(t)$. By the current-divider rule,

$$\hat{\imath}_C = \left(\frac{1000}{1000 + 100}\right) 100\,\hat{\imath}_i(t) = 90.91\hat{\imath}_i(t)$$

which agrees closely with the result of Example 10.10. Because of the graphical determination of β and r_o, not more than the first two figures in the answer are meaningful. The incremental model for Example 10.11—that is, for Figure 10.16(a)—would be identical to that for Figure 10.19(b) except that the resistance R_L would be placed in parallel with R_2.

Note that we can divide the general analysis of transistor circuits into two parts. First we establish the operating point by letting voltage and current sources take on their constant values, with the time-varying input signal removed. If there are capacitors and inductors, we replace them by open and short circuits, respectively.

Then we develop the incremental model, replacing the transistor by the circuit shown in Figure 10.18. Because this model involves only the time-varying components of the variables, the constant voltage and current sources are not included. We can regard a sufficiently large capacitor or inductor as a short or open circuit, respectively, in the incremental model. However, in contrast to graphical solutions, this assumption is not necessary. Capacitors and inductors can retain their assigned values in the model, and we can solve the resulting linear RLC circuit by any of the methods we developed in Chapter 9.

10.3 TIME-VARYING AND NONLINEAR CAPACITORS AND INDUCTORS

Additional problems arise in the modeling process when the circuit includes time-varying or nonlinear capacitors and inductors. This section provides a brief introduction to such circuits, including a discussion about the appropriate choice of state variables.

Changes in device parameters can occur as the result of independently applied external means or in response to signals within the system. Moving the plates of a capacitor closer together increases the capacitance, while inserting an iron slug into an air-core inductor increases the inductance. In one type of microphone, the force of sound waves impinging on a diaphragm varies the spacing between plates of a capacitor. In parametric amplifiers, the value of a capacitance is made a periodic function of time. Magnetic amplifiers and various other magnetic devices make use of intentionally exaggerated nonlinearities in inductors.

Recall that the element laws for fixed linear capacitors and inductors are

$$i_C = C\dot{e}_C$$

$$e_L = L\frac{di_L}{dt} \tag{40}$$

Furthermore, e_C and i_L are directly related to the energy stored in the devices and are usually chosen to be state variables. If we can find algebraic equations for i_C and e_L in terms of the state variables and inputs, then the state-variable equations follow directly from (40). For linear elements, $q = Ce_C$ and $\Lambda = Li_L$, where q is the charge on the capacitor and Λ is the flux linkage for the inductor. If we have written the state-variable equations for a linear circuit in terms of e_C and i_L, we can easily write them in terms of q and Λ by using the substitutions $e_C = q/C$ and $i_L = \Lambda/L$.

LINEAR TIME-VARYING ELEMENTS

The charge on a linear time-varying capacitor is directly proportional to the voltage, but the coefficient of proportionality $C(t)$ varies with time. Using the general law $i_C = \dot{q}$, we have

$$i_C = \dot{q} \tag{41a}$$

$$= \frac{d}{dt}[C(t)e_C] \tag{41b}$$

$$= C(t)\dot{e}_C + \dot{C}e_C \tag{41c}$$

The general law for an inductor is $e_L = \dot{\Lambda}$, where Λ denotes the flux linkage. For a linear time-varying inductor, $\Lambda = L(t)i_L$ and

$$e_L = \dot{\Lambda} \tag{42a}$$

$$= \frac{d}{dt}[L(t)i_L] \tag{42b}$$

$$= L(t)\frac{di_L}{dt} + \dot{L}i_L \tag{42c}$$

Note that (41c) and (42c) reduce to (40) for the special case where $C(t)$ and $L(t)$ are constants.

If we choose q and Λ as state variables and if we can express i_C and e_L as algebraic functions of the state variables and inputs, then the state-variable equations follow from (41a) and (42a). On the other hand, if e_C and

i_L are the state variables and if i_C and e_L are expressed in terms of these variables and the inputs, (41c) and (42c) will yield the state-variable equations. We use both sets of state variables in the following example.

EXAMPLE 10.13   The circuit shown in Figure 10.20 contains a linear time-varying inductor and capacitor. Find the state-variable equations in terms of e_C and i_L and also in terms of q and Λ.

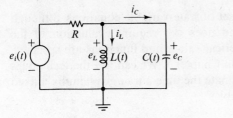

Figure 10.20 Circuit with time-varying elements for Example 10.13.

Solution   We note that

$$i_C = \frac{1}{R}[e_i(t) - e_C] - i_L$$

$$e_L = e_C \tag{43}$$

When the state variables are e_C and i_L, we insert (43) into (41c) and (42c) and solve for the derivatives of the state variables. The resulting state-variable equations are

$$\dot{e}_C = \frac{1}{C(t)}\left[\frac{1}{R}e_i(t) - \left(\dot{C} + \frac{1}{R}\right)e_C - i_L\right] \tag{44}$$

$$\frac{di_L}{dt} = \frac{1}{L(t)}[e_C - \dot{L}i_L]$$

When the state variables are q and Λ, we rewrite (43) as

$$i_C = \frac{1}{R}\left[e_i(t) - \frac{q}{C(t)}\right] - \frac{\Lambda}{L(t)} \tag{45}$$

$$e_L = \frac{q}{C(t)}$$

Inserting (45) into (41a) and (42a) and rearranging give the state-variable equations

$$\dot{q} = \frac{1}{R} e_i(t) - \frac{q}{RC(t)} - \frac{\Lambda}{L(t)}$$

$$\dot{\Lambda} = \frac{q}{C(t)}$$

(46)

Either (44) or (46) is an acceptable set of state-variable equations, although (46) is somewhat more concise and does not require evaluation of the derivatives of $C(t)$ and $L(t)$. If numerical values and functions are specified, we can use the numerical methods in Chapter 7. We can use the techniques of Section 5.4 if we want to approximate the time-varying system by a fixed linear model with incremental variables.

NONLINEAR ELEMENTS

For a nonlinear capacitor, there is an algebraic relationship between the charge q and the voltage e_C, as shown in Figure 10.21(a), but the two quantities are not proportional. For a nonlinear inductor, there is an

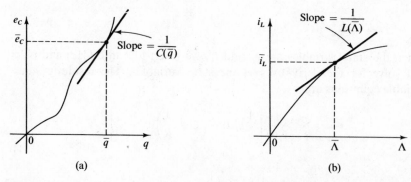

(a)   (b)

Figure 10.21 Typical characteristic curves for nonlinear elements. (a) Capacitor. (b) Inductor.

algebraic relationship between the flux linkage and the current, as shown in Figure 10.21(b), but not a direct proportionality. We shall assume for such nonlinear elements that

$$e_C = f_C(q)$$

$$i_L = f_L(\Lambda)$$

(47)

where f_C and f_L are known single-valued algebraic functions and where the curves pass through their respective origins.*

The general laws in (41a) and (42a) are still valid and

$$i_C = \dot{q}$$
$$e_L = \dot{\Lambda} \tag{48}$$

The usual choice of state variables for circuits with nonlinear capacitors and inductors is the charge on each capacitor and the flux linkage for each inductor. If we can use the element laws and Kirchhoff's laws to express i_C and e_L in terms of the state variables and inputs, then we can write the state-variable equations from (48).

For linear elements, the curves shown in Figure 10.21 would be straight lines with constant slopes of $1/C$ and $1/L$, respectively. For a nonlinear capacitor, we can define the capacitance corresponding to the point $(\bar{q}, \bar{e}_C)$ as the reciprocal of the slope of the tangent to the characteristic curve at that point. Because the value of the capacitance is a function of the charge, it is written $C(q)$. In a similar way, we can define the inductance of a nonlinear inductor as the reciprocal of the slope of the characteristic curve shown in Figure 10.21(b). Because it is a function of the flux linkage, it is written $L(\Lambda)$, and for the particular point $(\bar{\Lambda}, \bar{i}_L)$, it is $L(\bar{\Lambda})$. In the following example, we will see how $C(\bar{q})$ and $L(\bar{\Lambda})$ become the capacitance and inductance of elements in a linearized model that is valid about an operating point corresponding to $\bar{q}$ and $\bar{\Lambda}$.

EXAMPLE 10.14 Find a set of nonlinear state-variable equations for the circuit shown in Figure 10.22, which contains a nonlinear capacitor and a nonlinear inductor that are described by relationships of the form of (47). Find expressions for the operating point corresponding to $\bar{e}_i$, and derive a linearized model valid in the vicinity of the operating point.

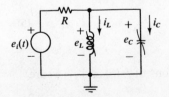

Figure 10.22 Circuit with nonlinear inductor and capacitor for Example 10.14.

* If the inductor exhibits hysteresis effects, the function f_L is not single-valued and the analysis becomes more difficult.

Solution From Kirchhoff's laws and the element laws,

$$i_C = \frac{1}{R}[e_i(t) - e_C] - i_L$$

$$= \frac{1}{R}[e_i(t) - f_C(q)] - f_L(\Lambda)$$

and

$$e_L = e_C = f_C(q)$$

Substituting these equations into (48) gives

$$\dot{q} = \frac{1}{R}[e_i(t) - f_C(q)] - f_L(\Lambda)$$

$$\dot{\Lambda} = f_C(q) \tag{49}$$

which are the nonlinear state-variable equations.

The derivatives of the state variables are zero at the operating point, so

$$f_C(\bar{q}) = 0 \tag{50a}$$

$$f_L(\bar{\Lambda}) = \frac{1}{R}[\bar{e}_i - f_C(\bar{q})] = \frac{1}{R}\bar{e}_i \tag{50b}$$

Since the nonlinear characteristic curves are assumed to pass through the origin, we see from (50a) that there is no voltage across the inductor-capacitor combination. From (50b), the current through the inductor is $\bar{e}_i/R$. This describes the circuit when the capacitor and inductor are replaced by open and short circuits, respectively, as should be expected at the operating point.

We define the incremental variables $\hat{q}$, $\hat{\Lambda}$, and $\hat{e}_i(t)$ by the equations

$$q = \bar{q} + \hat{q}$$

$$\Lambda = \bar{\Lambda} + \hat{\Lambda} \tag{51}$$

$$e_i(t) = \bar{e}_i + \hat{e}_i(t)$$

The first two terms in the Taylor-series expansions of $f_C(q)$ and $f_L(\Lambda)$ are

$$f_C(\bar{q}) + \left.\frac{df_C}{dq}\right|_{\bar{q}} \hat{q}$$

$$f_L(\bar{\Lambda}) + \left.\frac{df_L}{d\Lambda}\right|_{\bar{\Lambda}} \hat{\Lambda}$$

which can be written as

$$f_C(\bar{q}) + \frac{1}{C(\bar{q})}\hat{q}$$

$$f_L(\bar{\Lambda}) + \frac{1}{L(\bar{\Lambda})}\hat{\Lambda} \tag{52}$$

where $C(\bar{q})$ and $L(\bar{\Lambda})$ are the reciprocals of the slopes of curves such as those shown in Figure 10.21. Substituting (51), replacing the nonlinear terms by (52), and noting that $\dot{\bar{q}} = \dot{\bar{\Lambda}} = 0$, we can write (49) as

$$\dot{\hat{q}} = \frac{1}{R}\left[\bar{e}_i + \hat{e}_i(t) - f_C(\bar{q}) - \frac{1}{C(\bar{q})}\hat{q}\right] - f_L(\bar{\Lambda}) - \frac{1}{L(\bar{\Lambda})}\hat{\Lambda}$$

$$\dot{\hat{\Lambda}} = f_C(\bar{q}) + \frac{1}{C(\bar{q})}\hat{q}$$

Using (50) to cancel the constant terms in these equations gives the set of linearized state-variable equations

$$\dot{\hat{q}} = \frac{1}{R}\left[\hat{e}_i(t) - \frac{1}{C(\bar{q})}\hat{q}\right] - \frac{1}{L(\bar{\Lambda})}\hat{\Lambda} \tag{53}$$

$$\dot{\hat{\Lambda}} = \frac{1}{C(\bar{q})}\hat{q}$$

In these equations, $L(\bar{\Lambda})$ is a constant whose value depends on the operating point and hence on $\bar{e}_i$. Because of (50a), $\bar{q}$ is zero and we can replace the coefficient $C(\bar{q})$ by $C(0)$.

If we wish, we can rewrite (53) in terms of the incremental variables $\hat{e}_C$ and $\hat{\imath}_L$. From Figure 10.21, $\hat{q} = C(\bar{q})\hat{e}_C$ and $\hat{\Lambda} = L(\bar{\Lambda})\hat{\imath}_L$. Substituting these expressions into (53) and remembering that $C(\bar{q})$ and $L(\bar{\Lambda})$ are constants, we have the alternative set of linearized state-variable equations

$$\dot{\hat{e}}_C = \frac{1}{RC(\bar{q})}[\hat{e}_i(t) - \hat{e}_C] - \hat{\imath}_L$$

$$\frac{d\hat{\imath}_L}{dt} = \frac{1}{L(\bar{\Lambda})}\hat{e}_C$$

PROBLEMS

10.1 For the circuit shown in Figure P10.1, the nonlinear resistor obeys the relationship $e_2 = 2|i_2|i_2$.

　　a. Write the nonlinear input-output differential equation.

　　b. Find the operating point when $e_i(t) = 18 + \hat{e}_i(t)$, and draw the equivalent circuit corresponding to it.

　　c. Find the linearized element law relating $\hat{e}_2$ and $\hat{i}_2$ that is valid in the vicinity of the operating point you found in part b.

　　d. Find the linearized differential equation relating $\hat{e}_o$ and $\hat{e}_i(t)$.

　　e. Evaluate the time constant of the linearized model.

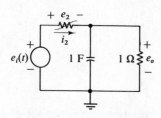

Figure P10.1

10.2 For the circuit shown in Figure P10.2, $e_1(t) = [15 + \cos 2t]\, U(t)$ and $e_2 = 3$ V for all t. The nonlinear resistor obeys the element law $e_o = 2i^3$.

　　a. Write the differential equation obeyed by $i(t)$.

　　b. Find $e_o(0+)$, assuming that steady-state conditions exist at $t = 0-$.

　　c. Find the operating-point values i and $\bar{e}_o$ for $t > 0$, when $\bar{e}_1 = 15$ V.

　　d. Derive the linearized model that is valid about the operating point you found in part c.

　　e. Evaluate the time constant of the linearized model.

　　f. Find $\hat{i}(0+)$ and $\hat{e}_o(0+)$.

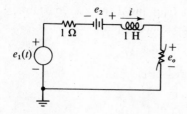

Figure P10.2

10.3 The nonlinear resistor in the circuit shown in Figure P10.3 obeys the element law $e_o = 3|i_L|i_L$.

   a. Write the nonlinear state-variable equation.

   b. Derive the linearized model when $e_i(t) = 5 + 0.4 \cos t$ for $t > 0$.

   c. Find the time constant of the linearized model.

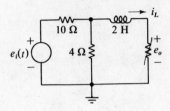

Figure P10.3

10.4 For the circuit shown in Figure P10.4, the element law for the nonlinear resistor is $i_2 = 2|e_C|e_C$.

   a. Write the nonlinear state-variable equations.

   b. Determine the values of $\bar{e}_C$ and $\bar{i}_L$ when $e_i(t) = 14 + \hat{e}_i(t)$.

   c. Derive the linearized state-variable equations that are valid about the operating point you found in part b.

   d. Find $\hat{e}_C(0+)$ and $\hat{i}_L(0+)$ when $e_C(0-) = 3$ V and $i_L(0-) = 1$ A.

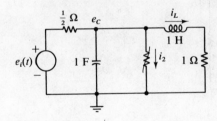

Figure P10.4

10.5 For the circuit shown in Figure P10.5, $e_2 = 0.5i_L^3$. The input voltage $e_i(t)$ is 4 V for $t \le 0$ and is $4 + \hat{e}_i(t)$ for $t > 0$.

   a. Write the nonlinear state-variable equations and an algebraic output equation for e_o.

   b. Find $e_C(0+)$ and $i_L(0+)$. Assume that steady-state conditions exist at $t = 0-$.

c. Find the values of the state variables at the operating point, and derive a set of linearized state-variable equations.

d. Find $\hat{e}_C(0+)$, $\hat{\imath}_L(0+)$, and $\hat{e}_o(0+)$.

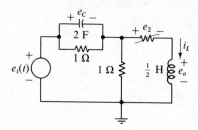

Figure P10.5

10.6 The element law for the nonlinear resistor in the circuit shown in Figure P10.6 is $i_o = |e_o|e_o$. The source voltage is $e_i(t) = \bar{e}_i + \hat{e}_i(t)$.

a. Find the operating-point values $\bar{e}_o$, $\bar{e}_C$, and i_L when $\bar{e}_i = 5.5$ V.

b. Find the linearized state-variable model that is valid about the operating point found in part a. Take $\hat{e}_C$ and $\hat{\imath}_L$ as the state variables and $\hat{e}_o$ as the output.

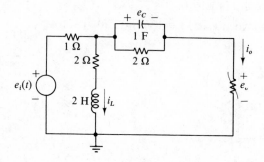

Figure P10.6

10.7 Sketch e_o and e_R for Figure 10.9(a) when the orientation of the diode is reversed.

10.8 For the partial circuit shown in Figure P10.8(a), $i = (e - A)/R$. Thus the plot of i vs e is a straight line with a slope of $1/R$ and a vertical intercept of $-A/R$.

a. Sketch i vs e for the partial circuit shown in Figure P10.8(b), assuming that the diode has the ideal characteristic shown in Figure 10.8(b).

b. Repeat part a when the diode orientation is reversed.

c. Sketch the voltage across the resistor in Figure P10.8(b), if $e(t) = 10 \sin \omega t$ volts and if $A = 5$ V.

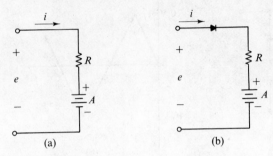

(a)   (b)

Figure P10.8

10.9 a. Sketch i vs e for the partial circuit shown in Figure P10.9(a), assuming that the diode has the ideal characteristic shown in Figure 10.8(b).

b. Repeat part a for the partial circuit shown in Figure P10.9(b).

c. Repeat part b when the diode orientation is reversed.

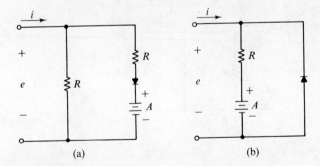

(a)   (b)

Figure P10.9

10.10 a. Sketch a curve of e_o vs $e_i(t)$ for the circuit shown in Figure P10.10(a). Assume that the diodes are ideal.

b. Sketch e_o vs t for the input shown in Figure P10.10(b) when $A = 5$ V.

c. Repeat part b if a 2-Ω resistor is added in series with each diode. Assume that $R = 10$ Ω.

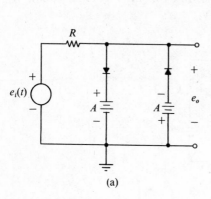

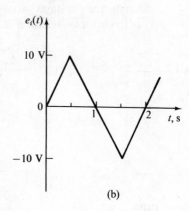

(a)

(b)

Figure P10.10

10.11 a. Sketch i vs e for the partial circuit shown in Figure P10.11(a). Assume that the diodes are ideal.

b. Repeat part a for the partial circuit shown in Figure P10.11(b).

c. Repeat part b when the diode orientation is reversed.

d. Repeat part b when a resistance R is added in series with the diode.

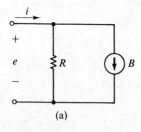

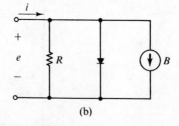

(a)

(b)

Figure P10.11

10.12 a. For the circuit shown in Figure P10.12(a), find e_o when $e_i(t) = A$, where A is positive and the diodes are ideal. Under these conditions, the right diode will be a short circuit and the left diode an open circuit.

b. Find e_o when $e_i(t) = -A$, where $A > 0$.

c. Sketch e_o vs e_i.

d. Sketch e_o vs t when $e_i(t)$ is the input shown in Figure P10.12(b). Assume that $R_1 = 1 \ \Omega$ and $R_2 = 4 \ \Omega$.

e. Repeat part d when $R_2 \gg R_1$.

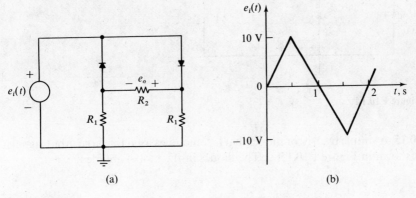

(a)   (b)

Figure P10.12

10.13 a. Sketch e_o vs $e_i(t)$ for the circuit shown in Figure P10.13(a). Assume that the diodes are ideal. Sketch e_o vs t when $e_i(t)$ is the input shown in Figure P10.12(b) and when $R_2 \gg R_1$.

b. Sketch e_o vs e_i for the circuit shown in Figure P10.13(b). Sketch e_o vs t when $e_i(t)$ is the input shown in Figure P10.12(b).

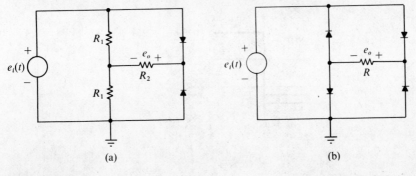

(a)   (b)

Figure P10.13

10.14 The switch in the circuit shown in Figure P10.14 is open for $t < 0$ and closed for $t > 0$. The source current $i_i(t)$ is 2 A for all values of t. If $e_C(0-) = 8$ V, find and sketch e_C vs t.

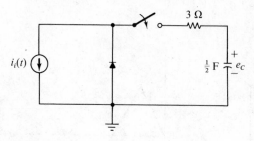

Figure P10.14

10.15 A simple transistor model that is sometimes used for large input signals is shown in Figure P10.15(a). The diodes in the model are ideal.

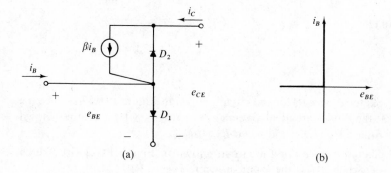

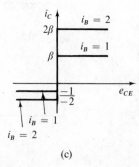

Figure P10.15

a. Show that the assumed model results in the i_B vs e_{BE} characteristic shown in Figure P10.15(b).

b. Show that the model results in the i_c vs e_{CE} characteristics shown in Figure P10.15(c). Indicate which of the lines comprising the collector characteristics correspond to the following cases: (1) only diode D_1 acting like a short circuit, (2) only diode D_2 acting like a short circuit, and (3) both diodes acting like short circuits.

10.16 Replace the transistor in the circuit shown in Figure 10.14(a) by the large-signal model shown in Figure P10.15(a), with $\beta = 100$.

a. Sketch a curve of i_C vs i_i for -2 mA $\le i_i \le 2$ mA.

b. Sketch e_{CE} vs i_i for -2 mA $\le i_i \le 2$ mA.

c. Use the result of part a to sketch i_C vs t when $i_i(t)$ is the input shown in Figure 10.14(c).

d. Use the result of part a to sketch i_C when $i_i(t)$ is four times the input shown in Figure 10.14(c).

10.17 Verify (34) by first finding the input-output differential equation for the circuit shown in Figure 10.15. Then assume that the particular solution for the input $e_i(t) = \sin \omega t$ has the form $y_P(t) = A \sin \omega t + B \cos \omega t$, and find A and B by the method used in Example 8.1. Use Table 8.1 to convert the expression to the form given in (34).

10.18 The circuit shown within the dashed lines in Figure P10.18 is the incremental transistor model shown in Figure 10.18 with $r_n = 100 \ \Omega$, $\mu = 0$, $\beta = 100$, and r_o infinite (an open circuit).

a. Find $\hat{i}_o$ when R_f is infinite.

b. Find $\hat{i}_o$ when $R_f = 1$ kΩ. (*Hint*: Use the node-equation method and write current-law equations at nodes B and C, with $\hat{i}_B = \hat{e}_B/100$.)

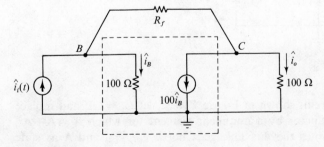

Figure P10.18

10.19 a. Write a set of state-variable equations and an algebraic output equation for the circuit shown in Figure 9.29 when the state variables are chosen to be the charge q for the capacitor and the flux linkage Λ for the inductor.

b. Repeat part a when the fixed capacitor is replaced by a time-varying linear capacitor for which $C(t) = 2 + \epsilon^t$ farads.

10.20 Repeat Problem 10.19 for the circuit shown in Figure P9.30(b).

10.21 a. Write the state-variable equations for the circuit shown in Figure P9.21 when the $\frac{1}{2}$-F capacitor is replaced by one for which $C(t) = 2 + \sin 3t$ farads. Choose the inductor current and the capacitor voltage as state variables.

b. Repeat part a when the state variables are chosen to be the inductor current and the capacitor charge.

c. Repeat part a when the state variables are the capacitor charge and the inductor flux linkage.

10.22 Find a set of state-variable equations for the circuit shown in Figure P10.22 for the conditions listed in parts a and b. Define your choice of state variables and their positive senses.

a. $L(t) = te^{-t}$ henries and $C_1 = 1$ F.

b. $L = 1$ H and $C_1(t) = te^{-t}$ farads.

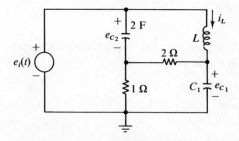

Figure P10.22

10.23 a. For the circuit shown in Figure P10.22, let $C_1 = 1$ F and replace the linear inductor by a nonlinear element for which $i_L = \Lambda + \Lambda^3$, where Λ denotes the flux linkage. Choose e_{C_1}, e_{C_2}, and Λ as state variables and find the state-variable equations.

b. Find the operating point corresponding to $\bar{e}_i = 30$ V.

c Derive a linearized model that is valid about the operating point you found in part b.

10.24 For the circuit shown in Figure P10.24, choose as state variables the capacitor charge q and the inductor flux linkage Λ. Find the state-variable equations and an output equation for e_o for the following conditions.

    a. $L = 2$ H and C is a nonlinear capacitor for which $e_c = \sqrt{q}$.

    b. $C = 0.5$ F and L is a nonlinear inductor for which $i_L = \Lambda^3$.

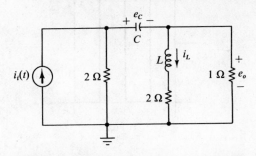

Figure P10.24

10.25 The element law for the nonlinear resistor in the circuit shown in Figure P10.25(a) is $e_A = 3|i_A|i_A$.

    a. Write the nonlinear state-variable equation when the state variable is i_L. Also write an output equation for e_o.

    b. Find the operating point corresponding to $\bar{e}_i = 9$ V, and derive the linearized model that is valid in the vicinity of the operating point.

    c. Prepare a computer simulation to calculate e_o for $t > 0$ for both the nonlinear and the linearized models. Use the input voltage shown in Figure P10.25(b), which is 9 V for $t \le 0$ and which is a square wave with a period of T_p for $t > 0$.

    d. Select a suitable step size and run the program long enough to determine the important features of the output. You should run the program for $T_p = 0.5$ and 2.0 s. For both of these cases, the output should include plots of e_o vs t for the nonlinear model and plots of $\hat{e}_o + \bar{e}_o$ vs t for the linear model. Obtain both plots on the same axes and comment on the accuracy of the response of the linearized model.

    e. Repeat part d for the input shown in Figure P10.25(c).

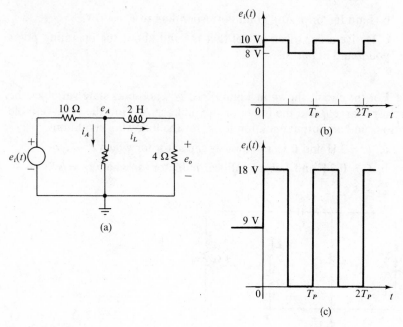

Figure P10.25

10.26 Repeat Problem 10.25 using the inputs shown in Figure P10.26. Because $e_i(t) = 0$ for $t < 0$, use $\bar{e}_i = 0$.

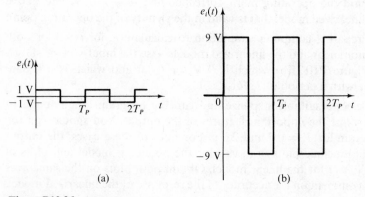

Figure P10.26

10.27 a. With $e_2 = 0.5i_L^3$ for the nonlinear resistor in the circuit shown in Figure P10.5, prepare a digital simulation that will solve the nonlinear and linearized state-variable equations and calculate e_o when the input, initial conditions, and operating point are given.

b. Use the program to determine the response for $t > 0$ of the linear and nonlinear models when $e_i(t) = 4 + [\sin 2t] U(t)$.

c. Repeat part b when $e_i(t) = [4 + \sin 2t] U(t)$.

d. Repeat part b when $e_i(t)$ is the input voltage shown in Figure P10.25(b).

e. Repeat part b when $e_i(t) = U(t) - U(t - 0.5)$.

CHAPTER 11
ELECTROMECHANICAL SYSTEMS

We can construct a wide variety of very useful devices by combining electrical and mechanical elements. Such electromechanical systems include potentiometers, the galvanometer, the microphone and loudspeaker, and motors and generators.

This chapter is organized according to the type of electrical element involved in the coupling. We shall consider couplings that are achieved by mechanically varying a resistance, by moving a current-carrying conductor in a magnetic field, or by varying a capacitance that has an electrical field between the plates. Additional types of electromechanical coupling exist, but we shall not consider them here. Variable resistors can be studied without any new element or interconnection laws. However, we shall have to introduce new laws in order to model systems coupled by magnetic and electrical fields.

11.1 RESISTIVE COUPLING

We can control a variable resistance by mechanical motion either continuously by moving an electrical contact or discretely by opening and closing a switch. Since resistors cannot store energy, this method of coupling electrical and mechanical parts of a system does not involve mechanical forces that depend on electrical variables, in contrast to coupling by magnetic and electrical fields.

Figure 11.1(a) shows a strip of conducting material having resistivity ρ and cross-sectional area A. One terminal is fixed to the left end of the

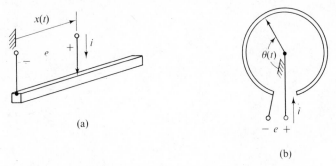

(a)

(b)

Figure 11.1 Variable resistors. (a) Translational. (b) Rotational.

conductor. The other terminal, known as the *wiper*, is free to slide along the bar while maintaining a good electrical connection at all times, i.e., zero resistance at the contact point. Since the resistance per unit length of the bar is ρ/A, the resistance between the two terminals is

$$R = \left[\frac{\rho}{A}\right] x(t) \tag{1}$$

where $x(t)$, the displacement of the wiper, can vary with time. The rotational equivalent of this device is shown in Figure 11.1(b). Here the resistance between the two terminals is a function of the angle $\theta(t)$, which defines the angular orientation of the wiper with respect to the fixed terminal.

A very useful device known as a *potentiometer* is obtained by adding a third terminal to the right ends of the variable resistors shown in Figure 11.1. For the translational potentiometer shown in Figure 11.2(a), the two end terminals are normally connected across a voltage source, and the voltage at the wiper is considered the output. The resistances R_1 and R_2 depend on the wiper position and are therefore labeled $R_1(x)$ and $R_2(x)$ in the figure. In our mathematical development, however, we shall omit the x in

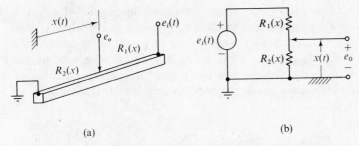

(a)   (b)

Figure 11.2 (a) Translational potentiometer. (b) Equivalent circuit.

parentheses. We can regard either the position $x(t)$ or the voltage source $e_i(t)$ or both as inputs.

The circuit diagram for the potentiometer is shown in Figure 11.2(b). Provided that no current flows through the wiper, we can apply the voltage-divider rule of (9.27) to obtain

$$e_o = \left[\frac{R_2}{R_1 + R_2}\right] e_i(t) \qquad (2)$$

If the distance and the resistance between the fixed terminals are denoted by x_{max} and R_T, respectively, then $R_2 = [R_T/x_{max}] x(t)$ and $R_1 + R_2 = R_T$. Substituting these two expressions into (2), we can write the output voltage as

$$e_o = \left[\frac{1}{x_{max}}\right] x(t) e_i(t) \qquad (3)$$

where the ratio $x(t)/x_{max}$ must be between 0 and 1, inclusive.

Thus the output voltage is the input voltage multiplied by a gain of $x(t)/x_{max}$, which is dependent on the mechanical variable $x(t)$. A simulation diagram for this interpretation is shown in Figure 11.3(a). Alternatively, we can interpret (3) as saying that the output voltage is proportional to the product of the input voltage $e_i(t)$ and the mechanical variable $x(t)$, as indicated in Figure 11.3(b).

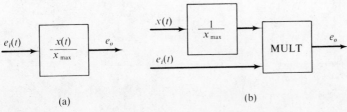

(a)   (b)

Figure 11.3 Simulation diagrams for a translational potentiometer.

EXAMPLE 11.1   The rotary potentiometer shown in Figure 11.4 has a constant voltage E applied across its fixed terminals, and the wiper is connected to a constant resistance R_o, which might represent the resistance of a voltmeter or a recording device. Find the relationship between the output voltage e_o and the angular orientation of the mechanical rotor attached to the wiper. Assume that the resistance per unit length of the potentiometer is constant.

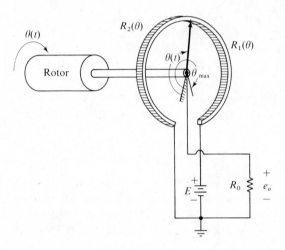

Figure 11.4 Potentiometer used to measure the angular orientation of a rotor.

Solution   The wiper contact divides the total potentiometer resistance R_T into R_1 and R_2, such that in the circuit shown in Figure 11.5(a),

$$R_2 = \left[\frac{R_T}{\theta_{max}}\right]\theta(t)$$

$$R_1 + R_2 = R_T \tag{4}$$

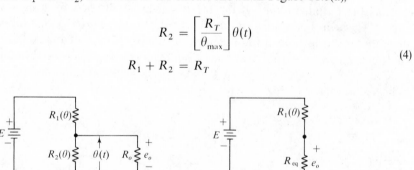

(a)   (b)

Figure 11.5 Equivalent circuits for Example 11.1.

The values of R_1 and R_2 depend on $\theta(t)$, as we emphasize by labeling them $R_1(\theta)$ and $R_2(\theta)$ in the figure. Since R_o and R_2 are connected in parallel, they can be replaced by a single equivalent resistance $R_{eq} = R_o R_2/(R_o + R_2)$, giving the circuit shown in Figure 11.5(b).

By the voltage-divider rule, the output voltage is

$$
\begin{aligned}
e_o &= \left(\frac{R_{eq}}{R_{eq} + R_1}\right) E \\
&= \left(\frac{R_o R_2}{R_o R_2 + R_o R_1 + R_1 R_2}\right) E \\
&= \left(\frac{R_o R_2}{R_o R_T + R_1 R_2}\right) E \\
&= \left[\frac{1}{1 + \left(\dfrac{R_T}{R_o}\right)\left(\dfrac{R_1}{R_T}\right)\left(\dfrac{R_2}{R_T}\right)}\right] \left(\frac{R_2}{R_T}\right) E
\end{aligned}
\tag{5}
$$

Substituting (4) into (5) gives

$$
e_o = \left(\frac{E}{\theta_{max}}\right) \left[\frac{1}{1 + \left(\dfrac{R_T}{R_o}\right)\left[1 - \dfrac{\theta(t)}{\theta_{max}}\right]\left[\dfrac{\theta(t)}{\theta_{max}}\right]}\right] \theta(t)
\tag{6}
$$

which is the desired result.

Unfortunately, (6) is a nonlinear relationship between the output voltage and the rotor angle because of the presence of $\theta(t)$ in the coefficient of $\theta(t)$. Although we could still use the potentiometer to measure $\theta(t)$, the voltmeter or recording device would require a nonlinear scale, which would be inconvenient.

However, if the output resistance R_o is large compared to the total potentiometer resistance R_T, then the coefficient in the brackets is approximately unity since $0 \le \theta(t)/\theta_{max} \le 1$. Then (6) reduces to the linear relationship

$$
e_o = \left[\frac{E}{\theta_{max}}\right] \theta(t) \qquad \text{for } R_o \gg R_T
\tag{7}
$$

Equation (7) indicates that in a simulation diagram we can treat the potentiometer as a gain of E/θ_{max} volts per radian with $\theta(t)$ as its input and e_o as its output. This simplification is equivalent to saying that the current through R_o is negligible compared to that through R_2, i.e., that the parallel combination of R_o and R_2 has a resistance equal to R_2. When R_o is not sufficiently large to allow us to use (7), we say that the resistance R_o *loads* the potentiometer, and we should use the nonlinear expression in (6).

11.2 COUPLING BY A MAGNETIC FIELD

A great variety of electromechanical devices contain current-carrying wires that can move within a magnetic field.* The physical laws governing this type of electromechanical coupling are given in introductory physics textbooks such as Halliday and Resnick. They state that (1) a wire in a magnetic field that carries a current will have a force exerted on it, and (2) a voltage will be induced in a wire that moves relative to the magnetic field. The variables we need to model such devices are

f_e,   the force on the conductor in newtons (N)

v,   the velocity of the conductor in meters per second (m/s)

ϕ,   the flux in webers (Wb)

$\mathscr{B}$,   the flux density of the magnetic field in webers per square meter (Wb/m^2)

i,   the current in the conductor in amperes (A)

e_m,   the voltage induced in the conductor in volts (V)

The force on a conductor of differential length $d\ell$ carrying a current i in a magnetic field of flux density $\mathscr{B}$ is

$$d\mathbf{f_e} = i(d\ell \times \mathscr{B}) \tag{8}$$

where the boldface type indicates vectors that can have any spacial orientation. The cross in (8) represents the vector cross product. To obtain the total electrically induced force $\mathbf{f_e}$, we must integrate (8) along the length of the conductor.

In our applications, the wires will be either straight conductors that are perpendicular to a unidirectional magnetic field or circular conductors in a radial magnetic field. In either case, the differential length $d\ell$ will be perpendicular to a uniform flux density $\mathscr{B}$. Then (8) simplifies to the scalar

---

* A number of useful devices achieve electromechanical coupling by having a mechanical displacement vary the characteristics of the flux path, as in a solenoid. We shall not consider such devices, but the interested reader may consult Fitzgerald and Kingsley.

relationship

$$f_e = \mathcal{B}\ell i \tag{9}$$

where the direction of the force is perpendicular to both the wire and the magnetic field and can be found by the following rule. If the bent fingers of the right hand are pointed from the positive direction of the current toward the positive direction of the magnetic field (through the right angle), the thumb will point in the positive direction of the force. Figure 11.6(a) shows that for a positive current into the page and a positive flux to the left, the force on a straight conductor is upward. The position of the right hand corresponding to this situation is shown in Figure 11.6(b).

The voltage induced in a conductor of differential length $d\ell$ moving with velocity **v** in a field of flux density $\mathcal{B}$ is

$$de_m = (\mathbf{v} \times \mathcal{B}) \cdot d\ell \tag{10}$$

where the dot denotes the scalar product, or dot product, used with vector notation. We obtain the total induced voltage for a conductor by integrating (10) between the ends of the conductor.

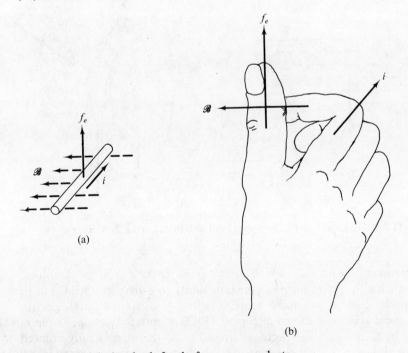

(a)

(b)

Figure 11.6 Right-hand rule for the force on a conductor.

In our applications, the three vectors in (10) are mutually perpendicular, and integrating (10) yields the scalar relationship

$$e_m = \mathscr{B}\ell v \tag{11}$$

If the bent fingers of the right hand are pointed from the positive direction of the velocity toward the positive direction of the magnetic field (through the right angle), the thumb will point in the direction in which the current caused by the induced voltage tends to flow. In Figure 11.7(a), a straight conductor is shown moving downward in a magnetic field directed to the left. As indicated by the polarity signs and by the sketch shown in Figure 11.7(b), the positive sense of the induced voltage is into the page. If the conductor were part of a complete circuit with no external sources, the current in the conductor would be into the page. According to Figure 11.6, this current would cause an upward force to be exerted on the conductor which would oppose the downward motion.

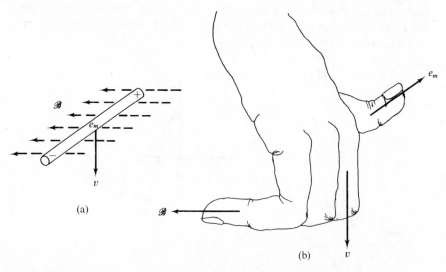

Figure 11.7 Right-hand rule for the voltage induced in a moving conductor.

Equations (9) and (11), which describe the force and induced voltage associated with a wire moving perpendicularly to a magnetic field, can be incorporated in a schematic representation of a translational electro-mechanical system as shown in Figure 11.8. The induced voltage is repre-sented by a source in the electrical circuit, while the magnetically induced force is shown acting on the mass M to which the conductor is attached.

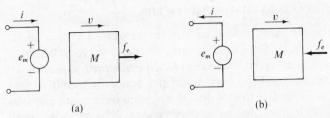

Figure 11.8 Representations of a translational electromechanical system. (a) Electrical to mechanical power flow. (b) Mechanical to electrical power flow.

We can determine the proper polarity marks for e_m and the reference direction for f_e by examining the specific device under consideration, but the figure indicates two combinations of reference directions that are consistent with the conservation of energy.

In Figure 11.8(a), the polarity of the electrical source is such that power is absorbed from the remainder of the circuit when both e_m and i are positive. Likewise, when f_e and v are positive, there is a transfer of power into the mechanical part of the system. Conversely, for Figure 11.8(b), power flows from the mechanical side to the electrical side when all four variables are positive.

It is instructive to evaluate the power involved in the coupling mechanism. The external power delivered to the electrical part of Figure 11.8(a) is

$$p_e = e_m i = (\mathcal{B}\ell v)i$$

while the power available to whatever mechanical elements are attached to the coil is

$$p_m = f_e v = (\mathcal{B}\ell i)v$$

Hence,

$$p_m = p_e$$

which says that any power delivered to the coupling mechanism in electrical form will be passed on undiminished to the mechanical portion. Of course, any practical system has losses resulting from the resistance of the conductor and the friction between the moving mechanical elements. However, any such dissipative elements can be modeled separately by a resistor in the electrical circuit or a viscous-friction element acting on the mass. In a similar way, you can demonstrate that for the coupling shown in Figure 11.8(b), the mechanical power supplied by the force f_e is transmitted to whatever electrical elements are connected across e_m.

11.3 DEVICES COUPLED BY MAGNETIC FIELDS

Having introduced the basic laws that govern the behavior of a single conductor in a magnetic field, we shall describe several of the most common types of electromechanical systems and derive their mathematical models. In each case, we consider idealized versions of the device that omit certain aspects that may be important from a design viewpoint but are not essential to understanding their operation as part of an overall system. We shall examine in turn the galvanometer, a microphone, and a motor.

THE GALVANOMETER

The galvanometer is a device that produces an angular deflection dependent on the current passing through a coil attached to a pointer. It is widely used in electrical measurement devices. As shown in Figure 11.9(a), a permanent magnet supplies a radial magnetic field and the flux passes through a stationary iron cylinder between the poles of the magnet. A coil of wire whose terminals can be connected to an external circuit is suspended

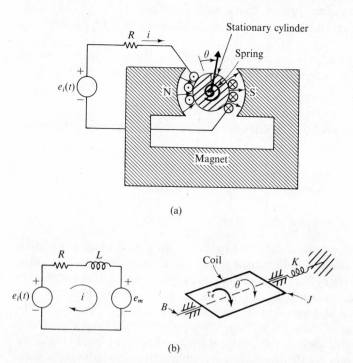

(a)

(b)

Figure 11.9 Galvanometer. (a) Physical device. (b) Diagram used for analysis.

by bearings so that it can rotate about a horizontal axis passing through the center of the cylinder. A torsional spring mounted on its axis restrains the coil.

The magnet provides a uniform flux density $\mathscr{B}$ within the air gaps between it and the iron cylinder directed from the north to the south pole. The moment of inertia of the coil is J, and the combination of bearing friction and damping due to the air is represented by the viscous-damping coefficient B. The torsional spring has the rotational spring constant K. It is assumed that the electrical connection between the external circuit and the movable coils is made in such a way that the connection exerts no torque on the coil.

The coil consists of N rectangular turns, each of which has a radius of a and a length of ℓ along the direction of the axis of rotation, and it has a total inductance L. The dots in the wires on the left side of the coil and the crosses on the right side indicate that when i is positive, the current flows out of the page in the left conductors and into the page in the right conductors. The remainder of the circuit consists of a voltage source $e_i(t)$ and a resistance R that accounts for any resistance external to the galvanometer as well as the resistance of the coils.

For purposes of analysis, we represent the idealized system by the circuit and mechanical diagrams shown in Figure 11.9(b). We have used the rotational equivalent of Figure 11.8(a) rather than Figure 11.8(b) to represent the electromechanical coupling, because the purpose of the device is to convert an electrical variable (current) into a mechanical variable (angular displacement). It remains to use (9) and (11), with the appropriate right-hand rules, to determine the expressions for e_m and τ_e, the electrically induced torque.

Assume that the rotation of the coil is sufficiently small that all the conductors remain in the region of constant flux density. Then we can obtain the torque acting about the axis of the coil by summing the torques on the $2N$ individual conductors of length ℓ. Because the magnetic field is confined to the iron cylinder as it passes between the air gaps, there is no contribution to the torque from the ends of the coil outside the gaps. If the current is positive, each conductor on the left side of the coil has a force of $f_e = \mathscr{B}\ell i$ acting upward, while the conductors on the right side have forces of the same magnitude acting downward. Because the arrow denoting the electrically induced torque τ_e was taken as clockwise in Figure 11.9(b), the expression for the torque is

$$\tau_e = (2N\mathscr{B}\ell a)i \tag{12}$$

Next we express the voltage induced in the coil in terms of the angular velocity $\dot{\theta}$. Since there are $2N$ conductors in series, which move with the velocity $a\dot{\theta}$ with respect to the magnetic field, a voltage of $2N\mathcal{B}\ell a\dot{\theta}$ is induced in the coil. We find the sign of the mechanically induced voltage e_m by applying the right-hand rule illustrated in Figure 11.7(b). The conductors on the right side of the coil in Figure 11.9(a) move downward when $\dot{\theta} > 0$ and the flux-density vector $\mathcal{B}$ points to the right, toward the south pole. Thus the positive sense of e_m is directed toward the viewer, and the corresponding current direction is out of the paper, opposite to the reference arrow for i in the figure. Since the circuit shown in Figure 11.9(b) was drawn with this assumed polarity for e_m, we have

$$e_m = (2N\mathcal{B}\ell a)\dot{\theta} \tag{13}$$

Summing torques on the coil and using (12) give

$$J\ddot{\theta} + B\dot{\theta} + K\theta = (2N\mathcal{B}\ell a)i \tag{14}$$

Using (13) in a voltage-law equation for the single loop comprising the electrical part of the system gives

$$L\frac{di}{dt} + Ri + (2N\mathcal{B}\ell a)\dot{\theta} = e_i(t) \tag{15}$$

Equations (14) and (15) comprise the complete model of the galvanometer, which is a third-order model. In practical situations, the inductance of the coil is often sufficiently small to allow us to neglect the term $L\,di/dt$ in (15). This simplification is particularly helpful in solving for the response to an input voltage, because the model becomes second-order. When this term is dropped, we can solve (15) for the current, obtaining

$$i = \frac{1}{R}[e_i(t) - (2N\mathcal{B}\ell a)\dot{\theta}]$$

By substituting this expression into (14), we have for the input-output equation

$$\ddot{\theta} + \left(\frac{B}{J} + \frac{\alpha^2}{JR}\right)\dot{\theta} + \frac{K}{J}\theta = \frac{\alpha}{JR}e_i(t) \tag{16}$$

where $\alpha = 2N\mathcal{B}\ell a$ is the electromechanical coupling coefficient. Comparing the result with (8.43), we find that the undamped natural frequency ω_n and the damping ratio ζ of the galvanometer are

$$\omega_n = \sqrt{\frac{K}{J}}$$

$$\zeta = \frac{1}{2\sqrt{KJ}}\left(B + \frac{\alpha^2}{R}\right)$$

Hence, the undamped natural frequency depends on the mechanical parameters K and J, and the damping ratio depends on both the mechanical and electrical parameters and the electromechanical coupling coefficient α.

We can find the sensitivity of the galvanometer in radians per volt by solving for the forced response to the constant excitation $e_i(t) = A$. The steady-state value of θ will be the particular solution of (16), namely

$$\theta_{ss} = \left(\frac{\alpha}{KR}\right)A \tag{17}$$

which is a constant. The galvanometer sensitivity is $\theta_{ss}/A = \alpha/KR$. Thus a higher flux density will make the device more sensitive, and increasing either the spring stiffness or the electrical resistance will reduce its sensitivity.

We can arrive at the same result for the sensitivity by the following argument. Since in the steady state the rotor will be stationary, no voltage will be induced in the coils. Thus the steady-state current will be $i_{ss} = A/R$. The steady-state torque exerted on the rotor through the action of the magnetic field will be $(\tau_e)_{ss} = \alpha A/R$, which must be balanced entirely by the torsional spring that exerts the steady-state torque $K\theta_{ss}$. Equating these two torques gives (17).

If we wish to characterize the system by a set of state-variable equations, we can choose θ and ω as the state variables and use (16) to write

$$\dot{\theta} = \omega$$

$$\dot{\omega} = -\frac{K}{J}\theta - \left(\frac{B}{J} + \frac{\alpha^2}{JR}\right)\omega + \frac{\alpha}{JR}e_i(t)$$

Had we retained the inductance of the coil, the current i would have become the third state variable. You are asked to write the equations and draw the simulation diagram for this case in Problem 11.4.

A MICROPHONE

The microphone shown in cross section in Figure 11.10 consists of a diaphragm attached to a circular coil of wire that moves back and forth through a magnetic field when sound waves impinge on the diaphragm. The magnetic field is supplied by a cylindrical permanent magnet having concentric north and south poles, which result in radial lines of flux directed inward toward the axis of the magnet. The coil has N turns with a radius of a and is connected in series with an external resistor R, across which the output voltage is measured. In the figure, the resistance of the coil has been neglected, but the inductance of the coil is represented by the inductor L located externally to the coil. If the coil resistance were not negligible,

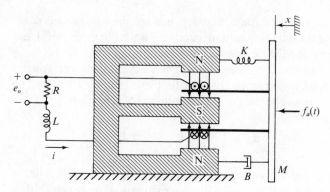

Figure 11.10 Representation of a microphone.

another resistor could be added in series with the external resistor in order to account for it.

A single stiffness element K has been used to represent the stiffness of the entire diaphragm, and the viscous-friction element B has been used to account for the energy dissipation due to air resistance. The net force of the impinging sound waves is represented by $f_a(t)$, which is the input to the system. Although the forces acting on the diaphragm are certainly distributed in nature, it is a justifiable simplification to consider them as point forces associated with the lumped elements K, B, and M.

The first step in modeling the microphone is to construct idealized diagrams of the electrical and mechanical portions. In developing the equations for the galvanometer, we assumed at the outset positive senses for τ_e and e_m. Then we wrote expressions for these quantities, determining their signs by using the right-hand rules. This time, we shall first determine the positive directions of f_e, the electrically induced force on the diaphragm, and the mechanically induced voltage e_m when the other variables are positive. After that, we shall label the diagram with senses that agree with the positive directions of f_e and e_m. With this approach, we know in advance that the expressions for f_e and e_m will have plus signs.

Figure 11.11(a) shows the upper portion of a single turn of the coil as viewed from the diaphragm looking toward the magnet. Since the flux arrow points downward from the north to the south pole and the current arrow points to the left, the right-hand rule shown in Figure 11.6(b) indicates that the positive sense of f_e is toward the diaphragm. Because the velocity arrow points away from the diaphragm, the right-hand rule shown in Figure 11.7(b) indicates that the positive sense of the induced voltage is the same as that of the current. We can reach similar conclusions by examining any other

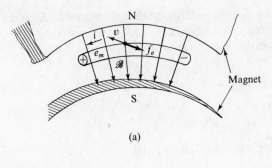

(a)

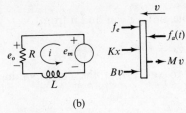

(b)

Figure 11.11 Microphone. (a) Relationships for a portion of a single coil. (b) Diagram used for analysis.

part of the coil. Thus we can draw the complete diagram of the system, as shown in Figure 11.11(b).

Because of the radial symmetry of the coil and the flux lines in the air gap, the entire coil of length $2\pi aN$ is perpendicular to the flux. Thus

$$f_e = \alpha i \tag{18a}$$

$$e_m = \alpha v \tag{18b}$$

where $\alpha = 2\pi aN\mathscr{B}$ is the electromechanical coupling coefficient for the system.

Summing forces on the free-body diagram for the diaphragm and using (18a), we obtain

$$M\dot{v} + Bv + Kx = -\alpha i + f_a(t) \tag{19}$$

We find the circuit equation by applying Kirchhoff's voltage law and using (18b), obtaining

$$L\frac{di}{dt} + Ri = \alpha v \tag{20}$$

Since the system has three independent energy-storing elements (L, K, and M), an appropriate set of state variables is i, x, and v. By rewriting (19) and (20), we obtain the state-variable equations

$$\frac{di}{dt} = \frac{1}{L}(-Ri + \alpha v)$$

$$\dot{x} = v$$

$$\dot{v} = \frac{1}{M}[-\alpha i - Kx - Bv + f_a(t)]$$

The output e_o is not a state variable, but we can find it from

$$e_o = Ri$$

In Problem 11.6, you are asked to verify that the corresponding input-output equation is

$$\dddot{e}_o + \left(\frac{R}{L} + \frac{B}{M}\right)\ddot{e}_o + \left(\frac{K}{M} + \frac{BR + \alpha^2}{ML}\right)\dot{e}_o + \frac{KR}{ML}e_o = \frac{\alpha R}{ML}\dot{f}_a \qquad (21)$$

It is interesting to note that since the right side of (21) is proportional to $\dot{f}_a$ rather than to $f_a(t)$, the forced response for a constant-force input is zero. Thus a constant force applied to the diaphragm yields no output voltage in the steady state.

A DIRECT-CURRENT MOTOR

A direct-current (dc) motor is somewhat similar to the galvanometer but has several significant differences. In all but the smallest motors, the magnetic field is established not by a permanent magnet, but by a current in a separate field winding on the iron core that comprises the stationary part of the motor, the *stator*. Figure 11.12 indicates the manner in which the field winding establishes the flux when the field current i_F flows. Because of the saturation effects of magnetic fields in iron, the flux ϕ is not necessarily proportional to i_F at high currents.

In a dc motor, the iron cylinder between the poles of the magnet is free to rotate and is called the *rotor*. The rotating coils are embedded in the surface of the rotor and are known as the *armature winding*. There is no restraining torsional spring, and the rotor is free to rotate through an indefinite number of revolutions. However, this fact requires a significant deviation from the galvanometer in the construction of the rotor and the armature winding. First, if the rotor shown in Figure 11.12 rotates through

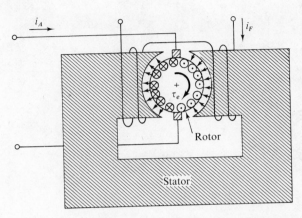

Figure 11.12 DC motor showing field and armature windings.

180° without a change in the direction of the currents in the individual armature conductors, then the torque exerted on it through the magnetic field will undergo a change in direction. Second, if there were a direct connection of the external armature circuit to the rotating armature, the wires would soon become tangled and halt the machine.

To solve both these problems, we use a *commutator*, which consists of a pair of low-resistance carbon brushes that are fixed with respect to the stator and make contact with the ends of the armature windings on the rotor (see Fitzgerald and Kingsley for details). As indicated in Figure 11.12, when a conductor is located to the right of the commutator brushes, a positive value of i_A implies that the individual conductor current will be directed toward the reader. When the conductor is located to the left of the brushes, its current flows away from the reader when $i_A > 0$. As the ends of a particular conductor pass under the brushes, the direction of the current in that conductor changes sign. Under the arrangement just described, each conductor exerts a unidirectional torque on the rotor as it passes through a complete revolution. Hence, the sliding contact at the brushes solves the mechanical problem of connecting the stationary and rotating parts of the armature circuit.

For modeling purposes, it is convenient to represent the important characteristics of the motor as shown in Figure 11.13. Representing the armature by a stationary circuit having resistance R_A, inductance L_A, and induced-voltage source e_m is justified by the presence of the commutator, which makes the armature behave as if it were stationary even though the individual conductors are indeed rotating. Likewise, the field circuit has resistance R_F and inductance L_F, but no induced voltage. The rotor has

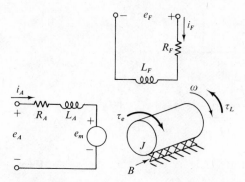

Figure 11.13 Diagram of the dc motor used for analysis.

moment of inertia J, rotational viscous-damping coefficient B, a driving torque τ_e caused by the forces acting on the individual conductors, and a load torque τ_L. The electrical inputs to the motor may be considered to be the currents i_A and i_F or the applied voltages e_A and e_F. The output is ω, the angular velocity of the rotor.

We begin the modeling process by expressing the voltage e_m and the torque τ_e in terms of the other system variables. We then write voltage-law equations for both the armature and the field circuits, unless i_A or i_F is an input, and apply D'Alembert's law to the rotor.

When we are modeling dc motors and generators, it is convenient to express the flux density $\mathscr{B}$ as

$$\mathscr{B} = \frac{1}{A}\phi(i_F) \tag{22}$$

where $\phi(i_F)$ is the total flux established by the field current and A is the effective cross-sectional area of the flux path in the air gap between the rotor and stator. If ℓ denotes the total effective length of the armature conductors within the magnetic field and a denotes the radius of the armature, the electromechanical torque exerted on the rotor is

$$\tau_e = \left(\frac{\phi}{A}\right)\ell a\, i_A \tag{23}$$

Since the parameters ℓ, a, and A depend only on the geometry of the motor, we can define the motor parameter

$$\gamma = \frac{\ell a}{A} \tag{24}$$

and rewrite (23) in terms of the flux and armature current as

$$\tau_e = [\gamma\phi(i_F)]\,i_A \tag{25}$$

Similarly, the voltage induced in the armature is

$$e_m = [\gamma\phi(i_F)]\omega \tag{26}$$

Keep in mind that the flux $\phi(i_F)$ in (25) and (26) is a function of the field current i_F and, for that matter, is generally a nonlinear function.

Having established the model for the internal behavior of a dc motor represented by Figure 11.13 and (24), (25), and (26), we are prepared to develop the models of complete electromechanical systems involving dc motors. We shall consider two such systems in the following examples.

EXAMPLE 11.2   Derive the state-variable equations for the dc motor shown in Figure 11.14, which has the constant field voltage E_F, an applied armature voltage $e_i(t)$, and a load torque $\tau_L(t)$. Also obtain the input-output equation with ω as the output, and determine the steady-state angular velocities corresponding to the following sets of inputs: (1) $e_i(t) = E$, $\tau_L(t) = 0$ and (2) $e_i(t) = 0$, $\tau_L(t) = T$.

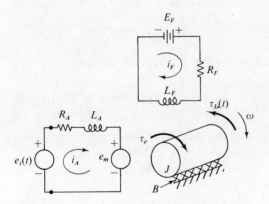

Figure 11.14 DC motor with constant field current.

Solution   Since the field voltage E_F is constant, the field current will be the constant $i_F = E_F/R_F$. We can write the electromechanical driving torque τ_e and the induced voltage e_m as

$$\tau_e = \alpha i_A$$
$$e_m = \alpha\omega \tag{27}$$

where α is a constant defined by

$$\alpha = \gamma\phi(i_F) \qquad (28)$$

and γ is given by (24). We select i_A and ω as the state variables and write a loop equation for the armature circuit and a torque equation for the rotor. Then, using (27) and solving for the derivatives of the state variables, we find the state-variable equations to be

$$\frac{di_A}{dt} = \frac{1}{L_A}[-R_A i_A - \alpha\omega + e_i(t)]$$

$$\dot{\omega} = \frac{1}{J}[\alpha i_A - B\omega - \tau_L(t)] \qquad (29)$$

To find the input-output equation for the angular velocity ω, we use the operator notation $p = d/dt$ to rewrite (29) as

$$(L_A p + R_A)i_A = -\alpha\omega + e_i$$

$$(Jp + B)\omega = \alpha i_A - \tau_L$$

Combining these equations to eliminate i_A yields the single operator equation

$$[(Jp + B)(L_A p + R_A) + \alpha^2]\omega = \alpha e_i - (L_A p + R_A)\tau_L$$

which yields the second-order input-output differential equation

$$\ddot{\omega} + \left(\frac{R_A}{L_A} + \frac{B}{J}\right)\dot{\omega} + \left(\frac{R_A B + \alpha^2}{JL_A}\right)\omega = \frac{\alpha}{JL_A}e_i(t) - \frac{1}{J}\dot{\tau}_L - \frac{R_A}{JL_A}\tau_L(t) \qquad (30)$$

As expected, both the electrical and the mechanical parameters contribute to the system's undamped natural frequency ω_n and damping ratio ζ.

To solve for the steady-state motor speed when the voltage source has the constant value $e_i(t) = E$ and when $\tau_L(t) = 0$, we omit all derivative terms in (30) and substitute for $e_i(t)$ and $\tau_L(t)$, obtaining

$$\omega_{ss} = \frac{\alpha E}{R_A B + \alpha^2} \qquad (31)$$

In physical terms, the motor will run at a constant speed such that the driving torque $\tau_e = \alpha i_A$ exactly balances the viscous-frictional torque $B\omega_{ss}$. However, the steady-state armature current is $i_A = (E - e_m)/R_A$ where $e_m = \alpha\omega_{ss}$. Making the appropriate substitutions, we again obtain (31).

When $e_i(t) = 0$ and $\tau_L(t)$ has the constant value T, the steady-state solution to (30) is

$$\omega_{ss} = -\frac{R_A T}{R_A B + \alpha^2} \tag{32}$$

which indicates that the motor will be driven backward at a constant angular velocity. To understand the behavior of the motor under this condition, we observe that the electromechanical torque, $\tau_e = \alpha i_A$, must balance the sum of the load torque T and the viscous-frictional torque $B\omega_{ss}$. Thus an armature current must flow that will make $\alpha i_A = T + B\omega_{ss}$. Furthermore, since the applied armature voltage is zero and $e_m = \alpha\omega_{ss}$, it follows that $i_A R_A = -\alpha\omega_{ss}$. As anticipated, solving these two equations for ω_{ss} results in (32).

In this situation, the motor is acting as a generator connected to a load of zero resistance. Part of the mechanical power supplied by the load torque is being converted to electrical form and dissipated in the armature resistance R_A. Basically, we may think of a *generator* as a motor that is being driven mechanically and that delivers a portion of the power to an electrical load connected across the armature terminals.

You can verify that when the constant applied voltage is $e_i(t) = E$ and the constant load torque is $\tau_L(t) = \alpha E/R_A$, the steady-state motor speed is $\omega_{ss} = 0$. In this condition, the electromechanical driving torque τ_e exactly matches the load torque and the motor is stalled.

EXAMPLE 11.3 The motor shown in Figure 11.15(a) has a constant armature current i_A and a variable field-current source. The relationship between the flux ϕ and the field current $i_F(t)$ is nonlinear and is shown in Figure 11.15(b). A load having moment of inertia J_L and viscous-damping coefficient B_L is connected to the rotor by a rigid shaft. Find a linear model suitable for analyzing small perturbations about the operating point A indicated on the flux-versus-current curve.

Solution From (25), the electromechanical torque is

$$\tau_e = \gamma\phi i_A \tag{33}$$

where, in this case, γ and i_A are constants and $\phi = \phi(i_F)$. Because the armature current is constant and the field current is specified, there is no need to write an equation for either of the electrical circuits. Rather, we obtain the system model by summing torques on the rotor and load and using (33), obtaining

$$(J_R + J_L)\dot{\omega} + (B_R + B_L)\omega = \gamma i_A \phi(i_F) \tag{34}$$

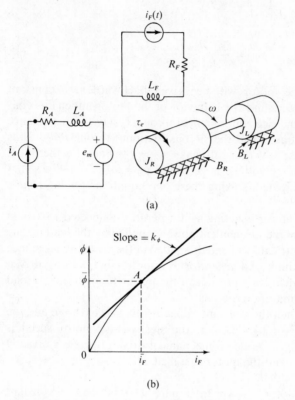

(a)

(b)

Figure 11.15 DC motor for Example 11.3. (a) Diagram used for analysis. (b) Nonlinear field characteristic.

To obtain the linearized model, we write the field current and angular velocity as $i_F(t) = \bar{i}_F + \hat{i}_F(t)$ and $\omega = \bar{\omega} + \hat{\omega}$, respectively. The operating-point values $\bar{i}_F$ and $\bar{\omega}$ must satisfy (34), which reduces to

$$\bar{\omega} = \frac{\gamma i_A \phi(\bar{i}_F)}{B_R + B_L} \tag{35}$$

The flux is approximated by the first two terms in the Taylor-series expansion for $\phi(i_F)$ about the operating point, namely

$$\phi(\bar{i}_F) + k_\phi \hat{i}_F \tag{36}$$

where

$$k_\phi = \left. \frac{d\phi}{di_F} \right|_{\bar{i}_F} \tag{37}$$

Substituting (36) into (34), writing $\omega = \bar{\omega} + \hat{\omega}$, and using (35) to cancel the constant terms, we find the linearized model to be

$$(J_R + J_L)\dot{\hat{\omega}} + (B_R + B_L)\hat{\omega} = \gamma i_A k_\phi \hat{i}_F(t) \tag{38}$$

where k_ϕ is the slope of the curve of ϕ vs i_F evaluated at the operating point.

Note that (38) is a first-order equation, implying that the system is only first-order. Although there are two inductors that can store energy, their currents are those of the current sources i_A and $i_F(t)$ and thus cannot be state variables.

Furthermore, it is apparent that the moments of inertia J_R and J_L and the friction coefficients B_R and B_L are merely added to obtain equivalent parameters for the system as a whole. If the motor had been connected to its load through a gear train (as would usually be the case), the equivalent parameters, when reflected to the motor, would be

$$J_{eq} = J_R + \frac{1}{N^2} J_L$$

$$B_{eq} = B_R + \frac{1}{N^2} B_L$$

The symbol N denotes the motor-to-load gear ratio; that is, the motor rotates at N times the angular velocity of the load.

11.4 COUPLING BY AN ELECTRICAL FIELD

In Section 9.2, we stated that the capacitance of two parallel plates of area A, separated a distance d by a dielectric material with permittivity ε, is $C = \varepsilon A/d$ provided that fringing effects are neglected. It follows that we can achieve electromechanical coupling by making either the distance d or the effective area A vary with time by mechanical means. For example, if the separation between the parallel plates shown in Figure 11.16 is the variable x, the device will obey the nonlinear equation

$$C = \frac{\varepsilon A}{x} \tag{39}$$

Provided that the permittivity of the dielectric material between the plates is constant, i.e., not affected by the voltage between the plates, the charge q on each plate and the voltage e between the plates are related by

$$q = C(x)e \tag{40}$$

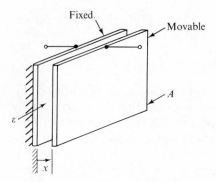

Figure 11.16 Capacitor with a movable plate.

The symbol $C(x)$ emphasizes that the capacitance C is a function of only the displacement variable x. If x varies, there will be variations in either q or e or both.

Energy is stored in the electrical field that exists between the plates of a charged capacitor, and a force is exerted on the plates by the electrical field. Thus any variation in x will involve a force acting on the movable capacitor plate, and there will be a flow of power between the mechanical and electrical parts. For a linear capacitor described by (40), the force due to the electrical field is given by

$$f_e = \frac{1}{2}e^2\frac{dC}{dx} \tag{41}$$

where the positive sense of f_e is the same as that of x (see Fitzgerald and Kingsley for a derivation).

For the variable capacitor shown in Figure 11.16, differentiating (39) with respect to x gives

$$\frac{dC}{dx} = -\frac{\varepsilon A}{x^2}$$

Thus the electrically induced force is

$$f_e = -\frac{1}{2}\varepsilon A\frac{e^2}{x^2} \tag{42}$$

We can express the force in terms of the charge rather than the voltage by using (39) and (40) to eliminate the voltage e in (42), giving

$$f_e = -\frac{1}{2\varepsilon A}q^2 \tag{43}$$

which does not explicitly involve the displacement x. Modeling a system that has a movable capacitor plate is illustrated in the following example.

EXAMPLE 11.4 The mass M shown in Figure 11.17 is attached to one plate of a capacitor, whose other plate is stationary. The capacitor is connected in series with a constant voltage source and a resistor, across which the output voltage e_o is measured. The system inputs are the applied force $f_a(t)$ and the voltage source E. Derive the mathematical model of the nonlinear system in terms of the total variables and in linearized form with incremental variables.

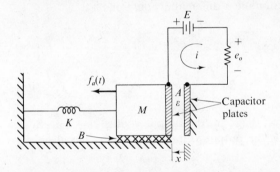

Figure 11.17 Variable capacitor connected to a mechanical system.

Solution The circuit diagram for the electrical part of the system is shown in Figure 11.18(a). Since the circuit is composed of a single loop, Kirchhoff's voltage law is

$$Ri + e_C = E \qquad (44)$$

where e_C denotes the voltage between the plates of the capacitor. In order to avoid mathematical complications due to the variable capacitor, we

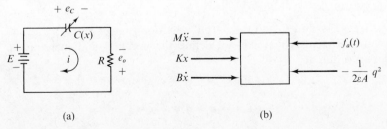

(a)   (b)

Figure 11.18 Capacitive measuring device. (a) Circuit diagram. (b) Free-body diagram

rewrite (44) in terms of the capacitor charge q. Using $i = \dot{q}$ and using (39) and (40) for e_C, we find

$$R\dot{q} + \left(\frac{1}{\varepsilon A}\right)xq = E \tag{45}$$

The free-body diagram for the mass is shown in Figure 11.18(b). The electrically induced force f_e is given by (43) in terms of the charge q, and the arrow denoting the positive sense of this force points to the left since it is in the same direction as the displacement arrow. Using the free-body diagram to obtain the mechanical equation and rewriting the electrical equation (45), we have the pair of coupled nonlinear differential equations

$$M\ddot{x} + B\dot{x} + Kx + \frac{1}{2\varepsilon A}q^2 = f_a(t) \tag{46a}$$

$$R\dot{q} + \frac{1}{\varepsilon A}xq = E \tag{46b}$$

From inspection of the circuit diagram, the output voltage is $e_o = -Ri = -R\dot{q}$. If we solve (46b) for $\dot{q}$, the output voltage is given by the algebraic equation

$$e_o = \left(\frac{1}{\varepsilon A}\right)xq - E \tag{47}$$

If we want a state-variable model, we choose x, $v = \dot{x}$, and q as state variables, in which case

$$\dot{x} = v$$

$$\dot{v} = \frac{1}{M}\left[-Kx - Bv - \frac{1}{2\varepsilon A}q^2 + f_a(t)\right]$$

$$\dot{q} = \frac{1}{R}\left(-\frac{1}{\varepsilon A}xq + E\right)$$

The inputs are $f_a(t)$ and E, and the output equation is given by (47).

To obtain a linearized model, we define the nominal and incremental variables according to

$$x = \bar{x} + \hat{x}$$

$$q = \bar{q} + \hat{q}$$

$$f_a(t) = \bar{f}_a + \hat{f}_a(t)$$

$$e_o = \bar{e}_o + \hat{e}_o$$

The operating point will be the equilibrium condition corresponding to the solution of the algebraic equations

$$K\bar{x} + \frac{1}{2\varepsilon A}\bar{q}^2 = \bar{f}_a \tag{48a}$$

$$\frac{1}{\varepsilon A}\bar{x}\bar{q} = E \tag{48b}$$

$$\bar{e}_o = 0 \tag{48c}$$

which we obtain by replacing all variables by their nominal values in (46) and (47). Combining (48a) and (48b) to eliminate $\bar{q}$ gives

$$K\bar{x} + \frac{\varepsilon A E^2}{2}\frac{1}{\bar{x}^2} = \bar{f}_a$$

which is equivalent to the cubic expression

$$K\bar{x}^3 - \bar{f}_a\bar{x}^2 + \tfrac{1}{2}\varepsilon A E^2 = 0 \tag{49}$$

The existence of a real positive root of (49) will depend on the relative values of $\bar{f}_a$, K, and E. Actually (49) is based on the assumption that the free length of the spring is such that the spring force is zero when $x = 0$. We could relax this restriction by writing the spring force as $f_K = K(x - x_0)$ where x_0 is the displacement of the movable capacitor plate when the spring is undeflected. The value of x_0 has no effect on the form of the linearized model, although it does affect the values of those coefficients that are dependent on $\bar{q}$ and $\bar{x}$.

Rewriting (46) in terms of the nominal and incremental variables and noting that $\dot{\bar{x}} = \ddot{\bar{x}} = \dot{\bar{q}} = 0$, we have

$$M\ddot{\hat{x}} + B\dot{\hat{x}} + K(\bar{x} + \hat{x}) + \left(\frac{1}{2\varepsilon A}\right)(\bar{q}^2 + 2\bar{q}\hat{q} + \hat{q}^2) = \bar{f}_a + \hat{f}_a(t)$$

$$R\dot{\hat{q}} + \left(\frac{1}{\varepsilon A}\right)(\bar{x}\bar{q} + \bar{x}\hat{q} + \bar{q}\hat{x} + \hat{x}\hat{q}) = E \tag{50}$$

When we use the operating-point equations (48a) and (48b) to cancel the constant terms and when we drop the terms $\hat{q}^2$ and $\hat{x}\hat{q}$, which involve products of incremental variables, (50) reduces to the linear equations

$$M\ddot{\hat{x}} + B\dot{\hat{x}} + K\hat{x} + \frac{\bar{q}}{\varepsilon A}\hat{q} = \hat{f}_a(t)$$

$$R\dot{\hat{q}} + \left(\frac{1}{\varepsilon A}\right)(\bar{x}\hat{q} + \bar{q}\hat{x}) = 0 \tag{51}$$

where $\bar{x}$ and $\bar{q}$ must satisfy (48a) and (48b). We find the output equation in the linearized model by linearizing (47) using (48b). It is

$$\hat{e}_o = \left(\frac{1}{\varepsilon A}\right)(\bar{x}\hat{q} + \bar{q}\hat{x}) \tag{52}$$

Equations (51) and (52) comprise the complete linearized model, which is third-order.

PROBLEMS

11.1 a. Use (6) to plot curves of e_o/E vs θ/θ_{max} for a rotary potentiometer with the load resistance R_o when (1) $R_o = 0.5R_T$, (2) $R_o = R_T$, and (3) $R_o = 10R_T$.

b. Let θ_{app} denote the apparent angular displacement, as calculated when $R_T/R_o = 0$. From the plots drawn for part (a), sketch curves of $(\theta - \theta_{app})/\theta_{max}$ versus e_o/E.

11.2 The potentiometer model shown in Figure 11.2(b) includes the resistances R_1 and R_2, both of which depend on $x(t)$. However, many potentiometers contain some inductance because they are constructed by winding many turns of wire about a core. The circuit shown in Figure P11.2 represents this inductance by the lumped elements whose values also depend on $x(t)$.

a. Find the differential equation relating e_o to $x(t)$ and $e_i(t)$.

b. Assume that R_2 and L_2 are proportional to the displacement $x(t)$. Let $L_2 = L_T[x(t)/x_{max}]$ and $R_2 = R_T[x(t)/x_{max}]$, where $L_1 + L_2 = L_T$ and $R_1 + R_2 = R_T$. Find the differential equation relating e_o to $x(t)$ and $e_i(t)$. Compare your answer to (3).

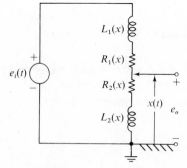

Figure P11.2

11.3 Two identical rotary potentiometers, similar to the one shown in Figure 11.4, are connected across a constant voltage source E as indicated in Figure P11.3. The output voltage is the difference in the voltages at the two wipers. The two wiper positions are denoted as $\theta_a(t)$ and $\theta_b(t)$. The various resistances satisfy the relationships

$$R_T = R_1 + R_2 = R_3 + R_4$$

$$R_2 = [\theta_a(t)/\theta_{max}] R_T$$

$$R_4 = [\theta_b(t)/\theta_{max}] R_T$$

a. Find an expression for e_o in terms of $\theta_a(t)$, $\theta_b(t)$, θ_{max}, and E when R_o is infinite (an open circuit).

b. Find an expression for e_o in terms of $\theta_a(t)$, $\theta_b(t)$, θ_{max}, E, R_T, and R_o that is valid for all values of R_o.

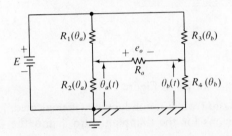

Figure P11.3

11.4 a. Find the input-output differential equation and also the state-variable equations for the galvanometer shown in Figure 11.9 when the inductance L of the coil is not neglected.

b. Draw simulation diagrams corresponding to your answers to part a.

11.5 Figure P11.5 shows a galvanometer whose flux is obtained from the same current i that passes through the movable coil, rather than from a permanent magnet as shown in Figure 11.9. The flux density in the air gaps is $\mathcal{B} = k_{\mathcal{B}} i$. The coil has moment of inertia J and viscous-damping coefficient B, and it is restrained by a torsional spring with spring constant K. The total length of the coil in the magnetic field is d, and its radius is a.

a. Take θ, ω, and i as state variables and write the state-variable equations.

b. Draw a simulation diagram.

c. Find θ corresponding to the operating point $\bar{e}_i$.

d. Find a set of linearized state-variable equations valid in the vicinity of the operating point you found in part c.

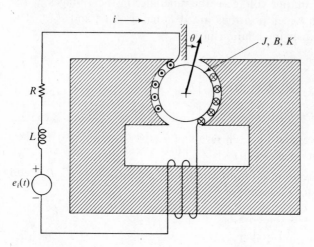

Figure P11.5

11.6 a. Verify (21) for the microphone shown in Figure 11.10.

b. Write the input-output equation when the inductance is neglected, i.e., when $L = 0$. Find expressions for the damping ratio ζ and the undamped natural frequency ω_n.

11.7 A loudspeaker produces sound waves by the movement of a diaphragm in response to an electrical input. In the cross-sectional view shown in Figure P11.7, $e_i(t)$ is the input and the output is the displacement x. A coil

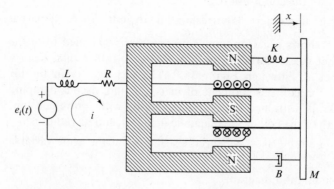

Figure P11.7

of wire with N turns and radius a is attached to the diaphragm. Let $\alpha = 2\pi a N \mathscr{B}$, where $\mathscr{B}$ is the flux density in the air gap of the permanent magnet.

    a. Define an appropriate set of state variables and write the state-variable equations.

    b. Draw a simulation diagram.

    c. Find the input-output differential equation.

11.8 A transducer to measure translational motion is shown in Figure P11.8. The permanent magnet produces a uniform magnetic field in the air gap with flux density $\mathscr{B}$ and can move with a displacement x and velocity v. A wire that is fixed in space has a length d within the magnetic field. The inductance and resistance of the wire are included in the lumped elements L and R, and the output of the system is the voltage e_o across the resistor R_o.

    a. Find the differential equation relating e_o and $x(t)$.

    b. If $i(0) = 0$, find the response for all $t > 0$ when $x(t) = U(t)$. What is the steady-state response?

    c. Repeat part b when $x(t) = tU(t)$. Assume that the wire remains within the magnetic field for the time period of interest.

    d. Find the steady-state response when $x(t) = A \sin \omega t$.

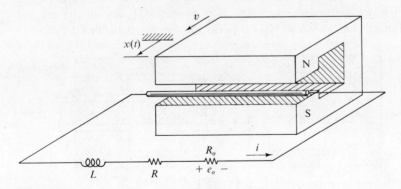

Figure P11.8

11.9 The conductor of mass M shown in Figure P11.9 can move vertically through a uniform magnetic field of flux density $\mathscr{B}$ whose positive sense is into the paper. There is no friction, the effective length of the conductor in the field is d, and $e_i(t)$ is a voltage source. The lumped elements $e_i(t)$, L, and R are outside the magnetic field.

a. Write a set of state-variable equations.

b. For what constant voltage $\bar{e}_i$ will the conductor remain stationary?

c. What will be the steady-state velocity of the conductor if $e_i(t)$ is always zero?

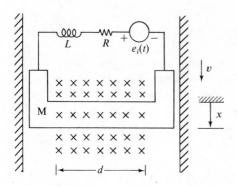

Figure P11.9

11.10 A plunger is made to move horizontally through the center of a fixed cylindrical permanent magnet by the application of a voltage source $e_i(t)$. Attached to the plunger is a coil having N turns and radius a. Figure P11.10 shows the system, including a cross-sectional view of the magnet and plunger, and indicates typical paths for the magnetic flux by dashed

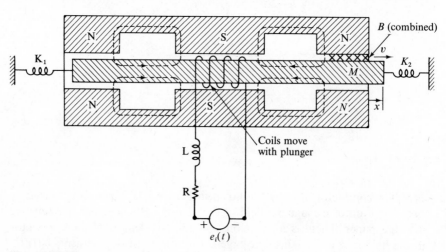

Figure P11.10

lines. The magnetic field between the plunger and the south pole is assumed to have a constant flux density $\mathscr{B}$. The resistance and inductance of the coil are represented by the lumped elements R and L, and the plunger has mass M.

    a. Define an appropriate set of state variables, indicating their positive senses, and write the state-variable equations.

    b. Draw a simulation diagram for the equations in part a.

    c. Determine the direction in which the plunger will move if there is no initial stored energy and if $e_i(t) = U(t)$.

    d. Find the steady-state displacement of the plunger for the input in part c.

11.11 For the electric motor shown in Figure 11.15(a), let $i_F(t)$ have the constant value i_F.

    a. Write the differential equation describing the system and identify the time constant when $i_A(t)$ is the input and ω the output.

    b. If $i_A(t) = i_{A_1}$ for all $t < 0$ and $i_A(t) = i_{A_2}$ for all $t > 0$, sketch ω vs t. Assume that steady-state conditions exist at $t = 0-$.

    c. Repeat part b when i_{A_2} is replaced by $-i_{A_1}$. Find the value of t for which $\omega = 0$.

11.12   a. Write the differential equation describing the motor shown in Figure 11.15(a) when $i_A(t)$ and $i_F(t)$ are separate inputs and when $\phi = k_\phi i_F(t)$.

    b. Find an expression for $\bar{\omega}$ at the operating point corresponding to $\bar{i}_A$ and $\bar{i}_F$.

    c. If $i_A(t) = \bar{i}_A + \hat{i}_A(t)$ and $i_F(t) = \bar{i}_F + \hat{i}_F(t)$, find a linearized model that is valid in the vicinity of the operating point you found in part b.

11.13 Repeat Problem 11.12, except assume that ϕ is the nonlinear function of i_F shown in Figure 11.15(b).

11.14 Assume that $L_A = 0$ for the motor shown in Figure 11.15(a). Replace the current source i_A by a voltage source that has a constant value of E_A volts.

    a. Write the differential equation describing the system if the input is $i_F(t)$ and if $\phi = k_\phi i_F(t)$.

    b. Find the operating point corresponding to $\bar{i}_F$.

    c. Find a linearized model that is valid about the operating point you found in part b. Identify its time constant.

11.15 For the motor depicted in Figure 11.14, replace $e_i(t)$ by a constant voltage source E_A, and replace the source in the field winding by a time-varying voltage $e_F(t)$. Assume that $\phi = k_\phi i_F$.

   a. Using i_A, i_F, and ω as state variables, write the state-variable equations.

   b. Draw a simulation diagram.

   c. Find $\bar{\omega}$ for the operating point corresponding to $e_F(t) = \bar{e}_F$ and $\tau_L(t) = 0$.

   d. Derive a linearized model valid about the operating point you found in part c.

11.16 The field and armature windings of an electric motor are connected in parallel directly across a voltage source $e_i(t)$, as shown in Figure P11.16. The resistances of the field and armature windings are R_F and R_A, respectively, and the inductances of both windings are negligible.

   a. Find the differential equation relating ω to $e_i(t)$ and $\tau_L(t)$.

   b. Find an expression for $\bar{\omega}$ at the operating point corresponding to $e_i(t) = \bar{e}_i$ and $\tau_L(t) = 0$.

   c. Derive a linearized differential equation that is valid about the operating point you found in part b. Identify the time constant of the linearized model.

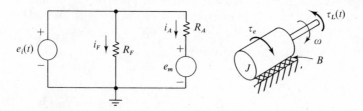

Figure P11.16

11.17 For the system shown in Figure P11.16, add an armature inductance L_A and a field inductance L_F in series with R_A and R_F, respectively.

   a. Write a set of state-variable equations.

   b. Find an expression for $\bar{\omega}$ at the operating point corresponding to $e_i(t) = \bar{e}_i$ and $\tau_L = 0$.

11.18 A Ward-Leonard speed-control system is shown in Figure P11.18. To the left of the dashed line a generator is represented whose rotor is driven at a constant angular velocity $\bar{\omega}_g$. The field is excited by a variable voltage source $e_i(t)$, and the armature terminals are connected to the armature of a motor. Assume that $\phi_g = k_\phi i_{F_g}$. The motor, shown to the right of the dashed line, has its field excited by a constant voltage source. In addition to the moment of inertia J_m and the viscous-friction coefficient B_m associated with the rotor, there is a variable load torque $\tau_L(t)$. Assume that the inductances L_{F_g} and L_{F_m} of the field windings are small enough to be neglected, so that neither i_{F_g} nor i_{F_m} is a state variable.

a. Choose an appropriate set of state variables and write the state-variable equations. Define any new symbols that are introduced.

b. Draw a simulation diagram.

c. Find the input-output differential equation relating ω_m to $e_i(t)$ and $\tau_L(t)$.

d. Determine the steady-state value of ω_m when $e_i(t) = U(t)$ and $\tau_L(t) = 0$.

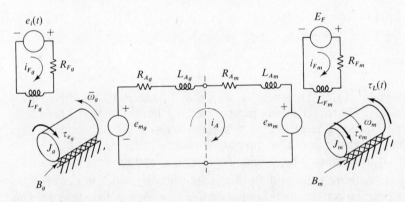

Figure P11.18

11.19 Figure P11.19 shows a tachometer with an input that is the angular velocity $\omega(t)$ of a cylindrical rotor. A coil embedded in the rotor has N rectangular turns, each with a radius a and a length ℓ in the direction of the axis of rotation. The coil moves through a uniform magnetic field of flux density $\mathscr{B}$ and is connected by a commutator to the external electrical circuit. The resistance and inductance of the coil are represented by the lumped elements R and L, respectively. A linear potentiometer, similar to the one shown in Figure 11.2, has a total resistance R_T and is placed across the terminals of the tachometer. The position of the potentiometer wiper is $x(t)$, treated as another input, and the maximum displacement of the wiper is $x_{\max}$.

a. Find the differential equation relating the output voltage e_o to $\omega(t)$ and $x(t)$. Let $\alpha = 2\mathcal{B}N\ell a$.

b. Find $\bar{e}_o$ at the operating point corresponding to $\bar{\omega}$ and $\bar{x}$.

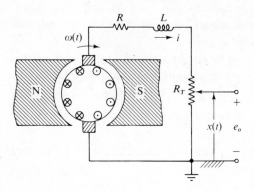

Figure P11.19

11.20 The rotor of the generator shown in Figure P11.20 is driven by a torque $\tau_a(t)$. The rotor has a moment of inertia J, but the friction is negligible. The field winding is excited by a constant voltage source, resulting in a constant magnetic flux. The armature resistance is denoted by R_A, and the armature inductance is negligible. Use (27) for e_m and τ_e.

a. With the rotor initially at rest and with the switch in the left-hand position connecting the armature to a short circuit, the applied torque is $\tau_a(t) = U(t)$. Find and sketch ω as a function of time.

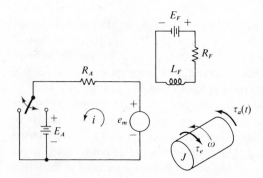

Figure P11.20

b. After the rotor has reached a steady-state speed under the conditions of part a, the switch is thrown to the right, thereby connecting the armature to the battery. Find and sketch ω vs t if a constant unit torque continues to be applied to the rotor. Also find the power supplied to the battery.

c. After a new steady-state speed has been reached under the conditions of part b, the applied torque is removed, with the switch still in the right-hand position. Again find and sketch ω vs t.

11.21 Use (51) to find an input-output differential equation relating the incremental variables $\hat{x}$ and $\hat{f}_a(t)$ for the system considered in Example 11.4.

CHAPTER 12
THE LAPLACE TRANSFORM

Frequently, we can transform a mathematical problem that is difficult to solve as it stands into an equivalent problem whose solution is much easier. If the solution of the transformed problem can be converted back to the framework of the original problem, then the process of transformation, solution, and conversion of the solution may be more attractive than attempting a direct solution of the original problem.

A familiar example of such a process is the use of logarithms. While logarithms transform numbers into other numbers, there are also procedures for transforming functions of one variable into functions of another variable. Among these are the Fourier series, the Fourier transform, and the Laplace transform. Since the Laplace transform converts fixed linear differential equations into algebraic equations, it is of considerable value in analyzing the response of a dynamic system whose model is fixed and linear.

In this chapter, we shall first define the Laplace transform and derive the transforms of several common functions of time. We shall develop properties that are useful in finding the transforms of specific time functions and in transforming the equations describing dynamic models. We shall complete the initial mathematical development by showing how to convert a transform back to the corresponding time function. In the following section, we use Laplace transforms to solve for the response of a variety of models. We conclude the chapter by introducing and illustrating several additional transform properties.

12.1 TRANSFORMS OF FUNCTIONS

The *Laplace transform* converts a function of a real variable, which will always be time t in our applications, into a function of a complex variable that is denoted by s. The transform of $f(t)$ is represented symbolically by either $\mathscr{L}[f(t)]$ or $F(s)$, where the symbol $\mathscr{L}$ stands for "the Laplace transform of." One can think of the Laplace transform as providing a means of transforming a given problem from the *time domain*, where all variables are functions of t, to the *complex-frequency domain*, where all variables are functions of s.

The defining equation for the Laplace transform* is

$$F(s) = \int_0^\infty f(t)\,\epsilon^{-st}\,dt \tag{1}$$

Some authors take the lower limit of integration as $t = 0-$ rather than $t = 0$. Either convention is acceptable provided that the transform properties are developed in a manner consistent with the defining integral.

Since the integration in (1) with respect to t is carried out between the limits of zero and infinity, the resulting transform will not be a function of t. In the factor ϵ^{-st} appearing in the integrand, we treat s as a constant in carrying out the integration.

The variable s is a complex quantity, which we can write as

$$s = \sigma + j\omega$$

where σ and ω are the real and the imaginary parts of s, respectively. We can write the factor ϵ^{-st} in (1) as

$$\epsilon^{-st} = \epsilon^{-\sigma t}\epsilon^{-j\omega t}$$

---

* Equation (1) defines the one-sided Laplace transform. There is a more general two-sided Laplace transform, which is useful for theoretical work but is seldom used for solving for system responses. The interested reader may consult DeRusso, Roy, and Close, Section 3.3, or Schwarz and Friedland for a discussion of the subject.

Since the magnitude of $\epsilon^{-j\omega t}$ is always unity, $|\epsilon^{-st}| = \epsilon^{-\sigma t}$. By placing appropriate restrictions on σ, we can ensure that the integral converges in most cases of practical interest, even if the function $f(t)$ becomes infinite as t approaches infinity. For all time functions considered in this book, convergence of the transform integral can be achieved.

To gain familiarity with the use of the transform definition and to begin the development of a table of transforms, we shall derive the Laplace transforms of several common functions. The results, along with others, are included in Appendix B.

STEP FUNCTION

Since the unit step function $U(t)$ is unity for all $t > 0$, it follows from (1) that

$$\mathcal{L}[U(t)] = \int_0^\infty \epsilon^{-st} \, dt$$

$$= \frac{\epsilon^{-st}}{-s} \bigg|_{t=0}^{t=\infty} \tag{2}$$

In order for the integral to converge, $|\epsilon^{-st}|$ must approach zero as t approaches infinity. Since $|\epsilon^{-st}| = \epsilon^{-\sigma t}$, the integral will converge provided that $\sigma > 0$. Hence, the expression for the transform of the unit step function converges for all values of s in the right half of the complex plane, and (2) becomes

UNIT STEP
FUNCTION

$$\mathcal{L}[U(t)] = \frac{1}{s} \tag{3}$$

Note that the variable t has disappeared in the integration process and that the result is strictly a function of s.

For all functions of time with which we shall be concerned, the transform definition will converge for all values of s to the right of some vertical line in the complex s-plane. Although knowledge of the region of convergence is needed in some advanced applications, we do not need such knowledge for the applications in this book. When evaluating Laplace transforms, we shall assume that σ, the real part of s, is sufficiently large to ensure convergence. If you are interested in regions of convergence, you should consult one of the more advanced references on system theory cited in Appendix D.

Because $U(t) = 1$ for all $t > 0$, the constant 1 is equivalent to $U(t)$ over the interval $t > 0$ and has the same transform as the unit step function. Thus

$$\mathscr{L}[1] = \frac{1}{s} \tag{4}$$

EXPONENTIAL FUNCTION

Depending on the value of the parameter a, the function $f(t) = \epsilon^{-at}$ represents an exponentially decaying function, a constant, or an exponentially growing function for $t > 0$, as shown in Figure 12.1. In any case, the Laplace transform of the exponential function is

$$\mathscr{L}[\epsilon^{-at}] = \int_0^\infty \epsilon^{-at}\epsilon^{-st}\, dt$$

$$= \int_0^\infty \epsilon^{-(s+a)t}\, dt$$

$$= \left.\frac{\epsilon^{-(s+a)t}}{-(s+a)}\right|_{t=0}^{t=\infty}$$

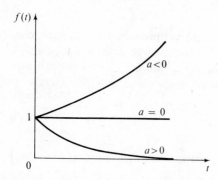

Figure 12.1 The exponential function ϵ^{-at} for various values of a.

The upper limit vanishes when σ, the real part of s, is greater than $-a$, so

$$\mathscr{L}[\epsilon^{-at}] = \frac{1}{s+a} \tag{5}$$

Note that if $a = 0$, the exponential function ϵ^{-at} reduces to the constant value of 1. Likewise, (5) reduces to (4), as it must for this value of a.

RAMP FUNCTION

The unit ramp function is defined to be $f(t) = t$ for $t > 0$. Substituting for $f(t)$ in (1), we have

$$\mathscr{L}[t] = \int_0^\infty t\epsilon^{-st}\, dt \tag{6}$$

To evaluate the integral in (6), we use the formula for integration-by-parts, namely

$$\int_a^b u\, dv = uv \Big|_a^b - \int_a^b v\, du \tag{7}$$

where the limits a and b apply to the variable t. Making the identifications $u = t$ and $v = \epsilon^{-st}/(-s)$, $a = 0$, and $b = \infty$, we can rewrite (6) as

$$\mathscr{L}[t] = \frac{t\epsilon^{-st}}{(-s)}\Big|_0^\infty - \frac{1}{(-s)}\int_0^\infty \epsilon^{-st}\, dt \tag{8}$$

It can be shown that

$$\lim_{t\to\infty} t\epsilon^{-st} = 0 \tag{9}$$

provided that σ, the real part of s, is positive. Thus

$$\mathscr{L}[t] = 0 - 0 + \frac{1}{s}\int_0^\infty \epsilon^{-st}\, dt \tag{10}$$

Since the integral in (10) is $\mathscr{L}[U(t)]$, which from (3) has the value $1/s$, (10) simplifies to

$$\mathscr{L}[t] = \frac{1}{s^2} \qquad \text{RAMP FUNCTION} \\ \qquad\qquad\qquad f(t) = t \tag{11}$$

RECTANGULAR PULSE

The rectangular pulse shown in Figure 12.2 has unit height and a duration of L. Its Laplace transform is

$$F(s) = \int_0^L \epsilon^{-st}\, dt + \int_L^\infty 0\epsilon^{-st}\, dt$$

$$= \frac{1}{(-s)}\epsilon^{-st}\Big|_0^L + 0$$

$$= \frac{1}{s}(1 - \epsilon^{-sL}) \tag{12}$$

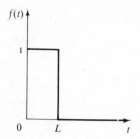

Figure 12.2 A rectangular pulse of unit height and duration L.

As the pulse duration L becomes infinite, the pulse approaches the unit step function $U(t)$, and the term ϵ^{-sL} in (12) approaches zero for all values of s having a positive real part. Thus as L approaches infinity, $F(s)$ given by (12) approaches $1/s$, which is $\mathscr{L}[U(t)]$.

THE IMPULSE

From (1), we can write the transform of the unit impulse as

$$\mathscr{L}[\delta(t)] = \int_{0}^{\infty} \delta(t)\,\epsilon^{-st}\,dt$$

This expression presents a dilemma because when the impulse was introduced in Section 8.3, we adopted the convention that $\delta(t)$ is an impulse at $t = 0$, which is the lower limit of integration in (1). To avoid this problem, we shall now adopt the convention that when we are working with Laplace transforms the unit impulse $\delta(t)$ occurs just after $t = 0$, namely at $t = 0+$. With this stipulation, (8.30) is applicable with $t_a = 0+$, $b = 0$, and $c = \infty$, and we can say that

$$\int_{0}^{\infty} \delta(t)\,\epsilon^{-st}\,dt = \epsilon^{-st}\big|_{t=0+} = 1$$

so

$$\mathscr{L}[\delta(t)] = 1 \tag{13}$$

In books where the lower limit of the transform integral in (1) is taken to be $t = 0-$, the impulse can be considered to occur at $t = 0$. This convention also yields (13).

TRIGONOMETRIC FUNCTIONS

When $f(t)$ is replaced by $\sin \omega t$ or $\cos \omega t$ in the definition of the Laplace transform, we can evaluate the resulting integral by using the identities given in Table 8.1, by integration-by-parts, or by consulting a table of integrals. Using an integral table, we find that

$$\mathscr{L}[\sin \omega t] = \int_0^\infty \sin \omega t \, \epsilon^{-st} \, dt$$

$$= \frac{\epsilon^{-st}}{(-s)^2 + \omega^2}(-s \sin \omega t - \omega \cos \omega t)\Big|_0^\infty$$

$$= \frac{\omega}{s^2 + \omega^2} \tag{14}$$

By a similar procedure, we can show that

$$\mathscr{L}[\cos \omega t] = \frac{s}{s^2 + \omega^2} \tag{15}$$

12.2 TRANSFORM PROPERTIES

The Laplace transform has a number of properties that are useful in finding the transforms of functions in terms of known transforms and in solving for the responses of dynamic models. We shall state, illustrate, and in most cases derive those properties that will be useful in later work. They are tabulated in Appendix B.

Throughout this section, the symbols $F(s)$ and $G(s)$ denote the Laplace transforms of the arbitrary time functions $f(t)$ and $g(t)$, and a and b denote arbitrary constants. Our object is to express the transforms of various functions of $f(t)$ and $g(t)$ in terms of $F(s) = \int_0^\infty f(t)\epsilon^{-st} \, dt$ and $G(s) = \int_0^\infty g(t)\epsilon^{-st} \, dt$.

MULTIPLICATION BY A CONSTANT

To express $\mathscr{L}[af(t)]$ in terms of $F(s)$, where a is a constant and where $F(s) = \mathscr{L}[f(t)]$, we use (1) to write

$$\mathscr{L}[af(t)] = \int_0^\infty af(t)\epsilon^{-st} \, dt$$

$$= a \int_0^\infty f(t)\epsilon^{-st} \, dt$$

$$= aF(s) \tag{16}$$

Thus multiplying a function of time by a constant multiples its transform by the same constant.

SUPERPOSITION

The transform of the sum of the two functions $f(t)$ and $g(t)$ is

$$\mathcal{L}[f(t) + g(t)] = \int_0^\infty [f(t) + g(t)] \epsilon^{-st} dt$$

$$= \int_0^\infty f(t) \epsilon^{-st} dt + \int_0^\infty g(t) \epsilon^{-st} dt$$

$$= F(s) + G(s) \tag{17}$$

Using (16) and (17), we have the general superposition property

$$\mathcal{L}[af(t) + bg(t)] = aF(s) + bG(s) \tag{18}$$

for any constants a and b and any transformable functions $f(t)$ and $g(t)$. As an illustration of the superposition property, we can evaluate $\mathcal{L}[2 + 3\sin 4t]$ by using (18) with (4) and (14) to write

$$\mathcal{L}[2 + 3\sin 4t] = \frac{2}{s} + 3\left(\frac{4}{s^2 + 4^2}\right)$$

$$= \frac{2s^2 + 12s + 32}{s^3 + 16s}$$

MULTIPLICATION BY AN EXPONENTIAL

If we replace $f(t)$ in (1) by the function $f(t)\epsilon^{-at}$, we have

$$\mathcal{L}[f(t)\epsilon^{-at}] = \int_0^\infty f(t)\epsilon^{-at}\epsilon^{-st} dt$$

$$= \int_0^\infty f(t)\epsilon^{-(s+a)t} dt$$

$$= F(s + a) \tag{19}$$

In words, (19) states that multiplying a function $f(t)$ by ϵ^{-at} is equivalent to replacing the variable s by the quantity $s + a$ wherever it occurs in $F(s)$.

With this property, we can derive several of the transforms in Appendix B rather easily from other entries in the table. Specifically, since $\mathcal{L}[\cos \omega t] =$

$s/(s^2 + \omega^2)$ and $\mathcal{L}[\sin \omega t] = \omega/(s^2 + \omega^2)$, we can write

$$\mathcal{L}[\epsilon^{-at} \cos \omega t] = \frac{s+a}{(s+a)^2 + \omega^2}$$

$$\mathcal{L}[\epsilon^{-at} \sin \omega t] = \frac{\omega}{(s+a)^2 + \omega^2}$$

Also, since $\mathcal{L}[t] = 1/s^2$, it follows that

$$\mathcal{L}[t\epsilon^{-at}] = \frac{1}{(s+a)^2}$$

MULTIPLICATION BY TIME

We obtain the transform of the product of $f(t)$ and the variable t by differentiating the transform $F(s)$ with respect to the complex variable s and then multiplying by -1:

$$\mathcal{L}[tf(t)] = -\frac{d}{ds}F(s) \tag{20}$$

To prove (20), we note that

$$\frac{d}{ds}F(s) = \frac{d}{ds}\left[\int_0^\infty f(t)\epsilon^{-st}\,dt\right]$$

$$= -\int_0^\infty tf(t)\epsilon^{-st}\,dt$$

$$= -\mathcal{L}[tf(t)] \tag{21}$$

Multiplying both sides of (21) by -1 results in (20).

We can illustrate the use of this property by deriving the entry in Appendix B for $\mathcal{L}[t^n]$ where n is any positive integer. Since $\mathcal{L}[1] = \mathcal{L}[U(t)] = 1/s$, it follows that

$$\mathcal{L}[t] = -\frac{d}{ds}\left(\frac{1}{s}\right) = \frac{1}{s^2}$$

$$\mathcal{L}[t^2] = -\frac{d}{ds}\left(\frac{1}{s^2}\right) = \frac{2}{s^3}$$

$$\mathcal{L}[t^3] = -\frac{d}{ds}\left(\frac{2}{s^3}\right) = \frac{2\cdot 3}{s^4}$$

and, for the general case,

$$\mathscr{L}[t^n] = \frac{n!}{s^{n+1}} \tag{22}$$

DIFFERENTIATION

Because we shall need to take the Laplace transform of each term in a differential equation when solving system models for their responses, we must derive expressions for the transforms of derivatives of arbitrary order. We shall first develop and illustrate the formula for obtaining the transform of df/dt in terms of $F(s) = \mathscr{L}[f(t)]$. Then we shall use this result to derive expressions for the transforms of higher derivatives.

First derivative From the transform definition (1), we can write the transform of df/dt as

$$\mathscr{L}[\dot{f}] = \int_0^\infty \left(\frac{df}{dt}\right) \epsilon^{-st}\, dt \tag{23}$$

We can rewrite (23) by using the formula for integration-by-parts given by (7), with $a = 0$ and $b = \infty$. The result is

$$\int_0^\infty u\, dv = uv \Big|_0^\infty - \int_0^\infty v\, du$$

If we let $u = \epsilon^{-st}$ and $dv = (df/dt)\, dt$, then $du = -s\epsilon^{-st}\, dt$, $v = f(t)$, and (23) becomes

$$\mathscr{L}[\dot{f}] = \epsilon^{-st}f(t)\Big|_0^\infty - \int_0^\infty f(t)(-s\epsilon^{-st})\, dt$$

For all the functions we shall encounter, there will be values of s for which $\epsilon^{-st}f(t)$ approaches zero as t approaches infinity, so

$$\mathscr{L}[\dot{f}] = [0 - f(0)] + s\int_0^\infty f(t)\epsilon^{-st}\, dt$$

$$= sF(s) - f(0) \tag{24}$$

To illustrate the application of (24), let $f(t) = \sin \omega t$. Then $F(s) = \omega/(s^2 + \omega^2)$ from (14) and $f(0) = \sin 0 = 0$, so

$$\mathscr{L}[\dot{f}] = s\left(\frac{\omega}{s^2 + \omega^2}\right) - 0 = \frac{s\omega}{s^2 + \omega^2} \tag{25}$$

We can verify this result by noting that $\dot{f} = \omega \cos \omega t$ and that

$$\mathscr{L}[\omega \cos \omega t] = \omega \mathscr{L}[\cos \omega t] = \omega \left(\frac{s}{s^2 + \omega^2}\right)$$

which agrees with (25).

   Deriving and using (24) are straightforward when $f(t)$ is continuous at $t = 0$. However, when $f(t)$ is discontinuous at $t = 0$, we must take care in applying (24) because of the potential ambiguity in evaluating $f(0)$. To avoid such problems, we shall adopt the convention that $f(0) = f(0-)$, which is equivalent to saying that any discontinuity at the time origin is considered to occur just after $t = 0$, i.e., at $t = 0+$. In such cases $f(0) \neq f(0+)$, and df/dt will contain an impulse of weight $f(0+) - f(0)$ occurring at $t = 0+$. This impulse is within the integration interval $0 \leq t \leq \infty$ we used for (1), and thus it will contribute to the expression for $\mathscr{L}[\dot{f}]$. This approach is consistent with our definition of the unit step function in Chapter 8 as

$$U(t) = \begin{cases} 0 & \text{for } t \leq 0 \\ 1 & \text{for } t > 0 \end{cases}$$

and with our derivation of the transform of $\delta(t)$.

   To illustrate the application of (24) when $f(t)$ has a discontinuity at $t = 0+$, consider the product of $\cos \omega t$ and the unit step function $U(t)$, namely

$$f(t) = [\cos \omega t] \, U(t)$$

$$= \begin{cases} 0 & \text{for } t \leq 0 \\ \cos \omega t & \text{for } t > 0 \end{cases} \tag{26}$$

which is shown in Figure 12.3(a). Since $f(t)$ differs from $\cos \omega t$ only at the

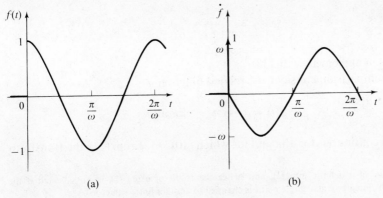

(a) (b)

Figure 12.3 The function $f(t) = [\cos \omega t] \, U(t)$ and its derivative.

point $t = 0$ within the interval $0 \leq t \leq \infty$, its transform is the same as that of $\cos \omega t$, namely*

$$F(s) = \frac{s}{s^2 + \omega^2} \tag{27}$$

We see from (26), however, that $f(0) = 0$ because of the presence of the unit step function. Thus (24) gives

$$\mathcal{L}[\dot{f}] = s\left(\frac{s}{s^2 + \omega^2}\right) - 0$$

$$= \frac{s^2}{s^2 + \omega^2} \tag{28}$$

To check this result, consider $\dot{f}$, the derivative of (26), which is shown in Figure 12.3(b). Because of the discontinuity in $f(t)$ at $t = 0+$, the function $\dot{f}$ contains a unit impulse at $t = 0+$. Consistent with the facts that $\dot{f}(t) = 0$ for all $t < 0$ and that the impulse occurs at $t = 0+$, we define $\dot{f}(0) = 0$. Thus

$$\dot{f} = \begin{cases} 0 & \text{for } t \leq 0 \\ \delta(t) - \omega \sin \omega t & \text{for } t > 0 \end{cases} \tag{29}$$

Transforming (29) gives

$$\mathcal{L}[\dot{f}] = 1 - \omega\left(\frac{\omega}{s^2 + \omega^2}\right)$$

$$= \frac{s^2}{s^2 + \omega^2}$$

which is in agreement with (28).

For comparison, consider the related function

$$g(t) = \cos \omega t \qquad \text{for all } t$$

which is continuous for all t and for which $g(0) = 1$. From (24), the transform

---

* The value of a definite integral, and hence the result of using (1), is not affected if the value of the integrand at a single point is changed to another finite value.

of its derivative is

$$\mathscr{L}[\dot{g}] = s\left(\frac{s}{s^2 + \omega^2}\right) - 1$$

$$= -\frac{\omega^2}{s^2 + \omega^2}$$

The derivative of $g(t)$ is $\dot{g} = -\omega \sin \omega t$ for all t. Taking the transform of this expression gives $-\omega^2/(s^2 + \omega^2)$, which agrees with the foregoing result.

In summary, whenever we use (24) we shall consider any discontinuities in $f(t)$ at the time origin to occur at $t = 0+$. We shall now develop formulas for finding the transforms of second and higher derivatives of $f(t)$ in terms of $F(s)$.

Second and higher derivatives   If we write (24) in terms of the function $g(t)$, it becomes

$$\mathscr{L}[\dot{g}] = sG(s) - g(0) \tag{30}$$

Now let $g(t) = \dot{f}(t)$. Then $\dot{g}(t) = \ddot{f}(t)$, and it follows from (30) and (24) that

$$\mathscr{L}[\ddot{f}] = s\mathscr{L}[\dot{f}(t)] - \dot{f}(0)$$

$$= s[sF(s) - f(0)] - \dot{f}(0)$$

$$= s^2 F(s) - sf(0) - \dot{f}(0) \tag{31}$$

If $\dot{f}$ is discontinuous at the time origin, we must use for $\dot{f}(0)$ in (31) the limit of $\dot{f}$ as t approaches zero from the left. Consider, for example,

$$f(t) = [\sin \omega t] \, U(t)$$

which is shown in Figure 12.4(a). The first and second derivatives of this function are

$$\dot{f} = [\omega \cos \omega t] \, U(t) \tag{32a}$$

$$\ddot{f} = \omega\delta(t) - [\omega^2 \sin \omega t] \, U(t) \tag{32b}$$

which are shown in Figure 12.4(b) and Figure 12.4(c). To be consistent with our previous development, we consider the discontinuity in $\dot{f}$ to occur at

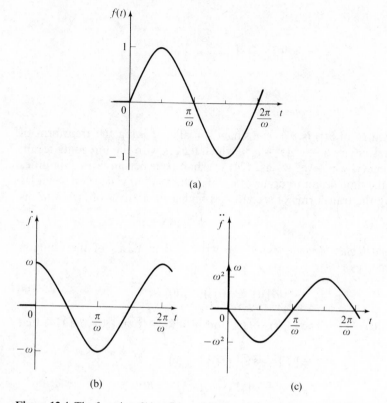

Figure 12.4 The function $f(t) = [\sin \omega t] \, U(t)$ and its first two derivatives.

$t = 0+$ and assume that $\dot{f}(0) = 0$. Then (31) gives

$$\mathcal{L}[\ddot{f}] = s^2 \left(\frac{\omega}{s^2 + \omega^2} \right) - s \cdot 0 - 0$$

$$= \frac{\omega s^2}{s^2 + \omega^2}$$

By a direct transformation of the right side of (32b), we obtain

$$\mathcal{L}[\ddot{f}] = \omega - \omega^2 \left(\frac{\omega}{s^2 + \omega^2} \right)$$

$$= \frac{\omega s^2}{s^2 + \omega^2}$$

which is the same result.

Equation (31) can be generalized to give the formula for the transform of the nth derivative of $f(t)$. The result is

$$\mathscr{L}\left[\frac{d^n f}{dt^n}\right] = s^n F(s) - s^{n-1} f(0) - \cdots - f^{(n-1)}(0) \tag{33}$$

where $f^{(n-1)}(0)$ denotes $d^{n-1} f/dt^{n-1}$ evaluated at $t = 0$.

INTEGRATION

The definite integral $\int_0^t f(\lambda)\, d\lambda$ will be a function of t because of the upper limit. From the transform definition given by (1), we can write

$$\mathscr{L}\left[\int_0^t f(\lambda)\, d\lambda\right] = \int_0^\infty \left[\int_0^t f(\lambda)\, d\lambda\right] \epsilon^{-st}\, dt$$

To evaluate the double integral on the right side of this expression, we use integration-by-parts with $u = \int_0^t f(\lambda)\, d\lambda$ and $dv = \epsilon^{-st}\, dt$. Then $du = f(t)\, dt$, $v = \epsilon^{-st}/(-s)$, and

$$\mathscr{L}\left[\int_0^t f(\lambda)\, d\lambda\right] = \left[\left(\frac{\epsilon^{-st}}{-s}\right)\int_0^t f(\lambda)\, d\lambda\right]\Bigg|_{t=0}^{t=\infty} - \int_0^\infty \left(\frac{\epsilon^{-st}}{-s}\right) f(t)\, dt$$

$$= [0 - 0] + \frac{1}{s}\int_0^\infty f(t)\epsilon^{-st}\, dt$$

$$= \frac{1}{s} F(s) \tag{34}$$

For example, if $f(t) = \cos \omega t$, then $F(s) = s/(s^2 + \omega^2)$, so

$$\mathscr{L}\left[\int_0^t \cos \omega\lambda\, d\lambda\right] = \frac{1}{s}\left(\frac{s}{s^2 + \omega^2}\right)$$

$$= \frac{1}{s^2 + \omega^2}$$

To check this result, we note that

$$\int_0^t \cos \omega\lambda\, d\lambda = \frac{1}{\omega}\sin \omega t$$

and

$$\mathcal{L}\left[\frac{1}{\omega}\sin \omega t\right] = \frac{1}{\omega}\left(\frac{\omega}{s^2+\omega^2}\right) = \frac{1}{s^2+\omega^2}$$

which agrees with the result obtained by using (34).

When solving dynamic models, we may need to transform a term such as $g(t) = g(0) + \int_0^t f(\lambda)\,d\lambda$. To do this, we note that $g(0)$ is a constant and thus has the transform $g(0)/s$, while we can transform the integral part of the term by using (34). The result is

$$G(s) = \frac{g(0)}{s} + \frac{F(s)}{s} \tag{35}$$

12.3 TRANSFORM INVERSION

When we use Laplace transforms to solve for the response of a system, we find the transform $F(s)$ of a particular variable, e.g., the output, first. The final step in the process, known as *transform inversion*, is to determine the corresponding time function $f(t)$ where $f(t) = \mathcal{L}^{-1}[F(s)]$, to be read "$f(t)$ is the inverse transform of $F(s)$."

For the types of problems encountered in this book, we do not need a completely general method of transform inversion. In this section, we present and illustrate a method that uses a partial-fraction expansion of $F(s)$. An extension of this method that permits handling an additional category of transformed functions appears in Section 12.5. More general techniques can be found in DeRusso, Roy, and Close or Schwarz and Friedland.

Assume that we can write the transform $F(s)$ as the ratio of two polynomials $N(s)$ and $D(s)$, such that

$$F(s) = \frac{N(s)}{D(s)} = \frac{b_m s^m + \cdots + b_0}{s^n + a_{n-1} s^{n-1} + \cdots + a_0} \tag{36}$$

Functions that can be written in the form of (36) are called *rational functions*. A *proper rational function* is one for which $m \leq n$, while a *strictly proper rational function* is one for which $m < n$.

The equation $D(s) = 0$ will have n roots denoted by $s_1, s_2, \ldots, s_n$, and $D(s)$ can be written in factored form as

$$D(s) = (s-s_1)(s-s_2)\cdots(s-s_n) \tag{37}$$

The quantities $s_1, s_2, ..., s_n$ are called the *poles* of $F(s)$ and are those values of s for which $F(s)$ becomes infinite.*

The *method of partial-fraction expansion* is applicable to any strictly proper rational function. Briefly, it allows us to express a known transform $F(s)$ as the sum of less complicated transforms. Using the table in Appendix B, we can identify the time functions corresponding to the individual transforms in the expansion and then use the superposition theorem to write $f(t)$.

We consider first the case where all the poles of $F(s)$ are distinct. We next modify the procedure to include repeated poles, where two or more of the quantities $s_1, s_2, ..., s_n$ are equal. Then we examine the case where the poles are complex numbers and discuss what to do if the degree of $N(s)$ is not less than that of $D(s)$. Throughout the section, we assume that the polynomial $D(s)$ has been factored, so that the values of the n poles are known.

DISTINCT POLES

The partial-fraction expansion theorem states that if $F(s)$ is a strictly proper rational function with distinct poles, it can be written as

$$F(s) = \frac{A_1}{s - s_1} + \frac{A_2}{s - s_2} + \cdots + \frac{A_n}{s - s_n} \tag{38}$$

where $A_1, A_2, ..., A_n$ are constants. We can write (38) with a summation sign as

$$F(s) = \sum_{i=1}^{n} A_i \left(\frac{1}{s - s_i} \right) \tag{39}$$

From (5), the term $1/(s - s_i)$ is the transform of the time function $\epsilon^{s_i t}$. Then, from the superposition formula given in (18), it follows that for $t > 0$,

$$f(t) = A_1 \epsilon^{s_1 t} + A_2 \epsilon^{s_2 t} + \cdots + A_n \epsilon^{s_n t}$$
$$= \sum_{i=1}^{n} A_i \epsilon^{s_i t} \tag{40}$$

In some cases, we can find the n poles s_i by factoring the denominator of $F(s)$ or, more generally, by finding the roots of $D(s) = 0$. We shall now develop a procedure for evaluating the n coefficients A_i so that we can write $f(t)$ as a sum of exponential time functions by using (40).

* It is assumed that $N(s) \neq 0$ at any of the poles.

Multiplying both sides of (38) by the term $(s - s_1)$ yields

$$(s - s_1)F(s) = A_1 + A_2 \frac{(s - s_1)}{(s - s_2)} + \cdots + A_n \frac{(s - s_1)}{(s - s_n)}$$

Since this equation must be a mathematical identity for all values of the variable s, we can set s equal to s_1 throughout the equation. Because the poles are distinct, $s_1 \neq s_j$ for $j = 2, 3, \ldots, n$. Thus each term on the right side will vanish except the term A_1, and we can write

$$A_1 = (s - s_1)F(s)|_{s = s_1} \tag{41}$$

To make it clear why (41) does not give a value of zero, we write $F(s)$ as $N(s)/D(s)$ with $D(s)$ in factored form. Then

$$A_1 = \left. \frac{(s - s_1)N(s)}{(s - s_1)(s - s_2)\cdots(s - s_n)} \right|_{s = s_1}$$

The term $(s - s_1)$ in the numerator will be canceled by the corresponding term in $D(s)$ before s is replaced by s_1.

Repeating the above process with s_1 replaced by s_2, we have

$$A_2 = (s - s_2)F(s)|_{s = s_2}$$

The general expression for the coefficients is

$$A_i = (s - s_i)F(s)|_{s = s_i} \qquad i = 1, 2, \ldots, n \tag{42}$$

In a numerical problem, we can check the calculation of $A_1, A_2, \ldots, A_n$ by combining the terms on the right side of (38) over a common denominator, which should give the original function $F(s)$.

EXAMPLE 12.1 Find the inverse transform of

$$F(s) = \frac{-s + 5}{(s + 1)(s + 4)}$$

Solution Comparing $F(s)$ to (36), we see that $N(s) = -s + 5$ so $m = 1$, and $D(s) = (s + 1)(s + 4)$ so $n = 2$. Since $D(s)$ is already in factored form, we see by inspection that the poles are $s_1 = -1$ and $s_2 = -4$. Thus we can rewrite the transform $F(s)$ in the form of (38) as

$$F(s) = \frac{A_1}{s + 1} + \frac{A_2}{s + 4}$$

Using (42) with $i = 1$ and 2, we find the coefficients of the partial-fraction expansion to be

$$A_1 = \left.\frac{(s+1)(-s+5)}{(s+1)(s+4)}\right|_{s=-1} = \frac{6}{3} = 2$$

$$A_2 = \left.\frac{(s+4)(-s+5)}{(s+1)(s+4)}\right|_{s=-4} = \frac{9}{-3} = -3$$

Hence, the partial-fraction expansion of the transform is

$$F(s) = \frac{2}{s+1} - \frac{3}{s+4}$$

and, from (40), the time function $f(t)$ is

$$f(t) = 2\epsilon^{-t} - 3\epsilon^{-4t} \qquad \text{for } t > 0$$

REPEATED POLES

If two or more of the n roots of $D(s) = 0$ are identical, these roots, which are poles of $F(s)$, are said to be *repeated*. When $F(s)$ contains repeated poles, (38) and (40) no longer hold. If $s_1 = s_2$, for example, the first two terms in (40) become $A_1 \epsilon^{s_1 t}$ and $A_2 \epsilon^{s_1 t}$, which are identical except for the multiplying constant, and (40) is not valid. If $s_1 = s_2$ and if the remaining poles are distinct, then (38) must be modified to be

$$F(s) = \frac{A_{11}}{(s-s_1)^2} + \frac{A_{12}}{s-s_1} + \frac{A_3}{s-s_3} + \cdots + \frac{A_n}{s-s_n} \tag{43}$$

Referring to Appendix B, we see that the first term on the right side of (43) is the transform of $A_{11} t\epsilon^{s_1 t}$, whereas the second term has the same form as the remaining ones. Thus the inverse transform is

$$f(t) = A_{11} t\epsilon^{s_1 t} + A_{12} \epsilon^{s_1 t} + A_3 \epsilon^{s_3 t} + \cdots + A_n \epsilon^{s_n t} \tag{44}$$

Note that the repeated pole at $s = s_1$ introduces into the time function a term of the form $t\epsilon^{s_1 t}$.

In order to evaluate any of the $n - 2$ coefficients $A_3, \ldots, A_n$ we can use the procedure given by (42). This formula cannot be used to find A_{11} and A_{12} because of the term $(s-s_1)^2$ in the denominator of $F(s)$. However, multiplying both sides of (43) by $(s-s_1)^2$ gives

$$(s-s_1)^2 F(s) = A_{11} + A_{12}(s-s_1) + A_3\frac{(s-s_1)^2}{(s-s_3)} + \cdots + A_n\frac{(s-s_1)^2}{(s-s_n)} \tag{45}$$

Setting s equal to s_1 throughout (45) results in

$$A_{11} = (s-s_1)^2 F(s)\big|_{s=s_1} \tag{46}$$

To find A_{12}, we note that the right side of (45) has the terms $A_{11} + A_{12}(s-s_1)$ and that all the remaining terms contain $(s-s_1)^2$ in their numerators. Thus if we differentiate both sides of (45) with respect to s, we have

$$\frac{d}{ds}[(s-s_1)^2 F(s)] = A_{12} + (s-s_1)G(s) \tag{47}$$

where $G(s)$ is a rational function without a pole at $s = s_1$. Note that in (47) the coefficient A_{11} is not present, A_{12} stands alone, and the function $G(s)$ that contains all the other coefficients is multiplied by the quantity $s - s_1$. Hence, setting s equal to s_1 in (47) gives

$$A_{12} = \left\{ \frac{d}{ds}[(s-s_1)^2 F(s)] \right\}\bigg|_{s=s_1} \tag{48}$$

where the differentiation must be performed before s is set equal to s_1.

If $F(s)$ has two or more pairs of identical poles, then each pair of poles contributes terms of the form $\epsilon^{s_i t}$ and $t\epsilon^{s_i t}$, and we can evaluate the coefficients of these terms by using (46) and (48) with the appropriate indices. If $F(s)$ has three or more identical poles, (43) must be modified further. For example, if $s_1 = s_2 = s_3$ and if the remaining poles are distinct, the partial-fraction expansion has the form

$$F(s) = \frac{A_{11}}{(s-s_1)^3} + \frac{A_{12}}{(s-s_1)^2} + \frac{A_{13}}{s-s_1} + \frac{A_4}{s-s_4} + \cdots + \frac{A_n}{s-s_n} \tag{49}$$

The formulas for evaluating the coefficients are

$$A_{11} = (s-s_1)^3 F(s)\big|_{s=s_1}$$

$$A_{12} = \left\{ \frac{d}{ds}[(s-s_1)^3 F(s)] \right\}\bigg|_{s=s_1}$$

$$A_{13} = \frac{1}{2!} \left\{ \frac{d^2}{ds^2}[(s-s_1)^3 F(s)] \right\}\bigg|_{s=s_1} \tag{50}$$

and

$$A_i = (s-s_i)F(s)\big|_{s=s_i} \qquad \text{for } i = 4, 5, \ldots, n$$

You will be asked to carry out the details of the derivation in Problem 12.14. The time function corresponding to (49) and (50) is

$$f(t) = \left(\frac{1}{2!} A_{11} t^2 + A_{12} t + A_{13} \right) \epsilon^{s_1 t} + A_4 \epsilon^{s_4 t} + \cdots + A_n \epsilon^{s_n t} \qquad (51)$$

EXAMPLE 12.2   Find the inverse Laplace transform of

$$F(s) = \frac{5s + 16}{(s + 2)^2 (s + 5)}$$

Solution   Since the denominator of $F(s)$ is given in factored form, we note that the poles are $s_1 = s_2 = -2$ and $s_3 = -5$. Because of the repeated pole, the partial-fraction expansion of $F(s)$ has the form

$$F(s) = \frac{A_{11}}{(s + 2)^2} + \frac{A_{12}}{s + 2} + \frac{A_3}{s + 5}$$

Using (46), (48), and (42) in order, we find that the coefficients are

$$A_{11} = (s + 2)^2 F(s) \Big|_{s = -2} = \frac{5s + 16}{s + 5} \Big|_{s = -2} = 2$$

$$A_{12} = \left\{ \frac{d}{ds} \left[\frac{5s + 16}{s + 5} \right] \right\} \Big|_{s = -2} = \frac{9}{(s + 5)^2} \Big|_{s = -2} = 1$$

$$A_3 = (s + 5) F(s) \Big|_{s = -5} = \frac{5s + 16}{(s + 2)^2} \Big|_{s = -5} = -1$$

Using numerical values for the coefficients gives the partial-fraction expansion of the transform as

$$F(s) = \frac{2}{(s + 2)^2} + \frac{1}{s + 2} - \frac{1}{s + 5}$$

The corresponding time function is

$$f(t) = 2t\epsilon^{-2t} + \epsilon^{-2t} - \epsilon^{-5t} \qquad \text{for } t > 0$$

COMPLEX POLES

The form of $f(t)$ given by (40) is valid for complex poles as well as for real poles. When (40) is used directly with complex poles, however, it has the disadvantage that the functions $\epsilon^{s_i t}$ and the coefficients A_i are complex. Hence, further developments are necessary in order for us to write $f(t)$ directly in terms of real functions and real coefficients. We shall present two methods for finding $f(t)$ that lead to slightly different but equivalent forms. First we show how mathematical identities are used to combine any complex terms in the partial-fraction expansion of $F(s)$ into real functions of time. Then we present the alternative method known as completing the square.

For simplicity, we shall assume that $F(s)$ has only two complex poles and that the order of its numerator is less than the order of its denominator. Any other poles, whether real or complex, will lead to additional terms in the partial-fraction expansion. By the superposition property, we can add the inverse transform of these additional terms to the inverse transform for the terms corresponding to the complex poles. With these restrictions, we can write the transform as

$$F(s) = \frac{N(s)}{(s+a-j\omega)(s+a+j\omega)} \tag{52}$$

which has poles at $s_1 = -a+j\omega$ and $s_2 = -a-j\omega$.

Partial-fraction expansion As indicated in (52), the complex poles of $F(s)$ will always occur in complex conjugate pairs. Using the partial-fraction expansion of (38) for distinct poles, we can write $F(s)$ as

$$F(s) = \frac{K_1}{s+a-j\omega} + \frac{K_2}{s+a+j\omega} \tag{53}$$

where

$$K_1 = (s+a-j\omega)F(s)|_{s=-a+j\omega} \tag{54a}$$

$$K_2 = (s+a+j\omega)F(s)|_{s=-a-j\omega} \tag{54b}$$

Since (54a) is identical to (54b) except for the sign of the imaginary term $j\omega$ wherever it appears, the coefficient K_2 is the complex conjugate of K_1, i.e., $K_2 = K_1^*$. Hence, we can write K_1 and K_2 in polar form as $K_1 = Ke^{j\phi}$ and $K_2 = Ke^{-j\phi}$ where K and ϕ are the magnitude and angle, respectively, of the complex number K_1. Thus both K and ϕ are real quantities and $K \geq 0$.

Rewriting (53) with K_1 and K_2 in polar form, we have

$$F(s) = \frac{Ke^{j\phi}}{s+a-j\omega} + \frac{Ke^{-j\phi}}{s+a+j\omega} \tag{55}$$

which has the form of (39), although the coefficients and poles are complex. From (40) with $n = 2$ and with the appropriate values substituted for A_i

and s_i, the complex form of the time function is

$$f(t) = K\epsilon^{j\phi}\epsilon^{(-a+j\omega)t} + K\epsilon^{-j\phi}\epsilon^{(-a-j\omega)t}$$

To obtain $f(t)$ as a real function of time, we factor the term $2K\epsilon^{-at}$ out of each term on the right side of this equation and combine the remaining complex exponentials to yield

$$f(t) = 2K\epsilon^{-at}\left[\frac{\epsilon^{j(\omega t + \phi)} + \epsilon^{-j(\omega t + \phi)}}{2}\right]$$

Recognizing from Table 8.1 that the term in the brackets is $\cos(\omega t + \phi)$, we can write

$$f(t) = 2K\epsilon^{-at}\cos(\omega t + \phi) \qquad \text{for } t > 0 \tag{56}$$

where the parameters a, ω, K, and ϕ are all real. Note that (55) and (56) constitute one of the entries in Appendix B.

Completing the square To develop the second of the two forms of $f(t)$ when $F(s)$ has complex poles, we rewrite (52) as

$$F(s) = \frac{B_1 s + B_2}{s^2 + 2as + a^2 + \omega^2} \tag{57}$$

where the denominator factors have been multiplied and where, since $m < n$, $N(s)$ has been replaced by a first-degree polynomial. Next we write the denominator of (57) as the sum of the perfect square $(s+a)^2$ and the constant ω^2, so that

$$F(s) = \frac{B_1 s + B_2}{(s+a)^2 + \omega^2}$$

Now we rewrite the numerator of $F(s)$ to have the term $(s+a)$ appear, giving

$$F(s) = \frac{B_1(s+a) + (B_2 - aB_1)}{(s+a)^2 + \omega^2}$$

$$= B_1\left[\frac{s+a}{(s+a)^2 + \omega^2}\right] + \left(\frac{B_2 - aB_1}{\omega}\right)\left[\frac{\omega}{(s+a)^2 + \omega^2}\right]$$

Referring to Appendix B, we see that the quantities within the brackets are the transforms of $\epsilon^{-at} \cos \omega t$ and $\epsilon^{-at} \sin \omega t$, respectively. Thus we can write

$$f(t) = B_1 \epsilon^{-at} \cos \omega t + \left(\frac{B_2 - aB_1}{\omega}\right)\epsilon^{-at} \sin \omega t$$

$$= \epsilon^{-at}[B_1 \cos \omega t + B_3 \sin \omega t] \tag{58}$$

where

$$B_3 = \frac{1}{\omega}(B_2 - aB_1) \tag{59}$$

Comparing (56) and (58), we see that we can express the parameters K and ϕ in terms of the coefficients B_1 and B_3, and vice versa. Using Table 8.1, you can verify that

$$B_1 = 2K \cos \phi \qquad B_3 = -2K \sin \phi$$

and

$$K = \tfrac{1}{2}\sqrt{B_1^2 + B_3^2} \qquad \phi = -\tan^{-1}\frac{B_3}{B_1}$$

EXAMPLE 12.3   Find $f(t)$ when

$$F(s) = \frac{4s + 8}{s^2 + 2s + 5}$$

Solution   Since the poles of $F(s)$ are complex, we can use either of the methods we have just described. The poles of $F(s)$ are the roots of $s^2 + 2s + 5 = 0$, namely $s_1 = -1 + j2$ and $s_2 = -1 - j2$. Hence $a = 1$, $\omega = 2$, and the partial-fraction expansion of $F(s)$ is

$$F(s) = \frac{K_1}{s + 1 - j2} + \frac{K_2}{s + 1 + j2}$$

Solving for K_1 according to (54a), we find

$$K_1 = \frac{(s + 1 - j2)(4s + 8)}{(s + 1 - j2)(s + 1 + j2)}\Bigg|_{s = -1 + j2}$$

$$= 2 - j1$$

$$= \sqrt{5}\,\epsilon^{-j0.4636}$$

Thus $K = \sqrt{5}$ and $\phi = -0.4636$ rad. Substituting the known values of a, ω, K, and ϕ into (56), we have

$$f(t) = 2\sqrt{5}\,\epsilon^{-t}\cos(2t - 0.4636) \qquad \text{for } t > 0 \qquad (60)$$

Having found $f(t)$ from the partial-fraction expansion of $F(s)$, we repeat the problem by the method of completing the square. Since the denominator of $F(s)$ is $s^2 + 2s + 5$, we can obtain a perfect-square term by adding and subtracting 1 to give

$$(s^2 + 2s + 1) + (5 - 1) = (s+1)^2 + 4$$

We must manipulate the numerator of $F(s)$ to contain the terms $(s + a)$ and ω, where $a = 1$ and $\omega = 2$. Thus we write

$$F(s) = \frac{4(s+1) + (8-4)}{(s+1)^2 + 2^2}$$

$$= 4\left[\frac{s+1}{(s+1)^2 + 2^2}\right] + 2\left[\frac{2}{(s+1)^2 + 2^2}\right]$$

$$= 4\mathscr{L}[\epsilon^{-t}\cos 2t] + 2\mathscr{L}[\epsilon^{-t}\sin 2t]$$

and

$$f(t) = 4\epsilon^{-t}\cos 2t + 2\epsilon^{-t}\sin 2t \qquad \text{for } t > 0$$

which we can convert to the form given by (60) by using the appropriate entry in Table 8.1.

Comparison of the two methods for finding $f(t)$ indicates that the partial-fraction expansion requires the manipulation of complex numbers. It is frequently employed, however, and we shall need it for a derivation in the next chapter. The method of completing the square has the advantage of avoiding the use of complex numbers.

Another difference between the two methods arises if $F(s)$ has more than the two poles specified by (52). For example, suppose

$$F(s) = \frac{N(s)}{(s^2 + 2s + 5)(s^2 + 4s + 13)}$$

where $N(s)$ is a polynomial of degree less than four. We can apply the partial-fraction expansion method directly, obtaining

$$F(s) = \frac{K_1\,\epsilon^{j\phi_1}}{s+1-j2} + \frac{K_1\,\epsilon^{-j\phi_1}}{s+1+j2} + \frac{K_3\,\epsilon^{j\phi_3}}{s+2-j3} + \frac{K_3\,\epsilon^{-j\phi_3}}{s+2+j3}$$

where K_1, K_3, ϕ_1, and ϕ_3 can be evaluated by equations similar to (54a) and where $f(t)$ can be written as the sum of two terms having the form of (56). To use the method of completing the square, we write

$$F(s) = \frac{As+B}{s^2+2s+5} + \frac{Cs+D}{s^2+4s+13} \tag{61}$$

where we must find the constants A, B, C, and D before we can use (58). To do this, we combine the two terms on the right side of (61) over a common denominator. Then we compare the coefficients of the numerator polynomial to those of $N(s)$ to give four equations that can be solved for the values of A, B, C, and D. You will be asked to do this for a particular $N(s)$ in Problem 12.16.

PRELIMINARY STEP OF LONG DIVISION

Remember that the techniques discussed so far are subject to the restriction that $F(s)$ is a strictly proper rational function, i.e., that m, the degree of the numerator polynomial $N(s)$, is less than n, the degree of the denominator polynomial $D(s)$. Otherwise, the partial-fraction expansion given for distinct poles by (39) or for a pair of repeated poles by (43) is not valid. In order to find the inverse transform of $F(s)$ when $m = n$, we must first write $F(s)$ as the sum of a constant* and a fraction whose numerator is of degree $n-1$ or less. We can accomplish this by dividing the numerator by the denominator so that the remainder is of degree $n-1$ or less. Then we may write

$$F(s) = A + F'(s) \tag{62}$$

where A is a constant and is the transform of $A\delta(t)$. The function $F'(s)$ is a ratio of polynomials having the same denominator as $F(s)$ but a numerator of degree less than n. We can find the inverse transform of $F'(s)$ by using the techniques described earlier in this section. Thus

$$f(t) = A\delta(t) + \mathcal{L}^{-1}[F'(s)] \tag{63}$$

EXAMPLE 12.4   Find $f(t)$ when

$$F(s) = \frac{2s^2+7s+8}{s^2+3s+2}$$

---

* If $m > n$, we can write $F(s)$ as the sum of a polynomial in s and a strictly proper rational function, but we shall not encounter such cases here.

Solution Since both the numerator and denominator of $F(s)$ are quadratic in s, we have $m = n = 2$. Before carrying out the partial-fraction expansion, we must rewrite $F(s)$ in the form of (62) by dividing its numerator by its denominator, as follows:

$$
\begin{array}{r}
2 \\
s^2 + 3s + 2 \overline{)\, 2s^2 + 7s + 8} \\
2s^2 + 6s + 4 \\
\hline
s + 4
\end{array}
$$

Because the poles of $F(s)$ are $s_1 = -1$ and $s_2 = -2$, we can write

$$
F(s) = 2 + \frac{s+4}{(s+1)(s+2)}
$$

$$
= 2 + \frac{A_1}{s+1} + \frac{A_2}{s+2}
$$

where

$$
A_1 = \left. \frac{(s+1)(s+4)}{(s+1)(s+2)} \right|_{s=-1} = 3
$$

$$
A_2 = \left. \frac{(s+2)(s+4)}{(s+1)(s+2)} \right|_{s=-2} = -2
$$

Thus

$$
F(s) = 2 + \frac{3}{s+1} - \frac{2}{s+2}
$$

and

$$
f(t) = 2\delta(t) + 3\epsilon^{-t} - 2\epsilon^{-2t} \qquad \text{for } t > 0
$$

which is the sum of an impulse at $t = 0+$ and two decaying exponentials.

12.4 SOLVING FOR THE RESPONSE

The preceding sections of this chapter have laid the groundwork for an efficient means of solving for the responses of fixed linear systems. The analysis of first-order systems is not difficult enough to warrant transform methods. However, for second- and higher-order systems, using transform methods is generally easier than using the methods discussed in Chapter 8. In addition, aquaintance with the transform approach will allow us to develop important concepts in Chapter 13. Before outlining the general technique, we shall consider a simple first-order example, which we would not normally solve by transform methods.

If a variable or one of its derivatives has a discontinuity at the time origin, we define its value at time zero to be the value approached through negative values of time. Thus $f(0) = f(0-)$ and $\dot{f}(0) = \dot{f}(0-)$, where the value at $t = 0-$ is the value at the left side of any discontinuity. This is consistent with the treatment in previous sections, e.g., our definition of $U(0) = 0$ and our interpretation of $\delta(t)$ as a unit impulse occurring at $t = 0+$. Any initial-condition terms, such as $f(0)$ and $\dot{f}(0)$, that appear in the transformed equations are due entirely to conditions for $t < 0$ and are not affected by a discontinuity in the input at time zero.

EXAMPLE 12.5   The capacitor in the circuit shown in Figure 12.5 is uncharged for $t < 0$, and the switch closes at $t = 0$. Using Laplace transforms, find $e_C(t)$ for $t > 0$.

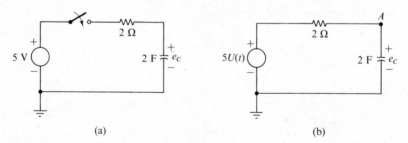

(a)   (b)

Figure 12.5 (a) Circuit for Example 12.5. (b) Equivalent circuit without the switch.

Solution   For convenience of analysis, we can redraw the original circuit as shown in Figure 12.5(b), where the switch and constant voltage source have been replaced by the step-function input $5U(t)$. Summing the currents leaving node A gives the first-order differential equation

$$\tfrac{1}{2}[e_C - 5U(t)] + 2\dot{e}_C = 0$$

Transforming this equation term by term, we have

$$\tfrac{1}{2}\{\mathscr{L}[e_C(t)] - \mathscr{L}[5U(t)]\} + 2\mathscr{L}[\dot{e}_C(t)] = 0$$

From Appendix B, we see that $\mathscr{L}[5U(t)] = 5/s$. Since $e_C(t)$ is unknown,

we cannot find its transform explicitly, but we use $E_C(s)$ to denote $\mathscr{L}[e_C(t)]$.*
Using the expression for the transform of a derivative, we rewrite the transformed equation as

$$\frac{1}{2}\left[E_C(s) - \frac{5}{s}\right] + 2[sE_C(s) - e_C(0)] = 0$$

Note that the original differential equation has been converted into an algebraic equation. Since the capacitor is uncharged for $t < 0$, $e_C(0) = 0$ and $e_C(t)$ must be continuous at $t = 0$. Solving this equation algebraically for $E_C(s)$ gives

$$E_C(s) = \frac{2.5/s}{2s + 0.5}$$

$$= \frac{1.25}{s(s + 0.25)}$$

To find the output voltage $e_C(t)$, we take the inverse transform by expanding $E_C(s)$ in its partial-fraction expansion and using Appendix B to identify the individual time functions. The expansion is

$$E_C(s) = \frac{A_1}{s} + \frac{A_2}{s + 0.25}$$

where

$$A_1 = sE_C(s)|_{s=0} = 5$$
$$A_2 = (s + 0.25)E_C(s)|_{s=-0.25} = -5$$

Thus

$$E_C(s) = \frac{5}{s} - \frac{5}{s + 0.25}$$

and

$$e_C(t) = 5 - 5\epsilon^{-0.25t} \qquad \text{for } t > 0$$

Since $e_C(t) = 0$ for all $t \leq 0$, we can write the solution in the alternative form

$$e_C(t) = 5[1 - \epsilon^{-0.25t}] U(t)$$

* As a rule, functions of time are denoted by lower-case letters and their Laplace transforms by the corresponding capital letters followed by the transform variable s in parentheses.

which is valid for all t. We can check this answer by solving the differential equation by the classical method of Chapter 8 or by using the short-cut method discussed in Section 9.6.

GENERAL PROCEDURE

Application of the Laplace transform to solve for the response of dynamic systems of any order consists of the following three steps:

1. Write and immediately transform the differential or integral-differential equations describing the system for $t > 0$, evaluating all the initial-condition terms that appear in the transformed equations.

2. Solve these algebraic equations for the transform of the output.

3. Evaluate the inverse transform to obtain the output as a function of time.

The first step transforms the original equations into a set of algebraic equations in the variable s. Then we may solve the algebraic equations by any convenient method to obtain the transform of the response. The inverse transform includes both the steady-state and the transient components and, unlike the classical approach, contains no unknown constants that must still be evaluated. A variety of examples illustrating this procedure follow.

EXAMPLE 12.6   The mechanical system shown in Figure 12.6 has the parameter values $M = 1$ kg, $B = 4$ N·s/m, and $K = 3$ N/m, and the applied force is $f_a(t) = 9$ N for all $t > 0$. The mass has no initial velocity, but it is released from a position 1 m to the right of its equilibrium position at the instant the force is applied. Find the displacement $x(t)$ for all $t > 0$.

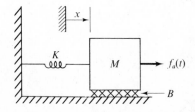

Figure 12.6 Mechanical system for Example 12.6.

Solution   The differential equation describing the system for $t > 0$ is $M\ddot{x} + B\dot{x} + Kx = f_a(t)$, which becomes

$$\ddot{x} + 4\dot{x} + 3x = 9$$

when the parameter and input values are substituted. Transforming the equation term by term gives

$$[s^2 X(s) - sx(0) - \dot{x}(0)] + 4[sX(s) - x(0)] + 3X(s) = \frac{9}{s}$$

Substituting the specified initial conditions $x(0) = 1$ and $\dot{x}(0) = 0$, we solve algebraically to obtain

$$X(s) = \frac{s + 4 + 9/s}{s^2 + 4s + 3} = \frac{s^2 + 4s + 9}{s(s+1)(s+3)}$$

Noting that $X(s)$ has three distinct real poles and that $m < n$, we write $X(s)$ in the form

$$X(s) = \frac{A_1}{s} + \frac{A_2}{s+1} + \frac{A_2}{s+3}$$

where

$$A_1 = sX(s)|_{s=0} = 3$$
$$A_2 = (s+1)X(s)|_{s=-1} = -3$$
$$A_3 = (s+3)X(s)|_{s=-3} = 1$$

Thus

$$X(s) = \frac{3}{s} - \frac{3}{s+1} + \frac{1}{s+3}$$

and the displacement is

$$x(t) = 3 - 3\epsilon^{-t} + \epsilon^{-3t} \qquad \text{for } t > 0$$

The response $x(t)$ reduces to the specified initial condition $x(0) = 1$ m and approaches a steady-state value of $x_{ss} = 3$ m. The two transient terms decay exponentially with time constants of 1 and $\frac{1}{3}$ seconds.

EXAMPLE 12.7 After steady-state conditions have been reached, the switch in Figure 12.7(a) opens at $t = 0$. Find the voltage e_o across the capacitor for all $t > 0$.

Solution The circuit for $t > 0$ is shown in Figure 12.7(b), with the switch open and with the node voltages e_o and e_A shown. The current-law equations

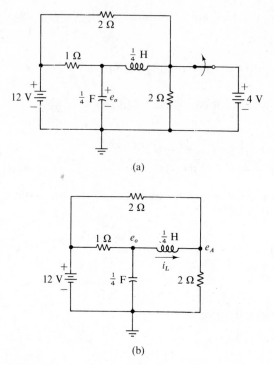

Figure 12.7 Circuit for Example 12.7. (a) Original circuit. (b) Circuit valid for $t > 0$.

at these two nodes are

$$e_o - 12 + \tfrac{1}{4}\dot{e}_o + i_L(0) + 4\int_0^t (e_o - e_A)\, d\lambda = 0$$

$$\tfrac{1}{2}(e_A - 12) - i_L(0) + 4\int_0^t (e_A - e_o)\, d\lambda + \tfrac{1}{2}e_A = 0$$

Transforming these equations, we have

$$E_o(s) - \frac{12}{s} + \frac{1}{4}[sE_o(s) - e_o(0)] + \frac{i_L(0)}{s} + \frac{4}{s}[E_o(s) - E_A(s)] = 0$$

$$\frac{1}{2}\left[E_A(s) - \frac{12}{s}\right] - \frac{i_L(0)}{s} + \frac{4}{s}[E_A(s) - E_o(s)] + \frac{1}{2}E_A(s) = 0$$

We must find the numerical values of the initial conditions $e_o(0)$ and $i_L(0)$ that appear in these transformed equations. Since e_o and i_L are measures of the energy stored in the capacitor and inductor, respectively, they cannot change instantaneously and do not have discontinuities at the time origin.

When the circuit is in the steady state with the switch closed, the capacitor and inductor may be replaced by open and short circuits, respectively. This is done in Figure 12.8(a), from which we find that $e_o(0) = 4$ V and $i_L(0) = 8$ A. Substituting these initial conditions into the transformed equations and collecting like terms, we get

$$\left[\frac{1}{4}s + 1 + \frac{4}{s}\right]E_o(s) - \frac{4}{s}E_A(s) = 1 + \frac{4}{s} \tag{64a}$$

$$-\frac{4}{s}E_o(s) + \left[1 + \frac{4}{s}\right]E_A(s) = \frac{14}{s} \tag{64b}$$

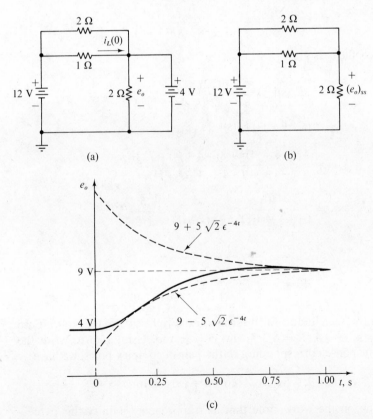

(a)

(b)

(c)

Figure 12.8 (a) Equivalent circuit for Example 12.7 just before the switch opens. (b) Equivalent circuit valid as t approaches infinity. (c) Complete response.

Since we want to find the capacitor voltage $e_o(t)$, the next step is to solve (64) for $E_o(s)$ by eliminating $E_A(s)$. Noting from (64b) that

$$E_A(s) = \frac{4E_o(s) + 14}{s + 4}$$

and substituting this expression into (64a), we find that

$$E_o(s) = \frac{4(s^2 + 8s + 72)}{s^3 + 8s^2 + 32s}$$

One pole of $E_o(s)$ is $s = 0$, and the remaining two are the roots of $s^2 + 8s + 32 = 0$. Hence the poles are $s_1 = 0$, $s_2 = -4 + j4$, and $s_3 = -4 - j4$, and we can expand the transformed output into the form

$$E_o(s) = \frac{A_1}{s} + \frac{A_2}{s + 4 - j4} + \frac{A_3}{s + 4 + j4} \tag{65}$$

where the coefficients are

$$A_1 = sE_o(s)|_{s=0} = \frac{(4)(72)}{32} = 9$$

$$A_2 = (s + 4 - j4)\,E_o(s)|_{s = -4 + j4}$$

$$= \frac{4[(-4 + j4)^2 + 8(-4 + j4) + 72]}{(-4 + j4)(-4 + j4 + 4 + j4)}$$

$$= \frac{4(40)}{(-4 + j4)(j8)} = \frac{5}{-(1 + j1)} = \frac{5}{\sqrt{2}}\epsilon^{j3\pi/4}$$

and

$$A_3 = A_2^* = \frac{5}{\sqrt{2}}\epsilon^{-j3\pi/4}$$

Comparing the second and third terms on the right side of (65) to (55), we see that $a = 4$, $\omega = 4$, $K = 5/\sqrt{2}$, and $\phi = \frac{3}{4}\pi$ rad. Using (56) to write the part of the response corresponding to the pair of complex poles, we have

$$e_o(t) = 9 + 5\sqrt{2}\,\epsilon^{-4t}\cos(4t + \tfrac{3}{4}\pi) \qquad \text{for } t > 0$$

As a check on the work, note that as t approaches infinity this expression gives a constant value of 9. We can also find the steady-state behavior with the switch open from Figure 12.8(b), where the capacitor and inductor have again been replaced by open and short circuits, respectively. The parallel combination of the 2-Ω and 1-Ω resistors is equivalent to $(2)(1)/(2+1) = \frac{2}{3}\,\Omega$. By the voltage-divider rule,

$$e_o(\infty) = \frac{2}{2+\frac{2}{3}}(12) = 9 \text{ V}$$

which agrees with the general equation for $t > 0$.

The complete response is shown in Figure 12.8(c). The transient component has an envelope that decays with a time constant of 0.25 s and has a period of $\frac{1}{2}\pi$ seconds.

EXAMPLE 12.8   Using Laplace transforms, find the impulse response $h(t)$ of a system that obeys the equation $\ddot{y} + 4\dot{y} + 4y = u(t)$.

Solution   Recall that the impulse response $h(t)$ is the zero-state response when $u(t) = \delta(t)$. If we rewrite the system equation as $\ddot{h} + 4\dot{h} + 4h = \delta(t)$ and assume that the impulse occurs at $t = 0+$, it follows that we should take $h(0)$ and $\dot{h}(0)$ as zero in evaluating the transforms of $\ddot{h}$ and $\dot{h}$.

Transforming both sides of the differential equation, using $\mathscr{L}[\delta(t)] = 1$ and denoting $\mathscr{L}[h(t)]$ by $H(s)$, we find that

$$[s^2 H(s) - sh(0) - \dot{h}(0)] + 4[sH(s) - h(0)] + 4H(s) = 1$$

which reduces to

$$(s^2 + 4s + 4)H(s) = 1$$

when the initial conditions are set equal to zero. Hence,

$$H(s) = \frac{1}{s^2 + 4s + 4} = \frac{1}{(s+2)^2}$$

Referring to Appendix B and noting that $h(t) = 0$ for all $t \leq 0$, we can write

$$h(t) = \begin{cases} 0 & \text{for } t \leq 0 \\ te^{-2t} & \text{for } t > 0 \end{cases} \tag{66}$$

which is shown in Figure 12.9.

Inspection of Figure 12.9 reveals that $\dot{h}(t)$ undergoes a discontinuity at $t = 0$. You can verify from (66) that $\dot{h}(0+) = 1$, whereas we used $\dot{h}(0) = 0$ in taking the transform of the original differential equation. This distinction between the initial slopes is consistent with our assertion that the impulse occurs at $t = 0+$.

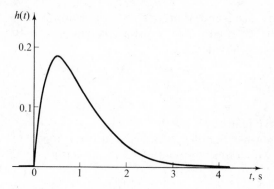

Figure 12.9 Impulse response found in Example 12.8.

EXAMPLE 12.9   Use the general procedure stated earlier in this section to solve for the unit step response of the circuit modeled in Example 9.10 and repeated in Figure 12.10.

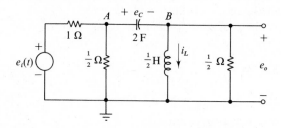

Figure 12.10 Circuit for Example 12.9.

Solution   Writing current-law equations at nodes A and B gives the pair of equations

$$e_A - e_i(t) + 2e_A + 2(\dot{e}_A - \dot{e}_o) = 0$$

$$2(\dot{e}_o - \dot{e}_A) + i_L(0) + 2\int_0^t e_o(\lambda)\, d\lambda + 2e_o = 0$$

(67)

Transforming (67) immediately, before performing any manipulations (e.g., differentiation), we obtain the pair of transformed equations

$$3E_A(s) + 2s[E_A(s) - E_o(s)] - e_C(0) = E_i(s)$$

$$2s[E_o(s) - E_A(s)] + e_C(0) + \frac{i_L(0)}{s} + \frac{2E_o(s)}{s} + 2E_o(s) = 0$$

where we have used $e_C(0)$ in place of $e_A(0) - e_o(0)$, which is involved in the transform of the term $\dot{e}_A - \dot{e}_o$. Since the definition of the step response requires zero initial stored energy, we take $i_L(0) = e_C(0) = 0$. Substitution of $1/s$ for $E_i(s)$ results in

$$(2s + 3) E_A(s) - 2sE_o(s) = \frac{1}{s} \tag{68a}$$

$$-2sE_A(s) + \left(2s + 2 + \frac{2}{s}\right) E_o(s) = 0 \tag{68b}$$

which we can solve for $E_o(s)$. Solving (68b) for $E_A(s)$, we get

$$E_A(s) = \left(\frac{s^2 + s + 1}{s^2}\right) E_o(s)$$

Substituting for $E_A(s)$ in (68a), we find that

$$\left[\frac{(2s + 3)(s^2 + s + 1)}{s^2} - 2s\right] E_o(s) = \frac{1}{s}$$

which reduces to

$$E_o(s) = \frac{s}{5s^2 + 5s + 3} \tag{69}$$

Since we have found the output transform $E_o(s)$, all that remains is to take the inverse transform to find the step response $e_o(t)$. The poles of $E_o(s)$ are $s_1 = -0.50 + j0.5916$ and $s_2 = -0.50 - j0.5916$. We can use either (56) or (58) for complex poles to show that the step response is

$$e_o(t) = 0.2619\epsilon^{-0.50t} \cos(0.5916t + 0.7017)$$

The following examples illustrate two alternative means of finding $E_o(s)$.

EXAMPLE 12.10 Starting with the state-variable equations derived in Example 9.10, use Laplace transforms to find the step response of the circuit discussed in Example 12.9.

Solution In Example 9.10, the state-variable and output equations were shown to be

$$\frac{di_L}{dt} = \tfrac{2}{5}[-i_L - 3e_C + e_i(t)]$$

$$\dot{e}_C = \tfrac{1}{10}[3i_L - 6e_C + 2e_i(t)]$$

and

$$e_o = \tfrac{1}{5}[-i_L - 3e_C + e_i(t)]$$

where the state variables i_L and e_C are shown in Figure 12.10. Transforming the state-variable and output equations gives the three algebraic transform equations

$$sI_L(s) - i_L(0) = \tfrac{2}{5}[-I_L(s) - 3E_C(s) + E_i(s)]$$
$$sE_C(s) - e_C(0) = \tfrac{1}{10}[3I_L(s) - 6E_C(s) + 2E_i(s)]$$

(70)

and

$$E_o(s) = \tfrac{1}{5}[-I_L(s) - 3E_C(s) + E_i(s)]$$ (71)

Although we could substitute $1/s = \mathscr{L}[U(t)]$ for $E_i(s)$ at this point, we shall retain $E_i(s)$ while solving the transform equations for $E_o(s)$. Doing so will result in a general relationship between $E_o(s)$ and $E_i(s)$ that is valid for any input that has a Laplace transform. Setting $i_L(0) = e_C(0) = 0$, we rearrange (70) in the form

$$(5s + 2)I_L(s) + 6E_C(s) = 2E_i(s)$$
$$(10s + 6)E_C(s) - 3I_L(s) = 2E_i(s)$$

These equations can be solved simultaneously for $I_L(s)$ and $E_C(s)$, giving

$$I_L(s) = \frac{2s}{5s^2 + 5s + 3} E_i(s)$$

$$E_C(s) = \frac{s+1}{5s^2 + 5s + 3} E_i(s)$$

Substituting these expressions into (71), the transformed output equation, and simplifying, we find that

$$E_o(s) = \frac{s^2}{5s^2 + 5s + 3} E_i(s)$$ (72)

We can now find the transform of the step response by merely substituting $E_i(s) = 1/s$ into (72), obtaining (69). However, (72) is a somewhat more general result than (69) in that it holds for any input, provided only that the input has a Laplace transform and that the initial stored energy is zero.

EXAMPLE 12.11   Repeat Example 12.9 by finding the input-output differential equation before using the Laplace transform.

Solution Starting with the node equations given by (67), we can use the p operator to show that the circuit's input-output equation is

$$5\ddot{e}_o + 5\dot{e}_o + 3e_o = \ddot{e}_i$$

We transform this equation, getting

$$5[s^2 E_o(s) - se_o(0) - \dot{e}_o(0)] + 5[sE_o(s) - e_o(0)] + 3E_o(s)$$
$$= s^2 E_i(s) - se_i(0) - \dot{e}_i(0) \qquad (73)$$

From (73), we see that the expression for $E_o(s)$ involves four different initial conditions, namely $e_o(0)$, $\dot{e}_o(0)$, $e_i(0)$, and $\dot{e}_i(0)$. Because there is zero initial stored energy and because the step-function input is assumed to occur just after $t = 0$ (that is, at $t = 0+$), it follows that each of these initial conditions is zero. Then (73) reduces to

$$(5s^2 + 5s + 3) E_o(s) = s^2 \left(\frac{1}{s}\right)$$

which is equivalent to (69).

In retrospect, using the input-output equation requires an assumption regarding precisely when the input commences and a careful evaluation of initial conditions. It is recommended that either the system's equations be transformed directly, without manipulation, or the state-variable form of the model be transformed.

12.5 ADDITIONAL TRANSFORM PROPERTIES

We now introduce several useful transform properties that were not needed for the examples in the Section 12.4. The first two of these concern functions that are shifted in time, and the others provide certain information about the time function directly from its transform, without our having to carry out the transform inversion.

TIME DELAY

If the function $f(t)$ is delayed by a units of time, we denote the delayed function as $f(t - a)$ where $a > 0$. In order to develop a general expression for the transform of the delayed function $f(t - a)$ in terms of the transform of $f(t)$, we must ensure that any part of $f(t)$ that is nonzero for $t < 0$ does not fall within the range $0 < t < \infty$ for the delayed function. Otherwise, a portion of the original time function will contribute to the transform of $f(t-a)$ but not to that of $f(t)$. To illustrate this point, consider the functions $f(t)$ and $f(t-a)$ shown in Figure 12.11. The shaded portion of $f(t)$ that is nonzero for $t < 0$ does not affect $\mathscr{L}[f(t)]$, because it is outside the limits of integration in (1). It does affect $\mathscr{L}[f(t-a)]$, however, because for $a > 0$, at least part of it falls within the interval $0 < t < \infty$, as shown in Figure 12.11(b).

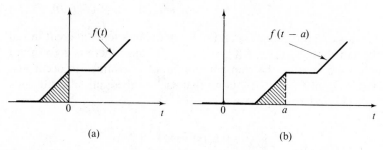

(a)   (b)

Figure 12.11 Illustration of a function for which the time-delay theorem is not applicable.

We shall consider functions of the form

$$f_1(t) = f(t)\, U(t)$$

which are the product of any transformable function $f(t)$ and the unit step function $U(t)$. Since $U(t) = 0$ for all $t \le 0$, the function $f_1(t)$ will be zero for all $t \le 0$, and the delayed function

$$f_1(t-a) = f(t-a)\, U(t-a)$$

will be zero for all $t \le a$. The functions $f_1(t)$ and $f_1(t-a)$ corresponding to the $f(t)$ defined in Figure 12.11(a) are shown in Figure 12.12.

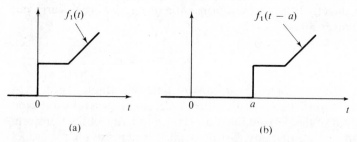

(a)   (b)

Figure 12.12 Illustration of a function for which the time-delay theorem is applicable.

To express the transform of $f(t-a)U(t-a)$, where $a > 0$, in terms of $F(s) = \mathscr{L}[f(t)]$, we start with the transform definition in (1) and write

$$\mathscr{L}[f(t-a)U(t-a)] = \int_0^\infty f(t-a)U(t-a)\epsilon^{-st}\,dt$$

Since

$$U(t-a) = \begin{cases} 0 & \text{for } t \le a \\ 1 & \text{for } t > a \end{cases}$$

we can rewrite the transform as

$$\mathscr{L}[f(t-a)U(t-a)] = \int_a^\infty f(t-a)\epsilon^{-st}\,dt$$

$$= \epsilon^{-sa}\int_a^\infty f(t-a)\epsilon^{-s(t-a)}\,dt$$

$$= \epsilon^{-sa}\int_0^\infty f(\lambda)\epsilon^{-s\lambda}\,d\lambda$$

$$= \epsilon^{-sa}F(s) \tag{74}$$

where $F(s)$ is the transform of $f(t)$ and where $a > 0$.

EXAMPLE 12.12   Use the time-delay theorem to derive the transform of the triangular pulse shown in Figure 12.13(a) and defined by the equation

$$f(t) = \begin{cases} 0 & \text{for } t \le 0 \\ At & \text{for } 0 < t \le L \\ 0 & \text{for } t > L \end{cases}$$

Solution   Any pulse that consists of straight lines can be decomposed into a sum of step functions and ramp functions. The triangular pulse shown in Figure 12.13(a) can be regarded as the superposition of the three functions shown in Figure 12.13(b), Figure 12.13(c), and Figure 12.13(d). These are a ramp starting at $t = 0$, a delayed ramp starting at $t = L$, and a delayed step function starting at $t = L$. Thus

$$f(t) = AtU(t) - A(t-L)U(t-L) - ALU(t-L) \tag{75}$$

From Appendix B, we note that

$$\mathscr{L}[AtU(t)] = \frac{A}{s^2} \tag{76}$$

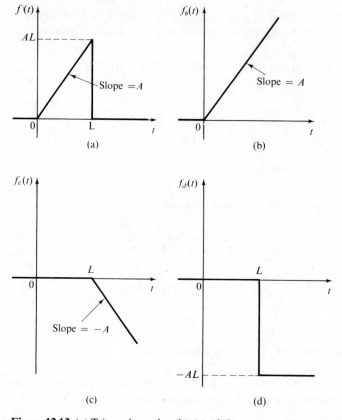

Figure 12.13 (a) Triangular pulse. (b), (c), (d) Its ramp and step components.

Using (74) with (76), we have

$$\mathscr{L}[-A(t-L)\,U(t-L)] = -\frac{A\epsilon^{-sL}}{s^2}$$

From (74) and the fact that $\mathscr{L}[U(t)] = 1/s$, it follows that

$$\mathscr{L}[-ALU(t-L)] = -\frac{AL\epsilon^{-sL}}{s}$$

Using the superposition theorem, we obtain the transform of the triangular pulse as

$$F(s) = \frac{A}{s^2}(1-\epsilon^{-sL}) - \frac{AL}{s}\epsilon^{-sL} \tag{77}$$

INVERSION OF SOME IRRATIONAL TRANSFORMS

The use of a partial-fraction expansion to find an inverse transform is restricted to transforms that are rational functions of s. However, the transform given by (77) is not a rational function because of the factor ϵ^{-sL}. For transforms that would be rational functions except for multiplicative factors in the numerator such as ϵ^{-sL}, we may use the time-delay theorem in (74).

Assume that an irrational transform can be written as

$$F(s) = F_1(s) + F_2(s)\epsilon^{-sa} \tag{78}$$

where $F_1(s)$ and $F_2(s)$ are rational functions and a is a positive constant. Then we can find the inverse transforms of $F_1(s)$ and $F_2(s)$ by using partial-fraction expansions and, if the order of the numerator is not less than that of the denominator, a preliminary step of long division. Denote the inverse transforms of $F_1(s)$ and $F_2(s)$ by $f_1(t)$ and $f_2(t)$, respectively. Then, by (74),

$$f(t) = f_1(t) + f_2(t-a)\,U(t-a) \tag{79}$$

EXAMPLE 12.13 Find the inverse transform of

$$F(s) = \frac{A}{s^2}(1 - \epsilon^{-sL}) - \frac{AL}{s}\epsilon^{-sL}$$

Solution We can rewrite the transform $F(s)$ in the form of (78) as

$$F(s) = \frac{A}{s^2} - \left(\frac{A}{s^2} + \frac{AL}{s}\right)\epsilon^{-sL}$$

Hence, $F_1(s) = A/s^2$ and $f_1(t) = At$ for $t > 0$. The rational portion of the remaining term is

$$F_2(s) = -\left(\frac{A}{s^2} + \frac{AL}{s}\right)$$

which has as its inverse transform

$$f_2(t) = -At - ALU(t) \qquad \text{for } t > 0$$

Using (79) with $a = L$, we can write the complete inverse transform as

$$f(t) = At - A(t-L)\,U(t-L) - ALU(t-L) \qquad \text{for } t > 0$$

which agrees with (75) and Figure 12.13(a).

INITIAL-VALUE AND FINAL-VALUE THEOREMS

It is possible to determine the limits of $f(t)$ as time approaches zero and infinity directly from its transform $F(s)$ without having to find $f(t)$ for all $t > 0$. First we consider the limit of $f(t)$ as time approaches zero through positive values (i.e., from the right). This limit of $f(t)$ is denoted by $f(0+)$. To evaluate this limit directly from $F(s)$, we use the *initial-value theorem*, which states that

$$f(0+) = \lim_{s \to \infty} sF(s) \tag{80}$$

where the limit exists.

If $F(s)$ is a rational function, (80) will yield a finite value provided that the degree of the numerator polynomial is less than that of the denominator, i.e., provided that $m < n$. If we attempt to use (80) when $m = n$, the result will be infinite. Recall from Section 12.3 that to find $f(t)$ when $m = n$, we must use a preliminary step of long division to write $F(s)$ as the sum of a constant and a transform for which $m < n$. Since the inverse transform of the constant is an impulse at $t = 0+$, the value of $f(0+)$ is undefined when $m = n$.

The *final-value theorem* states that

$$f(\infty) = \lim_{s \to 0} sF(s) \tag{81}$$

provided that $F(s)$ has no poles in the right half of the complex plane and, with the possible exception of a single pole at the origin, has no poles on the imaginary axis. The symbol $f(\infty)$ denotes the limit of $f(t)$ as t approaches infinity. Proofs of both these theorems are given in Cannon, Section 17.8.

To gain some insight into the effect of this restriction on the use of the final-value theorem, we recall that the forms of the terms in a partial-fraction expansion are dictated by the locations of the poles of $F(s)$. Suppose, for example, that

$$F(s) = \frac{A_1}{s} + \frac{A_2(s+\alpha)}{(s+\alpha)^2 + \beta^2} + \frac{A_3}{s-b} + \frac{A_4 \omega}{s^2 + \omega^2}$$

where α, β, b, and ω are positive real constants. The expansion implies that the poles of $F(s)$ are $s_1 = 0$, $s_2 = -\alpha + j\beta$, $s_3 = -\alpha - j\beta$, $s_4 = b$, $s_5 = j\omega$, and $s_6 = -j\omega$. The corresponding time function for $t > 0$ is

$$f(t) = A_1 + A_2 \epsilon^{-\alpha t} \cos \beta t + A_3 \epsilon^{bt} + A_4 \sin \omega t$$

The limits of the first two terms as t approaches infinity are A_1 and zero, respectively. However, $A_3 \epsilon^{bt}$ increases without limit, while $A_4 \sin \omega t$ oscillates continually without approaching a constant value. Thus, because of the poles of $F(s)$ at $s_4 = b$, $s_5 = j\omega$, and $s_6 = -j\omega$, the function $f(t)$ does not approach a limit as t approaches infinity. As another example, double poles of $F(s)$ at the origin will cause the partial-fraction expansion to have terms of the form

$$F(s) = \frac{A_{11}}{s^2} + \frac{A_{12}}{s} + \cdots$$

The corresponding time function for $t > 0$ is

$$f(t) = A_{11}t + A_{12} + \cdots$$

which again does not approach a limit.

The use of the initial-value and final-value theorems is illustrated in the following example and in the next chapter. Example 12.15 considers a transform for which neither theorem is applicable.

EXAMPLE 12.14   Use the initial-value and final-value theorems to find $f(0+)$ and $f(\infty)$ when

$$F(s) = \frac{s^2 + 2s + 4}{s^3 + 3s^2 + 2s} \tag{82}$$

Solution   From (80), the initial value of $f(t)$ is

$$f(0+) = \lim_{s \to \infty} \frac{s(s^2 + 2s + 4)}{s^3 + 3s^2 + 2s}$$

$$= \lim_{s \to \infty} \frac{s^3 + 2s^2 + 4s}{s^3 + 3s^2 + 2s} \tag{83}$$

Because $f(0+)$ is the limit of a ratio of polynomials in s as s approaches infinity, we need consider only the highest powers in s in both the numerator and denominator. Hence, (83) reduces to

$$f(0+) = \lim_{s \to \infty} \frac{s^3}{s^3} = 1$$

Before applying the final-value theorem, we must verify that the conditions necessary for it to be valid are satisfied. In this case, we can rewrite (82) with its denominator in factored form as

$$F(s) = \frac{s^2 + 2s + 4}{s(s+1)(s+2)} \tag{84}$$

which has distinct poles at $s = 0$, -1, and -2. Since $F(s)$ has a single pole at the origin with the remaining poles inside the left half of the s-plane, we can apply the final-value theorem. Using (81), we find that

$$f(\infty) = \lim_{s \to 0} \frac{s(s^2 + 2s + 4)}{s(s^2 + 3s + 2)}$$

$$= \lim_{s \to 0} \frac{s^2 + 2s + 4}{s^2 + 3s + 2} = 2$$

In this example, it is a simple task to evaluate $f(t)$ for all $t > 0$ by writing $F(s)$ as given by (84) in its partial-fraction expansion. The result is

$$f(t) = 2 - 3\epsilon^{-t} + 2\epsilon^{-2t} \qquad \text{for } t > 0$$

and it is apparent that the values we found for $f(0+)$ and $f(\infty)$ are correct.

EXAMPLE 12.15   Explain why the initial-value and final-value theorems are not applicable to the transform

$$F(s) = \frac{s^3 + 2s^2 + 6s + 8}{s^3 + 4s} \tag{85}$$

Solution   Attempting to apply the initial-value theorem, we would write

$$f(0+) = \lim_{s \to \infty} \frac{s(s^3 + 2s^2 + 6s + 8)}{s^3 + 4s}$$

$$= \lim_{s \to \infty} \frac{s^4 + 2s^3 + 6s^2 + 8s}{s^3 + 4s}$$

which is infinite. This is expected, because $m = n = 3$ in (85).

As for the final-value theorem, we see from (85) that we can write the denominator of $F(s)$ as $s(s^2 + 4) = s(s - j2)(s + j2)$. Hence, $F(s)$ has a pair of imaginary poles at $s_2 = j2$ and $s_3 = -j2$ and violates the requirements for the final-value theorem. If we carry out the partial-fraction expansion of $F(s)$, we find that

$$F(s) = 1 + \frac{2}{s} + \frac{2}{s^2 + 4}$$

which is the transform of the time function

$$f(t) = \delta(t) + 2 + \sin 2t \qquad \text{for } t > 0.$$

The initial-value theorem is invalid because of the impulse at $t = 0+$, and the final-value theorem is invalid because of the constant-amplitude sinusoidal term.

PROBLEMS

12.1 Using the definition of the Laplace transform in (1), evaluate the transforms of the following functions.

    a. $f_1(t) = t^2$

    b. $f_2(t) = \epsilon^{-at} \cos \omega t$

    c. The function $f_3(t)$ shown in Figure P12.1(a).

    d. The function $f_4(t)$ shown in Figure P12.1(b).

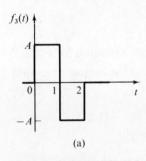
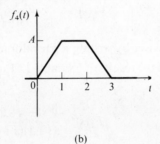

    (a)   (b)

Figure P12.1

12.2 a. Derive the expressions given in (14) and (15) for the transforms of $\sin \omega t$ and $\cos \omega t$ by using the identities in Table 8.1 and then applying (5).

    b. Using (14) and (15) with Table 8.1, derive expressions for the transforms of $\sin(\omega t + \phi)$ and $\cos(\omega t + \phi)$.

12.3 Use the properties tabulated in Appendix B to find the Laplace transform of each of the following functions of time.

    a. $f_1(t) = t\epsilon^{-2t} \cos 3t$

    b. $f_2(t) = t^2 \sin 2t$

    c. $f_3(t) = \dfrac{d}{dt}(t^2\epsilon^{-t})$

    d. $f_4(t) = \displaystyle\int_0^t \lambda^2 \epsilon^{-\lambda}\, d\lambda$

12.4 a. Use (31) and (24) to derive an expression for $\mathscr{L}[\ddot{f}]$.

b. Apply the result of part a to $f(t) = t \sin \omega t$ for all t. (*Hint*: Use (20) to show that $\mathscr{L}[t \sin \omega t] = 2\omega s/(s^2 + \omega^2)^2$.)

c. Solve part b by first finding an expression for $\ddot{f}$ and then taking its Laplace transform; that is, do not use the result of part a. Compare your answer to the one for part b and sketch $\ddot{f}$.

d. Repeat parts b and c for $f(t) = [t \sin \omega t] U(t)$.

12.5 a. Prove that $\mathscr{L}[f(t/a)] = aF(as)$.

b. Apply this property with $f(t) = \cos \omega t$ to find $\mathscr{L}[\cos 2\omega t]$.

12.6 a. Use (34) and the results of Example 12.1 to find the inverse transform of

$$F_1(s) = \frac{-s+5}{s(s+1)(s+4)}$$

Check your answer by writing a partial-fraction expansion for $F_1(s)$.

b. Use (24) and the results of Example 12.1 to find the inverse transform of

$$F_2(s) = \frac{s(-s+5)}{(s+1)(s+4)}$$

Check your answer by writing a partial-fraction expansion.

12.7 Using the results of Example 12.2, repeat Problem 12.6 for the following transforms.

a. $F_1(s) = \dfrac{5s+16}{s(s+2)^2(s+5)}$

b. $F_2(s) = \dfrac{s(5s+16)}{(s+2)^2(s+5)}$

For Problems 12.8 through 12.11, find $f(t)$ for the given $F(s)$.

12.8 a. $F(s) = \dfrac{2s^3 + 3s^2 + s + 4}{s^3}$

b. $F(s) = \dfrac{3s^2 + 9s + 24}{(s-1)(s+2)(s+5)}$

c. $F(s) = \dfrac{4}{s^2(s+1)}$

d. $F(s) = \dfrac{3s}{s^2+2s+26}$

12.9 a. $F(s) = \dfrac{s}{s^2+8s+16}$

b. $F(s) = \dfrac{1}{s(s^2+\omega^2)}$

c. $F(s) = \dfrac{8s^2+20s+74}{s(s^2+s+9.25)}$

d. $F(s) = \dfrac{2s^2+11s+16}{(s+2)^2}$

12.10 a. $F(s) = \dfrac{s^3+2s+4}{s(s+1)^2(s+2)}$

b. $F(s) = \dfrac{4s^2+10s+10}{s^3+2s^2+5s}$

c. $F(s) = \dfrac{3(s^3+2s^2+4s+1)}{s(s+3)^2}$

d. $F(s) = \dfrac{s^3-4s}{(s+1)(s^2+4s+4)}$

12.11 $F(s) = \dfrac{1}{(s^2+\omega^2)^2}$

12.12 a. Write the partial-fraction expansion for

$$F(s) = \frac{\alpha_1\alpha_2}{(s+\alpha_1)(s+\alpha_2)}$$

and find the inverse transform when α_1 and α_2 are positive constants and $\alpha_1 \neq \alpha_2$.

b. Find the inverse transform when $\alpha_1 = \alpha_2$.

c. Show that when $\alpha_1 = \alpha_2$, the answer for part a becomes an indeterminate form. Evaluate the indeterminate form by using L'Hôpital's rule, and compare the result to your answer to part b.

d. On the same axes, sketch $f(t)/\alpha_2$ vs $\alpha_2 t$ for $\alpha_1 = 0.5\alpha_2$ and for $\alpha_1 = \alpha_2$.

12.13 a. Find the inverse transform of

$$F(s) = \frac{s}{(s^2 + \omega_1^2)(s^2 + \omega_2^2)}$$

when $\omega_1 \neq \omega_2$.

b. Repeat part a when $\omega_1 = \omega_2$.

c. Show that when $\omega_1 = \omega_2$, the answer for part a becomes an indeterminate form. Evaluate the indeterminate form by using L'Hôpital's rule, and compare the result to your answer to part b.

12.14 a. Derive the expressions given in (50) for the coefficients in a partial-fraction expansion when $F(s)$ has a third-order pole.

b. Use (50) to find the inverse transform of

$$F(s) = \frac{4s^2 + 2s + 1}{s(s + 2)^3}$$

12.15 By comparing (56) and (58) and using Table 8.1, verify the four equations relating B_1 and B_3 to K and ϕ that follow (59).

12.16 a. Find the inverse transform of

$$F(s) = \frac{s^3 + 5s^2 + 21s + 29}{(s^2 + 2s + 5)(s^2 + 4s + 13)}$$

by evaluating the constants A, B, C, and D in (61).

b. Repeat part a by evaluating K_1, K_3, ϕ_1, and ϕ_3 in the equation preceding (61) and then using (56).

12.17 Find the inverse transform of $E_o(s)$ in Example 12.7 by first finding the constants A, B, and C in the expansion

$$E_o(s) = \frac{A}{s} + \frac{Bs + C}{s^2 + 8s + 32}$$

and then completing the square to evaluate the inverse transform of the term having complex poles.

12.18 Verify the expression given for $e_o(t)$ in Example 12.9 by taking the inverse transform of $E_o(s)$ given in (69).

12.19 For a system described by the equation $\ddot{y} + 3\dot{y} + 2y = u(t)$, use the Laplace transform to find $y(t)$ for $t > 0$ when the input is $u(t) = 5t$ for $t > 0$, and when the initial conditions are $y(0) = 1$ and $\dot{y}(0) = -1$. Sketch $y(t)$.

12.20 For the translational mechanical system shown in Figure P2.11, let $M = 1$ kg, $B_1 = 1$ N·s/m, $B_2 = 2$ N·s/m, $K_1 = 1$ N/m, $K_2 = 2$ N/m, and $K_3 = 3$ N/m. Find an expression for $X_1(s)$ in terms of the input transform $X_3(s)$ and the initial conditions $x_1(0)$, $x_2(0)$, and $v_1(0)$.

12.21 For the translational mechanical system shown in Figure P2.23, let both $f_a(t)$ and $x_2(t)$ be inputs. Use as parameter values $M_1 = 1$ kg, $B_1 = 2$ N·s/m, $K_1 = 3$ N/m, $K_2 = 6$ N/m, and $a = b$.

    a. Find an expression for $X_1(s)$ in terms of the input transforms $F_a(s)$ and $X_2(s)$ and the initial conditions $x_1(0)$ and $v_1(0)$.

    b. Find and sketch $x_1(t)$ for $t > 0$ for the inputs $f_a(t) = 10$ N and $x_2(t) = 0$ and for the initial conditions $x_1(0) = v_1(0) = 0$.

    c. Repeat part b when $f_a(t) = 0$ and $x_2(t) = t\epsilon^{-2t}$.

12.22 For the translational mechanical system shown in Figure 3.5, the parameter values are $M = 1$ kg, $B = 6$ N·s/m, $K_1 = 9$ N/m, and $K_2 = 4$ N/m.

    a. Find and sketch $x(t)$ for $t > 0$ if $x(0) = -1$ m, $v(0) = 0$, and $f_a(t) = 0$.

    b. Repeat part a with $f_a(t) = 2\epsilon^{-3t}$ and the same initial conditions.

12.23 For the translational mechanical system shown in Figure 3.6(a) and discussed in Example 3.4, let $M = 1$ kg, $B_1 = 2$ N·s/m, $B_2 = 1$ N·s/m, $K_1 = 1$ N/m, and $K_2 = 2$ N/m.

    a. Transform (3.7) and solve for $X_1(s)$ in terms of the input transform $F_a(s)$ and the initial conditions $x_1(0)$, $x_2(0)$, and $v_1(0)$.

    b. Find and sketch $x_1(t)$ for $t > 0$ when the initial conditions are $x_1(0) = x_2(0) = 0$ and $v_1(0) = -1$ m/s and when the input is $f_a(t) = 2$ N.

12.24 For the rotational system shown in Figure 4.17 and discussed in Example 4.5, let $J_1 = J_2 = 0.5$ kg·m², $B = 1$ N·m·s, and $K = 2$ N·m.

    a. Find an expression for $\Omega_2(s) = \mathscr{L}[\omega_2(t)]$ by transforming (4.34).

    b. If the system is initially at rest and the inputs are $\tau_L(t) = 0$ and $\tau_a(t) = \epsilon^{-2t}$, find and sketch ω_2 for $t > 0$.

For Problems 12.25 through 12.32, use the Laplace transform to solve for the response for $t > 0$ in the example or problem indicated.

12.25 Example 8.2.

12.26 Problem 8.2.

12.27 Problem 8.22.

12.28 Example 9.6.

12.29 Problem 9.12.

12.30 Problem 9.13.

12.31 Part e of Problem 9.25.

12.32 Problem 11.8.

12.33 Find and sketch e_o vs t for the circuit shown in Figure 9.17 and discussed in Example 9.4. The parameter values are $C = 0.5$ F, $R_1 = 2\ \Omega$, $R_2 = 1\ \Omega$, and $L = 4$ H. The inputs are $e_1(t) = 2$ V for all t and $e_2(t) = U(t)$ volts. Steady-state conditions exist at $t = 0$.

12.34 For the motor described by the state-variable equations in (11.29), let $L_A = 1$ H, $R_A = 5\ \Omega$, $J = 0.4$ kg·m², $B = 0.2$ N·m·s, and $\alpha = 2$ V·s. Assume that $e_i(t) = 100$ V for all t, that $\tau_L(t) = 20U(t)$ newton-meters, and that steady-state conditions exist at $t = 0$. Use the Laplace transform to find and sketch the angular velocity ω for $t > 0$.

12.35 For the electromechanical system shown in Figure P11.9, let $R = 40$ Ω, $L = 0.1$ H, $M = 0.1$ kg, and $\mathscr{B}d = 20$ V·s/m. If the initial conditions are $i(0) = v(0) = 0$ and the input is $e_i(t) = U(t)$ volts, find and sketch the velocity $v(t)$ for $t > 0$.

12.36 For the translational mechanical system shown in Figure P12.36, the parameter values are $M_1 = M_2 = 8$ kg, $B = 2$ N·s/m, and $K = 0.5$ N/m. Assume that at $t = 0$, the displacements and velocities are zero and the spring is neither stretched nor compressed.

    a. Draw appropriate free-body diagrams and derive the system equations.

    b. Transform the equations describing the system and solve for $X_1(s)$ in terms of $F_1(s)$ and $F_2(s)$.

    c. Find and sketch $v_1 = \dot{x}_1$ for $t > 0$ if $f_1(t) = 0$ and $f_2(t) = \delta(t)$.

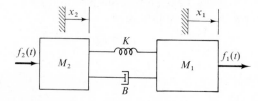

Figure P12.36

12.37 In the mechanical system shown in Figure P12.37, the two masses can move horizontally. They are released with $x_1(0) = x_{1_0}$ and $x_2(0) = x_{2_0}$ but with no initial velocity. Find an expression for $X_1(s)$.

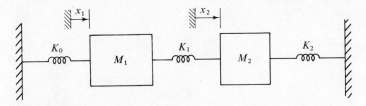

Figure P12.37

12.38 a. For the circuit shown in Figure P12.38, apply the node-equation method to nodes A and B and transform the resulting equations. Solve for $E_o(s)$ in terms of the input transforms $E_i(s)$ and $I_a(s)$ and the initial conditions $i_L(0)$ and $e_C(0)$.

b. Use the final-value theorem to find $e_o(\infty)$ when $e_i(t) = U(t)$ and $i_a(t) = 0$.

c. Repeat part b when $e_i(t) = 0$ and $i_a(t) = U(t)$.

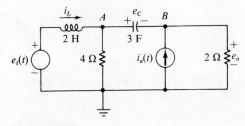

Figure P12.38

12.39 For the circuit shown in Figure P12.39, $e_i(t)$ is 4 V for all $t < 0$ and is 2 V for all $t > 0$. Use the Laplace transform to find and sketch e_o vs t for $t > 0$.

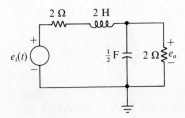

Figure P12.39

12.40 Repeat Problem 12.39 for the circuit shown in Figure P12.40.

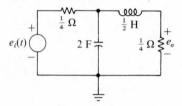

Figure P12.40

12.41 Steady-state conditions exist at $t = 0-$, with the switch closed, for the circuit shown in Figure P12.41. The switch opens at $t = 0$ and remains open thereafter. Find and sketch e_o and i_o for $t > 0$.

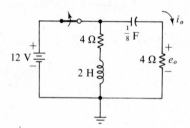

Figure P12.41

12.42 a. Use the time-delay theorem in (74) to derive the Laplace transform of the rectangular pulse shown in Figure 12.2.

b. Find the Laplace transform of the triangular pulse shown in Figure 12.13(a) by using (1) directly, rather than by decomposing the pulse into ramp and step components. Compare your result to (77).

12.43 Repeat part c and part d of Problem 12.1 by decomposing the functions into step functions and ramp functions, as appropriate.

12.44 Use (74) to sketch the inverse transform of $F(s) = A/[s(1 + \epsilon^{-s})]$. *Hint:* Use the infinite series

$$\frac{1}{1+\epsilon^{-s}} = 1 - \epsilon^{-s} + \epsilon^{-2s} - \epsilon^{-3s} + \cdots$$

12.45 Apply the initial-value and final-value theorems to find $f(0+)$ and $f(\infty)$ for each of the four transforms in Problem 12.10. If either theorem is not applicable to a particular transform, explain why this is so.

12.46 Repeat Problem 12.45 for the following transforms.

    a. $E_o(s)$ in Example 12.7.

    b. $E_o(s)$ in Example 12.9.

    c. The functions in part c and part d of Problem 12.9.

    d. $F(s) = \dfrac{4s^3}{(5s^2 + 3)^2}$

12.47 a. Using (24) and the initial-value theorem, show that

$$\dot{f}(0+) = \lim_{s \to \infty} [s^2 F(s) - sf(0)]$$

provided that the limit exists.

b. Use the property derived in part a to find $\dot{f}(0+)$ for the transform $F(s)$ given in (82). Assume that $f(t) = 1$ for $t \le 0$. If this property is not applicable, explain why this is so.

c. Repeat part b, assuming that $f(t) = 0$ for $t \le 0$.

d. Check your answers to part b and part c by differentiating the expression given for $f(t)$ in Example 12.14.

CHAPTER 13
TRANSFER-FUNCTION ANALYSIS

In Chapter 12 we presented the Laplace transform and used it to obtain the response of fixed linear dynamic systems. In this chapter, we shall examine further the transform solution of system models in the form of input-output differential equations.

We shall first consider the zero-input response and relate its transform to the system's characteristic polynomial that was defined in Chapter 8. Then we shall study the transform solution for the zero-state response and develop the concept of the transfer function. Following a discussion of the transform solution for the complete response, we shall examine in detail the responses to the unit impulse, the unit step function, and sinusoidal functions.

13.1 THE ZERO-INPUT RESPONSE

The mathematical model of an nth-order system with a single output can take any of the following forms:

1. A single nth-order input-output differential equation

2. An algebraic output equation and n simultaneous first-order differential equations

3. A set of differential equations at least one of which is of second or higher order, e.g., coupled second-order equations

In any of these cases, the Laplace transform and partial-fraction expansions provide a valuable tool for finding the system's response. Although the general mathematical development in this section is restricted to models in the form of an input-output equation, we shall illustrate the treatment of the multiequation forms by an example. We shall discuss the state-variable form in Chapter 17.

In Section 8.2, we defined the zero-input response to be the output when the input $u(t)$ is zero for all $t > 0$ and when the initial conditions are non-zero. Hence, the zero-input response of a fixed linear nth-order system satisfies a homogeneous nth-order differential equation of the form

$$a_n y^{(n)} + a_{n-1} y^{(n-1)} + \cdots + a_1 \dot{y} + a_0 y = 0 \qquad (1)$$

which is similar to (8.4) and where $y^{(n)}$ denotes $d^n y/dt^n$. The solution to (1) will involve the n initial conditions $y(0)$, $\dot{y}(0)$, ..., $y^{(n-1)}(0)$. Using the expressions for the transforms of derivatives given in Appendix B, we can transform (1) term by term to obtain

$$a_n [s^n Y(s) - s^{n-1} y(0) - \cdots - y^{(n-1)}(0)]$$
$$+ a_{n-1} [s^{n-1} Y(s) - s^{n-2} y(0) - \cdots - y^{(n-2)}(0)]$$
$$+ \cdots + a_1 [sY(s) - y(0)] + a_0 Y(s) = 0 \qquad (2)$$

where $Y(s) = \mathscr{L}[y(t)]$. If we retain the terms involving $Y(s)$ on the left side and collect those involving the initial conditions on the right side, (2) becomes

$$(a_n s^n + a_{n-1} s^{n-1} + \cdots + a_1 s + a_0) Y(s)$$
$$= a_n y(0) s^{n-1} + [a_n \dot{y}(0) + a_{n-1} y(0)] s^{n-2} + \cdots$$
$$+ [a_n y^{(n-1)}(0) + a_{n-1} y^{(n-2)}(0) + \cdots + a_1 y(0)]$$

Thus the transform of the zero-input response is

$$Y(s) = \frac{F(s)}{P(s)} \qquad (3)$$

where

$$F(s) = a_n y(0) s^{n-1} + [a_n \dot{y}(0) + a_{n-1} y(0)] s^{n-2} + \cdots$$
$$+ [a_n y^{(n-1)}(0) + a_{n-1} y^{(n-2)}(0) + \cdots + a_1 y(0)] \tag{4}$$

and

$$P(s) = a_n s^n + a_{n-1} s^{n-1} + \cdots + a_1 s + a_0 \tag{5}$$

Because $P(s)$ is of degree n and $F(s)$ is at most of degree $n-1$, $Y(s)$ is a strictly proper rational function. The numerator polynomial $F(s)$ depends on the initial conditions. The denominator polynomial $P(s)$ is identical to the characteristic polynomial in (8.5), except that it is written in terms of the complex variable s rather than r. When $P(s)$ is factored, it will have the form

$$P(s) = a_n(s - s_1)(s - s_2) \cdots (s - s_n)$$

As discussed in Section 12.3, the quantities $s_1, s_2, \ldots, s_n$ are the poles of the transformed output; hence, they determine the form of the zero-input response. They are also the roots of the characteristic equation given by (8.5). Thus the discussion in Chapter 8 about the roots of the characteristic equation applies equally well to the poles of the transformed output when the input is zero for $t > 0$.

For a first-order system, where $n = 1$,

$$P(s) = a_1 s + a_0$$
$$= a_1(s + 1/\tau)$$

where τ is the time constant. The zero-input response will have one of the forms shown in Figure 8.1.

For a second-order system, where $n = 2$,

$$P(s) = a_2 s^2 + a_1 s + a_0$$

Possible pole positions and the corresponding typical zero-input responses are shown in Figure 8.15 through Figure 8.18. When a_0 and a_2 have the same sign, we can rewrite $P(s)$ as

$$P(s) = a_2(s^2 + 2\zeta\omega_n s + \omega_n^2)$$

where ζ and ω_n are the damping ratio and the undamped natural frequency, respectively. We pointed out the significance of these two parameters in the discussion associated with Figure 8.19.

MODE FUNCTIONS

Since $Y(s)$ is a strictly proper rational function, we can expand it in partial fractions by the methods of Section 12.3. If the characteristic equation $P(s) = 0$ has the distinct roots $s_1, s_2, \ldots, s_n$, then

$$
\begin{aligned}
Y(s) &= \frac{F(s)}{a_n(s - s_1)(s - s_2)\cdots(s - s_n)} \\[2mm]
&= \frac{A_1}{s - s_1} + \frac{A_2}{s - s_2} + \cdots + \frac{A_n}{s - s_n}
\end{aligned}
\tag{6}
$$

where the coefficients $A_1, A_2, \ldots, A_n$ depend on the initial conditions $y(0)$, $\dot{y}(0), \ldots, y^{(n-1)}(0)$. Since $A_i/(s - s_i) = \mathscr{L}[A_i\,e^{s_i t}]$, the inverse transform of (6) is

$$
y(t) = A_1\,e^{s_1 t} + A_2\,e^{s_2 t} + \cdots + A_n\,e^{s_n t}
\tag{7}
$$

The exponential functions $e^{s_i t}$ that comprise $y(t)$ are referred to as the *mode functions* or *modes* of the system's zero-input response. Since the s_i are the roots of the characteristic equation, the mode functions are properties of the system. However, the coefficients A_i in (7) are the weightings of the individual mode functions, and they depend on the particular initial conditions. In fact, we can select specific initial conditions so as to eliminate any of the mode functions in the zero-input response by forcing the corresponding A_i to be zero. The following three examples will illustrate the manner in which we can use the Laplace transform to solve for a system's zero-input response and the manner in which the initial conditions affect the weighting of the system's modes.

EXAMPLE 13.1 The rotational mechanical system shown in Figure 4.11(a) is redrawn in Figure 13.1. For the parameter values $J = 1$ kg·m^2, $B = 5$ N·m·s/rad, and $K = 6$ N·m/rad, and for no applied torque, find and sketch the zero-input responses for the following sets of initial conditions:

1. $\theta(0) = \theta_0$ and $\dot{\theta}(0) = 0$
2. $\theta(0) = 0$ and $\dot{\theta}(0) = \dot{\theta}_0$

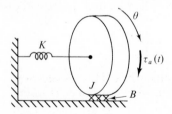

Figure 13.1 Rotational system for Example 13.1.

Solution   As derived in Example 4.1, the system model is

$$J\ddot{\theta} + B\dot{\theta} + K\theta = \tau_a(t)$$

which, for the given parameter values and no applied torque, becomes

$$\ddot{\theta} + 5\dot{\theta} + 6\theta = 0$$

Transforming each term in the differential equation and collecting those terms not involving $\Theta(s)$ on the right side, we get

$$(s^2 + 5s + 6)\Theta(s) = s\theta_0 + \dot{\theta}_0 + 5\theta_0$$

Dividing through by the characteristic polynomial

$$P(s) = s^2 + 5s + 6$$

gives $\Theta(s)$ for any combination of initial conditions, namely

$$\Theta(s) = \frac{s\theta_0 + \dot{\theta}_0 + 5\theta_0}{s^2 + 5s + 6} \tag{8}$$

The characteristic polynomial is quadratic and has a damping ratio $\zeta > 1$, so $\Theta(s)$ has two real distinct poles. To carry out the partial-fraction expansion in a general form before substituting specific values for θ_0 and $\dot{\theta}_0$, we write

$$\Theta(s) = \frac{s\theta_0 + \dot{\theta}_0 + 5\theta_0}{(s+2)(s+3)}$$

$$= \frac{A_1}{s+2} + \frac{A_2}{s+3}$$

where

$$A_1 = (s+2)\Theta(s)|_{s=-2} = 3\theta_0 + \dot{\theta}_0$$

$$A_2 = (s+3)\Theta(s)|_{s=-3} = -(2\theta_0 + \dot{\theta}_0)$$

Hence, the transform of the response is

$$\Theta(s) = \frac{3\theta_0 + \dot{\theta}_0}{s+2} - \frac{2\theta_0 + \dot{\theta}_0}{s+3}$$

and the corresponding time function is

$$\theta(t) = (3\theta_0 + \dot{\theta}_0)\epsilon^{-2t} - (2\theta_0 + \dot{\theta}_0)\epsilon^{-3t} \qquad \text{for } t > 0 \tag{9}$$

From an examination of (9), it is evident that the zero-input response is composed of the two exponentially decaying mode functions ϵ^{-2t} and ϵ^{-3t}, with the weighting of each mode function dependent on the initial angle θ_0 and the initial angular velocity $\dot{\theta}_0$. Specifically, the two desired responses are

1. $\theta(t) = \theta_0(3\epsilon^{-2t} - 2\epsilon^{-3t})$ for $\theta_0 \neq 0$ and $\dot{\theta}_0 = 0$
2. $\theta(t) = \dot{\theta}_0(\epsilon^{-2t} - \epsilon^{-3t})$ for $\theta_0 = 0$ and $\dot{\theta}_0 \neq 0$

The responses are shown in Figure 13.2, along with the respective mode functions.

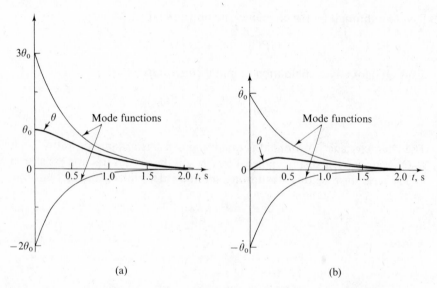

Figure 13.2 Responses of rotational system in Example 13.1 to specified initial conditions. (a) $\theta_0 \neq 0$ and $\dot{\theta}_0 = 0$. (b) $\theta_0 = 0$ and $\dot{\theta}_0 \neq 0$.

From inspection of (9), we can determine the relative values of θ_0 and $\dot{\theta}_0$ that will suppress either of the two modes in the zero-input response. For example, if $2\theta_0 + \dot{\theta}_0 = 0$, the response given by (9) reduces to

$$\theta(t) = (3\theta_0 + \dot{\theta}_0)\epsilon^{-2t}$$
$$= \theta_0 \epsilon^{-2t}$$

which would appear as the response of a first-order system having a time constant of $\tau = \frac{1}{2}$ s. Likewise, initial conditions satisfying $3\theta_0 + \dot{\theta}_0 = 0$

result in

$$\theta(t) = \theta_0 \epsilon^{-3t}$$

which is the response of a first-order system with $\tau = \frac{1}{3}$ s. Hence, one should not attempt to deduce the order of a system or the character of its mode functions based solely on a single sample of the zero-input response. Rather, one should be certain that each of the mode functions appears in the observed responses.

EXAMPLE 13.2   Find and sketch the zero-input response of the rotational system discussed in Example 13.1 for the parameter values $J = 1$ kg·m^2, $B = 2$ N·m·s/rad, and $K = 5$ N·m/rad and the initial conditions $\theta(0) = \theta_0$, $\dot{\theta}(0) = \dot{\theta}_0$.

Solution   Substituting the given parameter values into the model and transforming the resulting equation in terms of the arbitrary initial conditions θ_0 and $\dot{\theta}_0$ gives

$$\Theta(s) = \frac{s\theta_0 + \dot{\theta}_0 + 2\theta_0}{s^2 + 2s + 5} \tag{10}$$

which is similar to (8) but which has different coefficients in both the numerator and denominator. A check of the damping ratio indicates that $\zeta = 1/\sqrt{5}$, which is less than unity, implying that the two roots of the characteristic equation are complex. Since the mode functions $\epsilon^{s_i t}$ are themselves complex in such a case, it is preferable to leave the denominator in its quadratic form, rather than to expand $\Theta(s)$ into partial fractions with first-order denominators. Recalling that

$$\mathscr{L}\left[\epsilon^{-at}\cos \omega t\right] = \frac{s+a}{(s+a)^2 + \omega^2}$$

$$\mathscr{L}\left[\epsilon^{-at}\sin \omega t\right] = \frac{\omega}{(s+a)^2 + \omega^2} \tag{11}$$

we rewrite (10) as

$$\Theta(s) = \frac{\theta_0(s+1) + \dot{\theta}_0 + \theta_0}{(s+1)^2 + 2^2}$$

$$= \theta_0 \left[\frac{s+1}{(s+1)^2 + 2^2}\right] + \frac{\dot{\theta}_0 + \theta_0}{2}\left[\frac{2}{(s+1)^2 + s^2}\right] \tag{12}$$

From a comparison of (11) and (12), it is apparent that the zero-input response is

$$\theta(t) = \theta_0 \epsilon^{-t} \cos 2t + \tfrac{1}{2}(\dot{\theta}_0 + \theta_0)\epsilon^{-t}\sin 2t \qquad \text{for } t > 0 \qquad (13)$$

The responses for the two sets of initial conditions $\theta_0 \neq 0$, $\dot{\theta}_0 = 0$ and $\theta_0 = 0$, $\dot{\theta}_0 \neq 0$ are shown in Figure 13.3.

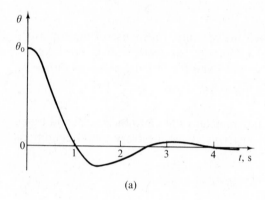

(a)

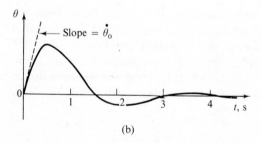

(b)

Figure 13.3 Responses of rotational system in Example 13.2 to specified initial conditions. (a) $\theta_0 \neq 0$ and $\dot{\theta}_0 = 0$. (b) $\theta_0 = 0$ and $\dot{\theta}_0 \neq 0$.

Examination of (13) suggests that in place of the complex mode functions $\epsilon^{(-1+j2)t}$ and $\epsilon^{(-1-j2)t}$ corresponding to the poles of $\Theta(s)$ at $s_1 = -1+j2$ and $s_2 = -1-j2$, it is appropriate to consider the functions $\epsilon^{-t}\cos 2t$ and $\epsilon^{-t}\sin 2t$. In more general terms, if a system has a pair of complex poles at $s_1 = \alpha + j\beta$ and $s_2 = \alpha - j\beta$, we can take the mode functions as either the complex functions $\epsilon^{(\alpha+j\beta)t}$ and $\epsilon^{(\alpha-j\beta)t}$ or the real functions $\epsilon^{\alpha t}\cos \beta t$ and $\epsilon^{\alpha t}\sin \beta t$.

EXAMPLE 13.3   The two-inertia rotational system modeled in Example 4.2 is shown in Figure 13.4, with the viscous damping in the original version omitted. Find and sketch the zero-input response $\theta_1(t)$ for the parameter values $J_1 = J_2 = 1$ kg·m², $K_1 = 1$ N·m/rad, and $K_2 = 2$ N·m/rad, with the initial conditions $\theta_1(0) = \theta_2(0) = 0.5$ rad and $\dot{\theta}_1(0) = \dot{\theta}_2(0) = 0$.

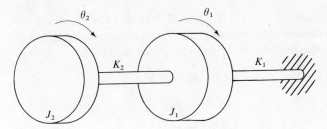

Figure 13.4 Rotational system for Example 13.3.

Solution   By substituting numerical values for J_1, J_2, K_1 and K_2 into (4.22) and setting $B_1 = B_2 = \tau_a(t) = 0$, we obtain the system model

$$\ddot{\theta}_1 + 3\theta_1 - 2\theta_2 = 0$$
$$\ddot{\theta}_2 + 2\theta_2 - 2\theta_1 = 0 \tag{14}$$

At this point, we could combine the pair of equations into a single homogeneous fourth-order differential equation for θ_1 and then transform the result to obtain $\Theta_1(s)$. However, that approach would require knowledge of $\ddot{\theta}_1(0)$ and $\dddot{\theta}_1(0)$. When working with coupled equations, including state-variable equations, it is more convenient to transform the differential equations immediately and then to solve the transformed equations for the transform of the desired output.

When transformed with the specified initial conditions, (14) becomes

$$(s^2 + 3)\Theta_1(s) - 2\Theta_2(s) = 0.5s$$
$$-2\Theta_1(s) + (s^2 + 2)\Theta_2(s) = 0.5s$$

Solving these two simultaneous algebraic equations for $\Theta_1(s)$ yields

$$\Theta_1(s) = \frac{0.5s^3 + 2s}{s^4 + 5s^2 + 2} \tag{15}$$

Although the denominator of $\Theta_1(s)$ is a polynomial of degree four in s, it contains only even powers of s and can be factored into the product

$(s^2 + 0.4384)(s^2 + 4.562)$ by means of the quadratic formula. Thus we can decompose $\Theta_1(s)$ into the sum of two terms having quadratic denominators. The expansion of $\Theta_1(s)$ is

$$\Theta_1(s) = \frac{A_1 s + B_1}{s^2 + 0.4384} + \frac{A_2 s + B_2}{s^2 + 4.562} \tag{16}$$

To evaluate the coefficients A_1, A_2, B_1, and B_2, we put the right side of (16) over a common denominator and compare the numerator coefficients to those of (15). Doing this, we obtain the following four equations:

$$A_1 + A_2 = 0.5$$

$$B_1 + B_2 = 0$$

$$4.562 A_1 + 0.4384 A_2 = 2$$

$$4.562 B_1 + 0.4384 B_2 = 0$$

Solving these equations and substituting the results into (16), we find that

$$\Theta_1(s) = \frac{0.4319s}{s^2 + 0.6622^2} + \frac{0.0681s}{s^2 + 2.136^2}$$

Referring to Appendix B, we see that the response is

$$\theta_1(t) = 0.4319 \cos 0.6622t + 0.0681 \cos 2.136t \qquad \text{for } t > 0$$

which is plotted in Figure 13.5. The figure indicates that the disk in question responds in a rather complicated fashion, which is the superposition of undamped oscillations at the two frequencies $\omega = 0.6622$ and 2.136 rad/s.

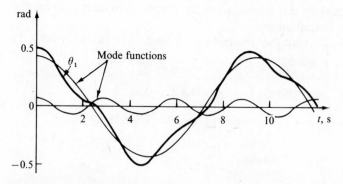

Figure 13.5 Response of the rotational system shown in Figure 13.4 to the initial conditions $\theta_1(0) = \theta_2(0) = 0.5$ rad and $\dot\theta_1(0) = \dot\theta_2(0) = 0$.

These cosine functions and the corresponding sine functions having the same frequencies can be considered the mode functions of the system. Any other combination of initial conditions would result in a response that is a weighted sum of these four mode functions.

13.2 THE ZERO-STATE RESPONSE

As defined in Section 8.2, the zero-state response is the response to a non-zero input when the initial stored energy is zero. We assume that the input starts at $t = 0+$ and that $y(0), \dot{y}(0), \ldots, y^{(n-1)}(0), u(0), \dot{u}(0), \ldots, u^{(m-1)}(0)$ are zero. The zero-state response will consist of both the forced response $y_P(t)$ and the free response $y_H(t)$. We shall consider the system model to be in the form of the general nth-order input-output equation

$$a_n y^{(n)} + a_{n-1} y^{(n-1)} + \cdots + a_1 \dot{y} + a_0 y = b_m u^{(m)} + \cdots + b_0 u \qquad (17)$$

INPUT - OUTPUT Eqn.

THE TRANSFER FUNCTION

Transforming both sides of (17) and collecting terms, we obtain the algebraic equation

$$(a_n s^n + a_{n-1} s^{n-1} + \cdots + a_1 s + a_0) Y(s) = (b_m s^m + \cdots + b_0) U(s)$$

which can be rearranged to give the transform of the output as

$$Y(s) = \left(\frac{b_m s^m + \cdots + b_0}{a_n s^n + a_{n-1} s^{n-1} + \cdots + a_1 s + a_0} \right) U(s) \qquad (18)$$

Hence, the output transform $Y(s)$ is the product of the input transform $U(s)$ and a rational function of the complex variable s whose coefficients are the coefficients in the input-output differential equation. This rational function of s is known as the system's *transfer function* and plays a key role in the analysis of linear systems. Denoting the transfer function by $H(s)$, we rewrite (18) as

$$Y(s) = H(s) U(s) \qquad (19)$$

Note that when we know a system's input-output differential equation, we can write its transfer function directly as

$$H(s) = \frac{b_m s^m + \cdots + b_0}{a_n s^n + a_{n-1} s^{n-1} + \cdots + a_1 s + a_0} \qquad (20)$$

Coefficients of INPUT-OUTPUT D.E.

Even if the system model is in a form other than the input-output differential equation, the transfer function can be found from (19) or its equivalent,

$$H(s) = \frac{Y(s)}{U(s)} \tag{21}$$

For example, if we know a system's state-variable and output equations, we can transform all the equations with zero for the initial values of the state variables and with an input that starts at $t = 0+$. We can then find the transfer function by manipulating the transformed equations into the form of (19) or (21).

POLES AND ZEROS

When the two polynomials in s that constitute $H(s)$ are factored, the transfer function will have the form

$$H(s) = K\frac{(s - z_1)(s - z_2)\cdots(s - z_m)}{(s - p_1)(s - p_2)\cdots(s - p_n)} \tag{22}$$

The quantities $z_1, z_2, ..., z_m$ are those values of s for which the numerator of $H(s)$ is zero; and they are called the *zeros* of the transfer function.* The quantities $p_1, p_2, ..., p_n$ are those values of s for which the denominator of $H(s)$ vanishes and for which $H(s)$ becomes infinite. They are the *poles* of the transfer function. From a comparison of (20) and (22), it follows that $K = b_m/a_n$, where we have assumed that neither b_m nor a_n may be zero. In the following discussion, we shall assume that none of the zeros coincides with any of the poles, i.e., that $z_i \neq p_j$ for all $1 \leq i \leq m$ and all $1 \leq j \leq n$. If a pole and a zero of $H(s)$ were coincident, some of the factors in (22) could be canceled, in which case we could not reconstruct the input-output differential equation from $H(s)$. Then it would not be possible to find the response of all the system's modes from the transfer function. For a treatment of such cases, see Chapter IV of Desoer.

From (22), it is apparent that a transfer function can be completely specified by its poles, its zeros, and the multiplying constant K. The poles and zeros may be complex numbers and can be represented graphically by points in a complex plane. This plane is called the *s-plane* or the *complex-frequency plane* and, since $s = \sigma + j\omega$, its real and imaginary axes are labeled σ and ω, respectively. The poles of $H(s)$ are indicated by crosses and the zeros by circles placed at the appropriate points.

Referring to (5), we see that the denominator of $H(s)$ in (20) is the characteristic polynomial of the system. Thus the poles $p_1, p_2, ..., p_n$ of

---

* If $m < n$, which is often the case, $H(s)$ will have a zero of multiplicity $n - m$ at infinity. However, one is usually interested in only the numerator zeros $z_1, z_2, ..., z_m$.

$H(s)$ are identical to the characteristic roots that appear in the zero-input response. As a consequence, we can write down the form of the zero-input response as soon as we know the poles of $H(s)$.

To lend further significance to the transfer-function concept, recall that the stability of a system is determined by the roots of its characteristic equation. Hence, the stability can also be characterized by the locations of the poles of the transfer function. If all the poles are inside the left half of the s-plane ($\sigma < 0$), the system is stable; and if at least one pole is in the right half of the s-plane ($\sigma > 0$), the system is unstable. If all the poles of $H(s)$ are in the left half of the s-plane except for distinct poles on the imaginary axis ($\sigma = 0$), the system is marginally stable. In addition, a system will be unstable if its transfer function has repeated poles on the imaginary axis. Before discussing the zero-state response in more detail, we shall illustrate some of these notions in the following example.

EXAMPLE 13.4   Find the transfer function and draw the corresponding pole-zero plot in the s-plane for the system described by the input-output equation

$$\dddot{y} + 7\ddot{y} + 15\dot{y} + 25y = 2\ddot{u} + 6\dot{u}$$

Also comment on the stability of the system and give the form of its zero-input response.

Solution   Transforming the input-output equation with zero initial conditions and with $u(0) = \dot{u}(0) = 0$, we have

$$(s^3 + 7s^2 + 15s + 25)\,Y(s) = (2s^2 + 6s)\,U(s)$$

which, when we solve for the ratio $Y(s)/U(s)$, gives

$$H(s) = \frac{2s^2 + 6s}{s^3 + 7s^2 + 15s + 25}$$

Alternatively, we could have used (20) to write $H(s)$ by inspection of the input-output equation.

In order to draw the pole-zero plot representing $H(s)$, we must factor its numerator and denominator. Although the numerator is readily factored, the denominator is a cubic in s. It turns out that

$$H(s) = \frac{2s(s+3)}{(s+5)(s^2+2s+5)} \tag{23a}$$

$$= 2\left[\frac{s(s+3)}{(s+5)(s+1-j2)(s+1+j2)}\right] \tag{23b}$$

Thus $H(s)$ has two real zeros (at $s = 0$ and $s = -3$) and three poles (a real one at $s = -5$ and a complex pair at $s = -1 + j2$ and $s = -1 - j2$), all of which can be represented by the pole-zero plot shown in Figure 13.6. In addition, the multiplying factor is $K = 2$.

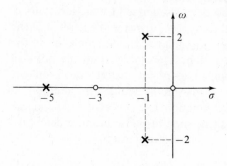

Figure 13.6 Pole-zero plot for $H(s)$ in Example 13.4.

Since the transfer function has all three of its poles in the left half of the s-plane, the system is stable. The fact that one of the zeros (at $s = 0$) is not inside the left half-plane has no bearing on the system's stability. For that matter, $H(s)$ can have zeros in the right half-plane and still correspond to a stable system. Knowing the three poles of the transfer function, we can immediately write down the form of the zero-input response in either of the following two equivalent forms:

$$y(t) = K_1 \epsilon^{-5t} + \epsilon^{-t}(K_2 \cos 2t + K_3 \sin 2t) \tag{24a}$$

$$y(t) = K_1 \epsilon^{-5t} + K_4 \epsilon^{-t} \cos(2t + \phi) \tag{24b}$$

EXAMPLE 13.5 For a linear system that contains no initial stored energy, suppose that we know the response to a certain input. Show that the new response is the derivative of the old response if we substitute a new input that is the derivative of the original input.

Solution If the original input and output are denoted by $u_1(t)$ and $y_1(t)$, then transforming (17) with

$$y(0) = \dot{y}(0) = \cdots y_1^{(n-1)}(0) = u_1(0) = \cdots u_1^{(m-1)}(0) = 0$$

gives

$$Y_1(s) = H(s) U_1(s)$$

For the new input $u_2(t)$, which is given by $u_2(t) = \dot{u}_1$, we have $U_2(s) = sU_1(s) - u_1(0) = sU_1(s)$. Thus with zero initial conditions,

$$Y_2(s) = sH(s)U_1(s)$$
$$= sY_1(s)$$

Thus

$$y_2(t) = \mathcal{L}^{-1}[sY_1(s)]$$

which, since $y_1(0) = 0$, yields $y_2(t) = \dot{y}_1$.

TRANSIENT AND STEADY-STATE COMPONENTS

Equation (19) is an expression for the transform of the system's zero-state response, namely the output when there is no initial stored energy. Recall from Chapter 8 that the transient response consists of those terms that decay to zero as t becomes large, while the remaining terms constitute the steady-state response. For a stable system and for an input that does not decay to zero, the steady-state response is the forced response* $y_P(t)$. We now show how this relationship is manifested in (19) and how to interpret it in terms of poles and zeros.

Assume that the transform of the input $u(t)$ can be written as a rational function of s, namely

$$(INPUT) = U(s) = \frac{N(s)}{D(s)} \tag{25}$$

where $N(s)$ and $D(s)$ are polynomials. For cases where the input contains a term delayed by a units of time, $N(s)$ will contain a factor ϵ^{-as}, which can be treated as in Example 12.13. Other inputs, such as $u(t) = 1/t$, that have Laplace transforms that do not fit (25) are beyond the scope of our consideration.

Rewriting (19) using (22) for $H(s)$ and (25) for $U(s)$, we have

$$y(s) = H(s) \, U(s)$$

$$Y(s) = \left(K \frac{(s-z_1)\cdots(s-z_m)}{(s-p_1)(s-p_2)\cdots(s-p_n)} \right) \cdot \frac{N(s)}{D(s)} \tag{26}$$

$$H(s)$$

* If a system is marginally stable, it can exhibit a steady-state response due to initial stored energy, even without an input. An unstable system can yield an unbounded response without an input. Hence, the designations "transient" and "steady-state" are most useful for stable systems.

Recognizing that $U(s)$ has its own poles and zeros, we see that the poles and zeros of $Y(s)$ are the combination of those of $H(s)$ and $U(s)$. If we expand $Y(s)$ in a partial-fraction expansion and all of its poles are distinct, it will have the form

$$Y(s) = \frac{A_1}{s-p_1} + \frac{A_2}{s-p_2} + \cdots + \frac{A_n}{s-p_n} + \frac{A_{n+1}}{s-p_{n+1}} + \cdots + \frac{A_q}{s-p_q} \quad (27)$$

where the poles $p_1, p_2, \ldots, p_n$ are the poles of $H(s)$ and the poles $p_{n+1}, \ldots, p_q$ are the poles of $U(s)$. Taking the inverse transform of each term, we have

$$y(t) = A_1 e^{p_1 t} + A_2 e^{p_2 t} + \cdots + A_n e^{p_n t} + A_{n+1} e^{p_{n+1} t} + \cdots + A_q e^{p_q t} \quad (28)$$

where the first n terms are the free response $y_H(t)$ and the last $q-n$ terms are the forced response $y_P(t)$. Hence,

$$y(t) = y_H(t) + y_P(t)$$

where

$$y_H(t) = A_1 e^{p_1 t} + \cdots + A_n e^{p_n t}$$
$$y_P(t) = A_{n+1} e^{p_{n+1} t} + \cdots + A_q e^{qt}$$

When the system is stable, the term $y_H(t)$ consists of the transients induced in having $y(t)$ start from zero initial conditions and reach the steady-state solution represented by $y_P(t)$. Keep in mind that when complex poles are present, it may be more efficient to include the corresponding second-order term (quadratic denominator and linear numerator) in the partial-fraction expansion, as in (12.61), rather than to factor it.

EXAMPLE 13.6   Find the form of the zero-state response of the system discussed in Example 13.4 for which

$$H(s) = \frac{2s(s+3)}{(s+5)(s^2+2s+5)}$$

when the input is $u(t) = 3 \cos 4t$ for $t > 0$. Identify the transient and steady-state terms.

Solution   The transform of the input is

$$U(s) = \frac{3s}{s^2+16}$$

When $H(s)$ and $U(s)$ are substituted into (19), it follows that

$$Y(s) = \frac{2s(s+3)}{(s+5)(s^2+2s+5)} \cdot \frac{3s}{(s^2+16)} \tag{29}$$

Referring to the procedures in Section 12.3, we can write the partial-fraction expansion for $Y(s)$ in either of the following two equivalent forms:

$$Y(s) = \frac{A_1}{s+5} + \frac{A_2}{s+1-j2} + \frac{A_2^*}{s+1+j2} + \frac{A_3}{s-j4} + \frac{A_3^*}{s+j4} \tag{30a}$$

$$Y(s) = \frac{A_1}{s+5} + \frac{A_4 s + A_5}{s^2+2s+5} + \frac{A_6 s + A_7}{s^2+16} \tag{30b}$$

where A_2 and A_3 are complex numbers and where the asterisk denotes the complex conjugate. If $A_2 = K_2 e^{j\phi_2}$ and $A_3 = K_3 e^{j\phi_3}$, we can use Appendix B to write from (30a)

$$y(t) = A_1 \epsilon^{-5t} + 2K_2 \epsilon^{-t} \cos(2t+\phi_2) + 2K_3 \cos(4t+\phi_3)$$

Alternatively, we may rewrite (30b) as

$$Y(s) = \frac{A_1}{s+5} + \frac{A_4(s+1)}{(s+1)^2+2^2} + \frac{A_5 - A_4}{(s+1)^2+2^2} + \frac{A_6 s}{s^2+4^2} + \frac{A_7}{s^2+4^2}$$

Then, by the entries in Appendix B, we have for the inverse transform

$$y(t) = A_1 \epsilon^{-5t} + A_4 \epsilon^{-t} \cos 2t + \left(\frac{A_5 - A_4}{2}\right) \epsilon^{-t} \sin 2t$$

$$+ A_6 \cos 4t + \frac{1}{4} A_7 \sin 4t$$

In Problem 13.8, you will be asked to calculate the coefficients and to show that when numerical values are inserted, the foregoing equations reduce to

$$y(t) = -0.3659\epsilon^{-5t} + 0.6975\epsilon^{-t} \cos(2t+3.118) + 1.378 \cos(4t-0.6894) \tag{31a}$$

$$y(t) = -0.3659^{-5t} - \epsilon^{-t}(0.6973 \cos 2t + 0.0162 \sin 2t)$$

$$+ (1.0632 \cos 4t + 0.8765 \sin 4t) \tag{31b}$$

For either form of (31), the first two terms constitute the transient response and have the same form as $y(t)$ given by (24), which was shown to be the form of the system's zero-input response. These terms come from the poles of $H(s)$, which are entirely within the left half of the s-plane. The last of the three terms in (31) is the steady-state response, a constant-amplitude oscillation resulting from the poles of $U(s)$, which are on the imaginary axis of the s-plane.

13.3 THE COMPLETE RESPONSE

When we are to evaluate the response of a fixed linear system to the combination of nonzero initial conditions and a nonzero input for $t > 0$, the principle of superposition allows us to add the zero-input and zero-state responses to give the complete response. Likewise, we can evaluate the Laplace transforms of the two responses separately and add them to give the transform of the complete response.

For the nth-order system described by (17), (3) gives the transform of the zero-input response, that portion due to the initial conditions. The transform of the zero-state response, that portion due to the input, is given by (19). Thus the transform of the total response is

$$Y(s) = \frac{F(s)}{P(s)} + H(s)\,U(s) \tag{32}$$

where $F(s)$ is the polynomial given by (4) and involves the initial conditions and where $P(s)$ is the characteristic polynomial given by (5). If we write $H(s)$ and $U(s)$ as ratios of polynomials, then (32) becomes

$$Y(s) = \frac{F(s)}{P(s)} + \frac{G(s)}{P(s)} \cdot \frac{N(s)}{D(s)} \tag{33}$$

where $G(s)$ is the numerator of $H(s)$ and where $U(s) = N(s)/D(s)$. The first term in (33) is the transform of the zero-input response, while the second is the transform of the zero-state response.

Let $p_1, p_2, \ldots, p_n$ denote the poles of $H(s)$, i.e., the roots of $P(s) = 0$, and let $p_{n+1}, \ldots, p_q$ denote the poles of the transformed input, i.e., the roots of $D(s) = 0$. If the poles are distinct, we can write the partial-fraction expansion of $Y(s)$ as

$$Y(s) = \left[\frac{K_1}{s - p_1} + \frac{K_2}{s - p_2} + \cdots + \frac{K_n}{s - p_n} \right]$$

$$+ \left[\frac{A_1}{s - p_1} + \frac{A_2}{s - p_2} + \cdots + \frac{A_n}{s - p_n} + \frac{A_{n+1}}{s - p_{n+1}} + \cdots + \frac{A_q}{s - p_q} \right]$$

where the expression in the first set of brackets is the transformed zero-input response and the expression in the second set is the transformed zero-state response. Only the constants $K_1, K_2, ..., K_n$ depend on the initial conditions. The complete time response is

$$y(t) = y_H(t) + y_P(t)$$

where

$$y_H(t) = (K_1 + A_1) \epsilon^{p_1 t} + \cdots + (K_n + A_n) \epsilon^{p_n t}$$

$$y_P(t) = A_{n+1} \epsilon^{p_{n+1} t} + \cdots + A_q \epsilon^{p_q t}$$

Comparing these expressions with (28), we see that the complete response is the same as the zero-state response, except that the terms in $y_H(t)$ have different coefficients. For a stable system with an input that does not decay to zero, $y_P(t)$ is the steady-state response, which is unaffected by the initial conditions.

We can rewrite (33) with a common denominator as

$$Y(s) = \frac{F(s)D(s) + G(s)N(s)}{P(s)D(s)} \tag{34}$$

The poles of $Y(s)$, the transform of the complete response, are the combined poles of the transfer function and the transformed input. The zeros of $Y(s)$ depend on the four polynomials $F(s)$, $D(s)$, $G(s)$, and $N(s)$.

EXAMPLE 13.7 The system we discussed in Example 13.4 and Example 13.6 obeys the differential equation

$$\dddot{y} + 7\ddot{y} + 15\dot{y} + 25y = 2\ddot{u} + 6\dot{u} \tag{35}$$

Find the complete response when the initial conditions are $y(0) = 2$, $\dot{y}(0) = 0$, $\ddot{y}(0) = 0$ and when the input is $u(t) = 3\epsilon^{-2t}$ for $t > 0$ and zero otherwise.

Solution Transforming (35) with the specified initial conditions and taking $u(0) = \dot{u}(0) = 0$, we obtain

$$s^3 Y(s) - 2s^2 + 7[s^2 Y(s) - 2s] + 15[sY(s) - 2] + 25Y(s) = (2s^2 + 6s) U(s)$$

where $U(s) = 3/(s+2)$. Solving for $Y(s)$, the transform of the complete response, we find that

$$Y(s) = \frac{2s^2 + 14s + 30}{P(s)} + \frac{(2s^2 + 6s)}{P(s)} \cdot \frac{3}{(s+2)}$$

where the characteristic polynomial is

$$P(s) = s^3 + 7s^2 + 15s + 25$$

which can be factored to give

$$P(s) = (s+5)(s^2+2s+5)$$

Combining the two terms on the right side of $Y(s)$ over a common denominator in order to express it as a rational function, we have

$$Y(s) = \frac{(2s^2+14s+30)(s+2)+6s^2+18s}{(s+2)P(s)}$$

$$= \frac{2s^3+24s^2+76s+60}{(s+2)(s+5)(s^2+2s+5)} \tag{36}$$

At this point, we expand $Y(s)$ in partial fractions in order to identify the individual time functions comprising $y(t)$. Since $Y(s)$ has real poles at $s = -2$ and $s = -5$ and a pair of complex poles, and since the order of the numerator is less than the order of the denominator, we can write

$$Y(s) = \frac{A_1}{s+2} + \frac{A_2}{s+5} + \frac{A_3 s + A_4}{s^2+2s+5} \tag{37a}$$

$$= \frac{A_1}{s+2} + \frac{A_2}{s+5} + \frac{A_3(s+1)}{(s+1)^2+2^2} + \left(\frac{A_4-A_3}{2}\right)\frac{2}{(s+1)^2+2^2} \tag{37b}$$

We can find the inverse transform of each term from Appendix B. The complete response in terms of the constants A_1, A_2, A_3 and A_4 is

$$y(t) = A_1 \epsilon^{-2t} + A_2 \epsilon^{-5t} + A_3 \epsilon^{-t}\cos 2t + \left(\frac{A_4-A_3}{2}\right)\epsilon^{-t}\sin 2t$$

$$\text{for } t > 0 \tag{38}$$

Note that each term in $y(t)$ decays to zero as t approaches infinity. This is because all the poles of $H(s)$ are in the left half of the s-plane and the input $u(t)$ decays to zero with increasing time. Hence, the steady-state response is zero.

To find $y(t)$ in numerical form, we find A_1 and A_2 in the usual manner. The results are $A_1 = -0.80$ and $A_2 = -0.50$. Using these values, we can combine the right side of (37a) over a common denominator and then compare the coefficients of its numerator polynomial to those of (36). This reveals that $A_3 = 3.30$ and $A_4 = 8.50$. Substituting these values into (38), we find that

$$y(t) = -0.80\epsilon^{-2t} - 0.50\epsilon^{-5t} + 3.30\epsilon^{-t}\cos 2t + 2.60\epsilon^{-t}\sin 2t \qquad \text{for } t > 0$$
$$(39)$$

We can evaluate $y(0+)$ by setting t equal to zero in (39). This gives $y(0+) = 2.0$, which has the same value as the specified initial condition $y(0) = 2$. If we differentiate (39) and then set t equal to zero, we find that $\dot{y}(0+) = 6$, which differs from the initial condition $\dot{y}(0) = 0$. The discontinuity in $\dot{y}$ results from the presence of the $\ddot{u}$ term on the right side of (35) and from the fact that the input is discontinuous.

To conclude this example, it is instructive to use the initial-value and final-value theorems to compute $y(0+)$ and the limit of $y(t)$ as t approaches infinity. We compute these values directly from the transform $Y(s)$, as given by (36). First, the initial value is

$$y(0+) = \lim_{s \to \infty} sY(s)$$

$$= \lim_{s \to \infty} \frac{2s^4 + 24s^3 + \cdots}{s^4 + 9s^3 + \cdots} = 2$$

The dots indicate low-order terms that do not affect the limiting process. The result agrees with that found by evaluating $y(t)$ at $t = 0+$. Next, the final value is

$$y(\infty) = \lim_{s \to 0} sY(s)$$

$$= \lim_{s \to 0} \frac{s(\cdots + 76s + 60)}{(\cdots + 55s + 50)} = 0$$

which is in agreement with (39), each term of which decays to zero as t approaches infinity. Here the dots indicate high-order terms that do not affect the limit when s approaches zero.

EXAMPLE 13.8   Repeat Example 13.7 when $y(0) = 2$, $\dot{y}(0) = 0$, $\ddot{y}(0) = 0$ and when $u(t) = 3\cos 4t$ for $t > 0$ and zero otherwise.

Solution   Transforming (35) with the specified conditions and then solving for the transform of the output, we obtain

$$Y(s) = \frac{2(s^2 + 7s + 15)}{(s+5)(s^2 + 2s + 5)} + \frac{2s^2 + 6s}{(s+5)(s^2 + 2s + 5)} \cdot \frac{3s}{s^2 + 16}$$

We could combine these terms over a common denominator. However, we recognize the second of the two terms as the transformed zero-state response, whose inverse transform we found in Example 13.6. The partial-fraction expansion for the first of the two terms is found to be

$$\frac{2(s^2+7s+15)}{(s+5)(s^2+2s+5)} = \frac{\frac{1}{2}}{s+5} + \frac{(3s+11)/2}{(s+1)^2+2^2}$$

whose inverse transform is

$$\tfrac{1}{2}\epsilon^{-5t} + \epsilon^{-t}(\tfrac{3}{2}\cos 2t + 2\sin 2t) \qquad \text{for } t > 0$$

By adding this expression to the answer for Example 13.6, we obtain for the complete response

$$y(t) = 0.1341\epsilon^{-5t} + \epsilon^{-t}(0.8027\cos 2t + 1.983\sin 2t)$$

$$+ (1.0632\cos 4t + 0.8765\sin 4t) \qquad \text{for } t > 0$$

This is, of course, the same as (31b) except that different coefficients multiply the transient terms.

13.4 RESPONSES TO SPECIAL INPUTS

In the analysis and design of dynamic systems, it is helpful to specify and evaluate the response of a system to a specific input in order to have a common point of reference. Three inputs that stand out because of their usefulness and acceptance are (1) the unit impulse $\delta(t)$, (2) the unit step function $U(t)$, and (3) the sinusoidal function $\sin \omega t$. We shall examine the responses of fixed linear systems to these three functions by using Laplace transforms and the system transfer function $H(s)$. In the case of the impulse and the step function, we shall be interested in the zero-state response, including both the transient and the steady-state portions, for both stable and unstable systems. For the sinusoidal input, on the other hand, we shall limit our attention to the steady-state response of stable systems.

The basis for our study of the responses to these special inputs will be (19), which is

$$Y(s) = H(s)\, U(s) \qquad\qquad (40)$$

where $Y(s)$ is the transform of the zero-state response, $H(s)$ is the system's transfer function, and $U(s)$ is the transform of the input.

UNIT IMPULSE RESPONSE

Since the transform of the unit impulse is unity, we have $U(s) = 1$ and (40) reduces to

$$Y(s) = H(s)$$

In words, the transform of the unit impulse response is merely the system's transfer function $H(s)$. In addition, $h(t)$ must be zero for all $t \le 0$, since it is the zero-state response and the impulse does not occur until $t = 0+$. Using the symbol $h(t)$ to denote the response to the unit impulse, we can write

$$h(t) = \begin{cases} 0 & \text{for } t \le 0 \\ \mathscr{L}^{-1}[H(s)] & \text{for } t > 0 \end{cases} \tag{41}$$

which is a significant result in spite of its simplicity. In particular, (41) serves to tie together the system's time-domain characterization in terms of $h(t)$ and its complex-frequency-domain characterization $H(s)$. For example, one can relate the locations in the s-plane of the poles and zeros of the transfer function to the character of the system's impulse response. Furthermore, although the subject is beyond the scope of this book, the impulse response can be related to the response to arbitrary inputs by an integral expression known as the convolution integral (for an example, see Chapter 4 of Frederick and Carlson).

In Section 8.3, we stated that the unit impulse response could be found by either (1) finding the appropriate initial conditions and solving the homogeneous differential equation, or (2) finding the response to a unit step function and differentiating the result. Now an easier and more direct method is available: We can find the transfer function and then take its inverse transform. From (17) and (20), we know that we can write $H(s)$ by inspection from the system's input-output differential equation, although we may encounter some computational difficulties in factoring its denominator if the system is of third or higher order. In the event that the system model is available as two or more simultaneous equations, e.g., as a set of state-variable equations and an output equation, we can find the transfer function by first transforming the simultaneous equations with zero initial conditions and then solving algebraically for the ratio $Y(s)/U(s)$, which is the transfer function.

EXAMPLE 13.9   Find the unit impulse response of the first-order system described by $\dot{y} + (1/\tau)y = u(t)$.

Solution The transfer function $Y(s)/U(s)$ is

$$H(s) = \frac{1}{s + 1/\tau}$$

Recalling that $\mathscr{L}[\epsilon^{-at}] = 1/(s + a)$, we can write

$$h(t) = \begin{cases} 0 & \text{for } t \le 0 \\ \epsilon^{-t/\tau} & \text{for } t > 0 \end{cases} \tag{42}$$

which is shown in Figure 13.7.

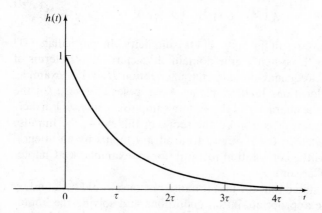

Figure 13.7 Impulse response for $\dot{y} + (1/\tau)y = u(t)$.

It is instructive to apply the initial-value theorem to $H(s)$ to evaluate $h(0+)$, getting

$$h(0+) = \lim_{s \to \infty} \frac{s}{s + 1/\tau} = 1$$

which agrees with (42) and Figure 13.7. It is easy to verify that the final-value theorem yields a value of zero.

EXAMPLE 13.10 Find and sketch the unit impulse response of the system whose input-output equation is

$$\ddot{y} + 4\dot{y} + 3\dot{y} = 2\dot{u} + u$$

Solution By inspection of the input-output differential equation, we see that the transfer function is

$$H(s) = \frac{2s+1}{s^3+4s^2+3s}$$

$$= \frac{2s+1}{s(s+1)(s+3)}$$

Carrying out the partial-fraction expansion, we find that

$$H(s) = \frac{\frac{1}{3}}{s} + \frac{\frac{1}{2}}{s+1} - \frac{\frac{5}{6}}{s+3}$$

Thus the unit impulse response is

$$h(t) = \begin{cases} 0 & \text{for } t \leq 0 \\ \frac{1}{3} + \frac{1}{2}\epsilon^{-t} - \frac{5}{6}\epsilon^{-3t} & \text{for } t > 0 \end{cases}$$

which is shown in Figure 13.8. In this case,

$$h(0+) = \lim_{s \to \infty} \frac{s(2s+1)}{s(s^2+4s+3)} = 0$$

and

$$\lim_{t \to \infty} h(t) = \lim_{s \to 0} \frac{s(2s+1)}{s(s^2+4s+3)} = \frac{1}{3}$$

both of which agree with the sketch of $h(t)$. Here $h(t)$ approaches a non-zero constant value because of the single pole at $s = 0$ in $H(s)$.

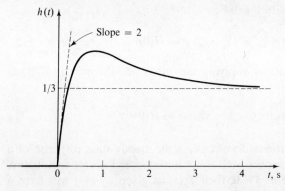

Figure 13.8 Impulse response for the system in Example 13.10.

It is often desirable to find the initial slope of the impulse response without determining the entire response. If we apply the initial-value theorem for the transform of a derivative,

$$\dot{h}(0+) = \lim_{s\to\infty} s\mathscr{L}[\dot{h}(t)]$$

$$= \lim_{s\to\infty} s[sH(s) - h(0)] \tag{43}$$

For the system in this example,

$$\dot{h}(0+) = \lim_{s\to\infty} \frac{2s^3 + s^2}{s^3 + 4s^2 + 3s} = 2$$

STEP RESPONSE

Since the transform of the unit step function is $1/s$, it follows from (40) that the transform of $y_U(t)$, the zero-state response of a system to a unit step-function input, is

$$Y_U(s) = H(s) \cdot \frac{1}{s} \tag{44}$$

Thus the unit step response $y_U(t)$ is zero for $t \le 0$, and for $t > 0$ it is the inverse transform of $H(s)/s$. The poles of $Y_U(s)$ consist of the poles of $H(s)$ and a pole at $s = 0$. Provided that $H(s)$ has no pole at $s = 0$, the time functions comprising $y_U(t)$ will be the mode functions of the response and a constant term resulting from the pole of $Y_U(s)$ at $s = 0$. In fact, if the system is stable, all the mode functions will decay to zero as t approaches infinity, and the steady-state portion of the step response will be due entirely to the pole at $s = 0$. The coefficient in the partial-fraction expansion of this steady-state term and thus $(y_U)_{ss}$ itself will be

$$(y_U)_{ss} = sY_U(s)|_{s=0} = H(0)$$

We can also get this result by applying the final-value theorem to $Y_U(s)$ as given by (44), namely

$$y_U(\infty) = \lim_{s\to 0} sY_U(s) = H(0)$$

Because the initial conditions do not affect the steady-state response of a stable system, the result applies even if the initial stored energy is not zero. If $H(s)$ has a single pole at $s = 0$, then $Y_U(s)$ as given by (44) will have a double pole at $s = 0$ and the step response will contain a ramp function in addition to a constant term. Then $y_U(t)$ will grow without bound as t approaches infinity.

To show that a system's unit step response is the integral of its unit impulse response, we use the theorem for the transform of an integral to write

$$\frac{1}{s} H(s) = \mathscr{L}\left[\int_0^t h(\lambda)\, d\lambda\right]$$

where λ is a dummy variable of integration. Since $\mathscr{L}[y_U(t)] = H(s)/s$, it follows that

$$y_U(t) = \int_0^t h(\lambda)\, d\lambda \tag{45}$$

In graphical terms (45) states that the unit step response $y_U(t)$ is the area under the curve of the unit impulse response $h(t)$. By the same token, the impulse response is the derivative of the step response, that is,

$$h(t) = \frac{d}{dt} y_U(t) \tag{46}$$

which agrees with (8.34).

EXAMPLE 13.11   Evaluate the unit step response of the first-order system $\dot{y} + (1/\tau)\, y = u(t)$.

Solution   From Example 13.9, the system's transfer function is $H(s) = 1/(s + 1/\tau)$. Thus (44) indicates that the transform of the step response is

$$Y(s) = \frac{1}{s(s + 1/\tau)}$$

$$= \frac{\tau}{s} - \frac{\tau}{s + 1/\tau}$$

Evaluating the inverse transform of $Y(s)$ and noting that the step response must be zero for $t \leq 0$, we find

$$y_U(t) = \begin{cases} 0 & \text{for } t \leq 0 \\ \tau(1 - \epsilon^{-t/\tau}) & \text{for } t > 0 \end{cases} \tag{47}$$

which is shown in Figure 13.9. You should verify that (47) is indeed the integral of the unit impulse response given by (42) and shown in Figure

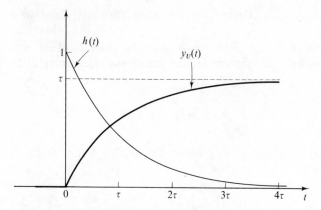

Figure 13.9 Step response for $\dot{y} + (1/\tau)y = u(t)$.

13.9. Finally, we note that the steady-state value of the unit step response is

$$(y_U)_{ss} = H(0) = \tau$$

which agrees with (47) and Figure 13.9.

EXAMPLE 13.12 Use Laplace transforms to find the unit step response of the system described by $\ddot{y} + 4\ddot{y} + 3\dot{y} = 2\dot{u} + u$, for which we found the impulse response in Example 13.10.

Solution From an inspection of the system's differential equation, the transfer function is

$$H(s) = \frac{2s + 1}{s^3 + 4s^2 + 3s}$$

Hence, the transform of the unit step response is

$$Y(s) = \frac{2s + 1}{s^2(s^2 + 4s + 3)}$$

whose partial-fraction expansion can be shown to be

$$Y(s) = \frac{\frac{1}{3}}{s^2} + \frac{\frac{2}{9}}{s} - \frac{\frac{1}{2}}{s + 1} + \frac{\frac{5}{18}}{s + 3}$$

Thus

$$y_U(t) = \tfrac{1}{3}t + \tfrac{2}{9} - \tfrac{1}{2}\epsilon^{-t} + \tfrac{5}{18}\epsilon^{-3t} \qquad \text{for } t > 0$$

As shown in Figure 13.10, the steady-state portion of the unit step response contains a ramp component with a slope of $\frac{1}{3}$, in addition to the constant of $\frac{2}{9}$. This is because $H(s)$ has a pole at $s = 0$. This pole combines with the pole at $s = 0$ due to the step-function input to give a double pole of $Y(s)$ at $s = 0$.

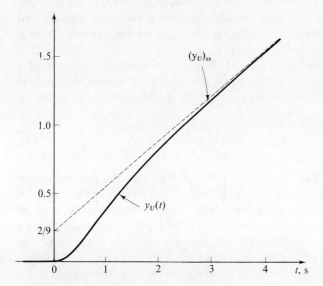

Figure 13.10 Step response for Example 13.12.

As an aid in sketching the unit step response, we note that in addition to $y_U(0) = 0$, the initial slope is zero, since $h(t)$ is the derivative of the step response and (from Figure 13.8) $h(0) = 0$.

FREQUENCY RESPONSE

In discussing the impulse and step responses, we were interested in both the transient and steady-state components. However, when considering the sinusoidal input

$$u(t) = \sin \omega t \tag{48}$$

we are normally interested only in the forced response. From our discussion of the particular solution in Section 8.1, we know that the forced response will have the same frequency as the input, but in general its amplitude and phase will differ from those of the input. For stable systems with a sinusoidal input, the forced response is also the steady-state response. Denoting the amplitude and phase angle of the steady-state response by A and ϕ, respectively, we can write

$$y_{ss}(t) = A \sin(\omega t + \phi) \tag{49}$$

which is referred to as the *sinusoidal-steady-state response*. It plays a key role in many aspects of system analysis, including electronic circuits and feedback control systems. We shall now show how we can express A and ϕ in terms of the transfer function $H(s)$.

Since the initial conditions do not affect the steady-state response of a stable system, we start with the transform of the zero-state response as given by (40). Since $\mathscr{L}[\sin \omega t] = \omega/(s^2 + \omega^2)$, the transformed response to (48) is

$$Y(s) = H(s)\left[\frac{\omega}{s^2 + \omega^2}\right] \tag{50}$$

Hence, the poles of $Y(s)$ will be the poles of $H(s)$ plus the two imaginary poles of $U(s)$ at $s = j\omega$ and $s = -j\omega$. Provided that the system is stable, all the poles of $H(s)$ will be in the left half of the complex plane, and all the terms in the response corresponding to these poles will decay to zero as time increases. Thus the steady-state response will result from the imaginary poles of $U(s)$ and will be a sinusoid at the frequency ω. To determine the steady-state component of the response, we can write (50) as

$$Y(s) = H(s)\left[\frac{\omega}{(s - j\omega)(s + j\omega)}\right]$$

$$= \frac{C_1}{s - j\omega} + \frac{C_2}{s + j\omega} + [\text{terms corresponding to poles of } H(s)] \tag{51}$$

where

$$C_1 = (s - j\omega)Y(s)\big|_{s = j\omega}$$

$$= H(s)\frac{\omega}{(s + j\omega)}\bigg|_{s = j\omega}$$

$$= \frac{H(j\omega)}{2j}$$

The constant C_2 is the complex conjugate of C_1; that is,

$$C_2 = -\frac{H(-j\omega)}{2j}$$

Since all the terms in (51) corresponding to the poles of $H(s)$ will decay if the system is stable, the transform of the steady-state response is

$$Y_{ss}(s) = \frac{H(j\omega)/2j}{s - j\omega} - \frac{H(-j\omega)/2j}{s + j\omega} \tag{52}$$

In general, $H(j\omega)$ is a complex number and may be written in polar form as

$$H(j\omega) = M(\omega)\epsilon^{j\theta(\omega)} \tag{53}$$

where $M(\omega)$ is the magnitude of $H(j\omega)$ and $\theta(\omega)$ is its angle. Both M and θ depend on the value of ω, as is emphasized by the ω within the parentheses. The quantity $H(-j\omega)$ is the complex conjugate of $H(j\omega)$, and we can write it as

$$H(-j\omega) = M(\omega)\epsilon^{-j\theta(\omega)} \tag{54}$$

Substituting (53) and (54) into (52) gives

$$Y_{ss}(s) = \frac{M}{2j}\left(\frac{\epsilon^{j\theta}}{s - j\omega} - \frac{\epsilon^{-j\theta}}{s + j\omega}\right)$$

Taking the inverse transforms of the two terms in $Y_{ss}(s)$, we find

$$y_{ss}(t) = \frac{M}{2j}(\epsilon^{j\theta}\epsilon^{j\omega t} - \epsilon^{-j\theta}\epsilon^{-j\omega t})$$

$$= \frac{M}{2j}[\epsilon^{j(\omega t + \theta)} - \epsilon^{-j(\omega t + \theta)}]$$

Using the exponential form of the sine function, as in Table 8.1, we obtain

$$y_{ss}(t) = M\sin(\omega t + \theta) \tag{55}$$

which has the form predicted in (49) where $A = M(\omega)$ and $\phi = \theta(\omega)$.

In Problem 13.31, you will use a similar derivation to show that for a stable system the steady-state response to $u(t) = B\sin(\omega t + \phi_1)$ is

$$y_{ss}(t) = BM\sin(\omega t + \phi_1 + \theta) \tag{56}$$

Likewise the steady-state response to $u(t) = B\cos(\omega t + \phi_2)$ is

$$y_{ss}(t) = BM\cos(\omega t + \phi_2 + \theta) \tag{57}$$

In summary, the sinusoidal-steady-state response of a stable linear system is a sinusoid having the same frequency as the input, an amplitude that is

$M(\omega)$ times that of the input, and a phase angle that is $\theta(\omega)$ plus the input angle, where $M(\omega)$ and $\theta(\omega)$ are the magnitude and angle of $H(j\omega)$. As a consequence, the function $H(j\omega)$, which is the transfer function evaluated for $s = j\omega$, is known as the *frequency-response function*.

Calculating and interpreting the frequency-response function is illustrated in the following examples. In the first example, we find the steady-state response to a sinusoidal input having a specified frequency. In the next two examples, we sketch curves of $M(\omega)$ and $\theta(\omega)$ as functions of ω. Such curves indicate how the magnitude and angle of the sinusoidal steady-state response change as the frequency of the input is changed.

EXAMPLE 13.13 Use (53) and (57) to find the steady-state response of the system described by the differential equation

$$\dddot{y} + 7\ddot{y} + 15\dot{y} + 25y = 2\ddot{u} + 6\dot{u}$$

to the input $u(t) = 3 \cos 4t$.

Solution As we found in Example 13.4, and as we can see by inspecting the differential equation,

$$H(s) = \frac{2s^2 + 6s}{s^3 + 7s^2 + 15s + 25}$$

We replace s by $j\omega$ with $\omega = 4$ rad/s, the frequency of the input, to form

$$H(j4) = \frac{-32 + j24}{-64 - 112 + j60 + 25}$$

$$= \frac{-32 + j24}{-87 - j4} = \frac{32 - j24}{87 + j4}$$

The magnitude of $H(j4)$ is the magnitude of its numerator divided by the magnitude of its denominator, namely

$$M = \frac{|32 - j24|}{|87 + j4|} = \frac{40.0}{87.09} = 0.4593$$

The angle of $H(j4)$ is the angle of its numerator minus the angle of its denominator, namely

$$\theta = \tan^{-1}(-24/32) - \tan^{-1}(4/87)$$

$$= -0.6435 - 0.0459 = -0.6894 \text{ rad}$$

Thus, by (57),

$$y_{ss}(t) = (3)(0.4593)\cos(4t - 0.6894)$$

$$= 1.378\cos(4t - 0.6894)$$

which agrees with the steady-state portion of the response found in Example 13.6.

EXAMPLE 13.14 Evaluate and sketch the magnitude and phase angle of the frequency-response function for the system described by the first-order equation $\dot{y} + (1/\tau)y = u(t)$.

Solution The system transfer function is $H(s) = 1/(s + 1/\tau)$. When we replace s by $j\omega$, we find the system's frequency-response function to be

$$H(j\omega) = \frac{1}{j\omega + 1/\tau} \tag{58}$$

In order to identify the magnitude $M(\omega)$ and angle $\theta(\omega)$ for (58), we convert it to polar form. To find $M(\omega)$, we divide the magnitude of the numerator of (58) by the magnitude of the denominator, obtaining

$$M(\omega) = \frac{1}{[\omega^2 + (1/\tau)^2]^{1/2}}$$

Subtracting the angle of the denominator from the numerator angle, we find that the phase angle* of $H(j\omega)$ is

$$\theta(\omega) = \arg[1] - \arg[j\omega + 1/\tau]$$

$$= 0 - \tan^{-1}\omega\tau$$

$$= -\tan^{-1}\omega\tau$$

To assist in sketching M vs ω and θ vs ω, we note that at $\omega = 0$ we have $M = \tau$ and $\theta = 0$. As ω approaches infinity, M diminishes monotonically to zero and θ decreases to $-\frac{1}{2}\pi$ rad. At $\omega = 1/\tau$, $M = \tau/\sqrt{2}$ and $\theta = -\frac{1}{4}\pi$ rad. The actual functions are shown in Figure 13.11.

EXAMPLE 13.15 For the circuit shown in Figure 13.12, evaluate the frequency-response function $H(j\omega)$ and sketch its magnitude versus frequency.

---

* The notation $\arg[z]$ denotes the angle of the complex quantity z. It is expressed in radians.

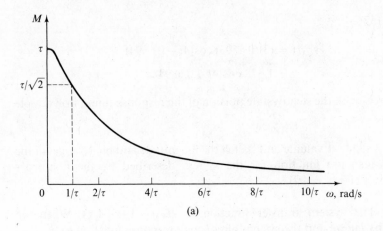

(a)

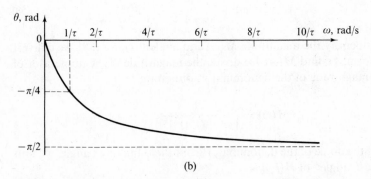

(b)

Figure 13.11 Frequency-response function for $\dot{y} + (1/\tau)y = u(t)$. (a) Magnitude $M(\omega)$. (b) Phase angle $\theta(\omega)$.

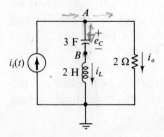

Figure 13.12 Circuit for Example 13.15.

Solution We start by evaluating the circuit's transfer function $H(s)$, which is the ratio $I_o(s)/I_i(s)$ when the initial stored energy is zero. Applying Kirchhoff's current law at nodes A and B gives

$$3(\dot{e}_A - \dot{e}_B) + \tfrac{1}{2}e_A - i_i(t) = 0 \tag{59a}$$

$$3(\dot{e}_B - \dot{e}_A) + i_L(0) + \tfrac{1}{2}\int_0^t e_B(\lambda)\,d\lambda = 0 \tag{59b}$$

The output current is given by

$$i_o = \tfrac{1}{2}e_A \tag{60}$$

Transforming (59) and (60) with zero initial voltage across the capacitor and zero initial current through the inductor gives

$$(3s + \tfrac{1}{2})E_A(s) - 3sE_B(s) = I_i(s)$$

$$-3sE_A(s) + \left(3s + \frac{1}{2s}\right)E_B(s) = 0$$

$$I_o(s) = \tfrac{1}{2}E_A(s)$$

We can solve this set of three transformed equations for $H(s) = I_o(s)/I_i(s)$. The result is

$$H(s) = \frac{6s^2 + 1}{6s^2 + 6s + 1}$$

We find the system's frequency-response function by setting s equal to $j\omega$ in $H(s)$. Noting that $(j\omega)^2 = -\omega^2$, we get

$$H(j\omega) = \frac{-6\omega^2 + 1}{-6\omega^2 + j6\omega + 1}$$

The magnitude function is

$$M(\omega) = \frac{|1 - 6\omega^2|}{|1 - 6\omega^2 + j6\omega|}$$

$$= \frac{|1 - 6\omega^2|}{[(1 - 6\omega^2)^2 + (6\omega)^2]^{1/2}}$$

which is plotted versus ω in Figure 13.13. The phase angle $\theta(\omega)$ is given by

$$\theta(\omega) = \arg[1 - 6\omega^2] - \arg[1 - 6\omega^2 + j6\omega]$$

which could also be plotted versus frequency.

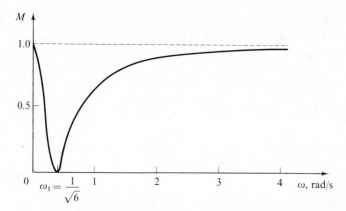

Figure 13.13 Frequency-response magnitude for the circuit in Example 13.15.

Note that when $\omega = 0$, $M(\omega)$ is unity. A sinusoidal function for which $\omega = 0$ reduces to a constant, so $H(0)$ is the steady-state response to the unit step function, which is consistent with our earlier discussion. We can check the fact that $H(0) = 1$ for Figure 13.12 by noting from Section 9.6 that we can replace the inductor and capacitor by short and open circuits, respectively, when finding the steady-state response to a constant input.

We see from Figure 13.13 that the sinusoidal-steady-state response is zero at a frequency of $\omega_1 = 1/\sqrt{6}$ rad/s. Because of the shape of the plot of $M(\omega)$ vs ω, such a circuit is often called a *notch filter*. For sinusoidal inputs with frequencies $\omega \ll \omega_1$ or $\omega \gg \omega_1$, the magnitude of the frequency-response function is approximately unity, which means that the amplitude of $(i_o)_{ss}$ will be close to that of $i_i(t)$. For inputs with $\omega \simeq 1/\sqrt{6}$ rad/s, however, the amplitude of $(i_o)_{ss}$ will be much less than that of $i_i(t)$. Hence, sinusoidal inputs in this frequency range are substantially attenuated by the circuit. Such a circuit can be used to filter out unwanted signals that have frequencies close to ω_1 without significantly affecting signals at other frequencies.

PROBLEMS

13.1 a. Repeat Example 13.1 for the parameter values $J = 1$ kg·m^2, $B = 4$ N·m·s/rad, and $K = 4$ N·m/rad and for no applied torque.

b. Identify the mode functions. For each mode function, give the restrictions on the initial conditions needed to eliminate it from the zero-input response.

13.2 a. For the circuit shown in Figure P13.2, find the transformed output $E_o(s)$ in terms of R, $e_C(0)$, $i_L(0)$, and $E_i(s)$.

b. Find the zero-input response and identify the mode functions when $R = \frac{2}{7} \Omega$. For each mode function, give the restrictions on the initial conditions needed to eliminate it from the zero-input response.

c. Repeat part b when $R = \frac{2}{3} \Omega$.

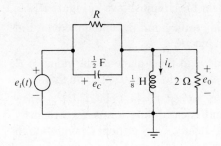

Figure P13.2

13.3 Consider the linearized model developed in Example 5.7 for a frictionless pendulum. Denote the initial values of the incremental angular displacement and velocity by $\hat{\theta}(0) = \theta_0$ and $\hat{\omega}(0) = \omega_0$.

a. Find an expression for the zero-input response in terms of θ_0 and ω_0 when the operating point is $\bar{\theta} = 0$, i.e., with the mass directly below the pivot.

b. Identify the two mode functions in part a. For each mode function, give the initial conditions needed to suppress that mode.

c. Repeat part a and part b when the operating point is $\bar{\theta} = \pi$ rad, i.e., with the mass directly above the pivot. Discuss in physical terms how, at least in principle, initial conditions can be selected that do not excite the unstable mode, thereby giving a decaying zero-input response.

13.4 a. Repeat Example 13.3 for the initial conditions $\theta_1(0) = \theta_2(0) = 0$ and $\dot{\theta}_1(0) = \dot{\theta}_2(0) = 0.5$ rad/s. Indicate which of the mode functions are not excited.

b. Repeat part a when $\theta_1(0) = \theta_2(0) = A$ and $\dot{\theta}_1(0) = \dot{\theta}_2(0) = B$, where A and B are arbitrary constants.

c. Verify that your answer to part b reduces to that of Example 13.3 for $A = 0.5$ and $B = 0$ and to that of part a for $A = 0$ and $B = 0.5$.

13.5 For the translational mechanical system shown in Figure P12.37, let $M_1 = M_2 = 1$ kg and $K_0 = K_1 = K_2 = 4$ N/m.

    a. Identify the mode functions of the zero-input response.

    b. Find the zero-input response of x_1 when the initial conditions are $x_1(0) = 0.1$ m and $\dot{x}_1(0) = x_2(0) = \dot{x}_2(0) = 0$.

    c. Repeat part b when $x_1(0) = x_2(0) = 0.1$ m and $\dot{x}_1(0) = \dot{x}_2(0) = 0$.

13.6 a. Find the transfer function $H(s)$ for the circuit shown in Figure P13.2.

    b. Find expressions for the damping ratio ζ and the undamped natural frequency ω_n in terms of the resistance R.

    c. Show how the poles of $H(s)$ move in the s-plane as the value of R varies. Show explicitly the pole positions for $R = 0$, $\frac{2}{7}\,\Omega$, and infinity.

    d. Find the zero-state response to $e_i(t) = [1 + \sin 2t]\,U(t)$ when $R = \frac{2}{7}\,\Omega$. Identify the transient and steady-state terms. Explain why there is no constant term in the steady-state response even though there is a constant term in the input.

13.7 a. Write the transfer function $H(s)$ for the translational mechanical system shown in Figure 3.5(a) when $M = 1$ kg, $K_1 = 1$ N/m, and $K_2 = 3$ N/m.

    b. Show how the poles of $H(s)$ move in the complex plane as the damping coefficient B is varied from zero to infinity.

    c. Give the form of the zero-input response when (1) $B = 2$ N·s/m, (2) $B = 4$ N·s/m, and (3) $B = 5$ N·s/m.

13.8 a. Verify the answer for Example 13.6 given in (13.31a) by evaluating the coefficients A_1, A_2, and A_3 in (13.30a).

    b. Determine the algebraic equations that the coefficients A_1, A_4, A_5, A_6, and A_7 in (13.30b) must obey. Then use (13.31b) to verify that the coefficients do satisfy these equations.

In Problems 13.9 through 13.19, find the transfer function for the system modeled in the indicated example or problem.

13.9 Problem 2.6 with the output x_1.

13.10 Problem 2.9 with the output z_2.

13.11 a. Problem 3.26 with the output v.

    b. Repeat part a with the output x.

13.12 Example 4.8 with the output x.

13.13 Example 4.9 with the output z.

13.14 a. Example 9.10 with the output e_o.

b. Repeat part a with the output i_C.

13.15 Problem 9.21.

13.16 a. Problem 9.23.

b. Repeat part a with the capacitor and left-hand resistor interchanged.

13.17 Problem 9.30 (for both circuits).

13.18 Problem 9.32.

13.19 Problem 11.7.

13.20 a. Find the transfer function $H(s) = \Theta(s)/E_i(s)$ for the galvanometer described by (11.16) where $L = 0$.

b. Repeat part a using (11.14) and (11.15), where L is included. Compare the result with your answer to part a.

13.21 a. Plot the pole-zero pattern for the transfer function

$$H(s) = \frac{s+1}{s^2 + 5s + 6}$$

b. Write the general form of the zero-input response and find the system's input-output differential equation.

c. Use the initial-value and final-value theorems, if they are applicable, to determine the values of the unit step response at $t = 0+$ and when t approaches infinity.

d. Find the unit step response $y_U(t)$ and the unit impulse response $h(t)$ using $H(s)$.

13.22 Repeat Problem 13.21 for

$$H(s) = \frac{s^2 + 2s + 2}{s^2 + 4s + 4}$$

13.23 Repeat Problem 13.21 for

$$H(s) = \frac{s+3}{s^3 + 7s^2 + 10s}$$

13.24 Repeat Problem 13.21 for

$$H(s) = \frac{12}{s(s^2 + 2s + 4)}$$

13.25 Repeat Problem 13.21 for

$$H(s) = \frac{s^2}{(2s + 1)(s^2 + 4)}$$

13.26 Solve part a of Problem 8.15 for the unit impulse response and the unit step response by using (41) and (45). Sketch both responses on the same axes.

13.27 Repeat Problem 13.26 for the differential equation in part b of Problem 8.15.

13.28 Repeat Problem 13.26 for the differential equation in Problem 8.16.

13.29 a. Find the transfer function for a system that obeys the equation

$$\ddot{y} + 4\dot{y} + 4y = u(t)$$

and use it to evaluate the unit step response.

b. Differentiate the answer to part a and compare the result to the unit impulse response found in Example 12.8.

13.30 Find the zero-state response of a system described by the transfer function in Problem 13.21 to the input $t\epsilon^{-t}$ for $t > 0$. Identify the free response $y_H(t)$, the forced response $y_P(t)$, and the transient terms.

13.31 a. By a derivation similar to the one used to obtain (55), show that the steady-state response of a stable system to the input $u(t) = B \cos \omega t$ is $y_{ss}(t) = BM \cos(\omega t + \theta)$, where M and θ are given by (53).

b. Derive the expression given in (56) for the steady-state response to $u(t) = B \sin(\omega t + \phi_1)$.

13.32 For each of the following transfer functions, plot the pole-zero pattern, draw curves of $M(\omega)$ vs ω and $\theta(\omega)$ vs ω, and comment briefly on your results. For the function in part c, include the numerical values for $\omega = 9.9$, 10.0, and 10.1 rad/s.

a. $H(s) = \dfrac{2}{s^2 + 2s + 1}$

b. $H(s) = \dfrac{2s^2}{s^2 + 2s + 1}$

c. $H(s) = \dfrac{s}{s^2 + 0.2s + 100}$

13.33 Repeat Problem 13.32 for the following transfer functions.

a. $H(s) = \dfrac{s-1}{s+1}$

b. $H(s) = \dfrac{(s+1)^2}{s(s+5)}$

c. $H(s) = \dfrac{s}{(s^2 + 0.2s + 100)^2}$

13.34 For the differential equation given in Example 8.1, find $H(j\omega)$ and $H(0)$. Also use (57) to verify the steady-state response to each of the inputs given in the example.

13.35 The steady-state response of a stable system to the input

$$u(t) = \sin 5t + \sin 10t + \sin 15t$$

has the form

$$y_{ss}(t) = A \sin(5t + \theta_1) + B \sin(10t + \theta_2) + C \sin(15t + \theta_3)$$

For a system described by the transfer function in part c of Problem 13.32, use (55) to find the values of A, B and C and calculate the ratios A/B and C/B.

13.36 The steady-state response of a system described by the transfer function in part a of Problem 13.32 to the input

$$u(t) = 1 + \sin t + \sin 10t$$

has the form

$$y_{ss}(t) = A + B \sin(t + \theta_2) + C \sin(10t + \theta_3)$$

Calculate A, B, C, and comment on the ratios B/A and C/A.

13.37 a. For the series RLC circuit shown in Figure 9.13, find the transfer function $H(s) = I(s)/E_i(s)$.

b. By examining $H(j\omega)$, determine the value of ω such that the steady-state response to $e_i(t) = \sin \omega t$ is $i_{ss}(t) = (1/R) \sin \omega t$.

13.38 a. Find the transfer function $H(s) = E_o(s)/I_i(s)$ for the parallel RLC circuit shown in Figure 9.14.

b. By examining $H(j\omega)$, determine the value of ω for which the steady-state response to $i_i(t) = \sin \omega t$ is $e_o(t) = R \sin \omega t$.

13.39 a. Find the transfer function $H_v(s) = V(s)/F_a(s)$ for the mechanical system shown in Figure P13.39.

b. Find the value of ω for which the mass M_2 will remain motionless in the steady state when the input is $f_a(t) = \sin \omega t$.

c. Let f_K denote the tensile force for the spring K and find the transfer function $H_K(s) = F_K(s)/F_a(s)$. Evaluate $H_K(j\omega)$ for the value of ω you found in part b and justify your answer to part b in physical terms.

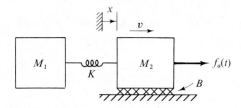

Figure P13.39

13.40 a. Find the transfer function $H(s) = E_o(s)/I_i(s)$ for the circuit shown in Figure P13.40.

b. Find the value of ω for which the steady-state voltage across C_2 is zero when the input is $i_i(t) = \sin \omega t$.

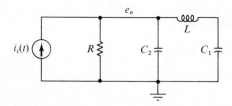

Figure P13.40

13.41 a. Verify that when $\alpha \leq 1$, the transfer function

$$H(s) = \frac{K}{s^2 + 2\alpha s + 100}$$

can be approximated by

$$H^*(s) = \frac{K}{(s+\alpha-j10)(s+\alpha+j10)}$$

b. Plot the pole-zero pattern for $H^*(s)$.

c. Using $H^*(j\omega)$ sketch to scale a curve of $M(\omega)/K$ vs ω, where M is defined by (53), when $\alpha = 0.2$ and $\alpha = 1.0$.

d. The transfer function $H^*(s)$ represents a band-pass filter having a center frequency of 10 rad/s and a bandwidth of 2α. Explain the appropriateness of these names in relation to your answers to part c.

e. Discuss the behavior of $M(\omega)$ as α approaches zero.

f. Show that the circuit in Figure 9.13 has $H(s)$ as its transfer function with $K = 100$ if the output is e_C. Find the value of α in terms of the circuit elements.

CHAPTER 14
THERMAL SYSTEMS

Thermal systems are ones in which the storage and flow of heat are involved. Their mathematical models are based on the fundamental laws of thermodynamics. Examples of thermal systems are a thermometer, an automobile engine's cooling system, an oven, and a refrigerator. Generally thermal systems are distributed, and thus they obey partial rather than ordinary differential equations. We shall restrict our attention to lumped mathematical models by making approximations where necessary. Our purpose is to obtain linear ordinary differential equations that are capable of describing the dynamic response to a good approximation. We shall not examine the steady-state analysis of thermodynamic cycles that might be required in the design of a chemical process. Neither shall we consider the evaluation of thermal properties, such as heat-transfer coefficients, in terms of physical parameters. Although systems that involve changes of phase such as boiling or condensation can be modeled, such treatment is beyond the scope of this book.

As in the previous chapters on modeling, we introduce the variables and the element laws used to describe the dynamic behavior of thermal systems first. Then we present a number of examples illustrating their application. A comprehensive example of the analysis of a thermal system appears in Chapter 18.

14.1 VARIABLES

The variables used to describe the behavior of a thermal system are

$$\theta = \text{temperature in kelvin (K)*}$$

$$q = \text{heat flow rate in joules per second (J/s) or in watts (W)}$$

where 1 watt $= 1$ joule per second.

The temperatures at various points in a distributed body generally differ from one another. For modeling and analysis, however, it is desirable to assume that all points in the body have the same temperature. This is presumably the average temperature of the body, and temperature deviations from the average at various points do not affect the validity of the single-temperature model. If this is not the case, the body may be partitioned into segments, each of which has an average temperature associated with it, as illustrated in a later example. Unless otherwise noted, we shall use average temperatures for individual bodies. Furthermore, because the temperature is a measure of the energy stored in a body (provided there are no changes of phase), we normally select the temperatures as the state variables of a thermal system.

For most thermal systems, an equilibrium condition exists that defines the nominal operation. Generally, only deviations of the variables from their nominal values are of interest from a dynamic point of view. In these cases, *incremental temperatures* and *incremental heat flow rates* are defined by relationships of the form

$$\hat{\theta}(t) = \theta(t) - \bar{\theta}$$

$$\hat{q}(t) = q(t) - \bar{q}$$

where $\bar{\theta}$ and $\bar{q}$ are the nominal values. The ambient temperature of the environment surrounding the system is considered constant and is denoted

---

* Although the kelvin is the SI temperature unit, degrees Celsius (°C) may be more familiar. A temperature expressed in kelvin can be converted to degrees Celsius by subtracting 273.15 from its value.

by θ_a. In some problems, the nominal values of the temperature variables may be equal to θ_a, in which case we may refer to the incremental temperatures as *relative temperatures*.

14.2 ELEMENT LAWS

A consequence of the laws of thermodynamics is that there are only two types of passive thermal elements, namely thermal capacitance and thermal resistance. Strictly speaking, thermal capacitance and thermal resistance are characteristics associated with bodies that are distributed in space and are not lumped elements. However, since we seek to describe the dynamic behavior of thermal systems by lumped models, we shall refer to them as elements. These elements are described below, and a brief discussion of thermal sources follows.

THERMAL CAPACITANCE

A physical body at a uniform temperature will have an algebraic relationship between its temperature and the heat stored within it. Provided that there is no change of phase and that the range of temperatures is not excessive, this relationship can be considered linear.

If $q_{in}(t) - q_{out}(t)$ denotes the net heat flow rate into the body as a function of time, then the net heat supplied between time t_0 and time t is

$$\int_{t_0}^{t} [q_{in}(\lambda) - q_{out}(\lambda)] \, d\lambda$$

We assume that the heat supplied during this time interval equals a constant C times the change in temperature. If the temperature of the body at the reference time t_0 is denoted by $\theta(t_0)$, then

$$\theta(t) = \theta(t_0) + \frac{1}{C} \int_{t_0}^{t} [q_{in}(\lambda) - q_{out}(\lambda)] \, d\lambda \tag{1}$$

The constant C is known as the *thermal capacitance* and has units of joules per kelvin (J/K). For a body having a mass M and specific heat σ, with units of joules per kilogram-kelvin, the thermal capacitance is $C = M\sigma$.

Differentiating (1), we have

$$\dot{\theta} = \frac{1}{C} [q_{in}(t) - q_{out}(t)] \tag{2}$$

which relates the rate of temperature change to the instantaneous net heat flow rate into the body. Because we generally select the temperatures of the bodies comprising a thermal system as the state variables, we shall use (2) extensively to write the state-variable equations. As indicated earlier, we can use (1) and (2) only when the temperature of the body is assumed to be uniform. If the thermal gradients within the body are so great that we cannot make this assumption, then the body should be divided into two or more parts with separate thermal capacitances.

THERMAL RESISTANCE

Heat can flow between points by three different mechanisms: conduction, convection, and radiation. We shall consider only conduction, whereby heat flows from one body to another through the medium connecting them at a rate proportional to the temperature difference between the points. Specifically, the flow of heat by conduction from a body at temperature θ_1 to a body at temperature θ_2 obeys the relationship

$$q(t) = \frac{1}{R}[\theta_1(t) - \theta_2(t)] \tag{3}$$

where R is the *thermal resistance* of the path between the bodies, with units of kelvin-seconds per joule (K·s/J) or kelvin per watt (K/W). For a path of cross-sectional area A and length d composed of material having a thermal conductivity α (with units of watts per meter-Kelvin), the thermal resistance is

$$R = \frac{d}{A\alpha} \tag{4}$$

We can use (3) only when the material or body being treated as a thermal resistance does not store any heat. Should it become important to account for the heat stored in the resistance, then we must also include a thermal capacitance in the model.

In developing lumped models of thermal systems, we often find it convenient to combine two or more thermal resistances into a single equivalent resistance. The following two examples illustrate the techniques for doing this for combinations of two resistances.

EXAMPLE 14.1   Figure 14.1(a) shows two bodies at temperatures θ_1 and θ_2 separated by two resistances R_1 and R_2. Heat flows through each of the resistances at the rate q but cannot flow through the perfect insulating

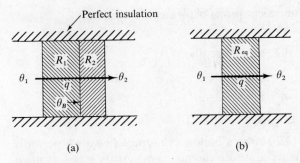

Perfect insulation

(a)   (b)

Figure 14.1 (a) Two thermal resistances in series. (b) Equivalent resistance.

material above and below the resistances. Find the value of the equivalent thermal resistance R_{eq} in Figure 14.1(b) and solve for the interface temperature θ_B.

Solution We can use (3) for each of the thermal resistances to express the heat flow rate q in terms of the resistance and the temperature difference. Specifically,

$$q = \frac{1}{R_1}(\theta_1 - \theta_B) \tag{5a}$$

$$q = \frac{1}{R_2}(\theta_B - \theta_2) \tag{5b}$$

Combining the equations to eliminate θ_B and arranging the result in the form of (3), we find that

$$q = \frac{1}{R_1 + R_2}(\theta_1 - \theta_2)$$

Hence the equivalent thermal resistance is

$$\boxed{R_{eq} = R_1 + R_2} \quad series \tag{6}$$

where R_1 and R_2 are said to be in *series* because the heat flow rate is the same through each.

To calculate the interface temperature θ_B, we combine (5a) and (5b) to eliminate q, getting

$$\theta_B = \frac{R_2 \theta_1 + R_1 \theta_2}{R_1 + R_2} \tag{7}$$

You should verify that equivalent forms of (7) are

$$\theta_B = \theta_1 - \frac{R_1}{R_{eq}}(\theta_1 - \theta_2)$$

$$\theta_B = \theta_2 + \frac{R_2}{R_{eq}}(\theta_1 - \theta_2)$$

EXAMPLE 14.2 Figure 14.2 shows a hollow cylindrical vessel whose walls have thickness T and a material whose thermal conductivity is α. Calculate the thermal resistances of the side of the cylinder (R_c) and of each end (R_e). Then find the equivalent resistance of the entire vessel in terms of R_c and R_e.

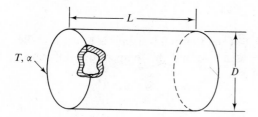

Figure 14.2 Cylindrical vessel for Example 14.2.

Solution From (4), with d replaced by T and A replaced by $\pi D^2/4$, the resistance of each end of the vessel is

$$R_e = \frac{4T}{\pi D^2 \alpha} \tag{8}$$

Similarly, the resistance of the cylindrical portion is

$$R_c = \frac{T}{\pi D L \alpha} \tag{9}$$

The rate at which heat flows through each end is

$$q_e = \frac{\Delta \theta}{R_e}$$

where $\Delta \theta$ denotes the interior temperature minus the exterior temperature.

Likewise, the heat flow rate through the cylindrical wall is

$$q_c = \frac{\Delta\theta}{R_c}$$

The total heat flow rate is

$$q_T = 2q_e + q_c$$

$$= \left(\frac{2}{R_e} + \frac{1}{R_c}\right)\Delta\theta \tag{10}$$

Since the equivalent thermal resistance R_{eq} must satisfy the relationship

$$q_T = \frac{\Delta\theta}{R_{eq}}$$

it follows from (10) that

$$\frac{1}{R_{eq}} = \frac{2}{R_e} + \frac{1}{R_c}$$

and

$$R_{eq} = \frac{R_c R_e}{2R_c + R_e} \tag{11}$$

Because the two ends and the cylindrical wall present independent paths for the heat to flow between the interior and exterior of the vessel, with the same temperature difference existing across each path, the three thermal resistances are said to be in *parallel*.

THERMAL SOURCES

There are two types of ideal thermal sources. Figure 14.3 represents a source that adds or removes heat at a specified rate. Heat is added to the system when $q_i(t)$ is positive and removed when $q_i(t)$ is negative. On occasion, we shall consider the temperature of a body to be an input, in which case the temperature is a known function of time regardless of the rate at which heat flows between that body and the rest of the system.

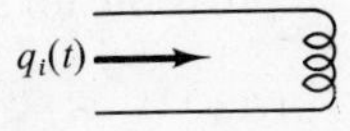

Figure 14.3 Representation of an ideal thermal source.

14.3 DYNAMIC MODELS OF THERMAL SYSTEMS

We shall demonstrate how to construct and analyze dynamic models of thermal systems by considering several examples. The general technique is to select the temperature of each thermal capacitance as a state variable and use (2) to obtain the corresponding state-variable equation. The net heat flow rate into a thermal capacitance depends on heat sources and heat flow rates through thermal resistances. By using (3), we can express the heat flow rates through the resistances in terms of the system's state variables, namely the temperatures of the thermal capacitances.

We first consider systems that have thermal capacitances from which heat can escape to the environment through thermal resistances. We illustrate approximating a distributed system by a lumped model by analyzing two possible models for the heating of a bar. In the final example, a thermal capacitance is heated by both a heater and an incoming liquid stream.

EXAMPLE 14.3   Figure 14.4 shows a thermal capacitance C enclosed by insulation that has an equivalent thermal resistance R. The temperature

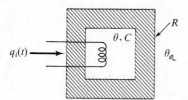

Figure 14.4 Thermal system with one capacitance.

within the capacitance is θ and is assumed to be uniform. Heat is added to the interior of the system at the rate $q_i(t)$. The nominal values of $q_i(t)$ and θ are denoted by $\bar{q}_i$ and $\bar{\theta}$, respectively. The ambient temperature surrounding the exterior of the insulation is θ_a, a constant. Find the system model in terms of θ, $q_i(t)$, and θ_a and also in terms of incremental variables. Solve for the transfer function and unit step response, and draw a simulation diagram.

Solution   The appropriate state variable is θ. We obtain an expression for its derivative by using (2) with

$$q_{\text{in}}(t) = q_i(t)$$

and, from (3),

$$q_{out}(t) = \frac{1}{R}(\theta - \theta_a)$$

Thus the state-variable model is

$$\dot{\theta} = \frac{1}{C}\left[q_i(t) - \frac{1}{R}(\theta - \theta_a)\right]$$

where we consider the ambient temperature θ_a an input to the system, along with $q_i(t)$. Rewriting the model, we have

$$\dot{\theta} + \frac{1}{RC}\theta = \frac{1}{C}q_i(t) + \frac{1}{RC}\theta_a \tag{12}$$

which is readily recognized as the differential equation of a linear first-order system with the time constant $\tau = RC$ and the inputs $q_i(t)$ and θ_a. At the operating point, (12) reduces to $\quad \bar{\theta} = c \quad \Rightarrow \quad \dot{\bar{\theta}} = 0$

$$\frac{1}{RC}\bar{\theta} = \frac{1}{C}\bar{q}_i + \frac{1}{RC}\theta_a \tag{13}$$

so

$$\bar{\theta} = \theta_a + R\bar{q}_i \tag{14}$$

When the system is in equilibrium, the temperature of the thermal capacitance is constant and the heat flow rate $\bar{q}_i$ supplied by the heater must equal the rate of heat flow through the thermal resistance. Then the temperature difference across the resistance is $R\bar{q}_i$, which agrees with (14).

To obtain a model in terms of incremental variables, we define

$$\hat{\theta}(t) = \theta(t) - \bar{\theta}$$

$$\hat{q}_i(t) = q_i(t) - \bar{q}_i$$

Substituting these expressions into (12) gives

$$\dot{\hat{\theta}} + \frac{1}{RC}(\hat{\theta} + \bar{\theta}) = \frac{1}{C}[\hat{q}_i(t) + \bar{q}_i] + \frac{1}{RC}\theta_a$$

which, by using (13), we reduce to

$$\dot{\hat{\theta}} + \frac{1}{RC}\hat{\theta} = \frac{1}{C}\hat{q}_i(t) \tag{15}$$

Examining (14) shows that if $\bar{q}_i > 0$, then $\bar{\theta} > \theta_a$ and the capacitance is being heated. If $\bar{q}_i < 0$, then $\bar{\theta} < \theta_a$ and the capacitance is being cooled. If $\bar{q}_i = 0$, then the nominal value of the temperature is $\bar{\theta} = \theta_a$ and $\hat{q}_i(t) = q_i(t)$. Because the system is linear, the incremental model given by (15) has the same coefficients regardless of the operating point.

Recall from Chapter 13 that the transfer function is $H(s) = Y(s)/U(s)$, where $U(s)$ is the transformed input and $Y(s)$ is the transform of the zero-state response. Thus we transform (15) with $\hat{\theta}(0) = 0$ to obtain

$$s\hat{\Theta}(s) + \frac{1}{RC}\hat{\Theta}(s) = \frac{1}{C}\hat{Q}_i(s)$$

We can rearrange this transformed equation to give the transfer function $H(s) = \hat{\Theta}(s)/\hat{Q}_i(s)$ as

$$H(s) = \frac{\dfrac{1}{C}}{s + \dfrac{1}{RC}}$$

which has a single pole at $s = -1/RC$. The response of $\hat{\theta}$ to a unit step function for $\hat{q}_i(t)$ will be

$$\hat{\theta} = R(1 - \epsilon^{-t/RC}) \qquad \text{for } t > 0$$

which approaches a steady-state value of R with the time constant RC. Note that we can also find the steady-state value of $\hat{\theta}$ by evaluating $H(s)$ at $s = 0$. To find the response in terms of the actual temperature θ, we merely add $\bar{\theta}$ to $\hat{\theta}$, getting

$$\theta = \bar{\theta} + R(1 - \epsilon^{-t/RC}) \qquad \text{for } t > 0$$

which is shown in Figure 14.5, along with the input $q_i(t)$.

To draw the simulation diagram, we use (12) to obtain $\dot{\theta}$ as the output of a summing junction and we use an integrator to get θ, resulting in Figure 14.6(a). We can get an alternative diagram by using (15) to form the derivative $\dot{\hat{\theta}}$ of the incremental temperature, integrating it, and then adding $\bar{\theta}$ to the result to give θ as the output. This version is shown in Figure 14.6(b). The diagrams include the initial conditions on the output of the integrators, as was first indicated in Figure 6.3. Note that the necessary initial condition is $\theta(0)$ for the integrator shown in Figure 14.6(a) and $\hat{\theta}(0)$ in Figure 14.6(b).

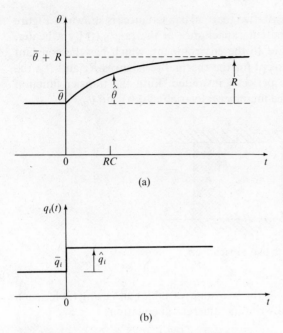

Figure 14.5 (a) Temperature response for Example 14.3. (b) Heat input rate.

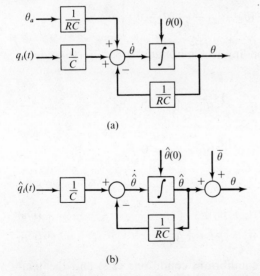

Figure 14.6 Simulation diagrams for Example 14.3. (a) θ as the integrator output. (b) $\hat{\theta}$ as the integrator output.

EXAMPLE 14.4   A system with two thermal capacitances is shown in Figure 14.7. Heat is supplied to the left capacitance at the rate $q_i(t)$ by a heater, and it is lost at the right end to the environment, which has the constant ambient temperature θ_a. Except for the thermal resistances R_1 and R_2, the enclosure is assumed to be perfectly insulated. Find the transfer function relating the transforms of the incremental variables $\hat{q}_i(t)$ and $\hat{\theta}_2$.

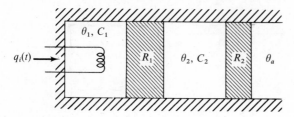

Figure 14.7 Thermal system with two capacitances.

Solution   Taking θ_1 and θ_2 as the two state variables and using (2) for each thermal capacitance, we can write the differential equations

$$\dot{\theta}_1 = \frac{1}{C_1}\left[q_i(t) - \frac{1}{R_1}(\theta_1 - \theta_2)\right]$$

$$\dot{\theta}_2 = \frac{1}{C_2}\left[\frac{1}{R_1}(\theta_1 - \theta_2) - \frac{1}{R_2}(\theta_2 - \theta_a)\right] \tag{16}$$

At the operating point corresponding to the nominal input $\bar{q}_i$, (16) reduces to

$$\bar{q}_i - \frac{1}{R_1}(\bar{\theta}_1 - \bar{\theta}_2) = 0$$

$$\frac{1}{R_1}(\bar{\theta}_1 - \bar{\theta}_2) - \frac{1}{R_2}(\bar{\theta}_2 - \theta_a) = 0$$

from which

$$\bar{\theta}_2 = \theta_a + R_2\bar{q}_i \tag{17a}$$

$$\bar{\theta}_1 = \theta_a + (R_1 + R_2)\bar{q}_i \tag{17b}$$

At the operating point, where equilibrium conditions exist and the temperatures are constant, the heat flow rates through R_1 and R_2 must both equal $\bar{q}_i$. Since the rates of heat flow are the same, we can regard the two resistances as being in series. The equivalent resistance is $R_{eq} = R_1 + R_2$, as in (6), so $\bar{\theta}_1$ must be the temperature difference $(R_1 + R_2)\bar{q}_i$ plus the ambient temperature θ_a, which agrees with (17b).

Defining the incremental temperatures $\hat{\theta}_1 = \theta_1 - \bar{\theta}_1$ and $\hat{\theta}_2 = \theta_2 - \bar{\theta}_2$, we can rewrite the two state-variable equations as

$$\dot{\hat{\theta}}_1 + \frac{1}{R_1 C_1}\hat{\theta}_1 = \frac{1}{R_1 C_1}\hat{\theta}_2 + \frac{1}{C_1}\hat{q}_i(t)$$

$$\dot{\hat{\theta}}_2 + \left(\frac{1}{R_1 C_2} + \frac{1}{R_2 C_2}\right)\hat{\theta}_2 = \frac{1}{R_1 C_2}\hat{\theta}_1$$

(18)

where the ambient temperature θ_a no longer appears. To find the transfer function $H(s) = \hat{\Theta}_2(s)/\hat{Q}_i(s)$, we transform (18) with $\hat{\theta}_1(0) = \hat{\theta}_2(0) = 0$, getting

$$\left(s + \frac{1}{R_1 C_1}\right)\hat{\Theta}_1(s) = \frac{1}{R_1 C_1}\hat{\Theta}_2(s) + \frac{1}{C_1}\hat{Q}_i(s)$$

$$\left[s + \left(\frac{1}{R_1 C_2} + \frac{1}{R_2 C_2}\right)\right]\hat{\Theta}_2(s) = \frac{1}{R_1 C_2}\hat{\Theta}_1(s)$$

Combining these equations to eliminate $\hat{\Theta}_1(s)$ and rearranging to form the ratio $\hat{\Theta}_2(s)/\hat{Q}_i(s)$, we get

$$H(s) = \frac{\dfrac{1}{R_1 C_1 C_2}}{s^2 + \left(\dfrac{1}{R_1 C_1} + \dfrac{1}{R_1 C_2} + \dfrac{1}{R_2 C_2}\right)s + \dfrac{1}{R_1 C_1 R_2 C_2}}$$

(19)

Note that the denominator of $H(s)$ is a quadratic function of s, implying that it will have two poles. In fact, for any combination of numerical values for R_1, R_2, C_1, and C_2, these poles will be real, negative, and distinct (as you will show in Problem 14.7). As a consequence, the transient response of the system consists of two decaying exponential functions.

To construct a simulation diagram, we rewrite (16) as

$$R_1 C_1 \dot{\theta}_1 = \theta_2 - \theta_1 + R_1 q_i(t)$$

$$R_2 C_2 \dot{\theta}_2 = \frac{R_2}{R_1}(\theta_1 - \theta_2) - \theta_2 + \theta_a$$

and then form the terms $R_1 C_1 \dot{\theta}_1$ and $R_2 C_2 \dot{\theta}_2$ as the outputs of summing junctions. Integrating the summing-junction outputs with the appropriate gains leads directly to the diagram shown in Figure 14.8. It is interesting to note that this diagram has three feedback loops, with the outer loop indicating that θ_2 directly affects the rate of change of θ_1. In Chapter 16, we shall demonstrate a technique for finding the transfer function $H(s)$ directly from the simulation diagram.

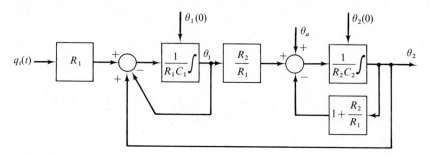

Figure 14.8 Simulation diagram for Example 14.4.

EXAMPLE 14.5   Consider a bar of length L and cross-sectional area A that is perfectly insulated on all its boundaries except at the left end, as shown in Figure 14.9. The temperature at the left end of the bar is $\theta_i(t)$, a known function of time that is the system's input. The interior of the bar is initially at the ambient temperature θ_a. The specific heat of the material is σ with units of joules per kilogram-kelvin, its density is ρ expressed in kilograms per cubic meter, and its thermal conductivity is α expressed in watts per meter-kelvin. Although the system is distributed and can be modeled exactly only by a partial differential equation, develop a lumped model consisting of a single thermal capacitance and resistance and then find the unit step response.

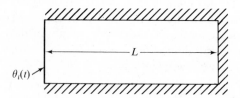

Figure 14.9 Insulated bar considered in Example 14.5.

Solution   Assume that all points within the bar have the same temperature θ except the left end, which has the prescribed temperature $\theta_i(t)$. Then the thermal capacitance is

$$C = \sigma \rho A L \tag{20}$$

with units of joules per kelvin. To complete the single-capacitance approximation, we assume the thermal resistance of the entire bar, which from (4) is

$$R = \frac{L}{A\alpha} \tag{21}$$

with units of kelvins per watt, separates the left end from the remainder of the bar. The lumped approximation is shown in Figure 14.10, with C and

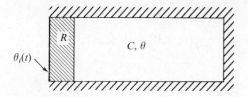

Figure 14.10 Single-capacitance approximation to the insulated bar shown in Figure 14.9.

R given by (20) and (21), respectively. From (2), with $q_{\text{out}} = 0$ because of the perfect insulation, and with

$$q_{\text{in}} = \frac{1}{R} [\theta_i(t) - \theta]$$

it follows that the single-capacitance model is

$$\dot{\theta} = \frac{1}{RC} [\theta_i(t) - \theta]$$

or

$$\dot{\theta} + \frac{1}{RC} \theta = \frac{1}{RC} \theta_i(t) \tag{22}$$

where $\theta(0) = \theta_a$.

If the input $\theta_i(t)$ is the sum of the constant ambient temperature θ_a and a step function of height B,

$$\theta_i(t) = \theta_a + BU(t)$$

In this example, it is convenient to define the nominal values of the temperatures to be the ambient temperature θ_a and to let the incremental variables be the temperatures relative to θ_a. The relative input temperature is

$$\hat{\theta}_i(t) = \theta_i(t) - \theta_a$$
$$= BU(t)$$

and the relative bar temperature is

$$\hat{\theta} = \theta - \theta_a$$

With these definitions, (22) becomes

$$\dot{\hat{\theta}} + \frac{1}{RC}(\hat{\theta} + \theta_a) = \frac{1}{RC}[\theta_a + BU(t)]$$

which reduces to

$$\dot{\hat{\theta}} + \frac{1}{RC}\hat{\theta} = \frac{B}{RC}U(t) \qquad (23)$$

with the initial condition $\hat{\theta}(0) = 0$. The response of the relative temperature is

$$\hat{\theta} = B(1 - \epsilon^{-t/RC}) \qquad \text{for } t > 0$$

which indicates that the temperature of the bar will rise from the ambient temperature to that of its left end with the time constant

$$RC = \frac{\sigma \rho L^2}{\alpha}$$

EXAMPLE 14.6 Analyze the step response of the insulated bar shown in Figure 14.9 using the two-capacitance approximation shown in Figure 14.11.

Solution As indicated in Figure 14.11, we take the value of each thermal capacitance as $0.5C$ where C is given by (20). Likewise, we take the value of each thermal resistance as $0.5R$ where R is given by (21). Applying (2) to the left capacitance with

$$q_{in} = \frac{1}{0.5R}[\theta_i(t) - \theta_1]$$

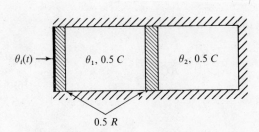

Figure 14.11 A two-capacitance approximation to the insulated bar shown in Figure 14.9.

and

$$q_{\text{out}} = \frac{1}{0.5R}(\theta_1 - \theta_2)$$

gives the state-variable equation

$$\theta_1 = \frac{1}{0.25RC}[\theta_i(t) - 2\theta_1 + \theta_2] \tag{24}$$

Applying (2) to the right capacitance with

$$q_{\text{in}} = \frac{1}{0.5R}(\theta_1 - \theta_2)$$

and $q_{\text{out}} = 0$ gives the second state-variable equation

$$\dot{\theta}_2 = \frac{1}{0.25RC}(\theta_1 - \theta_2) \tag{25}$$

To evaluate the responses of θ_1 and θ_2 to the input $\theta_i(t) = \theta_a + BU(t)$, we can define the relative temperatures $\hat{\theta}_1 = \theta_1 - \theta_a$, $\hat{\theta}_2 = \theta_2 - \theta_a$, and $\hat{\theta}_i(t) = \theta_i(t) - \theta_a$ and then derive the transfer functions $H_1(s) = \hat{\Theta}_1(s)/\hat{\Theta}_i(s)$ and $H_2(s) = \hat{\Theta}_2(s)/\hat{\Theta}_i(s)$. Knowing $H_1(s)$ and $H_2(s)$, we can write for the unit step responses

$$\theta_1(t) = \mathscr{L}^{-1}\left[H_1(s)\frac{B}{s} \right]$$

$$\theta_2(t) = \mathscr{L}^{-1}\left[H_2(s)\frac{B}{s} \right]$$

Rewriting (24) and (25) in terms of the relative temperatures, we have

$$\dot{\theta}_1 + \frac{8}{RC}\hat{\theta}_1 = \frac{4}{RC}\hat{\theta}_2 + \frac{4}{RC}\hat{\theta}_i(t)$$

$$\dot{\theta}_2 + \frac{4}{RC}\hat{\theta}_2 = \frac{4}{RC}\hat{\theta}_1$$

(26)

Transforming (26) with $\hat{\theta}_1(0) = \hat{\theta}_2(0) = 0$ gives the pair of algebraic equations

$$\left(s + \frac{8}{RC}\right)\hat{\Theta}_1(s) = \frac{4}{RC}[\hat{\Theta}_2(s) + \hat{\Theta}_i(s)]$$

(27a)

$$\left(s + \frac{4}{RC}\right)\hat{\Theta}_2(s) = \frac{4}{RC}\hat{\Theta}_1(s)$$

(27b)

Solving (27b) for $\hat{\Theta}_2(s)$, substituting the result into (27a), and rearranging terms, we find that

$$\left[\left(s + \frac{8}{RC}\right)\left(s + \frac{4}{RC}\right) - \left(\frac{4}{RC}\right)^2\right]\hat{\Theta}_1(s) = \left[\frac{4}{RC}s + \left(\frac{4}{RC}\right)^2\right]\hat{\Theta}_i(s)$$

Simplifying the quadratic factor on the left side and solving for $H_1(s) = \hat{\Theta}_1(s)/\hat{\Theta}_i(s)$, we obtain

$$H_1(s) = \frac{4}{RC}\left[\frac{s + \dfrac{4}{RC}}{s^2 + \dfrac{12}{RC}s + \dfrac{16}{(RC)^2}}\right]$$

$$= \frac{4}{RC}\left[\frac{s + \dfrac{4}{RC}}{\left(s + \dfrac{1.528}{RC}\right)\left(s + \dfrac{10.472}{RC}\right)}\right]$$

(28)

The transfer function has poles at $s_1 = -1.528/RC$ and $s_2 = -10.472/RC$, and the free response has two terms that decay exponentially with time constants

$$\tau_1 = \frac{RC}{1.528} = 0.6545RC$$

$$\tau_2 = \frac{RC}{10.472} = 0.0955RC$$

Since $H_1(0) = 1$ from (28), the steady-state value of $\hat{\theta}_1$ equals B when $\hat{\theta}_i(t) = BU(t)$. For this input,

$$
\hat{\Theta}_1(s) = \frac{4}{RC} \left[\frac{s + \dfrac{4}{RC}}{\left(s + \dfrac{1.528}{RC}\right)\left(s + \dfrac{10.472}{RC}\right)} \right] \cdot \frac{B}{s}
$$

$$
= \frac{A_1}{s} + \frac{A_2}{s + \dfrac{1.528}{RC}} + \frac{A_3}{s + \dfrac{10.472}{RC}} \tag{29}
$$

You can verify that the numerical values of the coefficients in the partial-fraction expansion are

$$
A_1 = B
$$

$$
A_2 = -0.7235B
$$

$$
A_3 = -0.2764B
$$

Substituting these values into (29) and taking the inverse transform, we find that for $t > 0$,

$$
\hat{\theta}_1(t) = B(1 - 0.7235\epsilon^{-1.528t/RC} - 0.2764\epsilon^{-10.472t/RC})
$$

From (27b),

$$
\hat{\Theta}_2(s) = \frac{4}{RCs + 4}\hat{\Theta}_1(s)
$$

$$
= \frac{16}{(RC)^2} \left[\frac{B}{s\left(s + \dfrac{1.528}{RC}\right)\left(s + \dfrac{10.472}{RC}\right)} \right]
$$

whose inverse transform can be shown to be

$$
\hat{\theta}_2(t) = B(1 - 1.1708\epsilon^{-1.528t/RC} + 0.1708\epsilon^{-10.472t/RC})
$$

The ratios $\hat{\theta}_1(t)/B$ and $\hat{\theta}_2(t)/B$ are shown in Figure 14.12 versus the normalized time variable t/RC. The exact solution for the response of the continuous bar is given in Section 10.3 of Doebelin (1972).

EXAMPLE 14.7 The insulated vessel shown in Figure 14.13 is filled with liquid at a temperature θ, which is kept uniform throughout the vessel by

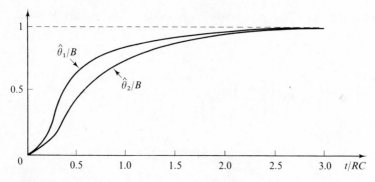

Figure 14.12 Response to a unit step-function in $\theta_i(t)$ for the two-capacitance approximation to the insulated bar shown in Figure 14.9.

perfect mixing. Liquid enters at a constant volumetric flow rate of $\bar{w}$, expressed in units of cubic meters per second, and at a temperature $\theta_i(t)$. It leaves at the same rate and at the temperature θ_o. Because of the perfect mixing, the exit temperature θ_o is the same as the liquid temperature θ. The thermal resistance of the vessel and its insulation is R, and the ambient temperature is θ_a, a constant.

Heat is added to the liquid in the vessel by a heater at a rate $q_h(t)$. The volume of the vessel is V, and the liquid has a density of ρ (with units of kilograms per cubic meter) and a specific heat of σ (with units of joules per kilogram-kelvin). Derive the system model and find the appropriate transfer functions.

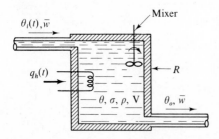

Figure 14.13 Insulated vessel with liquid flowing through it.

Solution   The thermal capacitance of the liquid is the product of the liquid's volume, density, and specific heat. The heat entering the vessel is the sum of that due to the heater and that contained in the incoming stream. The heat leaving the vessel is the sum of that taken out by the outgoing stream and that lost to the ambient through the vessel walls and insulation.

From (2),

$$\dot{\theta} = \frac{1}{C}(q_{\text{in}} - q_{\text{out}}) \tag{30}$$

where

$$q_{\text{in}} = q_h(t) + \bar{w}\rho\sigma\theta_i(t)$$

$$q_{\text{out}} = \bar{w}\rho\sigma\theta + \frac{1}{R}(\theta - \theta_a)$$

$$C = \rho\sigma V$$

Substituting these expressions into (30) and rearranging, we obtain the first-order differential equation

$$\dot{\theta} + \left(\frac{\bar{w}}{V} + \frac{1}{RC}\right)\theta = \frac{\bar{w}}{V}\theta_i(t) + \frac{1}{C}q_h(t) + \frac{1}{C}\theta_a \tag{31}$$

The time constant of the system is

$$\tau = \frac{1}{\dfrac{\bar{w}}{V} + \dfrac{1}{RC}} \tag{32}$$

which approaches RC as $\bar{w}$ approaches zero (no liquid flow). As R approaches infinity (perfect insulation), the time constant becomes $V/\bar{w}$, which is the time required to replace the total tank volume at the volumetric flow rate $\bar{w}$.

Since the initial values of the state variables must be zero when we calculate the transfer functions, we first rewrite the model in terms of incremental variables defined with respect to the operating point. Setting $\dot{\theta}$ equal to zero in (31) yields the relationship between the nominal values $\bar{q}_h$, $\bar{\theta}_i$, $\bar{\theta}$, and the ambient temperature θ_a, namely

$$\bar{q}_h + \frac{C\bar{w}}{V}(\bar{\theta}_i - \bar{\theta}) = \frac{1}{R}(\bar{\theta} - \theta_a) \tag{33}$$

Rewriting (31) in terms of the incremental variables $\hat{\theta} = \theta - \bar{\theta}$, $\hat{\theta}_i(t) = \theta_i(t) - \bar{\theta}_i$, and $\hat{q}_h(t) = q_h(t) - \bar{q}_h$ and using (33), we obtain for the incremental model

$$\dot{\hat{\theta}} + \frac{1}{\tau}\hat{\theta} = \frac{\bar{w}}{V}\hat{\theta}_i(t) + \frac{1}{C}\hat{q}_h(t) \tag{34}$$

The equilibrium condition, or operating point, corresponds to the initial condition $\hat{\theta}(0) = 0$ and to incremental inputs $\hat{q}_h(t) = \hat{\theta}_i(t) = 0$. To find the transfer function $H_1(s) = \hat{\Theta}(s)/\hat{Q}_h(s)$, we transform (34) with $\hat{\theta}(0) = 0$ and $\hat{\theta}_i(t) = 0$, obtaining

$$\left(s + \frac{1}{\tau}\right)\hat{\Theta}(s) = \frac{1}{C}\hat{Q}_h(s)$$

Solving for the ratio $\hat{\Theta}(s)/\hat{Q}_h(s)$ yields

$$H_1(s) = \frac{\dfrac{1}{C}}{s + \dfrac{1}{\tau}}$$

which has a single pole at $s = -1/\tau$. The steady-state value of $\hat{\theta}$ in response to a unit step change in the heater input is

$$H_1(0) = \frac{\tau}{C}$$

$$= \frac{1}{\bar{w}\rho\sigma + \dfrac{1}{R}}$$

Hence, either a high flow rate $\bar{w}$ or a low thermal resistance R tends to reduce the steady-state effect of a change in the heater input.

In similar fashion, you can verify that the transfer function $H_2(s) = \hat{\Theta}(s)/\hat{\Theta}_i(s)$ is

$$H_2(s) = \frac{\dfrac{\bar{w}}{V}}{s + \dfrac{1}{\tau}}$$

The steady-state response of $\hat{\theta}$ to a unit step change in the inlet temperature is

$$H_2(0) = \frac{\bar{w}\tau}{V} = \frac{\bar{w}\rho\sigma}{\bar{w}\rho\sigma + \dfrac{1}{R}}$$

$$= \frac{1}{1 + \dfrac{1}{\bar{w}\rho\sigma R}}$$

Thus a step change in the inlet temperature with constant heater input will affect the steady-state value of the liquid temperature by only some fraction of the change. Either a high flow rate or a high thermal resistance will tend to make that fraction approach unity.

In an actual process for which θ must be maintained constant in spite of variations in $\theta_i(t)$, a feedback control system may be used to sense changes in θ and make corresponding adjustments in q_h. The type of mathematical modeling and transfer-function analysis that we have demonstrated here plays an important part in the design of such control systems.

PROBLEMS

14.1 a. Find the equivalent thermal resistance for the three resistances shown in Figure P14.1(a). Also express the temperatures θ_A and θ_B at the interfaces in terms of θ_1 and θ_2.

b. The hollow enclosure shown in Figure P14.1(b) has four sides, each with resistance R_a, and two ends, each with resistance R_b. Find the equivalent thermal resistance between the interior and the exterior of the enclosure. Also calculate the total heat flow rate from the interior to the exterior when the interior of the enclosure is at a constant temperature θ_1 and the ambient temperature is θ_a.

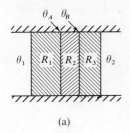
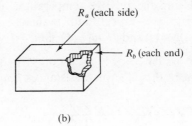

(a)   (b)

Figure P14.1

14.2 A perfectly insulated enclosure containing a heater is filled with a liquid having thermal capacitance C. The temperature of the liquid is assumed to be uniform and is denoted by θ. The heat supplied by the heater is $q_i(t)$.

a. Write the differential equation obeyed by the liquid temperature θ.

b. Solve the equation you found in part a for θ in terms of the arbitrary input $q_i(t)$ and the arbitrary initial temperature $\theta(0)$.

c. Sketch θ vs t when $q_i(t) = 1$ for $10 < t \le 20$ and zero otherwise.

14.3 Figure P14.3 shows a volume for which the temperature is θ_1 and the thermal capacitance is C. The volume is perfectly insulated from the environment except for the thermal resistances R_1 and R_2. Heat is supplied at the rate $q_i(t)$. The ambient temperature is θ_a.

    a. Write the system model.

    b. Find and sketch the response when $q_i(t) = AU(t)$ and $\theta_1(0) = \theta_a$. (*Hint*: First solve for the relative temperature $\hat{\theta}_1 = \theta_1 - \theta_a$.)

    c. Evaluate the transfer function $\hat{\Theta}_1(s)/Q_i(s)$ and sketch the magnitude of the frequency response versus ω.

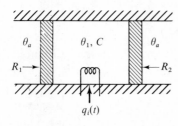

Figure P14.3

14.4 a. Repeat part a of Problem 14.3 when the temperature to the right of R_2 is $\theta_2(t)$, rather than the constant ambient temperature θ_a.

    b. Using the relative temperatures $\hat{\theta}_1 = \theta_1 - \theta_a$ and $\hat{\theta}_2 = \theta_2(t) - \theta_a$, rewrite the model and evaluate the transfer functions $H_1(s) = \hat{\Theta}_1(s)/\hat{\Theta}_2(s)$ and $H_2(s) = \hat{\Theta}_1(s)/Q_i(s)$.

14.5 The system shown in Figure P14.5 is composed of two thermal capacitances, three thermal resistances, and two heat sources.

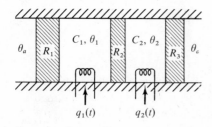

Figure P14.5

    a. Write the model in state-variable form, taking θ_1 and θ_2 as the state variables.

b. Derive the input-output differential equation relating θ_1, the inputs $q_1(t)$ and $q_2(t)$, and the ambient temperature θ_a.

c. Defining the relative temperature $\hat{\theta}_1$ as $\hat{\theta}_1 = \theta_1 - \theta_a$, write the transfer function $\hat{\Theta}_1(s)/Q_1(s)$ and evaluate the steady-state response to a unit step-function input. Identify the three other transfer functions associated with the system, but do not solve for them.

14.6 a. Repeat part a of Problem 14.5 when the temperature to the right of R_3 is $\theta_3(t)$, rather than the constant ambient temperature θ_a.

b. Rewrite the model in terms of the relative temperatures $\hat{\theta}_1 = \theta_1 - \theta_a$, $\hat{\theta}_2 = \theta_2 - \theta_a$, and $\hat{\theta}_3(t) = \theta_3(t) - \theta_a$, and derive the transfer functions $H_1(s) = \hat{\Theta}_1(s)/\hat{\Theta}_3(s)$ and $H_2(s) = \hat{\Theta}_2(s)/\hat{\Theta}_3(s)$.

14.7 The denominator of $H(s)$ in (19) has the form $s^2 + (a+b+c)s + ab$ where the parameters a, b, and c are positive. Show that the poles of $H(s)$ will always be real and negative. *Hint:* Show that the poles can be written as

$$ s = \tfrac{1}{2}\left[-(a+b+c) \pm \sqrt{(a-b)^2 + 2(a+b)c + c^2}\right] $$

14.8 Figure P14.8 shows an electronic amplifier and a fan that can be turned on to cool the amplifier. The electronic equipment has a thermal capacitance C and generates heat at the constant rate $\bar{q}$ when it is operating. The amplifier's thermal conductivity to the ambient due to convection is K with the fan off and $2K$ with it on. The ambient temperature is θ_a, and the temperature of the amplifier is θ_1.

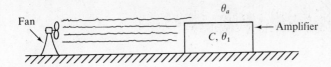

Figure P14.8

a. Write the differential equation obeyed by θ_1 with the fan off, and indicate how the equation must be modified when the fan is turned on.

b. Solve for and sketch θ_1 when the system undergoes the following sequence of operations. In each case, indicate the steady-state temperature and the time constant.

1. With $\theta_1(t_1) = \theta_a$ and the fan off, the amplifier is turned on at $t = t_1$.

2. After the system reaches steady-state conditions, the fan is turned on at $t = t_2$.

3. After the system reaches steady-state conditions, the amplifier is turned off at $t = t_3$ with the fan running.

c. Show how the response for $t > t_3$ in part b is affected if the fan is also turned off at $t = t_3$.

14.9 Figure P14.9 shows a piece of hot metal that has been immersed in a water bath to cool it. We can develop a simplified model by assuming that the temperatures of the metal and water, denoted by θ_m and θ_w, respectively, are uniform and that the rate of heat loss from the metal is proportional to the temperature difference $\theta_m - \theta_w$. The thermal capacitances are C_m and C_w, and the thermal resistance is R. We can neglect initially any heat lost to the environment at the surface.

a. Write the mathematical model of the system subject to these assumptions.

b. Taking the initial metal temperature and water temperature as $\theta_m(0)$ and $\theta_w(0)$, respectively, solve for and sketch θ_m vs t and θ_w vs t.

c. Modify your answer to part a to allow for heat flow from the water to the environment. Denote the ambient temperature by θ_a and the thermal resistance between the water and the environment by R_a.

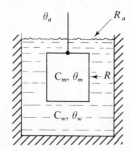

Figure P14.9

14.10 a. Solve part a and part b of Problem 14.9 in numerical form for the following parameter values and initial conditions. The parameter values are:

Metal mass = 50 kg
Metal specific heat = 460 J/(kg·K)
Liquid volume = 0.15 m^3
Liquid density = 1000 kg/m^3
Liquid specific heat = 4186 J/(kg·K)
Thermal resistance between the metal and the liquid = 10^{-3} s·K/J
Thermal resistance between the liquid and environment = 10^{-2} s·K/J

The initial conditions are:

Initial metal temperature = 750 K
Initial liquid temperature = 320 K
Ambient temperature = 300 K

b. Do part c of Problem 14.9 and solve for θ_m. Sketch θ_m vs t and show qualitatively the shape of the curve for θ_w.

14.11 A heat exchanger in a chemical process uses steam to heat a liquid flowing in a pipe. We can apply an approximate linear model to relate changes in the temperature of the liquid leaving the heat exchanger to changes in the rate of steam flow. This model is given by the transfer function

$$\frac{\hat{\Theta}(s)}{\hat{W}(s)} = \frac{A\epsilon^{-sT_d}}{(\tau_1 s + 1)(\tau_2 s + 1)}$$

where $\hat{\Theta}(s)$ and $\hat{W}(s)$ are the Laplace transforms of the incremental outlet temperature and the incremental steam flow rate, respectively. We can determine the coefficient A, the time delay T_d, and the time constants τ_1 and τ_2 experimentally by recording the response to a step change in the steam flow rate with all other process conditions held constant.

For parameter values $A = 0.5$ K·s/kg, $T_d = 20$ s, $\tau_1 = 15$ s, and $\tau_2 = 150$ s, evaluate and sketch the following:

a. The response to a step in $\hat{w}(t)$ of 20 kg/s
b. The unit impulse response $h(t)$
c. The magnitude of the frequency response $H(j\omega)$ vs ω.

14.12 Heat can be transferred from one body to another by radiation according to the relationship $q = \alpha_r(\theta_1^4 - \theta_2^4)$, where θ_1 and θ_2 are the temperatures of the bodies in kelvin and α_r is a coefficient dependent on various physical properties.

a. Using a Taylor-series expansion, obtain a linear mathematical model relating the incremental variables $\hat{q} = q - \bar{q}$, $\hat{\theta}_1 = \theta_1 - \bar{\theta}_1$, and $\hat{\theta}_2 = \theta_2 - \bar{\theta}_2$. Explain why this model does not yield an equivalent thermal resistance unless $\bar{\theta}_1 = \bar{\theta}_2$.

b. Rewrite the radiation element law in the form of a nonlinear resistive element by finding K_r where $q = (1/K_r)(\theta_1 - \theta_2)$. *Hint:* Note that $\theta_1^4 - \theta_2^4 = (\theta_1^2 + \theta_2^2)(\theta_1^2 - \theta_2^2)$.

14.13 Write the state-variable equations for a three-capacitance model of the insulated bar considered in Examples 14.5 and 14.6. Use three equal capacitances of value $C/3$ and three equal resistances of value $R/3$. Find the characteristic polynomial in terms of the parameter RC.

14.14 Repeat Problem 14.13 using nonuniform lengths for the three lumped segments of the bar. Specifically, take segments of length $0.2L$, $0.3L$, and $0.5L$, from left to right, where L is the length of the bar. Explain why such an approximation should yield greater accuracy that the equal-segment approximation of Problem 14.13.

14.15 Using a digital computer, simulate the response of the insulated bar considered in Examples 14.5 and 14.6 by using N elements of equal length, as shown in Figure P14.15. Normalize the model by taking $RC = 1$ and finding the response to $\hat{\theta}_i(t) = U(t)$ where the initial relative temperature of each element is taken as zero.

    a. Use $N = 3$ and plot θ_a, θ_b, and θ_c versus time.

    b. Use $N = 6$ and plot the temperatures

$$\theta_a^* = 0.5(\theta_1 + \theta_2)$$

$$\theta_b^* = 0.5(\theta_3 + \theta_4)$$

$$\theta_c^* = 0.5(\theta_5 + \theta_6)$$

Compare the results for the two cases and comment on the difference.

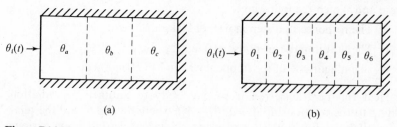

(a)   (b)

Figure P14.15

CHAPTER 15 —————————————————————
HYDRAULIC SYSTEMS

A hydraulic system is one in which liquids, generally considered incompressible, flow. Hydraulic systems commonly appear in chemical processes, automatic control systems, and actuators and drive motors for manufacturing equipment. Such systems are usually interconnected to mechanical systems through pumps, valves, and movable pistons. A turbine driven by water and used for driving an electric generator is an example of a system with interacting hydraulic, mechanical, and electrical elements. We will not discuss here the more general topic of fluid systems, which would include compressible fluids such as gases and air.

An exact analysis of hydraulic systems is usually not feasible because of their distributed nature and the nonlinear character of the resistance to flow. For our dynamic analysis, however, we can obtain satisfactory results by using lumped elements and linearizing the resulting nonlinear mathematical models. On the other hand, the design of chemical processes requires a more exact analysis using static rather than dynamic models.

In most cases, hydraulic systems operate with the variables remaining close to a specific operating point. Thus we are generally interested in models involving incremental variables. This fact is particularly helpful because such models are generally linear, although the model in terms of the total variables may be quite nonlinear.

In the next two sections, we shall define the variables to be used and introduce and illustrate the element laws. Then we will present a variety of examples to demonstrate the modeling process and the application of analytical techniques discussed in previous chapters, including Laplace transforms and transfer functions.

15.1 VARIABLES

Because hydraulic systems involve the flow and accumulation of liquid, the variables used to describe their dynamic behavior are

$$w = \text{flow rate in cubic meters per second (m}^3/\text{s})$$

$$v = \text{volume in cubic meters (m}^3)$$

$$h = \text{liquid height in meters (m)}$$

$$p = \text{pressure in newtons per square meter (N/m}^2)$$

Unless otherwise noted, a pressure will be the *absolute pressure*. In addition, we shall sometimes find it convenient to express pressures in terms of gauge pressures. A *gauge pressure*, denoted by p^*, is defined to be the difference between the absolute pressure and the atmospheric pressure p_a, namely

$$p^*(t) = p(t) - p_a \tag{1}$$

A pressure difference, denoted by Δp, is the difference between the pressures at two points.

15.2 ELEMENT LAWS

Hydraulic systems exhibit three types of characteristics that can be approximated by lumped elements: capacity, resistance to flow, and inertance. In this section, we shall discuss the first two. The inertance, which accounts for the kinetic energy of a moving fluid stream, is usually negligible, and we will not consider it (see Doebelin (1972), Section 4–4 for a treatment of this subject). A brief discussion of centrifugal pumps that act as hydraulic sources appears at the end of the section.

CAPACITANCE

If liquid is stored in an open vessel, there will be an algebraic relationship between the volume of the liquid and the pressure at the base of the vessel. If the cross-sectional area of the vessel is given by the function $A(h)$ where h is the height of the liquid level above the bottom of the vessel, then the liquid volume v is the integral of the area from the base of the vessel to the top of the liquid. Hence,

$$v = \int_0^h A(\lambda)\, d\lambda \tag{2}$$

where λ is a dummy variable of integration. For a liquid of density ρ expressed in kilograms per cubic meter, the absolute pressure p and the liquid height h are related by

$$p = \rho g h + p_a \tag{3}$$

where g is the gravitational constant (9.807 m/s^2) and where p_a is the atmospheric pressure, which is taken as $1.013 \times 10^5 \text{ N/m}^2$.

Equations (2) and (3) imply that for any vessel geometry, liquid density, and atmospheric pressure there will be a unique algebraic relationship between the pressure p and the liquid volume v. A typical characteristic curve describing this relationship is shown in Figure 15.1(a).

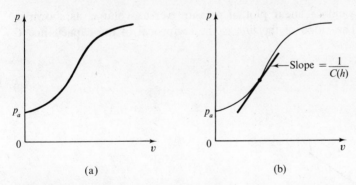

(a) (b)

Figure 15.1 Pressure versus liquid volume for a vessel with variable cross-sectional area $A(h)$.

If the tangent to the pressure-versus-volume curve is drawn at some point, as shown in Figure 15.1(b), then the reciprocal of the slope is defined to be the *hydraulic capacitance*, denoted by $C(h)$. As indicated by the h in parentheses, the capacitance depends on the point on the curve being considered and hence on the liquid height h. Now

$$C(h) = \frac{1}{\dfrac{dp}{dv}} = \frac{dv}{dp}$$

and, from the chain rule of differentiation,

$$C(h) = \frac{dv}{dh}\frac{dh}{dp}$$

We see that $dv/dh = A(h)$ from (2) and that $dh/dp = 1/\rho g$ from (3). Thus for a vessel of arbitrary shape,

$$C(h) = \frac{A(h)}{\rho g} \tag{4}$$

which has units of $m^4 \cdot s^2/kg$ or, equivalently, m^5/N.

For a vessel with constant cross-sectional area A, (2) reduces to $v = Ah$. We can substitute the height $h = v/A$ into (3) to obtain the pressure in terms of the volume, namely

$$p = \frac{\rho g}{A}v + p_a \tag{5}$$

Equation (5) yields a linear plot of pressure versus volume, as shown in Figure 15.2. The slope of the line is the reciprocal of the capacitance C where

$$C = \frac{A}{\rho g} \tag{6}$$

LINEAR ONLY
(A CONSTANT
WITH h)

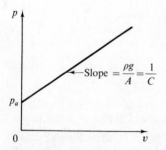

Figure 15.2 Pressure versus liquid volume for a vessel with constant A.

The volume of liquid in a vessel at any instant is the integral of the net flow rate into the vessel plus the initial volume. Hence, we can write

$$v(t) = v(0) + \int_0^t [w_{in}(\lambda) - w_{out}(\lambda)] \, d\lambda$$

which can be differentiated to give the alternative form

$$\dot{v} = w_{in}(t) - w_{out}(t) \tag{7}$$

To obtain expressions for the time derivatives of the pressure p and the liquid height h that are valid for vessels with variable cross-sectional areas, we use the chain rule of differentiation to write

$$\frac{dv}{dt} = \frac{dv}{dh}\frac{dh}{dt}$$

where dv/dt is given by (7) and where $dv/dh = A(h)$. Thus the rate of change of the liquid height depends on the net flow rate according to

$$\dot{h} = \frac{1}{A(h)}[w_{in}(t) - w_{out}(t)] \tag{8}$$

Alternatively, we can write dv/dt as

$$\frac{dv}{dt} = \frac{dv}{dp}\frac{dp}{dt}$$

where $dv/dp = C(h)$. Hence, the rate of change of the pressure at the base of the vessel is

$$\left(\begin{smallmatrix} AT\ base \\ OF\ VESSEL \end{smallmatrix}\right) \quad \dot{p} = \frac{1}{C(h)}[w_{in}(t) - w_{out}(t)] \tag{9}$$

where $C(h)$ is given by (4).

Because any of the variables v, h, and p can be used as a measure of the amount of liquid in a vessel, we generally select one of them as a state variable. Then (7), (8), or (9) will yield the corresponding state-variable equation when w_{in} and w_{out} are expressed in terms of the state variables and inputs. If the cross-sectional area of the vessel is variable, then the coefficient $A(h)$ in (8) will be a function of h and the system model will be nonlinear. To develop a linearized model, we must find the operating point, define the incremental variables, and retain the first two terms in the Taylor-series expansion. Likewise, the term $C(h)$ in (9) will cause the differential equation to be nonlinear because the capacitance varies with h, which in turn is a function of the pressure.

EXAMPLE 15.1   Consider a vessel formed by a circular cylinder of radius R and length L which contains a liquid of density ρ in units of kilograms per cubic meter. Find the hydraulic capacitance of the vessel when the cylinder is vertical, as shown in Figure 15.3(a). Then evaluate the capacitance when the cylinder is on its side, as shown in Figure 15.3(b).

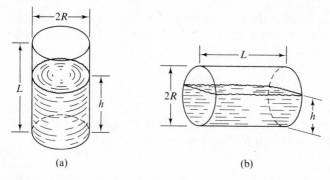

(a)   (b)

Figure 15.3 Cylindrical vessel for Example 15.1. (a) Cylinder vertical. (b) Cylinder horizontal.

Solution   For the configuration shown in Figure 15.3(a), the cross-sectional area is πR^2, independent of the liquid height. Thus we can use (6), and the vessel's hydraulic capacitance is $C_a = \pi R^2/\rho g$.

When the vessel is on its side as shown in Figure 15.3(b), the cross-sectional area is a function of the liquid height h. You can verify that the width of the liquid surface is $2\sqrt{R^2-(R-h)^2}$, which is zero when $h = 0$ and $h = 2R$ and which has a maximum value of $2R$ when $h = R$. Using (4), we find that the capacitance is

$$C_b = \frac{2L}{\rho g}\sqrt{R^2-(R-h)^2}$$

which is shown in Figure 15.4.

RESISTANCE

As liquid flows through a pipe, there is a drop in the pressure of the liquid over the length of pipe. There is likewise a pressure drop if the liquid flows through a valve or through an orifice. The change in pressure associated

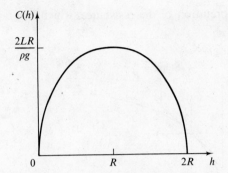

Figure 15.4 Capacitance of the vessel shown in Figure 15.3(b).

with a flowing liquid results from the dissipation of energy and usually obeys a nonlinear algebraic relationship between the flow rate w and the pressure difference Δp. The symbol for a valve is shown in Figure 15.5; it can also

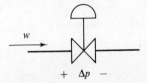

Figure 15.5 Symbol for a hydraulic valve.

be used for other energy-dissipating elements. A positive value of w indicates that liquid is flowing in the direction of the arrow, while a positive value of Δp indicates that the pressure at the end marked $+$ is higher than the pressure at the other end. The expression

$$w = k\sqrt{\Delta p} \tag{10}$$

describes an orifice and a valve and is a good approximation for turbulent flow through pipes (see Doebelin (1972), Section 4-2). We can treat all situations of interest to us by using a nonlinear element law of the form of (10). In this equation, k is a constant that depends on the characteristics of the pipe, valve, or orifice. A typical curve of flow rate versus pressure difference is shown in Figure 15.6(a).

Because (10) is a nonlinear relationship, we must linearize it about an operating point in order to develop a linear model of a hydraulic system. If we draw the tangent to the curve of w vs Δp at the operating point, the reciprocal of its slope is defined to be the *hydraulic resistance R*. Figure

15.6(b) illustrates the geometric interpretation of the resistance, which has units of newton-seconds per meter[5].

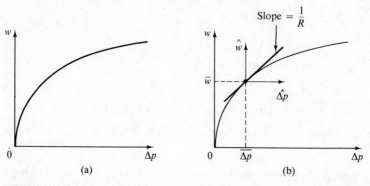

Figure 15.6 (a) Flow rate versus pressure difference given by (10). (b) Geometric interpretation of hydraulic resistance.

Expanding (10) in a Taylor series about the operating point gives

$$w = \bar{w} + \frac{dw}{d\,\Delta p}\bigg|_{\overline{\Delta p}}(\Delta p - \overline{\Delta p}) + \cdots$$

The incremental variables $\hat{w}$ and $\hat{\Delta p}$ are defined by

$$\hat{w} = w - \bar{w} \tag{11a}$$

$$\hat{\Delta p} = \Delta p - \overline{\Delta p} \tag{11b}$$

and the second- and higher-order terms in the expansion are dropped. Thus the incremental model becomes

$$\hat{w} = \frac{1}{R}\hat{\Delta p} \tag{12}$$

where

$$\frac{1}{R} = \frac{dw}{d\,\Delta p}\bigg|_{\overline{\Delta p}}$$

We can express the resistance R in terms of either $\overline{\Delta p}$ or $\bar{w}$ by carrying out the required differentiation using (10). Specifically,

$$\frac{1}{R} = \frac{d}{d\,\Delta p}(k\,\Delta p^{1/2})\Big|_{\overline{\Delta p}}$$

$$= \tfrac{1}{2}k(\overline{\Delta p})^{-1/2}$$

so

$$R = \frac{2\sqrt{\overline{\Delta p}}}{k} \qquad (13)$$

To express the resistance in terms of $\overline{w}$, we note from (10) that

$$\overline{w} = k\sqrt{\overline{\Delta p}} \qquad (14)$$

Substituting (14) into (13) gives the alternative equation for the hydraulic resistance as

$$R = \frac{2\overline{w}}{k^2} \qquad (15)$$

Because liquids typically flow through networks composed of pipes, valves, and orifices, we must often combine several relationships of the form of (10) into a single equivalent expression. Since we use linearized models in much of our analysis of hydraulic systems, it is important to develop rules for combining the resistances of linearized elements that occur in series and parallel combinations. In the following example, we consider the relationship of flow versus pressure difference and the equivalent resistance for two valves in series. You will treat the parallel-flow situation in Problem 15.3.

EXAMPLE 15.2   Figure 15.7(a) shows a series combination of two valves through which liquid flows at a rate of w and across which the pressure difference is Δp. An equivalent valve is shown in Figure 15.7(b). The two valves obey the relationships $w = k_a\sqrt{\Delta p_a}$ and $w = k_b\sqrt{\Delta p_b}$, respectively.

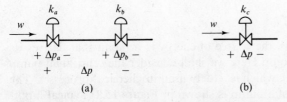

Figure 15.7 (a) Two valves in series. (b) Equivalent valve.

Find the coefficient k_c of an equivalent valve that obeys the relationship $w = k_c\sqrt{\Delta p}$. Also evaluate the resistance R_c of the series combination in terms of the individual resistances R_a and R_b.

Solution Because the two valves are connected in series, they have the same flow rate w, and the total pressure difference is $\Delta p = \Delta p_a + \Delta p_b$. To determine k_c, we write Δp in terms of w as

$$\Delta p = \Delta p_a + \Delta p_b = \left(\frac{1}{k_a^2} + \frac{1}{k_b^2}\right)w^2$$

and then solve for w in terms of Δp. After some manipulations, we find that

$$w = \left(\frac{k_a k_b}{\sqrt{k_a^2 + k_b^2}}\right)\sqrt{\Delta p} \tag{16}$$

Comparing (16) with (10), we see that the equivalent valve constant is

$$k_c = \frac{k_a k_b}{\sqrt{k_a^2 + k_b^2}} \tag{17}$$

Using (15) for the resistance of the linearized model of the equivalent valve, we can write

$$R_c = \frac{2\bar{w}}{k_c^2} = 2\bar{w}\left(\frac{1}{k_a^2} + \frac{1}{k_b^2}\right) \tag{18}$$

However, by applying (15) to the individual valves, we see that their resistances are $R_a = 2\bar{w}/k_a^2$ and $R_b = 2\bar{w}/k_b^2$, respectively. Using these expressions for R_a and R_b, we see that we can rewrite (18) as

$$R_c = R_a + R_b \tag{19}$$

which is identical to the result for a linear electrical circuit.

SOURCES

In most hydraulic systems, the source of energy is a pump that derives its power from an electric motor. Here we shall consider the centrifugal pump driven at a constant speed, which is widely used in chemical processes. The symbolic representation of a pump is shown in Figure 15.8. Typical input-output relationships for a centrifugal pump being driven at three different constant speeds are shown in Figure 15.9(a). Pump curves of Δp vs w are

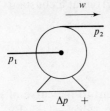

Figure 15.8 Symbolic representation of a pump.

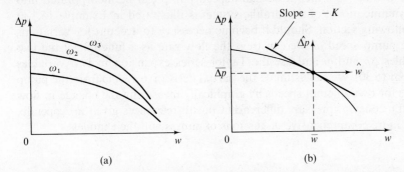

Figure 15.9 Typical centrifugal pump curves where $\Delta p = p_2 - p_1$. (a) For three different pump speeds ($\omega_1 < \omega_2 < \omega_3$). (b) Showing the linear approximation.

determined experimentally under steady-state conditions and are quite non-linear. To include a pump being driven at constant speed in a linear dynamic model, we first determine the operating point for the particular pump speed by calculating the values of $\overline{\Delta p}$ and $\overline{w}$. Then we find the slope of the tangent to the pump curve at the operating point and define it to be $-K$, which has units of newton-seconds per meter[5]. Having done this, we can express the incremental pressure difference $\hat{\Delta p}$ in terms of the incremental flow rate $\hat{w}$ as

$$\hat{\Delta p} = -K\hat{w} \tag{20}$$

where the constant K is always positive. Solving (20) for $\hat{w}$, we have

$$\hat{w} = -\frac{1}{K}\hat{\Delta p} \tag{21}$$

Figure 15.9(b) illustrates the relationship of the linearized approximation to the nonlinear pump curve.

We can write the Taylor-series expansion for a pump driven at a constant speed as

$$w = \bar{w} + \frac{dw}{d\,\Delta p}\bigg|_{\overline{\Delta p}} (\Delta p - \overline{\Delta p}) + \cdots$$

where the coefficient $(dw/d\,\Delta p)|_{\overline{\Delta p}}$ is the slope of the tangent to the curve of w vs Δp, measured at the operating point, and has the value $-1/K$. By dropping the second- and higher-order terms in the expansion and using the incremental variables $\hat{w}$ and $\hat{\Delta} p$, we obtain the linear relationship (21).

The manner in which a constant-speed pump can be incorporated into the dynamic model of a hydraulic system is illustrated in Example 15.4 in the following section. Should it become necessary to account for variations in the pump speed ω, we can treat the flow rate as a function of the two variables Δp and ω and use the Taylor-series expansion in two variables given by (5.30). We can estimate the partial derivative $\partial w/\partial\omega$ from the pump curves for two different speeds by graphically measuring the change in flow rate at a constant pressure difference. Consult references given in Appendix D for more comprehensive discussions of pumps and their models.

15.3 DYNAMIC MODELS OF HYDRAULIC SYSTEMS

In this section, we apply the element laws presented in Section 15.2 and use many of the analytical techniques from previous chapters. We will develop and analyze dynamic models for a single vessel with a valve, a combination of a pump, vessel, and valve, and finally two vessels with two valves and a pump. In each case, we will derive the nonlinear model and develop and analyze a linearized model.

EXAMPLE 15.3   Figure 15.10 shows a vessel that receives liquid at a flow rate $w_i(t)$ and loses liquid through a valve that obeys the nonlinear flow-pressure relationship $w_o = k\sqrt{p_1 - p_a}$. The cross-sectional area is A and the liquid density is ρ. Derive the nonlinear model obeyed by the absolute pressure p_1 at the bottom of the vessel. Then develop the linearized version valid in the vicinity of the operating point, and find the transfer function relating the transforms of the incremental input $\hat{w}_i(t)$ and the incremental pressure $\hat{p}_1$.

Having developed the system models in literal form, determine in numerical form the operating point, the transfer function, and the response to a 10% step-function increase in the input flow rate for the following parameter values:

$$A = 2 \, \text{m}^2$$

$$\rho = 1000 \, \text{kg/m}^3$$

$$k = 5.0 \times 10^{-5} \, \text{m}^4/\text{s} \cdot \text{N}^{1/2}$$

$$\bar{w}_i = 6.0 \times 10^{-3} \, \text{m}^3/\text{s}$$

$$p_a = 1.013 \times 10^5 \, \text{N/m}^2$$

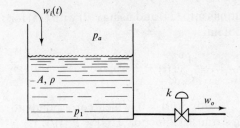

Figure 15.10 Hydraulic system for Example 15.3.

Solution Taking the pressure p_1 as the single state variable, we use (9) with $C(h)$ replaced by the constant $C = A/\rho g$ to write

$$\dot{p}_1 = \frac{1}{C} [w_{\text{in}}(t) - w_{\text{out}}(t)] \tag{22}$$

where

$$w_{\text{in}}(t) = w_i(t) \tag{23a}$$

$$w_{\text{out}}(t) = k\sqrt{p_1 - p_a} \tag{23b}$$

Substituting (23) into (22) gives the nonlinear system model as

$$\dot{p}_1 = \frac{1}{C} [-k\sqrt{p_1 - p_a} + w_i(t)] \tag{24}$$

To develop a linearized model, we rewrite (24) in terms of the incremental variables $\hat{p}_1 = p_1 - \bar{p}_1$ and $\hat{w}_i(t) = w_i(t) - \bar{w}_i$. The nominal values $\bar{p}_1$ and $\bar{w}_i$ must satisfy the algebraic equation

$$k\sqrt{\bar{p}_1 - p_a} = \bar{w}_i$$

or

$$\bar{p}_1 = p_a + \frac{1}{k^2} \bar{w}_i^2 \tag{25}$$

which corresponds to an outflow rate equal to the inflow rate, resulting in a constant liquid level and pressure. The nominal height of the liquid is

$$\bar{h} = \frac{(\bar{p}_1 - p_a)}{\rho g} \tag{26}$$

Making the appropriate substitutions into (24) and using (12) with (15) for the linearized valve equation, we obtain

$$\dot{\hat{p}}_1 + \frac{1}{RC}\hat{p}_1 = \frac{1}{C}\hat{w}_i(t) \tag{27}$$

where $R = 2\bar{w}_i/k^2$. Transforming (27) with $\hat{p}_1(0) = 0$, which corresponds to $p_1(0) = \bar{p}_1$, we find that the system's transfer function $H(s) = \hat{P}_1(s)/\hat{W}_i(s)$ is

$$H(s) = \frac{\dfrac{1}{C}}{s + \dfrac{1}{RC}} \tag{28}$$

which has a single pole at $s = -1/RC$.

For the parameter values specified, the operating point given by (25) reduces to

$$\bar{p}_1 = 1.013 \times 10^5 + \left(\frac{6.0 \times 10^{-3}}{5.0 \times 10^{-5}}\right)^2 = 1.157 \times 10^5 \text{ N/m}^2$$

and from (26), the nominal liquid height is

$$\bar{h} = \frac{1.440 \times 10^4}{1000 \times 9.807} = 1.468 \text{ m}$$

The numerical values of the hydraulic resistance and capacitance are, respectively,

$$R = \frac{2 \times 6.0 \times 10^{-3}}{(5.0 \times 10^{-5})^2} = 4.80 \times 10^6 \text{ N} \cdot \text{s/m}^5$$

$$C = \frac{2.0}{1000 \times 9.807} = 2.039 \times 10^{-4} \text{ m}^5/\text{N}$$

Substituting these values of R and C into (28), we obtain the numerical form of the transfer function as

$$H(s) = \frac{4904.}{s + 1.0216 \times 10^{-3}} \tag{29}$$

If $w_i(t)$ is originally equal to its nominal value of $\bar{w}_i = 6.0 \times 10^{-3}$ m^3/s and undergoes a 10% step-function increase, then $\hat{w}_i(t) = [0.60 \times 10^{-3}] U(t)$ m^3/s and $\hat{W}_i(s) = 0.60 \times 10^{-3} (1/s)$. Thus

$$\hat{P}_1(s) = \frac{4904. \times 0.60 \times 10^{-3}}{s(s + 1.0216 \times 10^{-3})}$$

$$= \frac{2.942}{s(s + 1.0216 \times 10^{-3})}$$

From the final-value theorem, the steady-state value of $\hat{p}_1$ is $s\hat{P}_1(s)$ evaluated at $s = 0$, namely

$$\lim_{t \to \infty} \hat{p}_1(t) = \frac{2.942}{1.0216 \times 10^{-3}} = 2880. \text{ N/m}^2$$

The time constant of the linearized model is $\tau = RC$, which becomes

$$\tau = (4.80 \times 10^6)(2.039 \times 10^{-4})$$

$$= 978.7 \text{ s}$$

which is slightly over 16 minutes. Thus the response of the incremental pressure is

$$\hat{p}_1 = 2880.(1 - \epsilon^{-t/978.7})$$

The change in the incremental level is $\hat{p}_1/\rho g$, which becomes

$$\hat{h} = 0.2937(1 - \epsilon^{-t/978.7})$$

To obtain the responses of the actual pressure and liquid level, we merely add the nominal values $\bar{p}_1 = 1.157 \times 10^5$ N/m^2 and $\bar{h} = 1.468$ m to these incremental variables. It is interesting to note that because of the nonlinear valve, the 10% increase in the flow rate results in 20% increases in both the gauge pressure $p_1 - p_a$ and the height h.

EXAMPLE 15.4 Find the linearized model of the hydraulic system shown in Figure 15.11(a), which consists of a constant-speed centrifugal pump feeding a vessel from which liquid flows through a pipe and valve obeying the relationship $w_o = k\sqrt{p_1 - p_a}$. The pump characteristic for the specified pump speed $\bar{\omega}$ is shown in Figure 15.11(b).

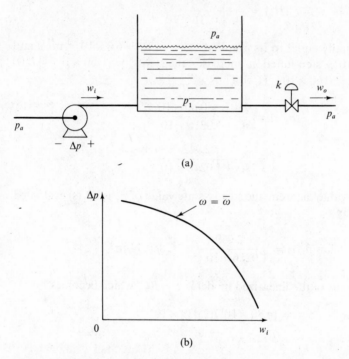

Figure 15.11 (a) System for Example 15.4. (b) Pump curve.

Solution   The equilibrium condition for the system corresponds to

$$\bar{w}_i = \bar{w}_o \tag{30}$$

where $\bar{w}_i$ and $\overline{\Delta p} = \bar{p}_1 - p_a$ must be one of the points on the pump curve in Figure 15.11(b), and where $\bar{w}_o$ obeys the nonlinear flow relationship

$$\bar{w}_o = k\sqrt{\overline{\Delta p}} \tag{31}$$

To determine the operating point, we find the solution to (30) graphically by plotting the valve characteristic (31) on the pump curve. Doing this gives Figure 15.12(a), where the operating point is the intersection of the valve curve and pump curve, designated as point A in the figure. Once we have located the operating point, we can draw the tangent to the pump curve as shown in Figure 15.12(b) and determine its slope $-K$ graphically.

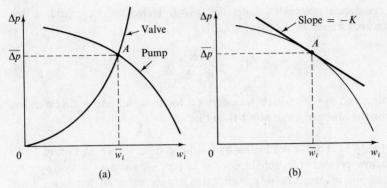

Figure 15.12 (a) Combined pump and valve curves for Example 15.4. (b) Pump curve with linear approximation.

Following this preliminary step, we can use (9) to write the model of the system as

$$\dot{p}_1 = \frac{1}{C}(w_i - w_o) \tag{32}$$

where, from (12), the approximate flow rate through the valve is

$$w_o = \bar{w}_o + \frac{1}{R}\hat{\Delta p} \tag{33}$$

and where, from (21), the approximate flow rate through the pump is

$$w_i = \bar{w}_i - \frac{1}{K}\hat{\Delta p} \tag{34}$$

Substituting (33) and (34) into (32), using $\dot{p}_1 = \dot{\hat{p}}_1$ and (30), and noting that $\hat{\Delta p} = \hat{p}_1$ since p_a is constant, we find the incremental model to be

$$\dot{\hat{p}}_1 = \frac{1}{C}\left(-\frac{1}{K} - \frac{1}{R}\right)\hat{p}_1$$

which we can write as the homogeneous first-order differential equation

$$\dot{\hat{p}}_1 + \frac{1}{C}\left(\frac{1}{K} + \frac{1}{R}\right)\hat{p}_1 = 0 \tag{35}$$

Inspection of (35) indicates that the magnitude of the slope of the pump curve at the operating point enters the equation in exactly the same manner as the resistance associated with the valve. Hence, if we evaluate the equivalent resistance R_{eq} according to

$$R_{eq} = \frac{RK}{R+K}$$

(35) is the same as (27), which was derived for a vessel and a single valve, except for the absence of an input flow rate.

EXAMPLE 15.5   The valves in the hydraulic system shown in Figure 15.13 obey the flow-pressure relationships $w_1 = k_1 \sqrt{p_1 - p_2}$ and $w_2 = k_2 \sqrt{p_2 - p_a}$. The atmospheric pressure is p_a, and the capacitances of the vessels are C_1 and C_2. Find the equations that determine the operating point, and show how the pump curve is used to solve them. Derive a linearized model that is valid about the operating point.

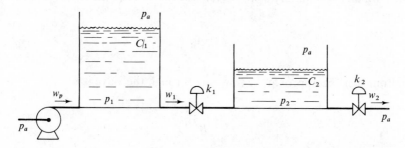

Figure 15.13 Hydraulic system with two vessels considered in Example 15.5.

Solution   Because the pump and the two vessels are in series at equilibrium conditions, we define the operating point by equating the three flow rates $\bar{w}_p$, $\bar{w}_1$, and $\bar{w}_2$. The flow rates through the two valves are given by

$$\bar{w}_1 = k_1 \sqrt{\bar{p}_1 - \bar{p}_2} \tag{36a}$$

$$\bar{w}_2 = k_2 \sqrt{\bar{p}_2 - p_a} \tag{36b}$$

The flow rate $\bar{w}_p$ through the pump and the pressure difference $\overline{\Delta p_1} = \bar{p}_1 - p_a$ must correspond to a point on the pump curve. Equations (36a) and (36b) are those of two valves in series, and, as shown in Example 15.2, we can replace such a combination of valves by an equivalent valve specified by (16). Hence,

$$\bar{w}_p = k_{eq} \sqrt{\overline{\Delta p_1}} \tag{37}$$

where, from (17),

$$k_{eq} = \frac{k_1 k_2}{\sqrt{k_1^2 + k_2^2}}$$

Plotting (37) on the pump curve as shown in Figure 15.12(a) will yield the values of $\overline{\Delta p}_1$ and $\overline{w}_p$, from which we can find the other nominal values.

With this information, we can develop the incremental model. Using (9), (12), and (21), we can write the pair of linear differential equations

$$\dot{\hat{p}}_1 = \frac{1}{C_1}\left[-\frac{1}{K}\hat{p}_1 - \frac{1}{R_1}(\hat{p}_1 - \hat{p}_2)\right]$$

$$\dot{\hat{p}}_2 = \frac{1}{C_2}\left[\frac{1}{R_1}(\hat{p}_1 - \hat{p}_2) - \frac{1}{R_2}\hat{p}_2\right] \tag{38}$$

where the valve resistances are given by $R_1 = 2\overline{w}_1/k_1^2$ and $R_2 = 2\overline{w}_2/k_2^2$, and where $-K$ is the slope of the pump curve at the operating point. As indicated by (38), the incremental model has no inputs and hence can respond only to nonzero initial conditions—that is, $\hat{p}_1(0) \neq 0$ and/or $\hat{p}_2(0) \neq 0$.

In practice, there might be additional liquid streams entering either vessel or the pump speed might be changed. It is also possible for either of the two valves to be opened or closed slightly. Such a change would modify the respective hydraulic resistance and could be modeled by using the techniques discussed in Chapter 5.

In the absence of an input, we can transform (38) to find the Laplace transform of the zero-input response. Doing this, we find that after rearranging,

$$\left[C_1 s + \left(\frac{1}{K} + \frac{1}{R_1}\right)\right]\hat{P}_1(s) = \frac{1}{R_1}\hat{P}_2(s) + C_1 \hat{p}_1(0) \tag{39a}$$

$$\left[C_2 s + \left(\frac{1}{R_1} + \frac{1}{R_2}\right)\right]\hat{P}_2(s) = \frac{1}{R_1}\hat{P}_1(s) + C_2 \hat{p}_2(0) \tag{39b}$$

We can find either $\hat{P}_1(s)$ or $\hat{P}_2(s)$ by combining these two equations into a single transform equation. The corresponding inverse transform will yield the zero-input response in terms of $\hat{p}_1(0)$ and $\hat{p}_2(0)$. The denominator of either $\hat{P}_1(s)$ or $\hat{P}_2(s)$ will be the characteristic polynomial of the system, which, as you may verify, is

$$s^2 + \left[\frac{1}{C_1}\left(\frac{1}{K} + \frac{1}{R_1}\right) + \frac{1}{C_2}\left(\frac{1}{R_1} + \frac{1}{R_2}\right)\right]s + \frac{1}{C_1 C_2}\left(\frac{1}{KR_1} + \frac{1}{KR_2} + \frac{1}{R_1 R_2}\right)$$

PROBLEMS

15.1 Figure P15.1 shows a conical vessel that has a circular cross section and contains a liquid. Evaluate and sketch the hydraulic capacitance as a function of the liquid height h. Also evaluate and sketch the gauge pressure $p^* = p - p_a$ at the base of the vessel as a function of the liquid volume v.

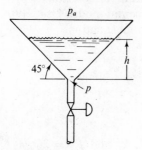

Figure P15.1

15.2 Find the equivalent hydraulic resistances for the linear models of hydraulic networks shown in Figure P15.2. Express your answers as single fractions.

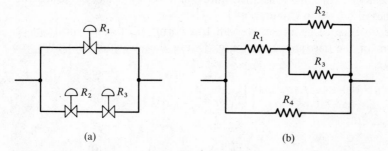

(a)   (b)

Figure P15.2

15.3 Two valves that obey the relationships $w_a = k_a \sqrt{\Delta p}$ and $w_b = k_b \sqrt{\Delta p}$ are connected in parallel, as indicated in Figure P15.3.

    a. Determine the equivalent valve coefficient k_c in terms of k_a and k_b such that the total flow rate is given by $w = k_c \sqrt{\Delta p}$.

    b. Show that the hydraulic resistance of the equivalent linearized model is $R_c = 2\sqrt{\overline{\Delta p}}/k_c$, where $\overline{\Delta p}$ denotes the nominal pressure drop across the valves.

c. Show that $R_c = R_a R_b/(R_a + R_b)$, where R_a and R_b are the hydraulic resistances of the individual valves evaluated at the nominal pressure difference $\overline{\Delta p}$.

d. Assume curves for w_a and w_b vs Δp and sketch the corresponding w_c. Indicate the linearized approximations at a typical value of $\overline{\Delta p}$.

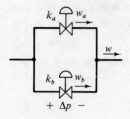

Figure P15.3

15.4 Consider the hydraulic system consisting of a single vessel and a valve that was modeled in Example 15.3.

a. Obtain the linearized model in terms of the incremental pressure that is valid for the nominal flow rate $\overline{w}_i = 3.0 \times 10^{-3}$ m^3/s, and evaluate the transfer function $\hat{P}_1(s)/\hat{W}_i(s)$. Also find $\bar{p}_1$.

b. Rewrite the model and transfer function you found in part a in terms of the incremental volume $\hat{v}$. Also find $\bar{v}$.

c. Solve for and sketch the incremental volume versus time when $\hat{w}_i(t) = 0.5 \times 10^{-3} U(t)$ cubic meters per second and when the system starts at the nominal conditions found in part a.

15.5 Write the state-variable equations for the system shown in Figure P15.5 using the incremental pressures $\hat{p}_1$ and $\hat{p}_2$ as state variables where the pump obeys the relationship

$$w_p = \overline{w}_p - \frac{1}{K}(\hat{p}_2 - \hat{p}_1).$$

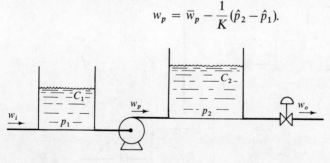

Figure P15.5

15.6 For the hydraulic system shown in Figure P15.6, the incremental input is $\hat{w}_i(t)$ and the incremental output is $\hat{w}_o$. The liquid density is ρ.

a. Taking $\hat{p}_1$ and $\hat{p}_2$ as incremental state variables, derive the state-variable and output equations.

b. Derive the system transfer function $\hat{W}_o(s)/\hat{W}_i(s)$.

c. Define the time constants $\tau_1 = R_1 C_1$ and $\tau_2 = R_2 C_2$, and evaluate the unit step response. Sketch the response for the case where $\tau_1 = 2\tau_2$.

d. Repeat part c for the case where $\tau_1 = \tau_2$.

e. Evaluate the transfer function relating the transform of $\hat{w}_i(t)$ and the transform of the incremental liquid height $\hat{h}_2$.

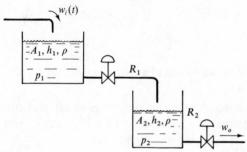

Figure P15.6

15.7 The hydraulic system shown in Figure P15.7 has the incremental pressure $\hat{p}_i(t)$ as its input and the incremental flow rate $\hat{w}_o$ as its output.

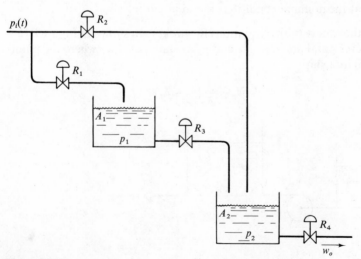

Figure P15.7

The liquid density is ρ.

　　a. Write the state-variable equations using the incremental pressures $\hat{p}_1$ and $\hat{p}_2$ as the state variables.

　　b. Find the transfer function $H(s) = \hat{W}_o(s)/\hat{P}_i(s)$.

　　c. Solve for the steady-state value of the unit step response. Justify your answer on physical grounds.

15.8 Figure P15.8 shows three identical tanks having capacitance C connected by identical lines having resistance R. An input stream flows into tank 1 at the incremental volumetric flow rate $\hat{w}_i(t)$, and an output stream flows from tank 3 at the incremental volumetric flow rate $\hat{w}_o(t)$.

　　a. Taking the incremental liquid volumes $\hat{v}_1$, $\hat{v}_2$, and $\hat{v}_3$ as the state variables, write the state-variable equations.

　　b. Let $RC = 1$ and find $\hat{V}_2(s)$ in terms of $\hat{W}_1(s)$ and $\hat{W}_o(s)$ by transforming the three state-variable equations and eliminating $\hat{V}_1(s)$ and $\hat{V}_3(s)$.

　　c. Solve for $\hat{v}_2$ as a function of time when $\hat{w}_i(t) = U(t)$ and $\hat{w}_o(t) = 0$. Repeat the solution when $\hat{w}_i(t) = 0$ and $\hat{w}_o(t) = U(t)$. Sketch the responses and comment on them.

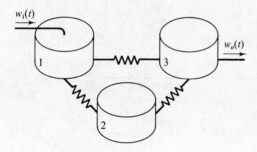

Figure P15.8

15.9 The valve and pump characteristics for the hydraulic system shown in Figure P15.9(a) are plotted in Figure P15.9(b). Curves are given for the two pump speeds 100 rad/s and 150 rad/s. The cross-sectional area of the vessel is 2.0 m^2, and the liquid density is $1000. \text{ kg/m}^3$.

　　a. Determine the steady-state flow rate, gauge pressure, and liquid height for each of the pump speeds for which curves are shown.

　　b. Derive the linearized models for each of the pump speeds in numerical form.

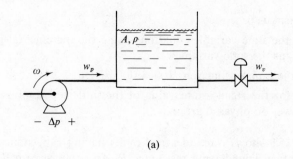

(a)

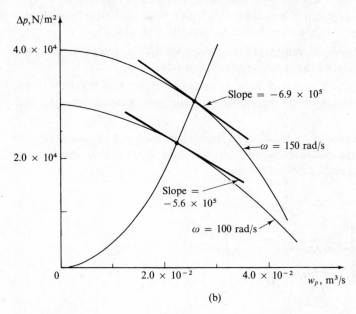

(b)

Figure P15.9

15.10 As shown in Figure P15.10(a), liquid can flow into a vessel through a constant-speed pump and valve, but it cannot leave the vessel. The assumed pump characteristic is shown in Figure P15.10(b), where α and β are the maximum flow rate and the maximum pressure difference, respectively. The valve is shut for all $t < 0$, is opened at $t = 0$, and presents no resistance to the flow of the liquid for $t > 0$.

    a. Write the differential equation obeyed by the liquid height h.

    b. Write expressions for the time constant and the steady-state height

    c. Solve for h and sketch it versus time.

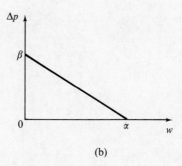

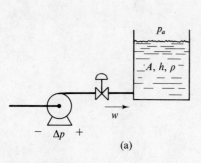

(a)   (b)

Figure P15.10

15.11 Consider the hydraulic system described in Problem 15.10 with a pump having the nonlinear pressure-flow relationship shown in Figure P15.11.

   a. Write the differential equation obeyed by the liquid height after the valve is opened.

   b. Using a calculator or a digital computer, solve for h and w and sketch them versus time. Use the following parameter values:

$$A = 1.50 \text{ m}^2$$

$$\rho g = 9807. \text{ N/m}^3$$

$$\alpha = 0.120 \text{ m}^3/\text{s}$$

$$\beta = 3.0 \times 10^4 \text{ N/m}^2$$

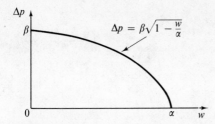

Figure P15.11

15.12 For the pump characteristic shown in Figure P15.9(b), use (5.31) to develop a linearized pump model in numerical form that is valid for small variations in both Δp and ω in the vicinity of the operating point. The operating point is defined by the values $\bar{\omega} = 100$ rad/s, $\bar{w} = 2.25 \times 10^{-2}$ m^3/s, and $\overline{\Delta p} = 2.28 \times 10^4$ N/m^2.

CHAPTER 16
FEEDBACK SYSTEMS

In the simulation diagrams of many dynamic systems, one or more closed paths or loops connect the diagram elements. Such systems are known as *feedback* or *closed-loop systems*, and they arise in a wide variety of situations. For example, feedback is inherent in the dynamic models of population and economic systems, and it is typically introduced into automatic control systems and electronic amplifiers in order to improve their performance.

This chapter combines the simulation-diagram and transfer-function concepts to yield the *block-diagram method* for representing and analyzing the models of feedback systems efficiently. We present several techniques for reducing the block diagrams of feedback systems to find the overall transfer function of the system from its input to its output. The chapter concludes with a comprehensive example that illustrates the modeling and analysis of an electromechanical feedback control system.

16.1 ELEMENTARY BLOCK DIAGRAMS

In Chapter 6, we developed the use of simulation diagrams for representing the mathematical models of dynamic systems. The dynamic element in such diagrams is always the integrator, and at that time we did not develop procedures for simplifying the diagrams. The basic difference between simulation and block diagrams is that in the latter, the dynamic elements are represented by their transfer functions and can be considerably more complex than integrators. Because the block diagram is merely a pictorial representation of a set of algebraic Laplace transform equations, it is possible to combine blocks by calculating equivalent transfer functions and thereby simplify the diagram. Following a description of the basic blocks, we will present procedures for combining series and parallel combinations of blocks. Diagrams involving feedback will be discussed in the next section.

BASIC BLOCKS

The basic elements of which block diagrams are composed are summing junctions, gains, and transfer-function blocks. As indicated in Figure 16.1(a) and Figure 16.1(b), summing junctions and gains are drawn exactly as we draw simulation diagrams, except that the variables are denoted by their Laplace transforms rather than by their time-domain representations, e.g., by $X(s)$ rather than by $x(t)$.

An integrator is represented by the transfer function $1/s$, as depicted in

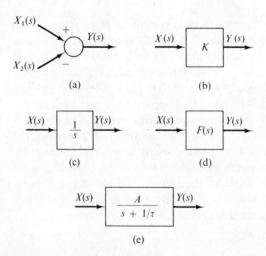

Figure 16.1 Basic block diagrams. (a) Summing junction. (b) Gain. (c) Integrator. (d) Arbitrary transfer function. (e) First-order system.

Figure 16.1(c). This representation results from the fact that an integrator having input $x(t)$ and output $y(t)$ obeys the relationship

$$y(t) = y(0) + \int_0^t x(\lambda)\, d\lambda$$

where λ is the dummy variable of integration. Setting $y(0)$ equal to 0 and transforming the equation give

$$Y(s) = \frac{1}{s} X(s)$$

Hence, the transfer function of the integrator is $Y(s)/X(s) = 1/s$.

Any system or combination of elements can be represented by a block containing its transfer function $F(s)$, as indicated in Figure 16.1(d). For example, the first-order system that obeys the input-output equation

$$\dot{y} + \frac{1}{\tau} y = Ax(t)$$

has as its transfer function

$$F(s) = \frac{A}{s + \dfrac{1}{\tau}}$$

Thus it could be represented by the block diagram shown in Figure 16.1(e).

SERIES COMBINATIONS

Two blocks are said to be in *series* when the output of one goes only to the input to the other, as shown in Figure 16.2(a). The transfer functions of the individual blocks in the figure are $F_1(s) = V(s)/X(s)$ and $F_2(s) = Y(s)/V(s)$.

When we evaluate the individual transfer functions, it is essential that we take any *loading effects* into account. This means that $F_1(s)$ is the ratio

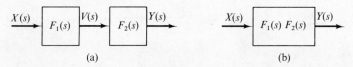

(a)   (b)

Figure 16.2 (a) Two blocks in series. (b) Equivalent diagram.

$V(s)/X(s)$ when the two subsystems are connected, so that any effect the second subsystem has on the first is accounted for in the mathematical model. The same statement holds for calculating $F_2(s)$. For example, the input-output relationship for a linear potentiometer loaded by a resistor connected from its wiper to the ground node was shown in Example 11.1 to differ from that of the unloaded potentiometer. Problem 16.22 considers a simple electrical circuit in which we must account for loading effects when finding the transfer function.

In Figure 16.2(a), $Y(s) = F_2(s)V(s)$ and $V(s) = F_1(s)X(s)$. It follows that

$$Y(s) = F_2(s)[F_1(s)X(s)]$$
$$= [F_1(s)F_2(s)]X(s)$$

Thus the transfer function relating the input transform $X(s)$ to the output transform $Y(s)$ is $F_1(s)F_2(s)$, the product of the individual transfer functions. The equivalent block diagram is shown in Figure 16.2(b).

PARALLEL COMBINATIONS

Two systems are said to be in *parallel* when they have a common input and their outputs are combined by a summing junction. If, as indicated in Figure 16.3(a), the individual blocks have the transfer functions $F_1(s)$ and $F_2(s)$ and the signs at the summing junction are both positive, the overall transfer function $Y(s)/X(s)$ will be the sum $F_1(s) + F_2(s)$, as shown in Figure 16.3(b). To prove this statement, we note that

$$Y(s) = V_1(s) + V_2(s)$$

where $V_1(s) = F_1(s)X(s)$ and $V_2(s) = F_2(s)X(s)$. Substituting for $V_1(s)$ and $V_2(s)$, we have

$$Y(s) = [F_1(s) + F_2(s)]X(s)$$

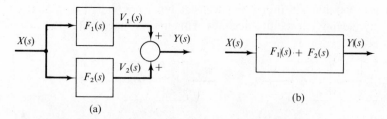

(a)

(b)

Figure 16.3 (a) Two blocks in parallel. (b) Equivalent diagram.

If either of the summing-junction signs associated with $V_1(s)$ or $V_2(s)$ is negative, we must change the sign of the corresponding transfer function in forming the overall transfer function. The following example illustrates the rules for combining blocks that are in parallel or series.

EXAMPLE 16.1   Evaluate the transfer functions $Y(s)/U(s)$ and $Z(s)/U(s)$ for the block diagram shown in Figure 16.4, giving the results as rational functions of s.

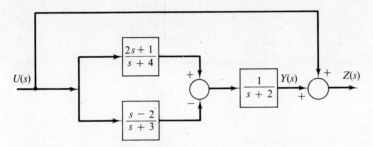

Figure 16.4 Block diagram for Example 16.1.

Solution   Since $Z(s)$ can be viewed as the sum of the outputs of two parallel blocks, one of which has $Y(s)$ as its output, we first evaluate the transfer function $Y(s)/U(s)$. To do this, we observe that $Y(s)$ can be considered the output of a series combination of two parts, one of which is a parallel combination of two blocks. Starting with this parallel combination, we write

$$\frac{2s+1}{s+4} - \frac{s-2}{s+3} = \frac{3s^2 + 9s - 5}{s^2 + 7s + 12}$$

and redraw the block diagram as shown in Figure 16.5(a). The series combination in this version has the transfer function

$$\frac{Y(s)}{U(s)} = \frac{3s^2 + 9s - 5}{s^2 + 7s + 12} \cdot \frac{1}{s + 2}$$

$$= \frac{3s^2 + 9s - 5}{s^3 + 9s^2 + 26s + 24}$$

which leads to the diagram shown in Figure 16.5(b). We can reduce the final parallel combination to the single block shown in Figure 16.5(c) by

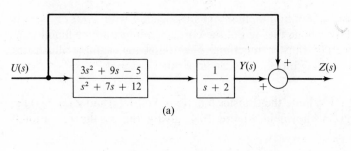

(a)

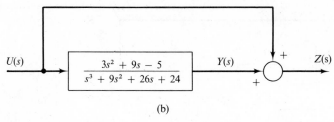

(b)

$$U(s) \rightarrow \boxed{\dfrac{s^3 + 12s^2 + 35s + 19}{s^3 + 9s^2 + 26s + 24}} \rightarrow Z(s)$$

(c)

Figure 16.5 Equivalent block diagrams for the diagram shown in Figure 16.4.

writing

$$\frac{Z(s)}{U(s)} = 1 + \frac{Y(s)}{U(s)}$$

$$= 1 + \frac{3s^2 + 9s - 5}{s^3 + 9s^2 + 26s + 24}$$

$$= \frac{s^3 + 12s^2 + 35s + 19}{s^3 + 9s^2 + 26s + 24}$$

In general, it is desirable to reduce the transfer functions of combinations of blocks to rational functions of s in order to simplify the subsequent analysis. This is particularly important in the next section when we are reducing feedback loops to obtain an overall transfer function.

16.2 BLOCK DIAGRAMS OF FEEDBACK SYSTEMS

Figure 16.6(a) shows the block diagram of a general feedback system that has a forward path from the summing junction to the output and a feedback path from the output back to the summing junction. The transforms of the system's input and output are $U(s)$ and $Y(s)$, respectively. The transfer function $G(s) = Y(s)/V(s)$ is known as the *forward transfer function*, while $H(s) = Z(s)/Y(s)$ is called the *feedback transfer function*. We must evaluate both these transfer functions with the system elements connected in order to properly account for the loading effects of the interconnections. The product $G(s)H(s)$ is referred to as the *open-loop transfer function*. The sign associated with the feedback signal from the block $H(s)$ at the summing junction is shown as minus because a minus sign naturally occurs in the majority of feedback systems, particularly in control systems.

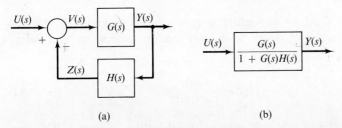

(a)

(b)

Figure 16.6 (a) Block diagram of a feedback system. (b) Equivalent diagram.

Given the model of a feedback system in terms of its forward and feedback transfer functions $G(s)$ and $H(s)$, it is often necessary to determine the *closed-loop transfer function* $Y(s)/U(s)$. We do this by writing the algebraic transform equations corresponding to the block diagram shown in Figure 16.6(a) and solving them for the ratio $Y(s)/U(s)$. We can write the following transform equations directly from the block diagram:

$$V(s) = U(s) - Z(s)$$

$$Y(s) = G(s)V(s)$$

$$Z(s) = H(s)Y(s)$$

If we combine these equations in such a way as to eliminate $V(s)$ and $Z(s)$, we find that

$$Y(s) = G(s)[U(s) - H(s)Y(s)]$$

which can be rearranged to give

$$[1 + G(s)H(s)]Y(s) = G(s)U(s)$$

Hence, the closed-loop transfer function $T(s) = Y(s)/U(s)$ is

$$T(s) = \frac{G(s)}{1 + G(s)H(s)} \qquad (1)$$

where it is implicit that the sign of the feedback signal at the summing junction is negative. It is readily shown that when a plus sign is used at the summing junction for the feedback signal, the closed-loop transfer function becomes

$$T(s) = \frac{G(s)}{1 - G(s)H(s)} \qquad (2)$$

A commonly used simplification occurs when the feedback transfer function is unity, i.e., $H(s) = 1$. Such a system is referred to as a *unity-feedback system*, and (1) reduces to

$$T(s) = \frac{G(s)}{1 + G(s)} \qquad (3)$$

In the remainder of this section, we consider three examples. The first two illustrate determining the closed-loop transfer function by reducing the block diagram and show the effects of feedback gains on the closed-loop poles, damping ratio, and undamped natural frequency. In the third example, a block diagram is drawn directly from the system's state-variable equations and then reduced to give the system's transfer functions.

EXAMPLE 16.2   Find the closed-loop transfer function for the feedback system shown in Figure 16.7(a) and compare the locations of the poles of the open-loop and closed-loop transfer functions in the s-plane.

Solution   By comparing the block diagram shown in Figure 16.7(a) with that shown in Figure 16.6(a), we see that $G(s) = 1/(s+\alpha)$ and $H(s) = \beta$. Substituting these expressions into (1) gives

$$T(s) = \frac{\dfrac{1}{s+\alpha}}{1 + \left(\dfrac{1}{s+\alpha}\right)\beta}$$

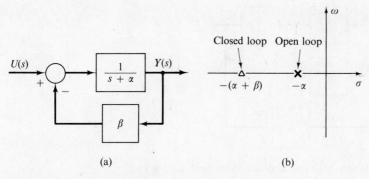

Figure 16.7 Single-loop feedback system for Example 16.2.

which we can write as a rational function of s by multiplying the numerator and denominator by $s + \alpha$. Doing this, we obtain the closed-loop transfer function

$$T(s) = \frac{1}{s + \alpha + \beta}$$

This result illustrates an interesting and useful property of feedback systems, namely that the poles of the closed-loop transfer function will differ from the poles of the open-loop transfer function $G(s)H(s)$. In this case, the single open-loop pole is at $s = -\alpha$, whereas the single closed-loop pole is at $s = -(\alpha + \beta)$. These pole locations are indicated in Figure 16.7(b) for positive α and β. Hence, in the absence of feedback, the pole of the transfer function $Y(s)/U(s)$ is at $s = -\alpha$ and the free response will be of the form $\epsilon^{-\alpha t}$. With feedback, however, the free response will be $\epsilon^{-(\alpha + \beta)t}$. Thus the time constant of the open-loop system is $1/\alpha$, whereas that of the closed-loop system is $1/(\alpha + \beta)$.

EXAMPLE 16.3 Find the closed-loop transfer function of the two-loop feedback system shown in Figure 16.8. Also express the damping ratio and undamped natural frequency of the closed-loop system in terms of the gains a_0 and a_1.

Solution Since the system's block diagram contains one feedback path inside another, we cannot use (1) directly to evaluate $Y(s)/U(s)$. However, we can redraw the block diagram such that the summing junction is split into two summing junctions, as shown in Figure 16.9(a). Then it is possible to use (1) to eliminate the inner loop by calculating the transfer function

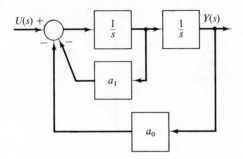

Figure 16.8 System with two feedback loops for Example 16.3.

$W(s)/V(s)$. Taking $G(s) = 1/s$ and $H(s) = a_1$ in (1), we obtain

$$\frac{W(s)}{V(s)} = \frac{\dfrac{1}{s}}{1 + \left(\dfrac{a_1}{s}\right)} = \frac{1}{s + a_1}$$

Redrawing Figure 16.9(a) with the inner loop replaced by a block having $1/(s + a_1)$ as its transfer function gives Figure 16.9(b). The two blocks in the forward path of this version are in series and can be combined by multiplying their transfer functions, giving the block diagram shown in Figure 16.9(c). Then we can apply (1) again to give the overall closed-loop transfer function $T(s) = Y(s)/U(s)$ as

$$T(s) = \frac{\dfrac{1}{s(s + a_1)}}{1 + \dfrac{1}{s(s + a_1)} \cdot a_0} = \frac{1}{s(s + a_1) + a_0}$$

$$= \frac{1}{s^2 + a_1 s + a_0} \tag{4}$$

The block-diagram representation of the feedback system corresponding to (4) is shown in Figure 16.9(d).

The poles of the closed-loop transfer function are the roots of the equation

$$s^2 + a_1 s + a_0 = 0 \tag{5}$$

which we obtain by setting the denominator of $T(s)$ equal to zero and which is the characteristic equation of the closed-loop system. Equation (5) has two roots, which may be real or complex depending on the sign of the quantity

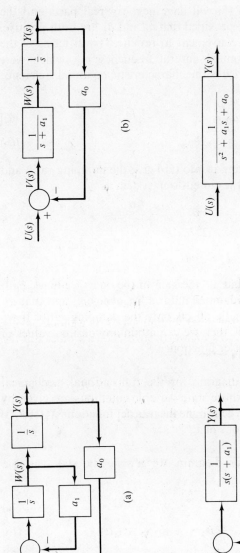

Figure 16.9 Equivalent block diagrams for the system shown in Figure 16.8.

$a_1^2 - 4a_0$. However, the roots of (5) will have negative real parts and the closed-loop system will be stable provided that a_0 and a_1 are both positive.

If the poles are complex, it is convenient to rewrite $T(s)$ in terms of the damping ratio ζ and the undamped natural frequency ω_n, which were introduced in Section 8.4. Since (5) is the characteristic equation, reference to (8.44) indicates that

$$r^2 + 2\zeta\omega_n r + \omega_n^2 = 0$$

$$a_0 = \omega_n^2 \tag{6a}$$

$$a_1 = 2\zeta\omega_n \tag{6b}$$

Solving (6a) for ω_n and substituting it into (6b) give the damping ratio and undamped natural frequency of the closed-loop system as

By changing GAINS a_1 & a_0, we can change ζ & ω_n For Closed-Loop Transfer function

$$\zeta = \frac{a_1}{2\sqrt{a_0}}$$

$$\omega_n = \sqrt{a_0}$$

We see from these expressions that a_0, the gain of the outer feedback path in Figure 16.8, determines the undamped natural frequency ω_n and that a_1, the gain of the inner feedback path, affects only the damping ratio. If we can specify both a_0 and a_1 at will, then we can attain any desired values of ζ and ω_n for the closed-loop transfer function.

EXAMPLE 16.4 Draw a block diagram for the translational mechanical system studied in Example 3.4 whose state-variable equations are given by (3.7). Reduce the block diagram to determine the transfer functions $X_1(s)/F_a(s)$ and $X_2(s)/F_a(s)$ as rational functions of s.

Solution To construct the block diagram, we rewrite the state-variable equations as

$$\dot{x}_1 = v_1 \tag{7a}$$

$$M\dot{v}_1 = -K_1 x_1 - B_1 v_1 + K_2 x_2 + f_a(t) \tag{7b}$$

$$B_2 \dot{x}_2 = K_1 x_1 - (K_1 + K_2)x_2 \tag{7c}$$

Then we use (7b) to draw a summing junction having $MsV_1(s)$, the transform of $M\dot{v}_1$, as its output. After the summing junction, we insert the transfer function $1/Ms$ to get $V_1(s)$, which, from (7a), equals $sX_1(s)$. Thus an integrator whose input is $V_1(s)$ has $X_1(s)$ as its output. Using (7c), we form a second summing junction having $B_2 sX_2(s)$ as its output. Following this summing junction by the transfer function $1/B_2 s$, we get $X_2(s)$ and can complete the four feedback paths required by the summing junctions. The result of these steps is the block diagram shown in Figure 16.10(a).

To simplify the block diagram, we use (1) to reduce each of the three inner feedback loops, obtaining the version shown in Figure 16.10(b). To

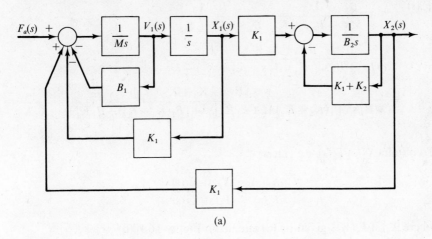

(a)

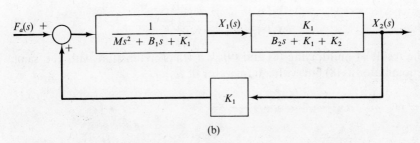

(b)

Figure 16.10 Block diagrams for the system in Example 16.4. (a) As drawn from (7). (b) With the three inner feedback loops eliminated.

evaluate the transfer function $X_1(s)/F_a(s)$, we can apply (2) to this single-loop diagram since the sign associated with the feedback signal at the summing junction is positive rather than negative. Doing this with

$$G(s) = \frac{1}{Ms^2 + B_1 s + K_1}$$

and

$$H(s) = \frac{K_1^2}{B_2 s + K_1 + K_2}$$

we find

$$\frac{X_1(s)}{F_a(s)} = \frac{\dfrac{1}{Ms^2 + B_1 s + K_1}}{1 - \dfrac{1}{Ms^2 + B_1 s + K_1} \cdot \dfrac{K_1^2}{B_2 s + K_1 + K_2}}$$

$$= \frac{B_2 s + K_1 + K_2}{(Ms^2 + B_1 s + K_1)(B_2 s + K_1 + K_2) - K_1^2}$$

$$= \frac{B_2 s + K_1 + K_2}{MB_2 s^3 + [(K_1 + K_2)M + B_1 B_2] s^2 + [B_1(K_1 + K_2) + B_2 K_1]s + K_1 K_2} \tag{8}$$

To obtain $X_2(s)/F_a(s)$, we can write

$$\frac{X_2(s)}{F_a(s)} = \frac{X_1(s)}{F_a(s)} \frac{X_2(s)}{X_1(s)}$$

where $X_1(s)/F_a(s)$ is given by (8) and, from Figure 16.10(b),

$$\frac{X_2(s)}{X_1(s)} = \frac{K_1}{B_2 s + K_1 + K_2} \tag{9}$$

The result of multiplying (8) and (9) is a transfer function with the same denominator as (8) but with a numerator of K_1.

16.3 APPLICATION TO A CONTROL SYSTEM

Most control systems use feedback to force the output variable to follow a reference input, while being relatively insensitive to the effects of one or more disturbance inputs. A common type of control system is the *servomechanism*, in which a mechanical output variable such as the position or angular rotation of an element is required to follow a reference input, which might be the orientation of a knob or dial that is varied by a human operator. For example, a mechanical manipulator used for working with radioactive materials would require several servomechanisms to translate the operator's hand motion into equivalent physical motion at a remote location behind the shielding material.

In this section, we model and analyze a simple servomechanism whose purpose is to make the angular orientation of an output shaft follow that of a manually adjusted dial. First, we will develop a mathematical model of the feedback system by writing the algebraic and differential equations describing the individual elements, transforming these equations and drawing a block diagram, and then reducing the block diagram to give a single transfer function. Finally, we shall analyze the system's performance and consider means of improving it.

SYSTEM DESCRIPTION

The positional servomechanism we will consider is shown in Figure 16.11. The manually set input potentiometer at the left has its two ends connected to constant voltage sources, and its wiper voltage e_1 will obey the algebraic relationship

$$e_1 = -K_\theta \theta_i(t) \tag{10}$$

where K_θ is a constant, provided that the potentiometer is linear and that no current is drawn by the amplifier, i.e., there is no loading. The output potentiometer at the right of the figure is identical to the input potentiometer, except that its wiper is mechanically connected to the output shaft and the polarities of the voltage sources connected to its ends are reversed. Hence, the voltage of the wiper of the output potentiometer obeys the equation

$$e_2 = K_\theta \theta_o \tag{11}$$

If both potentiometers are constrained to one full revolution and if the constant voltage sources are $\pm A$ volts, then $K_\theta = A/\pi$ volts per radian.

The amplifier's output voltage is proportional to the sum of its input voltages. Hence, we write

$$e_a = -K_A(e_1 + e_2) \tag{12}$$

where $-K_A$ is the amplifier gain in volts per volt. It is assumed that (12) holds regardless of the current i flowing in the armature circuit of the motor and that the amplifier draws no current from the wipers of the input and output potentiometers. Note that the amplifier is described by an algebraic rather than a differential equation, which implies that (12) holds regardless of how rapidly e_1 or e_2 may vary.

The motor is assumed to have a constant field current and negligible inductance in the armature winding. The electromechanical driving torque τ_e and the induced voltage e_m are given, as in (11.27), by

$$\tau_e = \alpha i$$

$$e_m = \alpha \dot{\phi}$$

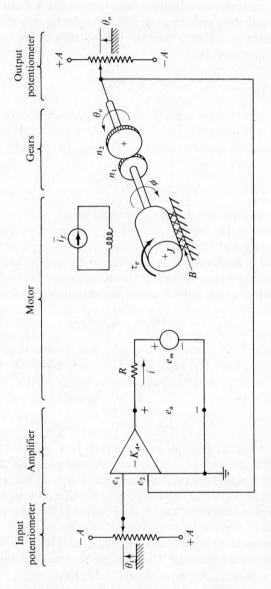

Figure 16.11 Servomechanism for Section 16.3.

where the coupling coefficient α has units of volt-seconds or, equivalently, newton-meters per ampere and is dependent on the field current i_f. The symbol ϕ denotes the angular orientation of the motor shaft. If the armature resistance is denoted by R, the viscous-friction coefficient by B, and the moment of inertia by J, the motor can be modeled by the pair of equations

$$i = \frac{1}{R}(e_a - \alpha\dot{\phi}) \tag{13a}$$

$$J\ddot{\phi} + B\dot{\phi} = \alpha i \tag{13b}$$

The motor shaft is connected to the output shaft through a pair of gears having the gear ratio $N = n_2/n_1$. Hence the motor angle ϕ and the output wiper angle θ_o are related by

$$\theta_o = \frac{1}{N}\phi \tag{14}$$

Although Figure 16.11 does not indicate a moment of inertia attached to the right gear or moments of inertia for the gears themselves, such moments of inertia could be referred to the motor shaft and incorporated in the value of J if they were not negligible. Likewise, any viscous friction associated with the output shaft could be referred to the motor and combined with B.

SYSTEM MODEL

In order to obtain a block diagram for the system, we transform (10) through (14), with zero initial conditions. For the combination of the input and output potentiometers and the amplifier, substituting (10) and (11) into (12) and taking the Laplace transform yield

$$E_a(s) = K_A K_\theta [\Theta_i(s) - \Theta_o(s)] \tag{15}$$

which is represented by the two gains of K_θ, the summing junction, and the gain of K_A within the dashed rectangle shown in Figure 16.12.

When (13) is transformed and $I(s)$ eliminated, we obtain the single transformed equation

$$\left[Js^2 + \left(B + \frac{\alpha^2}{R} \right)s \right]\Phi(s) = \frac{\alpha}{R}E_a(s)$$

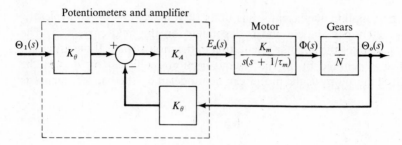

Figure 16.12 Servomechanism block diagram.

for the motor, which yields the transfer function

$$\frac{\Phi(s)}{E_a(s)} = \frac{\dfrac{\alpha}{RJ}}{s\left[s + \left(\dfrac{B}{J} + \dfrac{\alpha^2}{JR}\right)\right]}$$

If we define the parameters

$$K_m = \frac{\alpha}{RJ}$$

$$\tau_m = \frac{1}{\dfrac{B}{J} + \dfrac{\alpha^2}{JR}}$$

the transfer function of the motor becomes

$$\frac{\Phi(s)}{E_a(s)} = \frac{K_m}{s\left(s + \dfrac{1}{\tau_m}\right)} \tag{16}$$

which results in a single block located in the forward path of the diagram shown in Figure 16.12. Finally, we describe the gears by the gain $1/N$ according to

$$\Theta_o(s) = \frac{1}{N}\Phi(s) \tag{17}$$

and draw the feedback path from the output to the output-potentiometer block.

CLOSED-LOOP TRANSFER FUNCTION

To calculate $T(s) = \Theta_o(s)/\Theta_i(s)$, the transfer function of the closed-loop system, we note that the first block with gain K_θ is in series with the feedback-loop portion of the system. We can obtain the transfer function of the feedback loop by applying (1) with

$$G(s) = \frac{K_A K_m/N}{s(s + 1/\tau_m)}$$

$$H(s) = K_\theta$$

giving

$$T^*(s) = \frac{K_A K_m/N}{s^2 + (1/\tau_m)s + K_A K_m K_\theta/N}$$

To obtain the transfer function of the series combination, we multiply $T^*(s)$ by K_θ, getting

$$T(s) = \frac{K_A K_m K_\theta/N}{s^2 + (1/\tau_m)s + K_A K_m K_\theta/N} \tag{18}$$

By inspection of (18), we can observe several important aspects of the behavior of the closed-loop system. First, the system model is second-order and its closed-loop transfer function $T(s)$ has two poles and no zeros in the finite s-plane. Assuming that the values of all the parameters appearing in $T(s)$ are positive, the poles of $T(s)$ will be in the left half of the complex plane and the closed-loop system will be stable regardless of the specific numerical values of the parameters.

We obtain the steady-state value of the step response of the closed-loop system by setting s equal to zero in (18), obtaining

$$T(0) = \frac{K_A K_m K_\theta/N}{K_A K_m K_\theta/N} = 1$$

Hence, a step input for $\theta_i(t)$ will result in an output angle θ_o that is identical to the input angle in the steady state. This steady-state condition of zero error will occur regardless of the specific numerical values for the system parameters. This property is a result of the feedback structure of the servo-mechanism, whereby a signal that is proportional to the error signal $\theta_i(t) - \theta_o$ is used to drive the motor. Because the motor acts like an integrator (its transfer function $\Phi(s)/E_a(s)$ has a pole at $s = 0$), the armature voltage e_a must become zero if the system is to reach steady-state with a constant input; otherwise, ϕ would not be zero. Because of (12), the sum of the wiper voltages e_1 and e_2 must be zero in the steady state, and this condition requires that $(\theta_o)_{ss} = (\theta_i)_{ss}$.

DESIGN FOR SPECIFIED DAMPING RATIO

In practice, all the parameter values except one are often fixed and we must select the remaining one to yield some specific response characteristic such as damping ratio, undamped natural frequency, or steady-state value. Suppose that all the parameters except the amplifier gain K_A are fixed with the values listed in Table 16.1 and that we must select K_A to yield a value of 0.50 for the damping ratio.

Since the potentiometer gains are $K_\theta = A/\pi$, it follows that $K_\theta = 15.0/\pi = 4.775$ V/rad. Substituting the known parameter values into (18) results in the closed-loop transfer function

$$T(s) = \frac{(150 \times 4.775/10)\,K_A}{s^2 + (1/0.40)\,s + (150 \times 4.775/10)\,K_A}$$

$$= \frac{71.62 K_A}{s^2 + 2.50s + 71.62 K_A} \tag{19}$$

Comparing the denominator of (19) to the polynomial $s^2 + 2\zeta\omega_n s + \omega_n^2$, which is used to define the damping ratio ζ and the undamped natural frequency ω_n, we see that

$$2\zeta\omega_n = 2.50 \tag{20a}$$

$$\omega_n^2 = 71.62 K_A \tag{20b}$$

Setting ζ equal to its specified value of 0.50 in (20a), we get $\omega_n = 2.50$ rad/s. Substituting for ω_n^2 in (20b) gives the required amplifier-gain magnitude as

$$K_A = 0.08727 \text{ V/V}$$

Table 16.1 Numerical values of servomechanism parameters

Parameter	Value
A (magnitude of potentiometer voltages)	15.0 V
K_m (for the motor)	150. V·rad/s^2
τ_m (for the motor)	0.40 s
N (gear ratio)	10

Substituting this value of K_A into (19) yields the closed-loop transfer function in numerical form as

$$T(s) = \frac{6.25}{s^2 + 2.50s + 6.25} \tag{21}$$

You can verify that the denominator of $T(s)$ results in a pair of complex poles having $\zeta = 0.50$ as specified.

PROPORTIONAL-PLUS-DERIVATIVE FEEDBACK

The system designed in the foregoing paragraphs is constrained in several ways that turn out to be undesirable in practice. For example, although we obtain the specified damping ratio of 0.50, the value of ω_n was dictated by the requirement on ζ and could not have been specified independently. In practice, one might want to specify both ζ and ω_n for the closed-loop system. We shall demonstrate that this can be done, provided that a signal proportional to the angular velocity of the motor shaft (or output shaft) is fed back and added to the amplifier input.

Assume that a tachometer is attached directly to the motor and a potentiometer is placed across the output terminals of the tachometer, such that the voltage on the potentiometer wiper is

$$e_3 = K_T \dot{\phi}$$

where K_T has units of volt-seconds and is adjustable between zero and some positive maximum value. The signal e_3 is brought to the amplifier input, along with e_1 and e_2, such that the armature voltage is now

$$e_a = -K_A(e_1 + e_2 + e_3)$$
$$= K_A[K_\theta \theta_i(t) - K_\theta \theta_o - K_T \dot{\phi}]$$

With these additions, the transform of the armature voltage becomes

$$E_a(s) = K_A[K_\theta \Theta_i(s) - K_\theta \Theta_o(s) - K_T s\Phi(s)] \tag{22}$$

The modified system can be represented by the block diagram shown in Figure 16.13, which we obtain by separating the motor transfer function $K_m/[s(s+1/\tau_m)]$ into the pair of blocks in series shown in the figure and then adding the inner feedback path corresponding to the term $K_T s\Phi(s)$. Because we assume zero initial conditions when evaluating transfer functions, the input to the gain block K_T is $s\Phi(s) = \mathscr{L}[\dot{\phi}(t)]$. Thus a signal proportional to the angular velocity of the motor shaft is being fed through the inner feedback path to the summing junction.

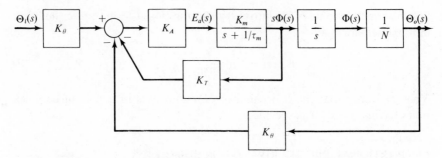

Figure 16.13 Block diagram of servomechanism with tachometer feedback added.

To determine the effect of the tachometer feedback on the closed-loop transfer function, we can reduce the inner loop shown in Figure 16.13 by applying (1) with $G(s) = K_A K_m/(s + 1/\tau_m)$ and $H(s) = K_T$ to give

$$T'(s) = \frac{K_A K_m}{s + (1/\tau_m) + K_A K_m K_T}$$

Applying (1) again with $G(s) = T'(s)/sN$ and $H(s) = K_\theta$ and then multiplying the result by K_θ, we obtain the overall transfer function

$$T(s) = \frac{K_A K_m K_\theta/N}{s^2 + [(1/\tau_m) + K_A K_m K_T]s + K_A K_m K_\theta/N} \tag{23}$$

Comparing (23) to (18) indicates that the only effect of the tachometer feedback is to modify the coefficient of s in the denominator of the transfer function. Hence, the damping ratio and undamped natural frequency now satisfy the relationships

$$\begin{aligned} 2\zeta\omega_n &= (1/\tau_m) + K_A K_m K_T \\ \omega_n^2 &= K_A K_m K_\theta/N \end{aligned} \tag{24}$$

If the values of both K_A and K_T can be selected with τ_m, K_m, K_θ, and N fixed as before, we can specify values for both ζ and ω_n. For example, using the parameter values given in Table 16.1 with $\zeta = 0.50$, we find that (24) reduces to

$$\omega_n = 2.50 + 150 K_A K_T \tag{25a}$$

$$\omega_n^2 = 71.62 K_A \tag{25b}$$

If we want ω_n to be 5.0 rad/s, it follows that

$$K_A = 0.3491 \text{ V/V}$$

$$K_T = 0.04775 \text{ V·s/rad}$$

When we substitute these parameter values and those listed in Table 16.1

into (23), the numerical form of the closed-loop transfer function with tachometer feedback is

$$T(s) = \frac{25.0}{s^2 + 5.0s + 25.0}$$

It is easy to verify that the two poles of $T(s)$ are complex and have $\zeta = 0.50$ and $\omega_n = 5.0$ rad/s, as required.

In concluding this example, we note that the control law given by (22) uses a combination of the output-shaft angle θ_o and the motor angular velocity $\dot{\phi}$ as the feedback signals. It is also possible to think in terms of having the output-shaft angular velocity $\dot{\theta}_o$ fed back by noting that the two angular velocities in question are proportional to one another, namely $\dot{\theta}_o = \dot{\phi}/N$. Hence we can rewrite (22) in the time domain and in terms of the output shaft angle as

$$e_a = K_A[K_\theta \theta_i(t) - K_\theta \theta_o - K_{\dot\theta} \dot{\theta}_o] \tag{26}$$

where $K_\theta = NK_T$, which is the gain of the angular-velocity term. The block diagram corresponding to this version of the control law can take either of the equivalent forms shown in Figure 16.14. If we combine the two parallel

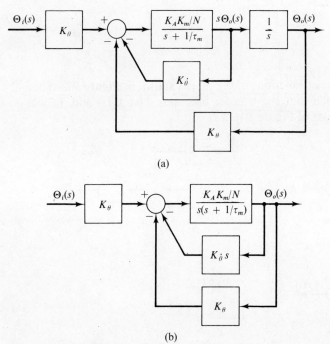

(a)

(b)

Figure 16.14 Equivalent block diagrams for servomechanism with tachometer feedback. (a) Feeding back $\dot{\theta}_o$. (b) Feeding back θ_o.

feedback paths shown in Figure 16.14(b) to yield the single feedback transfer function $K_\theta + K_{\dot\theta} s$, using (1) gives the same expression for $T(s) = \Theta_o(s)/\Theta_i(s)$ as (23).

PROBLEMS

16.1 Evaluate the transfer functions $T_1(s) = Y(s)/U(s)$ and $T_2(s) = Z(s)/U(s)$ as rational functions for the block diagram shown in Figure P16.1.

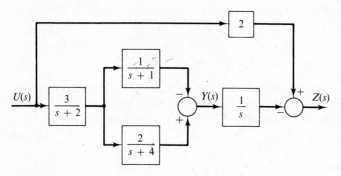

Figure P16.1

16.2 Find the transfer functions of the blocks shown in Figure P16.2(b) and Figure P16.2(c) such that the transfer functions $Y(s)/X_1(s)$ and $Y(s)/X_2(s)$ are identical to those of Figure P16.2(a).

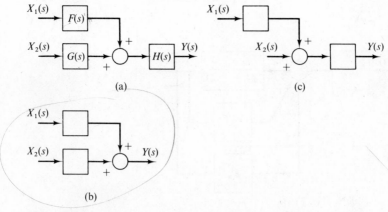

Figure P16.2

16.3 Find the transfer functions of the blocks shown in Figure P16.3(b) and Figure P16.3(c) such that the transfer functions $Y_1(s)/X(s)$ and $Y_2(s)/X(s)$ are identical to those of Figure P16.3(a).

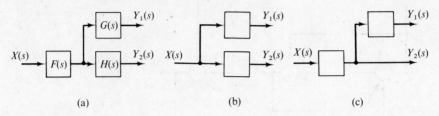

(a)   (b)   (c)

Figure P16.3

16.4 Draw a block diagram for the system described by (6.4) and use it to find $Y(s)/U(s)$ as a rational function of s.

16.5 For each of the block diagrams shown in Figure P16.5, determine the closed-loop transfer function $Y(s)/U(s)$ as a rational function of s.

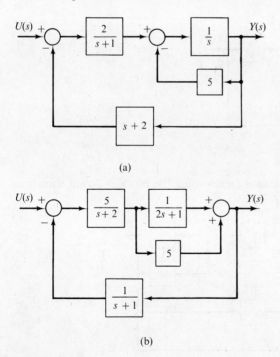

(a)

(b)

Figure P16.5

16.6 Find $Y(s)/U(s)$ as a rational function for the block diagram shown in Figure P16.6.

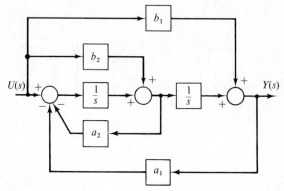

Figure P16.6

16.7 Find $Y(s)/U(s)$ as a rational function for the block diagram shown in Figure P16.7.

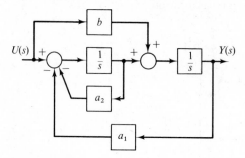

Figure P16.7

16.8 Find $Y(s)/U(s)$ as a rational function for the block diagram shown in Figure P16.8.

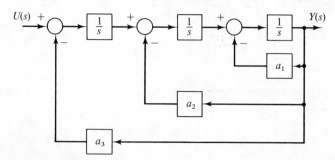

Figure P16.8

16.9 Find the closed-loop transfer functions $T_1(s) = Y(s)/U(s)$ and $T_2(s) = Z(s)/U(s)$ in terms of the individual transfer functions $A(s), ..., E(s)$ for the block diagram shown in Figure P16.9. Give your answer as a ratio of terms that involve only sums, differences, and products of the individual transfer functions.

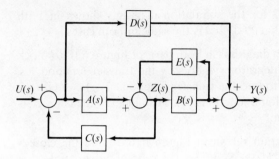

Figure P16.9

16.10 Repeat Problem 16.9 for the block diagram shown in Figure P16.10.

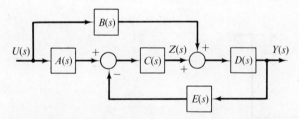

Figure P16.10

16.11 The state-variable and output equations of a general second-order system are given by (6.9) and (6.10).

  a. Draw a block diagram for the system.

  b. Evaluate the transfer functions $Q_1(s)/U(s)$, $Q_2(s)/U(s)$, and $Y(s)/U(s)$ as rational functions by transforming the equations with zero initial conditions and reducing them algebraically.

16.12 Convert the simulation diagram shown in Figure 6.10 to a block diagram by replacing the integrator symbols by their transfer function $1/s$. Then reduce the block diagram to find $Y(s)/U(s)$ as a rational function.

16.13 Repeat Problem 16.12 for the simulation diagram shown in Figure 6.11.

16.14 Repeat Problem 16.12 for the simulation diagrams shown in Figure 6.12 and Figure 6.13. Verify that $Y(s)/U(s)$ is the same in both cases.

16.15 Repeat Problem 16.12 for the simulation diagram shown in Figure 6.14.

16.16 Repeat Problem 16.12 for the simulation diagram shown in Figure 6.15(b), for the transfer function $X_2(s)/X_1(s)$.

16.17 Repeat Problem 16.12 for the simulation diagrams shown in both parts of Figure 6.16. Verify that $Y(s)/U(s)$ is the same in both cases.

16.18 Convert the simulation diagrams in both parts of Figure 6.18 to block diagrams by replacing the integrator symbols by the transfer function $1/s$. Reduce the block diagrams to find the transfer functions $T_x(s) = X(s)/\tau_a(s)$ and $T_\Theta(s) = \Theta(s)/\tau_a(s)$ as rational functions of s. Also identify the characteristic polynomial of the system.

16.19 a. For the block diagram shown in Figure P16.19, find the closed-loop transfer function $T(s) = Y(s)/U(s)$ as a ratio of polynomials.

  b. Determine the steady-state response to a unit step-function input in terms of K.

  c. Write the damping ratio ζ and undamped natural frequency ω_n in

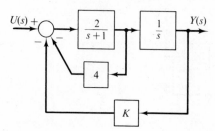

Figure P16.19

terms of the parameter K. Solve for the value of K for which $\zeta = 1/\sqrt{2}$ and find the corresponding numerical value of ω_n.

16.20 a. Find the closed-loop transfer function $T(s) = Y(s)/U(s)$ for the block diagram shown in Figure P16.20.

  b. Determine the undamped natural frequency ω_n and the damping ratio ζ of the closed-loop system. Solve for the value of K for which $\omega_n = 5$ and find the corresponding value of ζ.

  c. Find the steady-state value of the response when the input is a step function of height 5 and when K has the value determined in part b.

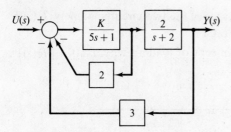

Figure P16.20

16.21 Consider the feedback system shown in Figure P16.21, which has inputs $U(s)$ and $V(s)$ and output $Y(s)$.

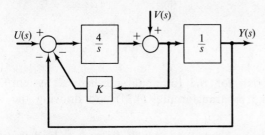

Figure P16.21

a. Find the transfer function $T_1(s) = Y(s)/U(s)$ as a ratio of polynomials. (Note that $V(s)$ should be set equal to zero for this calculation.)

b. Find the transfer function $T_2(s) = Y(s)/V(s)$ as a ratio of polynomials, taking $U(s) = 0$. *Hint:* Define the output of the right summing junction as $Z(s)$ and write the algebraic transform equations relating $V(s)$, $Z(s)$ and $Y(s)$.

c. Determine the damping ratio ζ and undamped natural frequency ω_n of the closed-loop system. Determine the value of K for which $\zeta = 1$.

d. Find the steady-state response when $u(t)$ is the unit step function and $v(t) = 0$. Repeat the solution when $u(t) = 0$ and $v(t)$ is the step function, and comment on the difference in the results.

16.22 Figure P16.22 shows two RC filters connected to a voltage source. The circuit shown in Figure P16.22(a) has a single pair of RC elements. The circuit shown in Figure P16.22(b) has two pairs of RC elements.

a. Evaluate the transfer functions of the two circuits, namely $T_1(s) = E_A(s)/E_i(s)$ and $T_2(s) = E_B(s)/E_i(s)$.

b. Explain in terms of the concept of loading why $T_2(s) \neq [T_1(s)]^2$.

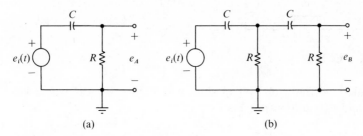

(a)   (b)

Figure P16.22

16.23 a. Draw a block diagram for the rotational mechanical system modeled in Example 4.5 by transforming (4.34) and drawing the corresponding diagram.

b. Set $\tau_L(s)$ equal to zero and reduce the block diagram to find the transfer function $T_1(s) = \Omega_2(s)/\tau_a(s)$.

c. Set $\tau_a(s)$ equal to zero and reduce the block diagram you found in part a to obtain the transfer function $T_2(s) = \Omega_2(s)/\tau_L(s)$.

d. Transform (4.35) and use the result to verify your answers to part b and part c.

16.24 a. Draw a block diagram for the electromechanical system modeled in Example 11.2 by transforming (11.29) and drawing the corresponding diagram.

b. Obtain the transfer functions $T_1(s) = \Omega(s)/E_i(s)$ and $T_2(s) = \Omega(s)/\tau_L(s)$ by reducing the block diagram, first with $\tau_L(s) = 0$ and then with $E_i(s) = 0$.

c. Transform (11.30) and use the result to verify your answers to part b.

16.25 Draw a block diagram for the electrical circuit modeled in Example 9.4 by transforming (9.23) and (9.24) and drawing the corresponding diagram. Define $\Delta e = e_1(t) - e_2(t)$ and evaluate the transfer function $T(s) = E_o(s)/\Delta E(s)$ by reducing the block diagram.

16.26 For the block diagram shown in Figure P16.26, the parameters α and β are positive constants.

　　a. Evaluate the closed-loop transfer function $Y(s)/U(s)$, and write an expression for its poles in terms of α, β, and K.

　　b. Show that the closed-loop poles are in the left half of the s-plane for all $K > 0$. Furthermore, show that the poles are real for $0 \le K \le 0.25(\alpha - \beta)^2$ and complex for $K > 0.25(\alpha - \beta)^2$.

　　c. Show that both poles are in the left half of the s-plane for $-\alpha\beta < K \le 0$ also and that at least one pole is in the right half of the s-plane for $K < -\alpha\beta$.

　　d. Take $\alpha = 1$ and $\beta = 3$ and plot the locus of the closed-loop poles for all values of K, both positive and negative. Indicate the pole locations for $K = -8, -3, 0, 1$, and 5.

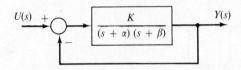

Figure P16.26

16.27 a. Find the closed-loop transfer function $Y(s)/U(s)$ in terms of the parameter K for the feedback system shown in Figure P16.27.

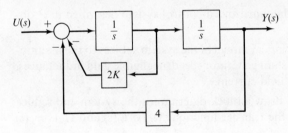

Figure P16.27

　　b. Write an expression for the closed-loop poles in terms of K and sketch the locus of these poles in the complex plane for $0 \le K \le 3$. Indicate the pole locations for $K = 0, 1, 2$, and 3 on the locus.

16.28 a. Find the closed-loop transfer function $Y(s)/U(s)$ in terms of the parameter K for the feedback system shown in Figure P16.28 (next page).

　　b. Write an expression for the closed-loop poles in terms of K, and sketch the locus of these poles in the complex plane for $-4 \le K \le 4$. Indicate the pole locations for $K = -4, -2, 0, 2$, and 4 on the locus.

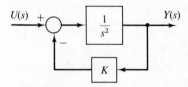

Figure P16.28

16.29 a. Verify that the block diagram shown in Figure P16.29 is a valid representation of the incremental transistor model described in Problem 10.18 with $R_f = 1 \text{ k}\Omega$. (Note that it has no dynamic elements.)

b. Find the transfer function $\hat{I}_o(s)/\hat{I}_i(s)$ by reducing the block diagram.

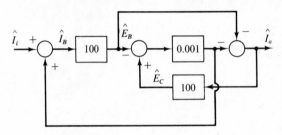

Figure P16.29

16.30 Consider the Ward-Leonard speed-control system described in Problem 11.18.

a. Write the mathematical model of the system subject to the assumptions stated, e.g., constant generator speed, negligible field inductances, and constant motor field current.

b. Using the model, draw a block diagram for the system and reduce the diagram to find the transfer functions $\Omega_m(s)/E_i(s)$ and $\Omega_m(s)/\tau_L(s)$ as rational functions of s.

c. Assume that a tachometer with gain K_T in units of volt-seconds per radian is attached to the motor shaft and that the generated voltage $K_T \omega_m$ is subtracted from an input voltage $K_T \omega_d(t)$ where $\omega_d(t)$ is the desired motor speed in radians per second. The difference between the voltages is applied to the generator field such that

$$e_i = K_T[\omega_d(t) - \omega_m]$$

Using the results of part b, draw a block diagram of the resulting feedback control system, and evaluate the closed-loop transfer functions $\Omega_m(s)/\Omega_d(s)$ and $\Omega_m(s)/\tau_L(s)$.

16.31 Figure P16.31 shows a thermal process with a feedback temperature controller. The controller receives the desired temperature $\theta_d(t)$ and the measured temperature θ_m as inputs and determines the heat q_h supplied by the heater. The uncontrolled process was modeled in Example 14.7, where we derived transfer functions for $\hat{\Theta}(s)/\hat{Q}_h(s)$ and $\hat{\Theta}(s)/\hat{\Theta}_i(s)$. Assume that the thermal resistance R is infinite and that the liquid flow rate is $\bar{w}$, a constant.

In terms of incremental variables, the controller is modeled by the relationship

$$\hat{Q}_h(s) = G_c(s)[\hat{\Theta}_d(s) - \hat{\Theta}_m(s)]$$

where $G_c(s)$ is the controller transfer function and the quantity $\hat{\Theta}_d(s) - \hat{\Theta}_m(s)$ is the transform of the measured temperature error. It is assumed that the sensor measures the actual temperature exactly—that is, $\hat{\theta}_m = \hat{\theta}$.

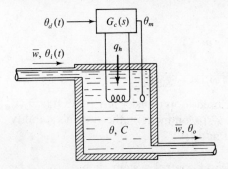

Figure P16.31

    a. Draw a block diagram representing the closed-loop system having incremental inputs of $\hat{\Theta}_d(s)$ and $\hat{\Theta}_i(s)$ and the incremental output $\hat{\Theta}(s)$.

    b. Evaluate the closed-loop transfer functions $\hat{\Theta}(s)/\hat{\Theta}_d(s)$ and $\hat{\Theta}(s)/\hat{\Theta}_i(s)$ in terms of the unspecified controller transfer function $G_c(s)$.

    c. Let $G_c(s) = K_c$, a constant, and evaluate the two closed-loop transfer functions $\hat{\Theta}(s)/\hat{\Theta}_d(s)$ and $\hat{\Theta}(s)/\hat{\Theta}_i(s)$. Show that for step inputs in $\hat{\theta}_d(t)$ and $\hat{\theta}_i(t)$ taken separately, the steady-state temperature error is not zero for finite values of K_c.

    d. Let $G_c(s) = K_c(1 + \alpha/s)$ and reevaluate the closed-loop transfer functions. Show that for step inputs in $\hat{\theta}_d(t)$ and $\hat{\theta}_i(t)$ taken separately, the steady-state temperature error is zero for all positive values of K_c and α.

16.32 Repeat the analysis of the temperature-control system described in Problem 16.31 when a dynamic model for the temperature sensor is included.

Ignore possible fluctuations in the inlet temperature. In terms of the incremental variables, the sensor is assumed to have the transfer function

$$\frac{\hat{\Theta}_m(s)}{\hat{\Theta}(s)} = \frac{1}{as+1}$$

where a is the sensor time constant, and the incremental inlet temperature $\hat{\theta}_i(t)$ is zero. Find the closed-loop transfer function $\hat{\Theta}(s)/\hat{\Theta}_d(s)$ for $G_c(s) = K_c$ and for $G_c(s) = K_c(1 + \alpha/s)$. Comment on the system's stability and steady-state error to a step input in $\hat{\theta}_d(t)$.

16.33 Repeat the analysis of the temperature-control system described in Problem 16.31 when a dynamic model is included for the heater, whose incremental input is denoted by $\hat{x}(t)$. Ignore possible fluctuations in the inlet temperature. Assume that

$$\frac{\hat{Q}_h(s)}{\hat{X}(s)} = \frac{1}{bs+1}$$

where

$$\hat{X}(s) = G_c(s)\left[\hat{\Theta}_d(s) - \hat{\Theta}_m(s)\right]$$

Take $\hat{\theta}_m = \hat{\theta}$ and $\hat{\theta}_i(t) = 0$. Find the closed-loop transfer function $\hat{\Theta}(s)/\hat{\Theta}_d(s)$ for $G_c(s) = K_c$ and for $G_c(s) = K_c(1 + \alpha/s)$. Comment on the system's stability and steady-state error to a step input in $\hat{\theta}_d(t)$.

16.34 Figure P16.34 shows a liquid-level control system such as might be found in a typical chemical process. The sensed level signal h_s is obtained by measuring the gauge pressure p_1^* at the bottom of the tank. The controller also receives a signal $h_d(t)$ indicating the desired level. The controller output x is used to position a linear control valve in a bypass line connected around a centrifugal pump. The bypass flow rate w_b is given by

$$w_b = k\sqrt{\Delta p}\left(\frac{x}{x_m}\right)$$

where x_m is the maximum valve opening, x is the actual valve opening, Δp is the pressure difference developed by the pump, and k is the valve coefficient. The pump is driven at a constant speed, and at the operating point the slope of the curve of Δp vs w is $-K$.

 a. Derive the linearized model of the control valve by finding the coefficients α and β in the expression $\hat{w}_b = \alpha\,\hat{\Delta p} + \beta\hat{x}$.

 b. Write the linearized system equations in terms of the incremental variables $\hat{h}$, $\hat{w}_o$, $\hat{w}_b$, $\hat{w}_p$, $\hat{\Delta p}$ and $\hat{p}_1$. Then draw the block diagram of the open-loop system with $\hat{X}(s)$ and $\hat{W}_o(s)$ as the inputs and $\hat{H}(s)$ as the output. Evaluate the transfer functions $\hat{H}(s)/\hat{X}(s)$ and $\hat{H}(s)/\hat{W}_o(s)$.

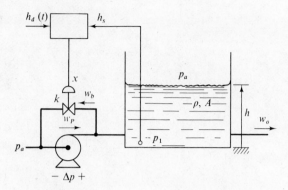

Figure P16.34

c. Taking $\hat{W}_o(s) = 0$, draw a block diagram of the closed-loop system when the controller is described by $\hat{X}(s) = K_c[\hat{H}_d(s) - \hat{H}_s(s)]$ and the sensor is described by $\hat{H}_s(s) = \hat{H}(s)$. Find the closed-loop transfer function $\hat{H}(s)/\hat{H}_d(s)$. Explain why the controller gain K_c should be negative.

d. Repeat part c using a dynamic model of the sensor such that $\hat{H}_s(s)/\hat{H}(s) = 1/(\tau s + 1)$.

16.35 Using a digital computer, simulate the response of the position-control system analyzed in Section 16.3 for the situations outlined in this problem. The program should implement (10) through (14) using the parameter values given in Table 16.1.

a. Taking $K_A = 0.08727$ V/V, simulate the response to a step input of 0.50 rad in $\theta_i(t)$, with zero initial conditions. Verify that the steady-state pointing error is zero.

b. Using the normalized step-response curves shown in Figure 8.20, estimate the damping ratio and undamped natural frequency of the closed-loop system. Compare your estimated values with the design values given in Section 16.3, namely $\zeta = 0.50$ and $\omega_n = 2.50$ rad/s.

c. Simulate the response to the ramp input $\theta_i(t) = 0.1t$ and determine the steady-state pointing error.

d. Repeat part a with amplifier gains of one-half and twice that used before. Comment on the effects on the damping ratio.

e. Change the control law to that specified by (22), using $K_A = 0.3491$ V/V and $K_T = 0.04775$ V·s/rad. Simulate the response to the step input used in part a. Comment on the effect of the revised control law on the damping ratio, undamped natural frequency, and steady-state error.

CHAPTER 17
MATRIX METHODS

One of the characteristics of state-variable models is that they are suitable for matrix notation and for the techniques of linear algebra. For example, we can represent any number of first-order state-variable equations by a single-matrix differential equation merely by making the appropriate definitions. Furthermore, when we are dealing with complex multi-input, multi-output systems, using matrix concepts and properties leads to an understanding of system behavior that would be difficult to achieve otherwise. An additional advantage of using matrices is that we can apply many of the theoretical properties of matrices that are taught in introductory linear-algebra courses to the study of dynamic systems once we have put their models into matrix form.

Section 17.1 presents the matrix forms of state-variable models for nonlinear and linear systems. In the remaining sections, we restrict our attention to fixed linear systems for which very general analytical results can be presented. First we obtain the zero-input response by using Laplace transforms. The state-transition matrix is introduced and several of its properties are derived. Then we study the zero-state response by means of the transfer-function matrix. The chapter concludes with an example in which matrix methods are used for solving for the complete response of a system.

We assume that the reader is somewhat familiar with matrix methods through an introductory course in linear algebra. Such basic matrix operations as multiplication, evaluation of determinants, and inversion are summarized in Appendix C.

17.1 MATRIX FORM OF SYSTEM MODELS

In this section, we introduce the notation we will use throughout the chapter and then use it to write the models of nonlinear, time-varying linear, and fixed linear systems.

Section 3.1 stated that we can write the mathematical model of a general nth-order system having m inputs in state-variable form as indicated in (3.1), which is repeated here, namely

$$
\begin{aligned}
\dot{q}_1 &= f_1(q_1, q_2, ..., q_n, u_1, ..., u_m, t) \\
\dot{q}_2 &= f_2(q_1, q_2, ..., q_n, u_1, ..., u_m, t) \\
&\vdots \\
\dot{q}_n &= f_n(q_1, q_2, ..., q_n, u_1, ..., u_m, t)
\end{aligned}
\tag{1}
$$

The variables $q_1, q_2, ..., q_n$ are the state variables, and $u_1, u_2, ..., u_m$ are the inputs. The functions $f_1, f_2, ..., f_n$ express the state-variable derivatives $\dot{q}_1, \dot{q}_2, ..., \dot{q}_n$ in terms of the state variables, the inputs, and time t. The initial conditions associated with (1) are the initial values of the state variables, $q_1(0), q_2(0), ..., q_n(0)$.

When dealing with the set of n state variables $q_1, q_2, ..., q_n$, we shall use the symbol $\mathbf{q}$ to denote the entire set.* Thus

$$
\mathbf{q} = \begin{bmatrix} q_1 \\ q_2 \\ \vdots \\ q_n \end{bmatrix}
$$

---

* Boldface symbols are used to denote matrices. The symbols are generally capitals, but lower-case letters may be used for vectors. Variables written in italic type are scalars.

which is a matrix having n rows and a single column, with each of its elements being one of the state variables. As such, its elements are functions of time. Matrices having a single column are commonly referred to as *column vectors* or, for short, *vectors*. Hence, the symbol $\mathbf{q}$ will be called the *state vector*, and it is understood that its ith element is the state variable q_i.

Likewise the m inputs $u_1, u_2, ..., u_m$ will be represented by the *input vector* $\mathbf{u}$, which is defined as

$$\mathbf{u} = \begin{bmatrix} u_1 \\ u_2 \\ \vdots \\ u_m \end{bmatrix}$$

NONLINEAR SYSTEMS

With this notation, we can write the system's state-variable model given by (1) as the single equation

$$\dot{\mathbf{q}} = \mathbf{f}(\mathbf{q}, \mathbf{u}, t)$$

where $\dot{\mathbf{q}}$ is the n-element vector whose elements are the derivatives of the state variables q_i, and where $\mathbf{f}$ is an n-element vector composed of the functions $f_i(\mathbf{q}, \mathbf{u}, t)$. The initial value of the state vector $\mathbf{q}$ is the vector

$$\mathbf{q}(0) = \begin{bmatrix} q_1(0) \\ q_2(0) \\ \vdots \\ q_n(0) \end{bmatrix}$$

If the system has p outputs $y_1, y_2, ..., y_p$, they will be defined by p algebraic equations of the form

$$
\begin{aligned}
y_1 &= g_1(\mathbf{q}, \mathbf{u}, t) \\
y_2 &= g_2(\mathbf{q}, \mathbf{u}, t) \\
&\vdots \\
y_p &= g_p(\mathbf{q}, \mathbf{u}, t)
\end{aligned}
\tag{2}
$$

which are represented by a single vector equation. Thus we can model a nonlinear or linear dynamic system of order n having m inputs and p outputs by the two equations

$$\dot{\mathbf{q}} = \mathbf{f}(\mathbf{q}, \mathbf{u}, t) \tag{3a}$$

$$\mathbf{y} = \mathbf{g}(\mathbf{q}, \mathbf{u}, t) \tag{3b}$$

where $\mathbf{f}$ and $\mathbf{g}$ are vector functions having n and p elements, respectively.

TIME-VARYING LINEAR SYSTEMS

For a linear system, the derivatives of the state variables must be linear combinations of the state variables and inputs. In this case, (1) becomes

$$
\begin{aligned}
\dot{q}_1 &= a_{11} q_1 + a_{12} q_2 + \cdots + a_{1n} q_n + b_{11} u_1 + \cdots + b_{1m} u_m \\
\dot{q}_2 &= a_{21} q_1 + a_{22} q_2 + \cdots + a_{2n} q_n + b_{21} u_1 + \cdots + b_{2m} u_m \\
&\vdots \\
\dot{q}_n &= a_{n1} q_1 + a_{n2} q_2 + \cdots + a_{nn} q_n + b_{n1} u_1 + \cdots + b_{nm} u_m
\end{aligned}
\tag{4}
$$

and (2) becomes

$$
\begin{aligned}
y_1 &= c_{11} q_1 + c_{12} q_2 + \cdots + c_{1n} q_n + d_{11} u_1 + \cdots + d_{1m} u_m \\
y_2 &= c_{21} q_1 + c_{22} q_2 + \cdots + c_{2n} q_n + d_{21} u_1 + \cdots + d_{2m} u_m \\
&\vdots \\
y_p &= c_{p1} q_1 + c_{p2} q_2 + \cdots + c_{pn} q_n + d_{p1} u_1 + \cdots + d_{pm} u_m
\end{aligned}
\tag{5}
$$

where any of the coefficients may be functions of time. If we define the $n \times n$ matrix

$$
\mathbf{A}(t) = \begin{bmatrix}
a_{11} & a_{12} & \cdots & a_{1n} \\
a_{21} & a_{22} & \cdots & a_{2n} \\
\vdots & \vdots & & \vdots \\
a_{n1} & a_{n2} & \cdots & a_{nn}
\end{bmatrix}
$$

the $n \times m$ matrix

$$
\mathbf{B}(t) = \begin{bmatrix}
b_{11} & b_{12} & \cdots & b_{1m} \\
b_{21} & b_{22} & \cdots & b_{2m} \\
\vdots & \vdots & & \vdots \\
b_{n1} & b_{n2} & \cdots & b_{nm}
\end{bmatrix}
$$

the $p \times n$ matrix

$$\mathbf{C}(t) = \begin{bmatrix} c_{11} & c_{12} & \cdots & c_{1n} \\ c_{21} & c_{22} & \cdots & c_{2n} \\ \vdots & \vdots & & \vdots \\ c_{p1} & c_{p2} & \cdots & c_{pn} \end{bmatrix}$$

and the $p \times m$ matrix

$$\mathbf{D}(t) = \begin{bmatrix} d_{11} & d_{12} & \cdots & d_{1m} \\ d_{21} & d_{22} & \cdots & d_{2m} \\ \vdots & \vdots & & \vdots \\ d_{p1} & d_{p2} & \cdots & d_{pm} \end{bmatrix}$$

where any element may be a function of time, we can write the state-variable and output equations of a general linear nth-order system with n state variables, m inputs, and p outputs as

$$\dot{\mathbf{q}} = \mathbf{A}(t)\mathbf{q} + \mathbf{B}(t)\mathbf{u} \tag{6a}$$

$$\mathbf{y} = \mathbf{C}(t)\mathbf{q} + \mathbf{D}(t)\mathbf{u} \tag{6b}$$

Reflecting a moment, we note that $\dot{q}_i$ is the ith element of $\dot{\mathbf{q}}$ and is obtained by multiplying the ith row of $\mathbf{A}(t)$ by the column vector $\mathbf{q}$ and adding to it the product of the ith row of $\mathbf{B}(t)$ and the column vector $\mathbf{u}$. Carrying out these operations gives

$$\dot{q}_i = a_{i1}q_1 + a_{i2}q_2 + \cdots + a_{in}q_n + b_{i1}u_1 + \cdots + b_{im}u_m$$

Likewise, we find the kth element of y by multiplying the kth row of $\mathbf{C}(t)$ by $\mathbf{q}$ and adding to it the product of the kth row of $\mathbf{D}(t)$ and the vector $\mathbf{u}$, obtaining

$$y_k = c_{k1}q_1 + c_{k2}q_2 + \cdots + c_{kn}q_n + d_{k1}u_1 + \cdots + d_{km}u_m$$

FIXED LINEAR SYSTEMS

When a system's model is fixed and linear, which is the case for nearly all the examples we shall consider, the elements of $\mathbf{A}$, $\mathbf{B}$, $\mathbf{C}$, and $\mathbf{D}$ are constants, so we can write (6) as

$$\dot{\mathbf{q}} = \mathbf{A}\mathbf{q} + \mathbf{B}\mathbf{u} \tag{7a}$$

$$\mathbf{y} = \mathbf{C}\mathbf{q} + \mathbf{D}\mathbf{u} \tag{7b}$$

Before we discuss solving the matrix form of the model for the system response, we shall consider two examples that illustrate the procedure for writing a state-variable model in matrix form.

EXAMPLE 17.1   Rewrite in matrix form the state-variable and output equations for the electrical circuit shown in Figure 9.29 and modeled in Example 9.10.

Solution   The state variables are the inductor current i_L and the capacitor voltage e_C, the input is the source voltage $e_i(t)$, and the single output is the voltage e_o. From (9.54), the state-variable equations are

$$\frac{di_L}{dt} = -\tfrac{2}{5}i_L - \tfrac{6}{5}e_C + \tfrac{2}{5}e_i(t)$$

$$\dot{e}_C = \tfrac{3}{10}i_L - \tfrac{3}{5}e_C + \tfrac{1}{5}e_i(t)$$

and the output equation was shown to be

$$e_o = -\tfrac{1}{5}i_L - \tfrac{3}{5}e_C + \tfrac{1}{5}e_i(t)$$

The state vector has two elements and is taken as

$$\mathbf{q} = \begin{bmatrix} i_L \\ e_C \end{bmatrix}$$

The input $\mathbf{u}$ is the voltage $e_i(t)$, and the output $\mathbf{y}$ is the voltage e_o. With these definitions, we can write the model in the form of (7) as

$$\dot{\mathbf{q}} = \underbrace{\begin{bmatrix} -\tfrac{2}{5} & -\tfrac{6}{5} \\ \tfrac{3}{10} & -\tfrac{3}{5} \end{bmatrix}}_{\mathbf{A}} \mathbf{q} + \underbrace{\begin{bmatrix} \tfrac{2}{5} \\ \tfrac{1}{5} \end{bmatrix}}_{\mathbf{B}} \mathbf{u}$$

$$\mathbf{y} = \underbrace{\begin{bmatrix} -\tfrac{1}{5} & -\tfrac{3}{5} \end{bmatrix}}_{\mathbf{C}} \mathbf{q} + \underbrace{\begin{bmatrix} \tfrac{1}{5} \end{bmatrix}}_{\mathbf{D}} \mathbf{u}$$

where the matrices **A**, **B**, **C**, and **D** are identified in the equations. Because there is only one input, **B** and **D** have only one column. Because there is only one output, **C** and **D** have only one row. Thus **B** is a column vector, **C** a row vector, and **D** a scalar. Such systems are referred to as *single-input, single-output systems* to distinguish them from the more general class of *multi-input, multi-output systems*.

EXAMPLE 17.2   Write in matrix form the linear model of the rotational mechanical system discussed in Example 4.5. Then replace the load-torque input $\tau_L(t)$ by the nonlinear function $\tau_L = \alpha|\omega_2|\omega_2$ and repeat the problem. In each case, take the output vector as the variables ω_1 and ω_2.

Solution   The state-variable equations are given by (4.34), which can be rewritten, with $\Delta\theta$ replaced by ϕ, as

$$\dot{\phi} = -\frac{K}{B}\phi + \omega_1 - \omega_2 \tag{8a}$$

$$\dot{\omega}_1 = -\frac{K}{J_1}\phi + \frac{1}{J_1}\tau_a(t) \tag{8b}$$

$$\dot{\omega}_2 = \frac{K}{J_2}\phi - \frac{1}{J_2}\tau_L(t) \tag{8c}$$

When we take the state vector as

$$\mathbf{q} = \begin{bmatrix} \phi \\ \omega_1 \\ \omega_2 \end{bmatrix}$$

the input vector as

$$\mathbf{u} = \begin{bmatrix} \tau_a(t) \\ \tau_L(t) \end{bmatrix}$$

and the output vector as

$$\mathbf{y} = \begin{bmatrix} \omega_1 \\ \omega_2 \end{bmatrix}$$

the matrix form of the system model is

$$\dot{\mathbf{q}} = \underbrace{\begin{bmatrix} -K/B & 1 & -1 \\ -K/J_1 & 0 & 0 \\ K/J_2 & 0 & 0 \end{bmatrix}}_{\mathbf{A}} \mathbf{q} + \underbrace{\begin{bmatrix} 0 & 0 \\ 1/J_1 & 0 \\ 0 & -1/J_2 \end{bmatrix}}_{\mathbf{B}}$$

$$\mathbf{y} = \underbrace{\begin{bmatrix} 0 & 1 & 0 \\ 0 & 0 & 1 \end{bmatrix}}_{\mathbf{C}} \mathbf{q} + \underbrace{\begin{bmatrix} 0 & 0 \\ 0 & 0 \end{bmatrix}}_{\mathbf{D}} \mathbf{u}$$

where the matrices $\mathbf{A}$, $\mathbf{B}$, $\mathbf{C}$, and $\mathbf{D}$ are identified in the equations. If the load torque $\tau_L(t)$, which is the second element of the input vector $\mathbf{u}$ and which appears in (8c), is changed to the nonlinear expression $\tau_L = \alpha |\omega_2| \omega_2$, the system model becomes nonlinear and these matrix equations are no longer valid. Because $\tau_L(t)$ is no longer an input, the input vector $\mathbf{u}$ becomes the scalar $u = \tau_a(t)$. Then, with $q_1 = \phi$, $q_2 = \omega_1$, and $q_3 = \omega_2$, the right-hand sides of the equations comprising (3) become

$$f_1(\mathbf{q}, \mathbf{u}, t) = -\frac{K}{B} q_1 + q_2 - q_3$$

$$f_2(\mathbf{q}, \mathbf{u}, t) = -\frac{K}{J_1} q_1 + \frac{1}{J_1} \tau_a(t)$$

$$f_3(\mathbf{q}, \mathbf{u}, t) = \frac{K}{J_2} q_1 - \frac{\alpha}{J_2} |q_3| q_3$$

and

$$g_1(\mathbf{q}, \mathbf{u}, t) = q_2$$

$$g_2(\mathbf{q}, \mathbf{u}, t) = q_3$$

17.2 SOLUTION FOR THE ZERO-INPUT RESPONSE

In the remainder of the chapter, we shall consider the response of fixed linear systems. We start with the *zero-input response* for which $\mathbf{u} = \mathbf{0}$ for all time and for which the initial state $\mathbf{q}(0)$ has at least one nonzero element. Then the matrix form of the system model given by (7) reduces to

$$\dot{\mathbf{q}} = \mathbf{A}\mathbf{q} \qquad (9a)$$

$$\mathbf{y} = \mathbf{C}\mathbf{q} \qquad (9b)$$

where $\mathbf{q}(0)$ is specified and $\mathbf{A}$ and $\mathbf{C}$ are constant. Because (9a) represents a set of simultaneous linear differential equations, we can use the Laplace transform to solve them for $\mathbf{q}$. Once we have found the state vector, we need only premultiply it by the matrix $\mathbf{C}$ to find the output vector $\mathbf{y}$.

TRANSFORM SOLUTION

The transform of a vector is defined to be that vector whose elements are the Laplace transforms of the corresponding time functions in the original vector.* For example, we denote the transform vector $\mathscr{L}[\mathbf{q}(t)]$ by $\mathbf{Q}(s)$ and define it to be

$$\mathbf{Q}(s) = \begin{bmatrix} Q_1(s) \\ Q_2(s) \\ \vdots \\ Q_n(s) \end{bmatrix}$$

where $Q_i(s) = \mathscr{L}[q_i(t)]$ for $i = 1, 2, ..., n$.

To transform the state-variable equation (9a), we first note that since the matrix $\mathbf{A}$ is constant, the transform of the right side is just $\mathbf{AQ}(s)$. As for the left side of the equation, $\mathscr{L}[\dot{\mathbf{q}}]$ is a vector whose elements are the transforms of the corresponding state-variable derivatives. Recalling from Chapter 12 that $\mathscr{L}[\dot{q}_i] = sQ_i(s) - q_i(0)$, we can write

$$\mathscr{L}[\dot{\mathbf{q}}] = s\mathbf{Q}(s) - \mathbf{q}(0)$$

where $\mathbf{q}(0)$ is the vector of initial conditions. Thus, when transformed, (9a) becomes

$$s\mathbf{Q}(s) - \mathbf{q}(0) = \mathbf{AQ}(s)$$

which can be rearranged to give

$$s\mathbf{Q}(s) - \mathbf{AQ}(s) = \mathbf{q}(0) \tag{10}$$

To combine the two terms on the left side of (10), we note that $s\mathbf{Q}(s)$ can be written as $s\mathbf{IQ}(s)$, where $\mathbf{I}$ is the identity matrix of order n. Hence, we can rewrite (10) as

$$(s\mathbf{I} - \mathbf{A})\mathbf{Q}(s) = \mathbf{q}(0) \tag{11}$$

where

$$s\mathbf{I} - \mathbf{A} = \begin{bmatrix} s - a_{11} & -a_{12} & \cdots & -a_{1n} \\ -a_{21} & s - a_{22} & \cdots & -a_{2n} \\ \vdots & \vdots & & \vdots \\ -a_{n1} & -a_{n2} & \cdots & s - a_{nn} \end{bmatrix}$$

We can find the transform of the state vector, $\mathbf{Q}(s)$, from (11) by premultiplying both of its sides by the inverse of the matrix $s\mathbf{I} - \mathbf{A}$, provided

---

* Capital letters followed by (s) are used for Laplace transforms, even when the time function is a vector.

that this inverse exists, i.e., provided that the matrix $sI - A$ is not singular. Since a square matrix is nonsingular if its determinant is nonzero, we can solve for $Q(s)$ provided that $|sI - A| \neq 0$. Hence, the inverse of the matrix $sI - A$ will exist for all points in the complex s-plane except those points that coincide with the characteristic values (eigenvalues) of the matrix A. This restriction is of no consequence in our solution for q because, as we shall see, these points (there are at most n of them) are the poles of the transform vector $Q(s)$. Since a Laplace transform is infinite when s equals one of its poles, we should not expect to obtain finite values for $Q(s)$ at its poles by matrix inversion or any other method.

Except for values of s that coincide with a characteristic value of A, we can write

$$Q(s) = (sI - A)^{-1} q(0)$$

where $(sI - A)^{-1}$ is an $n \times n$ matrix known as the *resolvent matrix*, each element of which is a function of s. Denoting the resolvent matrix by the symbol $\Phi(s)$, we have

$$Q(s) = \Phi(s) q(0) \tag{12}$$

where

$$\Phi(s) = (sI - A)^{-1} \tag{13}$$

All that remains is to take the inverse Laplace transform of $Q(s)$, which gives

$$q(t) = \mathscr{L}^{-1}\{Q(s)\}$$
$$= \mathscr{L}^{-1}\{\Phi(s) q(0)\}$$

Since the vector $q(0)$ does not involve the variable s, we may calculate the inverse transform of $\Phi(s)$ first and then multiply it by $q(0)$, obtaining

$$q(t) = \mathscr{L}^{-1}\{\Phi(s)\} q(0)$$

When we denote the inverse transform of $\Phi(s)$ by $\phi(t)$, which will be an $n \times n$ matrix, the desired expression for the zero-input response is

$$q(t) = \phi(t) q(0) \tag{14}$$

where

$$\phi(t) = \mathscr{L}^{-1}\{\Phi(s)\}$$

THE STATE-TRANSITION MATRIX

The matrix $\phi(t)$ is known as the *state-transition matrix* and has a number of useful properties, some of which we shall discuss. Since the inverse of a matrix can be written as its adjoint matrix divided by its determinant (see Appendix C), we can write the resolvent matrix from (13) as

$$\Phi(s) = \frac{1}{|s\mathbf{I} - \mathbf{A}|} \operatorname{adj}[s\mathbf{I} - \mathbf{A}] \tag{15}$$

where $\operatorname{adj}[s\mathbf{I} - \mathbf{A}]$ denotes the adjoint matrix of $s\mathbf{I} - \mathbf{A}$. Thus the state-transition matrix $\phi(t)$ can be written as

$$\phi(t) = \mathscr{L}^{-1}\left[\frac{1}{|s\mathbf{I} - \mathbf{A}|} \operatorname{adj}[s\mathbf{I} - \mathbf{A}]\right] \tag{16}$$

Evaluating $(s\mathbf{I} - \mathbf{A})^{-1}$ is difficult computationally because the matrix to be inverted is a function of s. Methods exist that are better suited for this task than using the adjoint matrix (see, for example, Melsa and Jones). However, our principal objective is to develop certain theoretical properties and to show the form of $\phi(t)$. For these tasks and for computational problems with matrices of order three or less, the adjoint method is acceptable.

As summarized in Appendix C, the adjoint of an $n \times n$ matrix is itself an $n \times n$ matrix that is the transposed matrix of cofactors, where the cofactor of an element is a determinant of order $n - 1$. It follows that $\operatorname{adj}[s\mathbf{I} - \mathbf{A}]$ will be an $n \times n$ matrix whose elements are polynomials in s of degree $n - 1$ or less. Furthermore, the determinant of $s\mathbf{I} - \mathbf{A}$ is

$$|s\mathbf{I} - \mathbf{A}| = \begin{vmatrix} s - a_{11} & -a_{12} & \cdots & -a_{1n} \\ -a_{21} & s - a_{22} & \cdots & -a_{2n} \\ \vdots & \vdots & & \vdots \\ -a_{n1} & -a_{n2} & \cdots & s - a_{nn} \end{vmatrix}$$

which will always reduce to a polynomial in s of degree n. This polynomial is the *characteristic polynomial* of the matrix $\mathbf{A}$. Hence, the matrix $\Phi(s)$ defined by (15) is an $n \times n$ matrix whose elements are rational functions of s having numerators of lower degree than that of the denominator. Specifically, a general element of $\Phi(s)$ will be of the form

$$\frac{\beta_{n-1} s^{n-1} + \cdots + \beta_0}{s^n + \alpha_{n-1} s^{n-1} + \cdots + \alpha_0}$$

where the numerator coefficients $\beta_{n-1}, ..., \beta_0$ depend on the specific row and column of the element. The denominator, however, which comes from the evaluation of $|s\mathbf{I} - \mathbf{A}|$, will be the characteristic polynomial and thus will be the same for each element of $\mathbf{\Phi}(s)$.

In principle, at least, the characteristic polynomial can be factored into the form

$$|s\mathbf{I} - \mathbf{A}| = s^n + \alpha_{n-1} s^{n-1} + \cdots + \alpha_0$$

$$= (s - s_1)(s - s_2) \cdots (s - s_n) \tag{17}$$

where the quantities $s_1, s_2, ..., s_n$ are known as the *characteristic values* (*eigenvalues*) of the matrix $\mathbf{A}$ (see Appendix C). Having factored the characteristic polynomial, we can evaluate the inverse transform of $\mathbf{\Phi}(s)$ by performing a partial-fraction expansion of each of its elements. If the characteristic values of $\mathbf{A}$ are distinct, the ijth term of the resolvent matrix will have as its expansion

$$\Phi_{ij}(s) = \frac{\gamma_{i1}}{s - s_1} + \frac{\gamma_{i2}}{s - s_2} + \cdots + \frac{\gamma_{in}}{s - s_n}$$

and the corresponding term in the state-transition matrix will be

$$\phi_{ij}(t) = \gamma_{i1} \epsilon^{s_1 t} + \gamma_{i2} \epsilon^{s_2 t} + \cdots + \gamma_{in} \epsilon^{s_n t} \tag{18}$$

Thus the state-transition matrix $\mathbf{\phi}(t)$ is an $n \times n$ matrix each of whose elements is a linear combination of the n mode functions $\epsilon^{s_1 t}, ..., \epsilon^{s_n t}$, where the s_i are the characteristic values of the matrix $\mathbf{A}$.

EXAMPLE 17.3   Evaluate the state-transition matrix for a system that obeys the state-variable equation $\dot{\mathbf{q}} = \mathbf{A}\mathbf{q}$ where

$$\mathbf{A} = \begin{bmatrix} 0 & 1 \\ -6 & -5 \end{bmatrix}$$

and identify the corresponding mode functions.

Solution   First we write the matrix $s\mathbf{I} - \mathbf{A}$ in numerical form and evaluate its determinant and inverse. Thus

$$s\mathbf{I} - \mathbf{A} = \begin{bmatrix} s & -1 \\ 6 & s+5 \end{bmatrix}$$

and

$$|s\mathbf{I} - \mathbf{A}| = s^2 + 5s + 6 = (s+2)(s+3)$$

Before taking the inverse of $s\mathbf{I} - \mathbf{A}$, we note that the characteristic polynomial associated with $\mathbf{A}$ is $P(s) = (s+2)(s+3)$. Hence, the characteristic values, which must satisfy $P(s) = 0$, are $s_1 = -2$ and $s_2 = -3$. This means that the mode functions of which the zero-input response is composed are ϵ^{-2t} and ϵ^{-3t}.

The next step is to find the resolvent matrix $\mathbf{\Phi}(s) = (s\mathbf{I} - \mathbf{A})^{-1}$. Following the steps outlined here and illustrated in Appendix C for taking the inverse of a matrix, we find the adjoint of $s\mathbf{I} - \mathbf{A}$ to be

$$\text{adj}\,[s\mathbf{I} - \mathbf{A}] = \begin{bmatrix} s+5 & 1 \\ -6 & s \end{bmatrix}$$

Dividing each element of the adjoint matrix by $|s\mathbf{I} - \mathbf{A}|$, we find that

$$\mathbf{\Phi}(s) = \begin{bmatrix} \dfrac{s+5}{(s+2)(s+3)} & \dfrac{1}{(s+2)(s+3)} \\ \dfrac{-6}{(s+2)(s+3)} & \dfrac{s}{(s+2)(s+3)} \end{bmatrix} \tag{19}$$

Carrying out a partial-fraction expansion of each of the four elements on the right side of (19), we get

$$\mathbf{\Phi}(s) = \begin{bmatrix} \dfrac{3}{s+2} - \dfrac{2}{s+3} & \dfrac{1}{s+2} - \dfrac{1}{s+3} \\ -\dfrac{6}{s+2} + \dfrac{6}{s+3} & -\dfrac{2}{s+2} + \dfrac{3}{s+3} \end{bmatrix}$$

By taking the inverse transform of each element of $\mathbf{\Phi}(s)$, we find the state-transition matrix to be

$$\phi(t) = \begin{bmatrix} 3\epsilon^{-2t} - 2\epsilon^{-3t} & \epsilon^{-2t} - \epsilon^{-3t} \\ -6\epsilon^{-2t} + 6\epsilon^{-3t} & -2\epsilon^{-2t} + 3\epsilon^{-3t} \end{bmatrix} \tag{20}$$

THE STATE AND OUTPUT RESPONSES

Having examined the state-transition matrix in some detail, we return to (14) to determine the general form of the state vector $\mathbf{q}$. Since $\mathbf{q}$ is the product of an $n \times n$ matrix of time functions, $\boldsymbol{\phi}(t)$, and an $n \times 1$ vector of constants, $\mathbf{q}(0)$, it follows that the elements of $\mathbf{q}$ are given by

$$q_i = \phi_{i1}(t)\, q_1(0) + \phi_{i2}(t)\, q_2(0) + \cdots + \phi_{in}(t)\, q_n(0) \qquad i = 1, 2, \ldots, n$$

From this expression, we see that q_i is a linear combination of the n functions of time $\phi_{ij}(t)$, $j = 1, 2, \ldots, n$, where the weightings of these functions are the corresponding initial values of the state variables, $q_1(0), q_2(0), \ldots, q_n(0)$. Furthermore, we have previously shown that the functions $\phi_{ij}(t)$ are themselves linear combinations of the n mode functions $\epsilon^{s_1 t}, \ldots, \epsilon^{s_n t}$. Hence, it follows that each of the state variables $q_i(t)$ is merely a linear combination of the system's mode functions, with the relative weightings dependent on the initial state vector $\mathbf{q}(0)$ and the partial-fraction coefficients $\gamma_{i1}, \gamma_{i2}, \ldots, \gamma_{in}$ in (18).

Once we have found the state vector $\mathbf{q}$, it is a straightforward matter to determine the response of the output variables. Since $\mathbf{y} = \mathbf{Cq}$ when the input is zero, the kth output is a linear combination of all the state variables, the weightings being given by the elements in the kth row of the $\mathbf{C}$ matrix. Specifically,

$$y_k = c_{k1}\, q_1 + c_{k2}\, q_2 + \cdots + c_{kn}\, q_n \qquad k = 1, 2, \ldots, p$$

Before discussing some of the properties of the state-transition matrix, we shall illustrate its use in evaluating both the state and output vectors for a second-order example in numerical form.

EXAMPLE 17.4 Determine the zero-input responses of the state and output vectors for the system described by the matrix equations

$$\dot{\mathbf{q}} = \begin{bmatrix} 0 & 1 \\ -6 & -5 \end{bmatrix} \mathbf{q} \qquad (21a)$$

$$\mathbf{y} = \begin{bmatrix} 2 & -1 \\ 0 & 1 \end{bmatrix} \mathbf{q} \qquad (21b)$$

Give the responses in terms of the elements of the initial state vector $\mathbf{q}(0)$.

Solution From inspection of (21), we see that the matrices $\mathbf{A}$ and $\mathbf{C}$ characterizing the system are

$$\mathbf{A} = \begin{bmatrix} 0 & 1 \\ -6 & -5 \end{bmatrix}$$

and

$$C = \begin{bmatrix} 2 & -1 \\ 0 & 1 \end{bmatrix}$$

The state-transition matrix corresponding to $\mathbf{A}$ was evaluated in Example 17.3 and is given by (20). Hence, to find $\mathbf{q}$ in terms of the initial states $q_1(0)$ and $q_2(0)$, we need only substitute (20) into (14) and carry out the matrix multiplication. For the first state variable, we find

$$q_1 = (3\epsilon^{-2t} - 2\epsilon^{-3t}) q_1(0) + (\epsilon^{-2t} - \epsilon^{-3t}) q_2(0)$$

which can be rewritten in terms of the mode functions ϵ^{-2t} and ϵ^{-3t} as

$$q_1 = [3q_1(0) + q_2(0)] \epsilon^{-2t} + [-2q_1(0) - q_2(0)] \epsilon^{-3t}$$

Similarly, you may verify that the response of the second state variable is

$$q_2 = [-6q_1(0) - 2q_2(0)] \epsilon^{-2t} + [6q_1(0) + 3q_2(0)] \epsilon^{-3t}$$

Finally we can obtain the response of the output variable $\mathbf{y}$ by substituting these expressions for q_1 and q_2 into the output equation given by (21b). For y_1, we get

$$y_1 = 2q_1 - q_2$$

which simplifies to

$$y_1 = [12q_1(0) + 4q_2(0)] \epsilon^{-2t} - [10q_1(0) + 5q_2(0)] \epsilon^{-3t} \tag{22}$$

The second element of the output vector is just q_2, so

$$y_2 = [-6q_1(0) - 2q_2(0)] \epsilon^{-2t} + [6q_1(0) + 3q_2(0)] \epsilon^{-3t} \tag{23}$$

PROPERTIES OF THE STATE-TRANSITION MATRIX

Starting with (14), which states that

$$\mathbf{q}(t) = \mathbf{\phi}(t)\mathbf{q}(0) \tag{24}$$

we can develop several properties of $\mathbf{\phi}(t)$ that are useful in the analysis of dynamic systems. First, by setting t equal to zero in (24), we have

$$\mathbf{q}(0) = \mathbf{\phi}(0)\mathbf{q}(0)$$

which implies that

$$\phi(0) = I \tag{25}$$

Thus the state-transition matrix reduces to the $n \times n$ identity matrix when its argument equals zero.

The initial time in (24) need not be restricted to $t = 0$. More generally, we can rewrite (24) in terms of the initial time t_0 as

$$\mathbf{q}(t) = \phi(t - t_0)\mathbf{q}(t_0) \tag{26}$$

which reduces to (24) if $t_0 = 0$. Using (26), we can express the state vector at some time t_2 in terms of $\mathbf{q}(t_0)$ as

$$\mathbf{q}(t_2) = \phi(t_2 - t_0)\mathbf{q}(t_0)$$

or in terms of $\mathbf{q}(t_1)$ as

$$\mathbf{q}(t_2) = \phi(t_2 - t_1)\mathbf{q}(t_1)$$

By using (26) again to express $\mathbf{q}(t_1)$ in terms of $\mathbf{q}(t_0)$, we can write the second expression for $\mathbf{q}(t_2)$ as

$$\mathbf{q}(t_2) = \phi(t_2 - t_1)\phi(t_1 - t_0)\mathbf{q}(t_0)$$

Now, comparing the two expressions for $\mathbf{q}(t_2)$, we see that

$$\phi(t_2 - t_1)\phi(t_1 - t_0) = \phi(t_2 - t_0) \tag{27}$$

Hence, the product of two state-transition matrices (for the same $\mathbf{A}$ matrix, of course) is merely the state-transition matrix with its argument set equal to the sum of the arguments of the two matrices being multiplied.

If we write (27) with $t_0 = t_2 = 0$, we see that

$$\phi(-t_1)\phi(t_1) = \phi(0)$$

But we know from (25) that $\phi(0)$ is the identity matrix, so

$$\phi(-t_1)\phi(t_1) = I$$

which implies that

$$[\phi(t_1)]^{-1} = \phi(-t_1) \tag{28}$$

Thus the state-transition matrix is nonsingular for all values of its argument, and we can obtain its inverse by merely changing the sign of its argument.

One way of gaining some insight into the physical significance of the state-transition matrix is to consider (24) for some very specific values of $\mathbf{q}(0)$. For instance, suppose $q_1(0) = 1$ and all the other elements of $\mathbf{q}(0)$ are zero. It follows from the definition of matrix multiplication that the resulting response for the state vector $\mathbf{q}$ will be the first column of $\boldsymbol{\phi}(t)$. In similar fashion, setting the ith initial state to unity and the rest to zero will result in a response for $\mathbf{q}$ that consists of the ith column of $\boldsymbol{\phi}(t)$. Hence, we could use a computer simulation of a system repetitively to compute and plot each element of the state-transition matrix, one column at a time, until we had determined all n^2 time functions.

EXAMPLE 17.5   For the state-transition matrix found in Example 17.3 and given by (20), verify that (25) and (28) are true in general and show that (27) holds for $t_0 = 0$, $t_1 = 1$, and $t_2 = 2$.

Solution   Setting t equal to zero in (20) gives

$$\boldsymbol{\phi}(0) = \begin{bmatrix} 3\epsilon^0 - 2\epsilon^0 & \epsilon^0 - \epsilon^0 \\ -6\epsilon^0 + 6\epsilon^0 & -2\epsilon^0 + 3\epsilon^0 \end{bmatrix} = \begin{bmatrix} 1 & 0 \\ 0 & 1 \end{bmatrix}$$

which agrees with (25).

To check (28), we shall show that $\boldsymbol{\phi}(-t)\boldsymbol{\phi}(t) = \mathbf{I}$ where $\mathbf{I}$ is the 2×2 identity matrix. From (20), we have

$$\boldsymbol{\phi}(t) = \begin{bmatrix} 3\epsilon^{-2t} - 2\epsilon^{-3t} & \epsilon^{-2t} - \epsilon^{-3t} \\ -6\epsilon^{-2t} + 6\epsilon^{-3t} & -2\epsilon^{-2t} + 3\epsilon^{-3t} \end{bmatrix}$$

so

$$\boldsymbol{\phi}(-t) = \begin{bmatrix} 3\epsilon^{2t} - 2\epsilon^{3t} & \epsilon^{2t} - \epsilon^{3t} \\ -6\epsilon^{2t} + 6\epsilon^{3t} & -2\epsilon^{2t} + 3\epsilon^{3t} \end{bmatrix}$$

For example, the 2,2 element* of $\boldsymbol{\phi}(-t)\boldsymbol{\phi}(t)$ is the product of the second row of $\boldsymbol{\phi}(-t)$ and the second column of $\boldsymbol{\phi}(t)$, namely

$$\begin{bmatrix} -6\epsilon^{2t} + 6\epsilon^{3t} & -2\epsilon^{2t} + 3\epsilon^{3t} \end{bmatrix} \begin{bmatrix} \epsilon^{-2t} - \epsilon^{-3t} \\ -2\epsilon^{-2t} + 3\epsilon^{-3t} \end{bmatrix}$$

$$= (-6\epsilon^{2t} + 6\epsilon^{3t})(\epsilon^{-2t} - \epsilon^{-3t}) + (-2\epsilon^{2t} + 3\epsilon^{3t})(-2\epsilon^{-2t} + 3\epsilon^{-3t})$$

$$= 1$$

---

* The i,j element of a matrix is the one in the ith row and the jth column.

In similar fashion, you can verify that the 1,1 element is also unity and that the 1,2 and 2,1 elements are zero, giving the 2×2 identity matrix.

To verify (27) for the specified values of t_0, t_1, and t_2, we must show that

$$\phi(2-1)\phi(1-0) = \phi(2-0)$$

which is equivalent to

$$\phi(1)\phi(1) = \phi(2) \tag{29}$$

For example, we can write the 1,1 element of the left side of (29) as the matrix product

$$\begin{aligned}
&[\ 3\epsilon^{-2} - 2\epsilon^{-3} \quad \epsilon^{-2} - \epsilon^{-3}\ ]\begin{bmatrix} 3\epsilon^{-2} - 2\epsilon^{-3} \\ -6\epsilon^{-2} + 6\epsilon^{-3} \end{bmatrix} \\
&= (3\epsilon^{-2} - 2\epsilon^{-3})^2 + (\epsilon^{-2} - \epsilon^{-3})(-6\epsilon^{-2} + 6\epsilon^{-3}) \\
&= 3\epsilon^{-4} - 2\epsilon^{-6}
\end{aligned}$$

which is indeed equal to the 1,1 element of $\phi(2)$. By evaluating the remaining three elements comprising the left and right sides of (29), you can show that (27) does hold for the values of t_0, t_1, and t_2 selected. It would be possible to verify (27) for arbitrary t_0, t_1, and t_2, but the process would be more tedious than it is when specific values are selected.

17.3 SOLUTION FOR THE ZERO-STATE RESPONSE

Having analyzed the zero-input response of a fixed linear system described in state-variable form, we now consider the *zero-state response* for which $q(0) = 0$ but the input u is not zero. The system model is given by (7), which is repeated here, namely

$$\dot{q} = Aq + Bu \tag{30a}$$

$$y = Cq + Du \tag{30b}$$

If (30a) is transformed with $q(0) = 0$, we get

$$sQ(s) = AQ(s) + BU(s)$$

where $U(s) = \mathcal{L}[\mathbf{u}(t)]$. As in the previous section, we solve the transformed equation for $\mathbf{Q}(s)$, obtaining

$$\mathbf{Q}(s) = (s\mathbf{I} - \mathbf{A})^{-1}\mathbf{B}\mathbf{U}(s)$$

$$= \mathbf{\Phi}(s)\mathbf{B}\mathbf{U}(s) \tag{31}$$

where $(s\mathbf{I} - \mathbf{A})^{-1}$ is the resolvent matrix $\mathbf{\Phi}(s)$. Hence, we can write the zero-state response of $\mathbf{q}$ as

$$\mathbf{q}(t) = \mathcal{L}^{-1}[\mathbf{\Phi}(s)\mathbf{B}\mathbf{U}(s)]$$

As we shall see, the zero-state response consists of a combination of the free response and the forced response. Rather than proceed further with the response of the state vector $\mathbf{q}$, we shall obtain $\mathbf{Y}(s)$, the transform of the output vector. Transforming the output equation (30b) and substituting (31) give

$$\mathbf{Y}(s) = \mathbf{C}\mathbf{\Phi}(s)\mathbf{B}\mathbf{U}(s) + \mathbf{D}\mathbf{U}(s)$$

$$= [\mathbf{C}\mathbf{\Phi}(s)\mathbf{B} + \mathbf{D}]\mathbf{U}(s) \tag{32}$$

Recall that in Chapter 13 the system transfer function $H(s)$ was defined as the ratio $Y(s)/U(s)$ where $Y(s)$ is the transform of the zero-state response. As a consequence of this definition, we could write $Y(s) = H(s)U(s)$. When dealing with matrices as we are here, we can not divide $Y(s)$ by $U(s)$. However, we can define the *transfer-function matrix* $\mathbf{H}(s)$ to be the quantity in brackets on the right side of (32), which when multiplied by $\mathbf{U}(s)$ gives $\mathbf{Y}(s)$.

Thus the transfer-function matrix is

$$\mathbf{H}(s) = \mathbf{C}\mathbf{\Phi}(s)\mathbf{B} + \mathbf{D} \tag{33}$$

and the Laplace transform of the zero-state output vector is

$$\mathbf{Y}(s) = \mathbf{H}(s)\mathbf{U}(s) \tag{34}$$

Often the output $\mathbf{y}$ is a function of only the state vector $\mathbf{q}$, in which case $\mathbf{D} = 0$ and (33) simplifies to

$$\mathbf{H}(s) = \mathbf{C}\mathbf{\Phi}(s)\mathbf{B} \tag{35}$$

We now define the *impulse-response matrix* $\mathbf{h}(t)$ as the inverse Laplace transform of the transfer-function matrix $\mathbf{H}(s)$. Thus

$$\mathbf{h}(t) = \mathcal{L}^{-1}[\mathbf{H}(s)] \tag{36}$$

is a $p \times m$ matrix of time functions such that $h_{ij}(t)$ is the response of y_i when u_j is a unit impulse, when all other inputs are zero, and when the initial value of the state vector is $\mathbf{q}(0) = \mathbf{0}$.

EXAMPLE 17.6 Evaluate the transfer-function matrix $\mathbf{H}(s)$ for the system described by the matrix equations

$$
\dot{\mathbf{q}} = \begin{bmatrix} 0 & 1 \\ -6 & -5 \end{bmatrix} \mathbf{q} + \begin{bmatrix} 2 & 1 \\ -1 & 0 \end{bmatrix} \mathbf{u}
$$
$$
\mathbf{y} = \begin{bmatrix} 2 & -1 \\ 0 & 1 \end{bmatrix} \mathbf{q}
$$
(37)

Then use $\mathbf{H}(s)$ to find the zero-state response to the input vector

$$
\mathbf{u} = \begin{bmatrix} \epsilon^{-t}U(t) \\ U(t) \end{bmatrix}
$$
(38)

Solution The system described by (37) has the same matrix $\mathbf{A}$ for which we found $\mathbf{\Phi}(s)$ in Example 17.3. Using (19) for $\mathbf{\Phi}(s)$, identifying $\mathbf{B}$ and $\mathbf{C}$ from (37), and noting that $\mathbf{D} = 0$, we can write (35) as

$$
\mathbf{H}(s) = \begin{bmatrix} 2 & -1 \\ 0 & 1 \end{bmatrix} \begin{bmatrix} \dfrac{s+5}{(s+2)(s+3)} & \dfrac{1}{(s+2)(s+3)} \\ \dfrac{-6}{(s+2)(s+3)} & \dfrac{s}{(s+2)(s+3)} \end{bmatrix} \begin{bmatrix} 2 & 1 \\ -1 & 0 \end{bmatrix}
$$

$$
= \begin{bmatrix} 2 & -1 \\ 0 & 1 \end{bmatrix} \begin{bmatrix} \dfrac{2s+9}{(s+2)(s+3)} & \dfrac{s+5}{(s+2)(s+3)} \\ \dfrac{-s-12}{(s+2)(s+3)} & \dfrac{-6}{(s+2)(s+3)} \end{bmatrix}
$$

$$
= \begin{bmatrix} \dfrac{5s+30}{(s+2)(s+3)} & \dfrac{2s+16}{(s+2)(s+3)} \\ \dfrac{-s-12}{(s+2)(s+3)} & \dfrac{-6}{(s+2)(s+3)} \end{bmatrix}
$$
(39)

As indicated by (39), $\mathbf{H}(s)$ is a 2×2 matrix, each element of which is a transfer function having the same two poles (at $s_1 = -2$ and $s_2 = -3$) and zeros that depend on its position in the matrix.

To find $\mathbf{Y}(s)$, we transform the input vector defined by (38), getting

$$
\mathbf{U}(s) = \begin{bmatrix} \dfrac{1}{s+1} \\[2ex] \dfrac{1}{s} \end{bmatrix}
$$

and then use (34) and (39) to write

$$
\mathbf{Y}(s) = \begin{bmatrix} \dfrac{5s+30}{(s+2)(s+3)} & \dfrac{2s+16}{(s+2)(s+3)} \\[2ex] \dfrac{-s-12}{(s+2)(s+3)} & \dfrac{-6}{(s+2)(s+3)} \end{bmatrix} \begin{bmatrix} \dfrac{1}{s+1} \\[2ex] \dfrac{1}{s} \end{bmatrix}
$$

$$
= \begin{bmatrix} \dfrac{7s^2 + 48s + 16}{s(s+1)(s+2)(s+3)} \\[2ex] \dfrac{-(s^2 + 18s + 6)}{s(s+1)(s+2)(s+3)} \end{bmatrix}
$$

We can now evaluate the zero-state responses of the two outputs y_1 and y_2 by performing a partial-fraction expansion of each element of $\mathbf{Y}(s)$. You can verify that the expansions of the two elements of $\mathbf{Y}(s)$ are

$$
Y_1(s) = \frac{8/3}{s} + \frac{25/2}{s+1} - \frac{26}{s+2} + \frac{65/6}{s+3}
$$

$$
Y_2(s) = -\frac{1}{s} - \frac{11/2}{s+1} + \frac{13}{s+2} - \frac{13/2}{s+3}
$$

Thus the zero-state response of the output vector is

$$
\mathbf{y} = \begin{bmatrix} \frac{8}{3} + \frac{25}{2}\epsilon^{-t} - 26\epsilon^{-2t} + \frac{65}{6}\epsilon^{-3t} \\[2ex] -1 - \frac{11}{2}\epsilon^{-t} + 13\epsilon^{-2t} - \frac{13}{2}\epsilon^{-3t} \end{bmatrix} \tag{40}
$$

which holds for $t > 0$. Actually, (40) is valid for $t \geq 0$, since $\mathbf{y}(0) = \mathbf{y}(0+) = \mathbf{0}$.

17.4 SOLUTION FOR THE COMPLETE RESPONSE

By the superposition property, the response of a linear system to a nonzero initial state vector $\mathbf{q}(0)$ and a nonzero input vector $\mathbf{u}$ is the sum of the zero-input and zero-state responses. It is known as the *complete response*. As an alternative to finding the zero-input and zero-state responses as functions of time and adding them, we can add their respective Laplace transforms to give the transform of the complete response. We shall develop this approach for the case when $\mathbf{D} = 0$ and then apply the result to the system considered in Example 17.6.

The transform of the zero-input state-variable response is given by (12) as

$$\mathbf{Q}(s) = \boldsymbol{\Phi}(s)\mathbf{q}(0)$$

Since with $\mathbf{D} = 0$, the output vector is

$$\mathbf{y} = \mathbf{Cq}$$

the transform of the zero-input output response, denoted here by $\mathbf{Y}_{zi}(s)$, is

$$\mathbf{Y}_{zi}(s) = \mathbf{C}\boldsymbol{\Phi}(s)\mathbf{q}(0) \tag{41}$$

For the transform of the zero-state output response, denoted here by $\mathbf{Y}_{zs}(s)$, we can use (34) and (35) to write

$$\mathbf{Y}_{zs}(s) = \mathbf{C}\boldsymbol{\Phi}(s)\mathbf{B}\mathbf{U}(s) \tag{42}$$

Because of the superposition property for linear systems and the super-position theorem for Laplace transforms, the transform of the complete response is

$$\begin{aligned}
\mathbf{Y}(s) &= \mathbf{Y}_{zi}(s) + \mathbf{Y}_{zs}(s) \\
&= \mathbf{C}\boldsymbol{\Phi}(s)\mathbf{q}(0) + \mathbf{C}\boldsymbol{\Phi}(s)\mathbf{B}\mathbf{U}(s) \\
&= \mathbf{C}\boldsymbol{\Phi}(s)[\mathbf{q}(0) + \mathbf{B}\mathbf{U}(s)]
\end{aligned} \tag{43}$$

Having found $\mathbf{Y}(s)$ in numerical form by using (43), we can take its inverse Laplace transform element by element.

EXAMPLE 17.7   Evaluate $\mathbf{y}$ for the system defined in Example 17.6 for the initial state vector

$$\mathbf{q}(0) = \begin{bmatrix} -1 \\ 1 \end{bmatrix}$$

and the input vector

$$\mathbf{u} = \left[\begin{array}{c} \epsilon^{-t}U(t) \\ U(t) \end{array} \right]$$

Solution   Although the zero-input response of this system was found in general terms in Example 17.4 and the zero-state response for the specified input was found in Example 17.6, we shall use (43) to find $\mathbf{Y}(s)$ and take its inverse transform. You can use the functions of time found in Example 17.4 and Example 17.6 to verify that the result obtained here by transforms is correct.

Using (19) for $\mathbf{\Phi}(s)$ and transforming $\mathbf{u}$, we can substitute the appropriate matrices into (43) to write

$$\mathbf{Y}(s) = \left[\begin{array}{cc} 2 & -1 \\ 0 & 1 \end{array} \right] \left[\begin{array}{cc} \dfrac{s+5}{(s+2)(s+3)} & \dfrac{1}{(s+2)(s+3)} \\[3mm] \dfrac{-6}{(s+2)(s+3)} & \dfrac{s}{(s+2)(s+3)} \end{array} \right]$$

$$\times \left\{ \left[\begin{array}{c} -1 \\ 1 \end{array} \right] + \left[\begin{array}{cc} 2 & 1 \\ -1 & 0 \end{array} \right] \left[\begin{array}{c} \dfrac{1}{s+1} \\[3mm] \dfrac{1}{s} \end{array} \right] \right\}$$

$$= \left[\begin{array}{cc} \dfrac{2s+16}{(s+2)(s+3)} & \dfrac{-s+2}{(s+2)(s+3)} \\[3mm] \dfrac{-6}{(s+2)(s+3)} & \dfrac{s}{(s+2)(s+3)} \end{array} \right] \left[\begin{array}{c} \dfrac{-s^2+2s+1}{s(s+1)} \\[3mm] \dfrac{s^2}{s(s+1)} \end{array} \right]$$

$$= \left[\begin{array}{c} \dfrac{-3s^3 - 10s^2 + 34s + 16}{s(s+1)(s+2)(s+3)} \\[3mm] \dfrac{s^3 + 6s^2 - 12s - 6}{s(s+1)(s+2)(s+3)} \end{array} \right] \tag{44}$$

Using partial-fraction expansions of the two elements of $\mathbf{Y}(s)$ leads to the complete response $\mathbf{y}$, namely

$$\mathbf{y} = \left[\begin{array}{c} \frac{8}{3} + \frac{25}{2}\epsilon^{-t} - 34\epsilon^{-2t} + \frac{95}{6}\epsilon^{-3t} \\[2mm] -1 - \frac{11}{2}\epsilon^{-t} + 17\epsilon^{-2t} - \frac{19}{2}\epsilon^{-3t} \end{array} \right] \tag{45}$$

This result is valid for $t > 0$. If we set $t = 0$ in (45), we get -3 and 1 for the two elements of $\mathbf{y}$, which is also the result of the matrix product $\mathbf{Cq}(0)$. The fact that these results are the same indicates that, in this example, $\mathbf{y}$ is continuous at $t = 0$.

In closing, it is worthwhile to comment on the computational task of evaluating the response of a multi-input, multi-output sytem. The second-order model used in the examples in this chapter is simple enough to permit us to evaluate the response in analytic form by hand. To do the same operations by hand for a third- or higher-order system would probably not be feasible. In such cases, one must usually resort to the digital computer. A direct solution can apply the methods described here or other algorithms that are better suited for computation. Alternatively, we can obtain a numerical simulation of the system's differential equation by using the methods presented in Chapter 7.

PROBLEMS

17.1 Consider the translational mechanical system shown in Figure 3.2(a) and discussed in Example 3.2. Let the output be the tensile force in the spring K_2.

    a. Taking the state variables as x_1, v_1, x_2 and v_2, write the state-variable and output equations in matrix form. Identify the matrices $\mathbf{A}, \mathbf{B}, \mathbf{C}$, and $\mathbf{D}$.

    b. Repeat part a using the state variables x_1, v_1, Δx, and Δv where $\Delta x = x_2 - x_1$ and $\Delta v = v_2 - v_1$.

    c. Show why the model incorporating the nonlinear elements cannot be written in the form of (4).

17.2   a. For the rotational mechanical system discussed in Example 4.3, write the state-variable equations in matrix form. Show why we cannot obtain either θ_1 or θ_2 from this model.

    b. Add θ_1 as the fourth state variable and write the state-variable equations in matrix form.

17.3 Consider the combined translational and rotational mechanical system shown in Figure 4.23 and modeled in Example 4.9.

    a. Write (4.49) in matrix form. Note that the system has two inputs, namely the gravitational force Mg and the applied force $f_a(t)$.

    b. Write (4.52) in matrix form. Write a matrix output equation for the displacements θ and x defined in (4.51).

17.4 Write in matrix form the model for the electrical circuit shown in Figure 9.28 and discussed in Example 9.9. Also write the matrix output equation for the current through the resistor R_1, taking the positive sense to the left in Figure 9.28.

17.5 Consider the motor shown in Figure 11.14 and discussed in Example 11.2. Taking i_A and ω as the state variables, write the state-variable equations in matrix form. Also write a matrix output equation for the output vector $\mathbf{y} = [\tau_e \ e_m]^T$. Identify the matrices $\mathbf{A}$, $\mathbf{B}$, $\mathbf{C}$, and $\mathbf{D}$.

17.6 The temperature of a uniform bar is analyzed in Example 14.5 and Example 14.6, and extensions of the analysis are proposed in Problems 14.13 through 14.15. For each of the following lumped models, write the state-variable equations in matrix form and write an output equation for the average temperature of the bar.

    a. The two-capacitance model with uniform segment lengths analyzed in Example 14.6.

    b. The three-capacitance model with uniform segment lengths described in Problem 14.13.

    c. The three-capacitance model with nonuniform lengths described in Problem 14.14.

17.7 A hydraulic system composed of three interconnected tanks is described in Problem 15.8.

    a. Write the system model in matrix form using the three incremental volumes as the state variables.

    b. Write a matrix equation for the output vector $\mathbf{y}$ whose elements are
            y_1, the incremental flow rate from tank 1 to tank 3
            y_2, the incremental pressure at the base of tank 2
            y_3, the average of the three incremental heights

In Problems 17.8 through 17.14, a system obeying the state-variable equation $\dot{\mathbf{q}} = \mathbf{A}\mathbf{q}$ has the given matrix $\mathbf{A}$.

    a. Evaluate the resolvent matrix $\mathbf{\Phi}(s)$.

    b. Calculate the state-transition matrix $\mathbf{\phi}(t)$.

    c. Write each element of $\mathbf{q}$ in terms of the elements of the arbitrary initial state vector $\mathbf{q}(0)$.

    d. Use (28) to evaluate $[\mathbf{\phi}(t)]^{-1}$.

17.8 $\mathbf{A} = \begin{bmatrix} 0 & 1 \\ 0 & -2 \end{bmatrix}$

17.9 $\mathbf{A} = \begin{bmatrix} 0 & 1 \\ -2 & -3 \end{bmatrix}$

17.10 $\mathbf{A} = \begin{bmatrix} -2 & 1 \\ 0 & -2 \end{bmatrix}$

17.11 $\mathbf{A} = \begin{bmatrix} 0 & 1 \\ -4 & 0 \end{bmatrix}$

17.12 $\mathbf{A} = \begin{bmatrix} 0 & 1 \\ -5 & -2 \end{bmatrix}$

17.13 $\mathbf{A} = \begin{bmatrix} a & 0 & 0 \\ 0 & b & 0 \\ 0 & 0 & c \end{bmatrix}$

17.14 $\mathbf{A} = \begin{bmatrix} 0 & 1 & 0 \\ 0 & 0 & 1 \\ 0 & -2 & -3 \end{bmatrix}$

17.15 Draw either a simulation diagram or a block diagram for each of the systems described in Problems 17.9, 17.11, and 17.13.

17.16 Draw either a simulation diagram or a block diagram for each of the systems described in Problems 17.10, 17.12, and 17.14.

17.17 Consider the system defined by the matrices

$$\mathbf{A} = \begin{bmatrix} 0 & 1 \\ 0 & -2 \end{bmatrix} \qquad \mathbf{B} = \begin{bmatrix} 1 \\ -1 \end{bmatrix} \qquad \mathbf{C} = \begin{bmatrix} 1 & -2 \end{bmatrix} \qquad \mathbf{D} = 0$$

a. Indicate the numbers of state variables, inputs, and outputs.

b. Using $\phi(t)$ as found in Problem 17.8, evaluate the state $\mathbf{q}$ and the output $\mathbf{y}$ for the initial state $\mathbf{q}(0) = \begin{bmatrix} 2 & 1 \end{bmatrix}^T$ and zero input.

c. Find the transfer-function matrix $\mathbf{H}(s)$.

d. Evaluate and sketch the output $\mathbf{y}$ when $\mathbf{q}(0) = \begin{bmatrix} 2 & 1 \end{bmatrix}^T$ and $\mathbf{u}(t) = U(t)$.

17.18 Consider the system defined by the matrices

$$\mathbf{A} = \begin{bmatrix} 0 & 1 \\ -2 & -3 \end{bmatrix} \qquad \mathbf{B} = \begin{bmatrix} 1 & 0 \\ 0 & 1 \end{bmatrix}$$

$$\mathbf{C} = \begin{bmatrix} 1 & 0 \end{bmatrix} \qquad \mathbf{D} = \begin{bmatrix} 0 & 0 \end{bmatrix}$$

a. Indicate the numbers of state variables, inputs, and outputs.

b. Using $\Phi(s)$ as found in Problem 17.9, determine the transfer-function matrix $\mathbf{H}(s)$.

c. Evaluate the impulse-response matrix $\mathbf{h}(t)$ and sketch each element versus time.

17.19 Consider the system defined by the matrices

$$\mathbf{A} = \begin{bmatrix} -2 & 1 \\ 0 & -2 \end{bmatrix} \quad \mathbf{B} = \begin{bmatrix} 2 \\ 1 \end{bmatrix} \quad \mathbf{C} = \begin{bmatrix} -1 & 2 \\ 3 & 1 \end{bmatrix} \quad \mathbf{D} = \begin{bmatrix} 0 \\ 0 \end{bmatrix}$$

a. Indicate the numbers of state variables, inputs, and outputs.

b. Using $\Phi(s)$ as found in Problem 17.10, determine the transfer-function matrix $\mathbf{H}(s)$.

c. Evaluate and sketch the zero-input response when $\mathbf{q}(0) = [4 \ -1]^T$.

d. Evaluate the zero-state response when

$$\mathbf{u}(t) = \begin{cases} 0 & \text{for } t \le 0 \\ 5\epsilon^{-t} & \text{for } t > 0 \end{cases}$$

17.20 Consider the system defined by the matrices

$$\mathbf{A} = \begin{bmatrix} 0 & 1 & 0 \\ 0 & 0 & 1 \\ 0 & -2 & -3 \end{bmatrix} \quad \mathbf{B} = \begin{bmatrix} 1 & 0 \\ 0 & 0 \\ 0 & -1 \end{bmatrix}$$

$$\mathbf{C} = \begin{bmatrix} 0 & 0 & 1 \\ 0 & 1 & 0 \end{bmatrix} \quad \mathbf{D} = \begin{bmatrix} 0 & 0 \\ 0 & 0 \end{bmatrix}$$

a. Indicate the numbers of state variables, inputs, and outputs.

b. Using $\phi(t)$ as found in Problem 17.14, evaluate the state $\mathbf{q}$ and the output $\mathbf{y}$ for the initial state $\mathbf{q}(0) = \lceil 0 \ 1 \ 0 \rceil^T$.

c. Determine the transfer-function matrix $\mathbf{H}(s)$ and the impulse-response matrix $\mathbf{h}(t)$.

d. Draw a block diagram of the system and use it to verify the correctness of the elements of $\mathbf{H}(s)$ found in part c.

CHAPTER 18
CASE STUDIES

In this final chapter, we model and analyze five dynamic systems that are somewhat more involved than the examples in previous chapters. We are also able to draw on all the analytical methods that we have introduced throughout the book.

The first study considers an electromechanical device that, in principle, could be used to obtain an electrical measurement of the acceleration of a moving body. The analysis uses transfer functions and a plot of the device's frequency response. Based on the character of the frequency-response plot, we draw certain conclusions regarding appropriate parameter values and the potential usefulness of the device.

Next, we consider a nonlinear electrical element known as the tunnel diode. A linearized approximation and a graphical procedure for determining equilibrium points are presented. We analyze the stability of the linearized models associated with these equilibrium points for two simple circuit models and indicate some applications.

The third study extends the treatment of the position-control system given in Section 16.3 to a velocity-control sytem designed to control the velocity of a mass driven by an electric motor. We obtain the system model as a block diagram with feedback and determine and analyze the closed-loop transfer function.

Then a thermal system having two thermal capacitances is modeled in numerical form and analyzed. We find the system's transfer function in terms of variables relative to the ambient condition and use it to obtain the response. We take advantage of the fact that the two time constants differ greatly in magnitude to solve for the time at which the temperature reaches a predetermined value.

Finally, we model a sociological system, in which the state variables are groups of people with distinguishing characteristics rather than the energy-related variables we have considered throughout the book. We first model the system as a set of discrete-time equations and then convert them to differential equations. Although this case study is very brief and qualitative, it demonstrates that the techniques of modeling and analysis that we have studied in the traditional context of physical systems are applicable to a far broader class of systems. Economics, transportation, ecology, and public services are among the areas in which the methods of dynamic modeling and analysis can be applied.

18.1 A DEVICE FOR MEASURING ACCELERATION

In many areas of technology, it is important to be able to measure and record the acceleration of a moving body as a function of time. For example, in the testing of automobiles for crash resistance, deceleration measurements are required. Accelerometers are important components in ship, aircraft, and rocket navigation and guidance systems.

Figure 18.1 shows an electromechanical device whose response depends on the acceleration of its case relative to an inertially fixed reference frame. The device is not intended to be representative of commonly used accelerometers, but modeling and analyzing its response will provide useful practice with many techniques we discussed in previous chapters.

SYSTEM DESCRIPTION

Basically, the device shown in Figure 18.1 consists of a case that is attached to the body or material whose motion is to be measured, a circular coil fixed to the case, and a permanent magnet supported on the case by a spring, with viscous damping between the magnet and the case. As the case moves

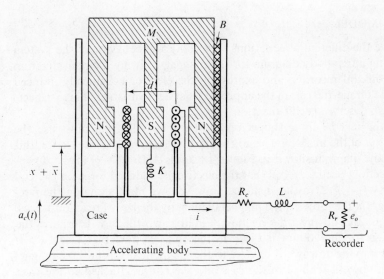

Figure 18.1 Acceleration-measuring device.

vertically due to motion of the body supporting it, a voltage is induced in the coil because of the relative motion of the coil and the magnetic field. A recorder is attached to the terminals of the coil and draws a graph of the coil voltage e_o as a function of time.

The magnet has mass M and provides a flux density of $\mathscr{B}$ in the annular space between its north and south poles. Although several springs would be used to support the mass, the single spring shown with spring constant K may be considered equivalent to whatever springs are present. The viscous-damping coefficient B accounts for all viscous effects between the magnet and the case.

The coil has N turns of diameter d, and the parameter $\ell = \pi d N$ denotes the length of the coil. The total coil resistance and inductance are modeled by lumped elements having values R_c and L, respectively. The dots and crosses associated with the coil in Figure 18.1 indicate that the assumed positive direction of current is clockwise as viewed from above. The recorder is attached directly to the terminals of the coil and is assumed to provide a resistance R_r in the coil circuit.

The acceleration of the case relative to a fixed inertial reference frame is denoted by $a_c(t)$ and is the input to the system. The vertical distance between the magnet and the case is $x + \bar{x}$, so the variable x denotes the incremental displacement of the magnet relative to the base, with the value $x = 0$ corresponding to the equilibrium condition. Likewise, the relative velocity between the magnet and case is equal to $\dot{x}$.

SYSTEM MODEL

We derive the differential equations describing the behavior of the system by drawing a free-body diagram for the magnet, drawing a circuit diagram for the coil and recorder, and expressing the electromechanically induced force and voltage in terms of the appropriate system variables. Three aspects of this task deserve specific mention.

First, the inertial force shown on the free-body diagram must use the acceleration of the mass relative to the inertial reference frame. Hence, this force on the diagram shown in Figure 18.2(a) is $M[a_c(t) + \ddot{x}]$ in the downward direction. Second, recall that the electrically induced force on the coil is given by $f_e = \mathscr{B}\ell i$. In this instance, however, we must show on the free-body diagram the force on the magnet, which, by the law of reaction forces, is $-\mathscr{B}\ell i$ in the downward direction. Finally, in order to determine the proper sign for the voltage induced in the coil, we note that when $\dot{x} > 0$ the coil is moving downward relative to the magnetic field, since the magnet is moving upward relative to the case and the coil is attached to the case.

Taking these points into consideration, we can draw the free-body and circuit diagrams shown in Figure 18.2. Summing forces on the magnet and writing a loop equation for the circuit, we obtain

$$L\frac{di}{dt} + (R_c + R_r)i = -\mathscr{B}\ell\dot{x} \tag{1a}$$

$$M\ddot{x} + B\dot{x} + Kx = \mathscr{B}\ell i - Ma_c(t) \tag{1b}$$

$$e_o = R_r i \tag{1c}$$

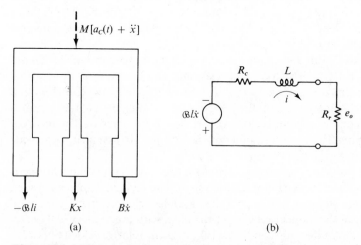

(a)   (b)

Figure 18.2 (a) Free-body diagram for the magnet. (b) Circuit diagram.

Using the model as expressed by (1) and treating the transform of the acceleration, $A_c(s)$, as the input, we can draw the block diagram shown in Figure 18.3. The diagram is grouped into two portions that are interconnected by paths having gains of $\mathscr{B}\ell$. On the left side is a second-order subsystem involving the mechanical parameters M, B, and K, and on the right side is a first-order subsystem involving the electrical parameters L, R_c, and R_r.

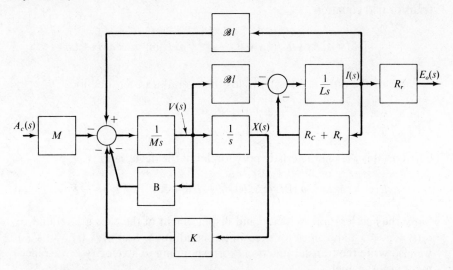

Figure 18.3 Block diagram of the acceleration-measuring device.

TRANSFER FUNCTION

For the purposes of analysis, it is desirable to determine a single overall transfer function from the input transform $A_c(s)$ to the output transform $E_o(s)$. Although we can reduce the block diagram to a single transfer function by using the technique discussed in Section 16.2, we shall transform (1) with zero initial conditions. Doing this, we obtain the following set of three algebraic transform equations

$$(Ls + R_c + R_r)I(s) = -\mathscr{B}\ell sX(s) \tag{2a}$$

$$(Ms^2 + Bs + K)X(s) = \mathscr{B}\ell I(s) - MA_c(s) \tag{2b}$$

$$E_o(s) = R_r I(s) \tag{2c}$$

Since we have three equations, we can combine them to eliminate the intermediate variables $X(s)$ and $I(s)$, obtaining a single equation relating the input transform $A_c(s)$ and the output transform $E_o(s)$.

First, if we combine (2a) and (2b) to eliminate $X(s)$, the result is

$$(Ms^2 + Bs + K)\left(\frac{Ls+R}{-\mathscr{B}\ell s}\right)I(s) - \mathscr{B}\ell I(s) = -MA_c(s)$$

where $R = R_c + R_r$. Combining the two terms involving $I(s)$ in this equation and then using (2c) to express $I(s)$ in terms of $E_o(s)$, we obtain the single transformed equation

$$\frac{1}{R_r}\left[(Ms^2 + Bs + K)(Ls + R) + \mathscr{B}^2\ell^2 s\right]E_o(s) = \mathscr{B}\ell MsA_c(s)$$

Solving for the ratio $E_o(s)/A_c(s)$ yields the overall transfer function

$$\frac{E_o(s)}{A_c(s)} = \frac{R_r\mathscr{B}\ell Ms}{P(s)} \tag{3}$$

where $P(s)$ is the characteristic polynomial of the device and is

$$P(s) = MLs^3 + (BL + MR)s^2 + (BR + KL + \mathscr{B}^2\ell^2)s + KR \tag{4}$$

Since the acceleration, velocity, and displacement of the case are related by $a_c(t) = \dot{v}_c = \ddot{x}_c$, and since for zero initial conditions $A_c(s) = sV_c(s) = s^2X_c(s)$, we also write the transfer functions corresponding to a velocity or displacement input, namely

$$\frac{E_o(s)}{V_c(s)} = \frac{R_r\mathscr{B}\ell Ms^2}{P(s)}$$

$$\frac{E_o(s)}{X_c(s)} = \frac{R_r\mathscr{B}\ell Ms^3}{P(s)}$$

Because the characteristic polynomial $P(s)$ is a cubic, it is difficult to be very specific about the behavior of the system as a measuring device without substituting numerical values for the parameters and calculating the frequency response or simulating the response to specific inputs, such as the impulse and step function.

Rather than doing this, we shall make the approximation that the coil inductance can be neglected. In terms of the frequency response, this approximation should be well justified for low frequencies but not for high frequencies. Since our interest is principally in the response at low frequencies, we are justified in setting L equal to zero in (4). With this change $P(s)$ becomes quadratic. In essence, we have eliminated the current as a state variable, and (3) reduces to

$$\frac{E_o(s)}{A_c(s)} = \frac{R_r \mathcal{B}\ell Ms}{MRs^2 + (BR + \mathcal{B}^2\ell^2)s + KR}$$

$$= \frac{(\mathcal{B}\ell R_r/R)s}{s^2 + \left(\dfrac{B}{M} + \dfrac{\mathcal{B}^2\ell^2}{MR}\right)s + \dfrac{K}{M}} \tag{5}$$

Inspection of (5) indicates that the transfer function $E_o(s)/A_c(s)$ has a zero at $s = 0$ and a pair of poles that may be real or complex. To determine the undamped natural frequency ω_n and the damping ratio ζ associated with these poles, we compare (5) to

$$H(s) = \frac{Cs}{s^2 + 2\zeta\omega_n s + \omega_n^2} \tag{6}$$

which can be viewed as the standard form for a transfer function having a zero at $s = 0$ and two poles. Comparing the coefficients in (5) and (6), we obtain

$$C = \frac{\mathcal{B}\ell R_r}{R}$$

$$\omega_n = \sqrt{\frac{K}{M}} \tag{7}$$

$$\zeta = \frac{1}{2}\sqrt{\frac{M}{K}}\left(\frac{B}{M} + \frac{\mathcal{B}^2\ell^2}{MR}\right)$$

If it were known what values of ζ and ω_n would result in a device that would perform well as a measuring instrument, a designer could attempt to select the physical parameters in order to achieve these values of ζ and ω_n. As indicated in the following section, frequency-response plots can be used to determine suitable values for ζ and ω_n.

FREQUENCY RESPONSE

The steady-state response to a sinusoidal input is found by examining $H(j\omega)$, as discussed in Section 13.4. A frequency-response analysis of (5) would be cumbersome unless specific numerical values were given for all but one or two of the physical parameters, so we shall work with (6). When s is replaced by $j\omega$, (6) becomes

$$H(j\omega) = \frac{jC\omega}{\omega_n^2 - \omega^2 + j2\zeta\omega_n\,\omega}$$

$$= \frac{j(C/\omega_n)(\omega/\omega_n)}{1 - (\omega/\omega_n)^2 + j2\zeta(\omega/\omega_n)} \tag{8}$$

where the second form of the equation results from dividing both the numerator and the denominator by ω_n^2. The quantity ω/ω_n can be thought of as a normalized frequency. Since the factor C/ω_n in the numerator of (8) is a multiplying constant that does not affect the variation of $H(j\omega)$ with ω, we shall also normalize the magnitude of the transfer function by defining

$$H_N(j\omega) = \frac{\omega_n}{C} H(j\omega)$$

$$= \frac{j(\omega/\omega_n)}{1 - (\omega/\omega_n)^2 + j2\zeta(\omega/\omega_n)} \tag{9}$$

We obtain the magnitude of $H_N(j\omega)$ by dividing the magnitude of its numerator by that of its denominator, resulting in

$$|H_N(j\omega)| = \frac{\omega/\omega_n}{\sqrt{(\omega/\omega_n)^4 + (4\zeta^2 - 2)(\omega/\omega_n)^2 + 1}} \tag{10}$$

Comparing $|H_N(j\omega)|$ for different values of ζ indicates the relative shapes of the frequency-response magnitudes corresponding to different values of the damping ratio. If the device is to be of value in measuring the acceleration of the case, there should be a range of frequencies for which $|H_N(j\omega)|$ is fairly flat, i.e., independent of frequency.

From (10), we see that $|H_N(j\omega)| \simeq \omega/\omega_n$ for small values of ω/ω_n and approaches $1/(\omega/\omega_n)$ for large values of ω/ω_n. For $\omega/\omega_n = 1$, $|H_N(j\omega)| = 1/2\zeta$. Using this information and calculating a few additional points, we can draw the plots shown in Figure 18.4. Logarithmic scales are commonly used for frequency response plots, and they allow us to include a wide range of values of $|H_N(j\omega)|$ and ω/ω_n.

It is apparent that the device must be heavily damped, i.e., $\zeta \gg 1$, if a range of frequencies is to be achieved for which $|H_N(j\omega)|$ is essentially constant. If $\zeta = 100$, $|H_N(j\omega)|$ will remain between 0.0045 and 0.0050 for $0.01 < \omega/\omega_n < 100$, which is a four-decade range of frequencies. In contrast, if $\zeta \leq 1$, there is no range of frequencies over which $|H_N(j\omega)|$ is essentially constant.

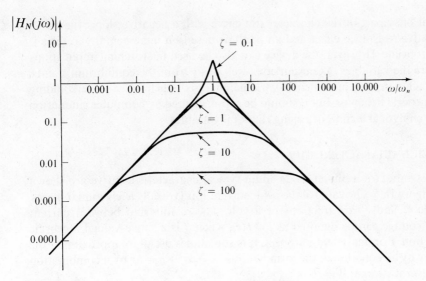

Figure 18.4 Frequency response of the acceleration-measuring device for several damping ratios.

Because $|H_N(j\omega)| \simeq \omega/\omega_n$ as ω approaches zero, a constant acceleration will result in a zero steady-state output. We can also see this from (6) by noting that the steady-state response to a unit step-function input is $H(0) = 0$. The reason for this effect is that a relative velocity of the coil with respect to the magnet is required to develop a voltage in the circuit. In order to read the relative displacement of the magnet with respect to the case, we could attach a pointer to the mass and put scale markings on the case. There would then be a steady-state output reading even for a constant acceleration of the case, and the device could sense accelerations of arbitrarily low frequencies. To obtain an electrical output for recording purposes, we can replace the pointer by the wiper of a potentiometer and use the voltage at the wiper arm as the output. Such an arrangement is discussed in Section 9.11 of Cannon.

18.2 THE TUNNEL DIODE

The *tunnel diode* is a nonlinear electrical element whose current-versus-voltage curve includes regions of both positive and negative slope. Hence, an increase in the voltage of the diode causes a decrease in the current under some conditions. When the variables remain close to a stable equilibrium point on a part of the characteristic curve with a negative slope, there is a negative resistance effect and an analysis based on incremental variables is appropriate. However, the device can also be used in switching applications where the variables take excursions well away from the equilibrium points. In such cases, an incremental-variable analysis is helpful in understanding the general behavior, but it should be supplemented by computer simulation and analytical techniques using the total variables.

DEVICE CHARACTERISTICS

The symbol for a tunnel diode and a typical characteristic curve are shown in Figure 18.5. The value of the peak current i_p is typically a few milliamperes, while e_p and e_v are fractions of a volt. The relationship between current and voltage can be denoted as $i = f(e)$, where f is a single-valued algebraic function. For analytical purposes, it is sometimes useful to approximate the curve by an equation of the form $i = a_3 e^3 + a_2 e^2 + a_1 e$ or by a combination of several straight lines.

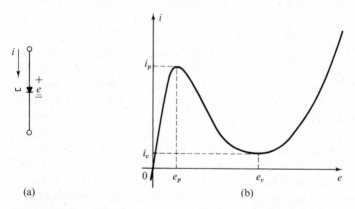

(a)   (b)

Figure 18.5 Tunnel diode. (a) Symbol. (b) Characteristic curve.

The algebraic relationship shown in Figure 18.5(b) describes a nonlinear resistor. However, a typical tunnel diode also exhibits some parasitic capacitance and inductance, and we may add external energy-storing elements purposely for a particular application. For a valid dynamic analysis, we must include the effects of the energy-storing elements.

LINEARIZED MODEL

When a linearized model valid in the vicinity of some equilibrium point is developed, the resistance in the model is determined by the slope of the characteristic curve, as in Example 10.5. Specifically, we denote the equilibrium point by $\bar{i}$, $\bar{e}$ and define the incremental variables $\hat{i}$ and $\hat{e}$ by

$$i = \bar{i} + \hat{i}$$
$$e = \bar{e} + \hat{e} \tag{11}$$

Using the first two terms of a Taylor-series expansion, as in (5.8), we can express the diode current i as

$$f(\bar{e}) + \frac{df}{de}\bigg|_{\bar{e}} (e - \bar{e}) = \bar{i} + \frac{1}{r}\hat{e} \tag{12}$$

where $1/r$ is the slope of the characteristic curve at the operating point. Thus, in terms of the incremental current and voltage, the model of the tunnel diode is

$$\hat{i} = \frac{1}{r}\hat{e} \tag{13}$$

Two possible operating points are shown in Figure 18.6(a). The resistance r_1 for the first point is positive, while the second resistance r_2 is negative.

We can also express the voltage of the device as an algebraic function of the current. To emphasize this, we redraw the characteristic curve with the axes interchanged as shown in Figure 18.6(b) and write $e = g(i)$. Note, however, that there is a range of currents for which g is not a single-valued function. By approximating $g(i)$ by the first two terms in its Taylor series,

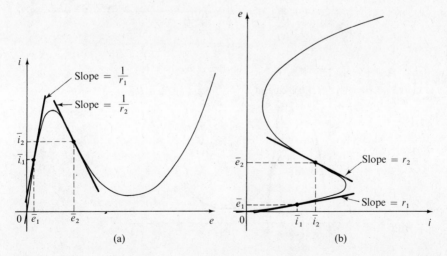

Figure 18.6 Determining the resistance for a linearized model. (a) $i = f(e)$. (b) $e = g(i)$.

we can write the diode voltage e as

$$g(i) + \frac{dg}{di}\bigg|_i (i - \bar{i}) = \bar{e} + r\hat{i} \tag{14}$$

where r is the slope of the curve in Figure 18.6(b), which is equal to the reciprocal of the slope at the corresponding point in Figure 18.6(a). Equation (14) yields the incremental model

$$\hat{e} = r\hat{i} \tag{15}$$

which is the same as (13), as it must be.

The operating point is typically determined by an external resistor and voltage source. If the circuit contains inductors and capacitors, we replace them by short and open circuits, respectively, when finding the operating point. For the circuit shown in Figure 18.7(a), we can use the same graphical

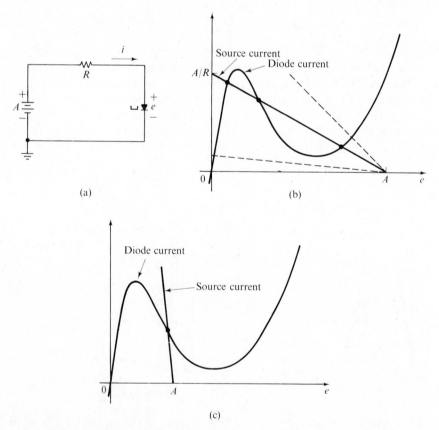

(a)

(b)

(c)

Figure 18.7 Determining the operating point.

approach as in Example 10.4. The equation describing the part of the circuit to the left of the tunnel diode is

$$i = \frac{1}{R}(A - e) \tag{16}$$

When plotted on the diode characteristic as shown in Figure 18.7(b) and Figure 18.7(c), (16) appears as a straight line called the *load line*. Since the current-voltage relationship must satisfy both (16) and the diode characteristic, the equilibrium points are at the intersections of the two curves. Note that when the load line is the solid line shown in Figure 18.7(b), there are three possible equilibrium points, indicated by the heavy dots. If the value of R is decreased so that the load line becomes the upper dashed curve or increased so that the load line becomes the lower dashed curve, there is only one intersection. Another case for which there is only one intersection is shown in Figure 18.7(c).

It is necessary to analyze the stability of each of the equilibrium points when all the inductors and capacitors are included in the model. We shall illustrate the technique for two cases.

A CAPACITOR IN PARALLEL WITH A TUNNEL DIODE

For the circuit shown in Figure 18.8(a), the current i_1 through the resistor R can differ from the diode current i when the circuit is not in a steady-state equilibrium condition. Since $i_C = C\dot{e}$,

$$\dot{e} = \frac{1}{C}(i_1 - i)$$

$$= \frac{1}{C}\left\{\frac{1}{R}\left[e_i(t) - e\right] - f(e)\right\} \tag{17}$$

where $i = f(e)$ describes the diode characteristic. At the operating point where $e_i(t)$ has the constant value $\bar{e}_i = A$, we have $\dot{e} = 0$, $i_1 = i$, and

$$\frac{1}{R}(A - \bar{e}) - i = 0 \tag{18}$$

Inserting the expressions $e_i(t) = A + \hat{e}_i(t)$, $e = \bar{e} + \hat{e}$, and $i = \bar{i} + \hat{i}$ into (17) and using (12) give

$$\dot{\hat{e}} = \frac{1}{C}\left\{\frac{1}{R}[A + \hat{e}_i(t) - \bar{e} - \hat{e}] - \bar{i} - \frac{1}{r}\hat{e}\right\}$$

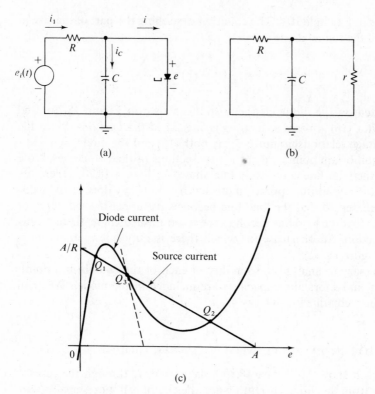

Figure 18.8 (a) Diode circuit with a capacitor. (b) Equivalent circuit for determining the time constant of the linearized model. (c) Possible equilibrium points.

Using (18) to cancel the constant terms, we obtain the linearized first-order model

$$\dot{\hat{e}} = \frac{1}{C}\left[\frac{1}{R}\hat{e}_i(t) - \left(\frac{1}{R} + \frac{1}{r}\right)\hat{e}\right] \qquad (19)$$

With $\hat{e}$, the incremental voltage across the capacitor, taken to be the state-variable, (19) is the linearized state-variable equation. It can be rewritten as

$$\dot{\hat{e}} + \frac{1}{R_{eq}C}\hat{e} = \frac{1}{RC}\hat{e}_i(t) \qquad (20)$$

where

$$R_{eq} = \frac{Rr}{R+r} \qquad (21)$$

By comparing (20) with (8.11), we see that the time constant is $\tau = R_{eq} C$. We reach the same conclusion by drawing the incremental model with the voltage source replaced by a short circuit, as shown in Figure 18.8(b), and then using the short-cut method described in Section 9.6.

For an equilibrium point such that the diode resistance r is positive, we see from (20) and (21) that the linearized model is stable. At an operating point for which r is negative, the model is stable if and only if $|r| > R$. Thus, for the solid load line shown in Figure 18.8(c), Q_1 and Q_2 represent stable equilibrium points and Q_3 an unstable equilibrium point. For the dashed load line, the only equilibrium point is Q_3, which this time is stable.

AN INDUCTOR IN SERIES WITH A TUNNEL DIODE

For the circuit shown in Figure 18.9(a), the voltage e_1 across the series combination of the source and external resistor may differ from the diode

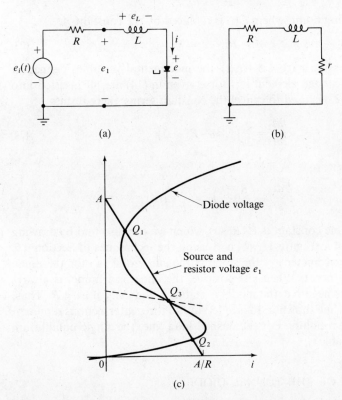

(a)   (b)

(c)

Figure 18.9 (a) Diode circuit with an inductor. (b) Equivalent circuit for determining the time constant of the linearized model. (c) Possible equilibrium points.

voltage e because of the presence of the inductor. Since $e_L = L\, di/dt$,

$$\frac{di}{dt} = \frac{1}{L}(e_1 - e)$$

$$= \frac{1}{L}[e_i(t) - Ri - g(i)] \tag{22}$$

where $e = g(i)$ describes the diode characteristic in Figure 18.9(c). If we choose i as a state variable, (22) has the appearance of a nonlinear first-order state-variable equation. Recall, however, that $g(i)$ is not a single-valued function of i (in contrast to the analysis of Figure 18.8(a), for which $f(e)$ was a single-valued function of e). Thus knowing the initial current $i(0)$ and the input $e_i(t)$ is not sufficient to allow us to solve for i by either analytical or numerical methods. We can, however, develop linearized models that are valid in the vicinity of any one of the three equilibrium points Q_1, Q_2, and Q_3.

At the operating point, where $e_i(t)$ is replaced by the constant A,

$$A - Ri - \bar{e} = 0 \tag{23}$$

When the expression $e_i(t) = A + \hat{e}_i(t)$, the incremental variables defined in (11), and the linearized element law for e given in (14) are all inserted into (22), and when (23) is used to cancel the constant terms, there results

$$\frac{d\hat{i}}{dt} = \frac{1}{L}[\hat{e}_i(t) - R\hat{i} - r\hat{i}] \tag{24}$$

or, equivalently,

$$\frac{d\hat{i}}{dt} + \left(\frac{R+r}{L}\right)\hat{i} = \frac{1}{L}\hat{e}_i(t) \tag{25}$$

From (25), the time constant is $L/(R + r)$, which we can also find by drawing the circuit shown in Figure 18.9(b) and using the techniques of Section 9.6. Recall that the parameter r is the slope of the curve of e vs i for the diode at the operating point. When r is positive, the linearized model is always stable. When r is negative, the model is stable if and only if $|r| < R$. Thus, for the solid load line shown in Figure 18.9(c), all three intersections represent stable equilibrium points. For the dashed load line, the single equilibrium point Q_3 is unstable.

APPLICATIONS OF THE TUNNEL DIODE

With this background, we could discuss a number of applications. We conclude this case study by presenting a sketchy outline of three of them.

Derivations and descriptions of the applications are given in books dealing with nonlinear circuits (see, for example, Anner and Section 19.1 of Alley and Atwood).

All three applications are based on the circuit shown in Figure 18.10(a), which includes an inductance of appropriate size. The operation is related to the stability of the equilibrium points, which in turn depends on the values of A and R. Instead of the e vs i characteristic diode curve shown in Figure 18.9(c), we use the more conventional i vs e curve.

Bistable operation We choose A and R to give a load line that intersects the diode characteristic in three places, as shown in Figure 18.10(b). All three intersections represent stable equilibrium points. Assume that $e_i(t)$ is zero at some reference time and that the diode variables correspond to the

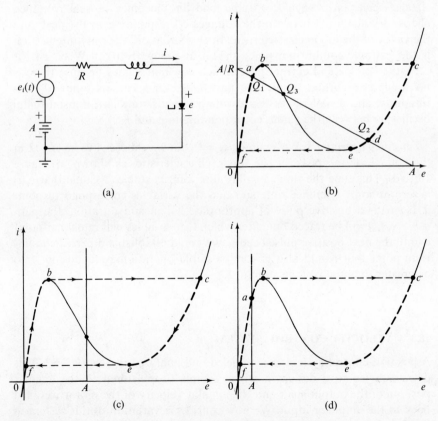

Figure 18.10 (a) Circuit used in switching applications. (b) Bistable operation. (c) Astable operation. (d) Monostable operation.

equilibrium point Q_1. We wish to trace out on the figure the path, called the *operating path*, that the diode variables describe when the signal $e_i(t)$ is applied.

If $e_i(t)$ is a positive pulse of appropriate height and duration, the operating path will be a, b, c, d. The path b, c is an instantaneous jump between branches of the characteristic curve. Note that because of the inductance, there is no instantaneous change in the current. If $e_i(t)$ is an appropriate negative pulse at a later time, the path of operation is d, e, f, a. In summary, the variables will move between the two stable equilibrium points Q_1 and Q_2 whenever the input contains the appropriate pulse. The equilibrium point in use depends on the sign of the last pulse; thus the circuit is said to have *digital memory*. The operating path never reaches the third equilibrium point Q_3.

Astable operation When $R = 0$, the load line becomes vertical. We then choose a value for A between the voltages corresponding to the peak and the valley of the diode characteristic. In this situation, the only intersection is the unstable equilibrium point shown in Figure 18.10(c). When $e_i(t)$ is zero, the voltage and current never reach constant values because there is no stable equilibrium point. The path $f, b, c, e, f, b, c, \ldots$ is followed continuously, and a plot of the diode voltage versus time is a nonsinusoidal oscillation between the values corresponding to c and f.

Monostable operation We again let $R = 0$, but we choose A to be less than the voltage at the peak in the diode characteristic, as shown in Figure 18.10(d). This time the single equilibrium point is stable. Assume that $e_i(t)$ is zero at some reference time and that the variables correspond to point a. If $e_i(t)$ is a positive pulse of appropriate height and duration, the path a, b, c, e, f, a will be traced out, after which the variables will remain at point a until the next positive pulse occurs. We could establish a different equilibrium point that would also give monostable operation by choosing A to be greater than the voltage at point e.

18.3 A VELOCITY-CONTROL SYSTEM

A position-control system was modeled and analyzed in Section 16.3. This discussion led to a design in which signals proportional to the angular position of the output shaft and the angular velocity of the motor were fed back to the amplifier input. We now consider a variation of this system to control the translational velocity of a mass driven by an electric motor through a rack and a pinion gear such as we analyzed in Example 4.8.

SYSTEM DESCRIPTION

The system is diagrammed in Figure 18.11. The voltage $e_1(t)$ is proportional to the desired velocity of the mass M and is supplied by an electronic signal generator, the details of which need not concern us. The amplifier, motor, and tachometer-potentiometer combination are identical to the corresponding elements in the positional servomechanism discussed in Section 16.3, except that the tachometer has the moment of intertia J_t. The rotor of the motor is connected by a rigid shaft to a pinion gear of radius r and negligible inertia. The pinion gear meshes with a rack that is attached to the mass, whose displacement and velocity are denoted by x and v, respectively. The mass is also subjected to a viscous-friction force and to an independently specified force $f_d(t)$ such as might be caused by a cutting tool in contact with a piece of material attached to the mass. We will consider the force $f_d(t)$ a disturbance input because we want the motion of the mass to be unaffected by this force.

SYSTEM MODEL

Taking the gear geometry into account, we can draw free-body diagrams for the combined motor and tachometer and for the mass as we did in Section 16.3 for the positional servomechanism. Writing the equations corresponding to the free-body diagrams and the equations for the electrical circuit, we obtain

$$(J_m + J_t)\dot{\omega} + B_1\omega + rf_c - \alpha i = 0 \tag{26a}$$

$$M\dot{v} + B_2 v + f_d(t) - f_c = 0 \tag{26b}$$

$$e_2 = K_T \omega \tag{26c}$$

$$e_a = -K_A[e_2 - e_1(t)] \tag{26d}$$

$$i = \frac{1}{R_a}(e_a - \alpha\omega) \tag{26e}$$

$$v = r\omega \tag{26f}$$

where f_c is the contact force between the rack and the pinion gear, defined as positive when it acts to the left on the mass, and where α is the electromechanical coupling coefficient defined in Section 16.3.

Although we can draw a block diagram directly from the six equations comprising (26), we shall first find an expression for the transform of ω in terms of the transforms of e_a and $f_d(t)$. Because the feedback signal is proportional to ω, the angular velocity must appear as one of the signals in the

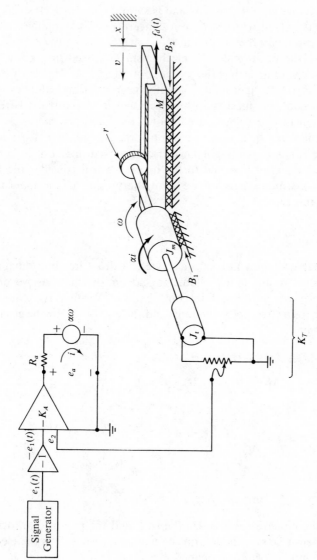

Figure 18.11 Mass with velocity controlled by a feedback system.

final diagram. Combining (26a) and (26b) to eliminate the contact force f_c, a variable that need not appear in the block diagram, we have

$$(J_m + J_t)\dot{\omega} + B_1\omega + r[M\dot{v} + B_2 v + f_d(t)] - \alpha i = 0 \tag{27}$$

By using (26f) to rewrite (27) in terms of the angular velocity, we have

$$(J_m + J_t + r^2M)\dot{\omega} + (B_1 + r^2B_2)\omega = \alpha i - rf_d(t)$$

or

$$J\dot{\omega} + B\omega = \alpha i - rf_d(t) \tag{28}$$

where the equivalent moment of inertia J and the equivalent damping coefficient B are defined by

$$J = J_m + J_t + r^2M \tag{29a}$$

$$B = B_1 + r^2B_2 \tag{29b}$$

Then we can use (26e) in (28) to express the armature current i in terms of e_a and ω, obtaining

$$J\dot{\omega} + \left(B + \frac{\alpha^2}{R_a}\right)\omega = \frac{\alpha}{R_a}e_a - rf_d(t) \tag{30}$$

Transforming (30) with zero initial conditions and solving for $\Omega(s) = \mathscr{L}[\omega(t)]$, we obtain

$$\Omega(s) = \frac{\left(\dfrac{\alpha}{R_aJ}\right)E_a(s) - \left(\dfrac{r}{J}\right)F_d(s)}{s + \dfrac{B}{J} + \dfrac{\alpha^2}{R_aJ}} \tag{31}$$

If we define the parameters

$$K_m = \frac{\alpha}{R_aJ}$$
$$\tag{32}$$
$$\tau_m = \frac{1}{\dfrac{B}{J} + \dfrac{\alpha^2}{R_aJ}}$$

which are identical to those defined in Section 16.3 except for the use of the equivalent moment of inertia and viscous friction and the use of R_a in place of R, we can write (31) as

$$\Omega(s) = \left(\frac{1}{s+1/\tau_m}\right)\left[K_m E_a(s) - \frac{r}{J}F_d(s)\right] \tag{33}$$

When converted to block-diagram form, (33) results in the two gains K_m and r/J with inputs $E_a(s)$ and $F_d(s)$, respectively, and a summing junction followed by the transfer function $1/(s+1/\tau_m)$ appearing in Figure 18.12. Note that the individual blocks labeled K_m, r/J, and $1/(s+1/\tau_m)$ do not correspond to separate physical components. A block diagram must represent the equations describing the system and is usually arranged in an order roughly comparable to the major subsystems, but an individual block need not be related to a separately identifiable physical entity.

To complete the block diagram, we combine (26c) and (26d) and transform them to give

$$E_a(s) = K_A[E_1(s) - K_T \Omega(s)]$$

which results in the feedback path, the left summing junction, and the amplifier gain K_A. Finally, since by transforming (26f) we have

$$V(s) = r\Omega(s)$$

we add the block with a gain of r at the right to complete the diagram shown in Figure 18.12.

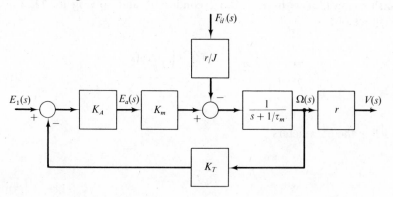

Figure 18.12 Block diagram of velocity control system showing both inputs.

CLOSED-LOOP TRANSFER FUNCTIONS

Because the system has the two inputs $e_1(t)$ and $f_d(t)$ and the single output v, it will possess two closed-loop transfer functions. These are the reference transfer function $V(s)/E_1(s)$ and the disturbance transfer function $V(s)/F_d(s)$.

We can find each of these transfer functions by setting the other input equal to zero, finding the closed-loop transfer function of the feedback loop, and multiplying the result by the gain r in the case of $V(s)/E_1(s)$ and by r^2/J for $V(s)/F_d(s)$.

To find $V(s)/E_1(s)$, we let $F_d(s)$ equal zero in Figure 18.12 and use (16.1) with $G(s) = K_A K_m/(s+1/\tau_m)$ and $H(s) = K_T$ to find $\Omega(s)/E_1(s)$. Multiplying the result by r, we find the reference transfer function to be

$$
\frac{V(s)}{E_1(s)} = \left[\frac{\dfrac{K_A K_m}{s+1/\tau_m}}{1 + \left(\dfrac{K_A K_m}{s+1/\tau_m}\right) K_T} \right] r
$$

$$
= \frac{K_A K_m r}{s+(1/\tau_m)+K_A K_m K_T} \tag{34}
$$

To find the disturbance transfer function $V(s)/F_d(s)$, we eliminate the reference input $E_1(s)$ shown in Figure 18.12. To account for the minus sign associated with the feedback path, we insert a block having a gain of -1. When we redraw the block diagram to show the input $F_d(s)$ at the left, as is customary, Figure 18.13 results. Inspection of Figure 18.13 indicates that the

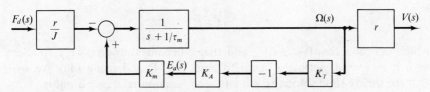

Figure 18.13 Equivalent block diagram when reference input is zero.

feedback loop is now characterized by the forward transfer function $G(s) = 1/(s+1/\tau_m)$, the feedback transfer function $H(s) = -K_m K_A K_T$, and a plus sign associated with the feedback path at the summing junction. When we use (16.2) with these expressions for $G(s)$ and $H(s)$ to reduce the feedback loop to a single transfer function and multiply the result by $-r^2/J$, we find the disturbance transfer function to be

$$
\frac{V(s)}{F_d(s)} = -\frac{r^2}{J} \left[\frac{\dfrac{1}{s+1/\tau_m}}{1 - \left(\dfrac{1}{s+1/\tau_m}\right)(-K_A K_m K_T)} \right]
$$

$$
= \frac{-(r^2/J)}{s+(1/\tau_m)+K_A K_m K_T} \tag{35}
$$

RESPONSE

Comparing (34) and (35), we see that the closed-loop system is characterized by a pair of first-order transfer functions having the same pole but different constants in their numerators. The pole of the closed-loop system is at

$$s = -[(1/\tau_m) + K_A K_m K_T]$$

The time constant of the closed-loop system, denoted by τ, is the reciprocal of the magnitude of this pole, namely

$$\tau = \frac{1}{(1/\tau_m) + K_A K_m K_T}$$

$$= \frac{\tau_m}{1 + K_A K_m K_T \tau_m}$$

If the parameter

$$\beta = K_A K_m K_T \tau_m \tag{36}$$

is defined, we can write the time constant of the closed-loop system as

$$\tau = \frac{\tau_m}{1 + \beta}$$

which indicates that $0 < \tau < \tau_m$ and that τ becomes progressively smaller as β is increased. Setting s equal to zero in (34) and using (36), we see that the steady-state velocity for a unit step-function reference input is

$$\frac{V(0)}{E_1(0)} = \frac{K_A K_m r}{(1/\tau_m) + K_A K_m K_T}$$

$$= \frac{r/K_T}{1 + \dfrac{1}{K_A K_m K_T \tau_m}}$$

$$= \frac{r/K_T}{1 + (1/\beta)} \tag{37}$$

Hence, as β is increased, the ratio of the steady-state value of the velocity v and a constant input signal $\bar{e}_1$ approaches a limit of r/K_T expressed in units of meters per second-volt. Note that this limiting value r/K_T depends only on the radius of the pinion gear and the gain of the tachometer-potentiometer combination, both of which can be precisely specified. This limiting value is independent of the values of K_A, K_m, and τ_m, each of which may not be known precisely and may be subject to variations with time, temperature, mechanical alignment, and other conditions. Thus, by maintaining a large value of β, we can make the steady-state performance of the system for the reference input $e_1(t)$ relatively insensitive to such parameter variations and uncertainties, which is a desirable situation.

Similarly, from (35) and (36), the steady-state velocity for a unit step-function disturbance force is

$$\frac{V(0)}{F_d(0)} = -\frac{r^2\tau_m/J}{1+\beta} \tag{38}$$

This expression approaches zero as β becomes large, provided that the increase in β is not due solely to an increase in τ_m. Because we do not want the velocity to be affected by the disturbance force, a large value of β is beneficial from this point of view also.

Since a large value of β is desirable in terms of dynamic response and steady-state behavior, one might be tempted to make its value very large. This can not be done in practice, however, because the system model we have analyzed does not take into account dynamic effects and nonlinearities that would no longer be negligible for very large values of β. For example, the amplifier has been modeled as a gain that is independent of frequency and not subject to saturation limits, and the armature inductance and the flexibility of the shaft connecting the motor to the pinion gear have been neglected. Including some of these additional effects would lead to a more complicated model that might become unstable if β were increased beyond some limiting value. The analytical tools necessary for investigating such questions are presented in books on feedback systems (see Dorf (1974), for example).

18.4 A THERMAL SYSTEM

Producing chemicals almost always requires control of the temperature of liquids contained in vessels. In a continuous process, a vessel within which a reaction is taking place typically has liquid flowing into and out of it continuously, and a control system is needed to maintain the liquid at a constant temperature. Such a system was modeled in Example 14.7. In a batch process, a vessel would typically be filled with liquid, sealed, and then heated to a prescribed temperature. In the design and operation of batch processes, it is important to be able to calculate in advance the time required for the liquid to reach the desired temperature. Such a batch system is modeled and analyzed in this case study.

SYSTEM DESCRIPTION

Figure 18.14 shows a closed, insulated vessel that is filled with liquid and contains an electrical heater immersed in the liquid. The heating element is contained within a metal jacket that has a thermal resistance of R_{HL}. The thermal resistance of the vessel and its insulation is R_{La}. The heater has a thermal capacitance of C_H, and the liquid has a thermal capacitance of C_L. The heater temperature is θ_H and that of the liquid is θ_L, which is assumed to be uniform because of the mixer in the vessel. The rate at which energy is supplied to the heating element is $q_i(t)$.

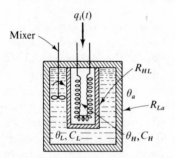

Figure 18.14 Vessel with heater.

The heater and the liquid are initially at the ambient temperature θ_a, with the heater turned off. At time $t = 0$, the heater is connected to an electrical source that supplies energy at a constant rate. We wish to determine the response of the liquid temperature θ_L and to calculate the time required for the liquid to reach a desired temperature, denoted by θ_d. The numerical values of the system parameters are:

Heater capacitance: $C_H = 20.0 \times 10^3$ J/K

Liquid capacitance: $C_L = 1.0 \times 10^6$ J/K

Heater-liquid resistance: $R_{HL} = 1.0 \times 10^{-3}$ s·K/J

Liquid-ambient resistance: $R_{La} = 5.0 \times 10^{-3}$ s·K/J

Ambient temperature: $\theta_a = 300$ K

Desired temperature: $\theta_d = 365$ K

We will derive the system model for an arbitrary input $q_i(t)$ and for an arbitrary ambient temperature θ_a, using θ_L and θ_H as the state variables. Then we will define a set of variables relative to the ambient conditions and find the transfer function relating the transforms of the relative liquid temperature and the input. Finally, we will calculate the time required for the liquid to reach the desired temperature.

SYSTEM MODEL

Since θ_H and θ_L represent the energy stored in the system, we can write the state-variable model as

$$\dot{\theta}_H = \frac{1}{C_H} [q_i(t) - q_{HL}]$$

$$\dot{\theta}_L = \frac{1}{C_L} [q_{HL} - q_{La}]$$

(39)

where $q_{HL} = (\theta_H - \theta_L)/R_{HL}$ and $q_{La} = (\theta_L - \theta_a)/R_{La}$. Substituting these expressions for q_{HL} and q_{La} and the appropriate numerical parameter values into (39) leads to the pair of state-variable equations

$$\dot{\theta}_H = -0.050\theta_H + 0.050\theta_L + [0.50 \times 10^{-4}] q_i(t)$$

$$\dot{\theta}_L = -[1.20 \times 10^{-3}] \theta_L + 10^{-3} \theta_H + [0.20 \times 10^{-3}] \theta_a$$

(40)

where the initial conditions are $\theta_H(0) = \theta_L(0) = \theta_a$.

At this point, we could transform (40) for a specific ambient temperature and a specific input $q_i(t)$. Then we could solve for $\Theta_L(s)$ and take its inverse transform to find $\theta_L(t)$. Instead, we define the relative variables

$$\hat{\theta}_H = \theta_H - \theta_a \tag{41a}$$

$$\hat{\theta}_L = \theta_L - \theta_a \tag{41b}$$

$$\hat{q}_i(t) = q_i(t) \tag{41c}$$

where the last equation implies that $\bar{q}_i = 0$. Using (41) to rewrite (40), we find that the system model in terms of the relative variables is

$$\dot{\hat{\theta}}_H = -0.050\hat{\theta}_H + 0.050\hat{\theta}_L + [0.50 \times 10^{-4}] \hat{q}_i(t)$$

$$\dot{\hat{\theta}}_L = -[1.20 \times 10^{-3}] \hat{\theta}_L + 10^{-3} \hat{\theta}_H$$

(42)

where the initial conditions are $\hat{\theta}_H(0) = \hat{\theta}_L(0) = 0$.

When transformed and rearranged, (42) becomes the pair of algebraic equations

$$(s+0.050)\hat{\Theta}_H(s) = 0.050\hat{\Theta}_L(s) + [0.50 \times 10^{-4}]\hat{Q}_i(s)$$
$$[s+(1.20 \times 10^{-3})]\hat{\Theta}_L(s) = 10^{-3}\hat{\Theta}_H(s) \tag{43}$$

Combining the two equations in (43) to eliminate $\hat{\Theta}_H(s)$, we find that the transfer function $H(s) = \hat{\Theta}_L(s)/\hat{Q}_i(s)$ is

$$H(s) = \frac{0.50 \times 10^{-4}}{1000s^2 + 51.20s + 0.010}$$

$$= \frac{0.50 \times 10^{-7}}{s^2 + 0.05120s + 10^{-5}}$$

$$= \frac{0.50 \times 10^{-7}}{(s+0.0510)(s+0.000196)} \tag{44}$$

SYSTEM RESPONSE

Having found the transfer function $\hat{\Theta}_L(s)/\hat{Q}_i(s)$ given by (44), we can now solve for the response to various inputs. Specifically, we shall find $\theta_L(t)$ for the ambient temperature $\theta_a = 300$ K, approximately 27°C or 80°F, and for a step-function input $\hat{q}_i(t) = 1.50 \times 10^4$ W for $t > 0$.

To find $\hat{\Theta}_L(s)$, we multiply $\hat{Q}_i(s) = [1.50 \times 10^4](1/s)$ by $H(s)$ as given by (44), getting

$$\hat{\Theta}_L(s) = \frac{0.750 \times 10^{-3}}{s(s+0.0510)(s+0.000196)} \tag{45}$$

Next we expand $\hat{\Theta}_L(s)$ in partial fractions to obtain

$$\hat{\Theta}_L(s) = \frac{75.0}{s} + \frac{0.289}{s+0.0510} - \frac{75.29}{s+0.000196}$$

Thus for $t > 0$, the relative liquid temperature is

$$\hat{\theta}_L = 75.0 + 0.289\epsilon^{-0.0510t} - 75.29\epsilon^{-0.000196t} \tag{46}$$

We obtain the actual liquid temperature by adding the ambient temperature of 300 K to $\hat{\theta}_L$, getting

$$\theta_L = 375.0 + 0.289\epsilon^{-0.0510t} - 75.29\epsilon^{-0.000196t} \tag{47}$$

which has a steady-state value of 375 K and is shown in Figure 18.15.

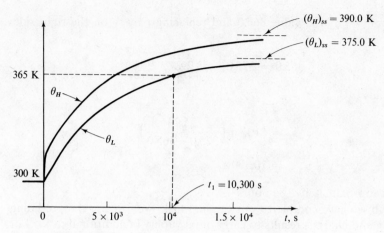

Figure 18.15 Response of liquid and heater temperatures.

From inspection of either (46) or (47), we see that the transient response of the system has two exponentially decaying modes with time constants of

$$\tau_1 = \frac{1}{0.0510} = 19.61 \text{ s}$$

$$\tau_2 = \frac{1}{0.000196} = 5102 \text{ s}$$

Because τ_2 exceeds τ_1 by several orders of magnitude, the transient having the longer time constant dominates the response. The principal effect of the shorter time constant is to give θ_L a zero slope at $t = 0+$, which it would not have if the transient response were a single decaying exponential.

Finally, we must calculate the value of t_1, the time required for the liquid to reach the desired temperature $\theta_d = 365$ K. Because two exponential terms are present, an explicit solution of (47) is not possible. However, by writing (47) in terms of the time constants τ_1 and τ_2 as

$$\theta_L = 375.0 + 0.289\epsilon^{-t/\tau_1} - 75.29\epsilon^{-t/\tau_2}$$

and noting that the solution for t_1 must be much greater than τ_1, we can see that the term $0.289\epsilon^{-t/\tau_1}$ is negligible when $t = t_1$. Hence, we can use the simpler approximate expression

$$\theta_L = 375.0 - 75.29\epsilon^{-t/5102} \tag{48}$$

where the value of 5102 s has been substituted for τ_2. Because (48) contains only one exponential function, we can solve it explicitly for t_1. Replacing the left side of (48) by the value 365.0 and replacing t by t_1 on the right side, we have

$$365.0 = 375.0 - 75.29\epsilon^{-t_1/5102}$$

which leads to

$$t_1 = 5102 \ln\left(\frac{75.29}{375.0 - 365.0}\right)$$

$$= 10{,}300 \text{ s}$$

which is approximately 2.86 hours.

Though we have completed our original task, we can use the preceding modeling and analysis results to carry out a variety of additional tasks. For instance, we can find the response of the heater temperature θ_H for the specified $q_i(t)$ by eliminating $\hat{\Theta}_L(s)$ from (43). The heater temperature θ_H is shown in Figure 18.15, and you are encouraged to evaluate the analytical expression for it. In practice, it might be important to evaluate the time t_1 for constant values of $\hat{q}_i(t)$ other than the value 1.50×10^4 W that we used. For example, if $\hat{q}_i(t)$ is doubled to 3.0×10^4 W, the desired liquid temperature $\theta_d = 365$ K will be reached in only 2916 s (48.6 minutes). On the other hand, if the constant energy-input rate is less than 1.30×10^4 W, the liquid temperature will never reach 365 K. If we want to investigate the effects of replacing the constant power source by a variable source, we can multiply the transfer function $H(s)$ by $\hat{Q}_i(s)$ to give the transform of the relative temperature. After we find the inverse transform, we can add it to the ambient temperature to get θ_L.

18.5 A SOCIOLOGICAL SYSTEM

It is possible to apply the techniques of modeling, analysis, and numerical solution to dynamic systems involving people and public services. In many such situations, the state variables are the numbers of people in particular groups, and the derivatives of the state variables are the rates at which people move into and out of the groups, e.g., birth and death rates. The inputs and some parameters of the model may be affected by policy decisions and the money available for various programs. The overall objective would be to select policies and allocate limited resources for best results. A properly constructed dynamic model may be of use in achieving this objective.

In this case study, we develop a very simple dynamic model of a system involving a population in which some members are afflicted with a hereditary disease. The model could be used to show the effects of various programs for screening the population for those with the disease before it reaches an advanced stage and for treating afflicted people. Our approach will be to describe the system variables and parameters and then to develop a discrete-time model. We will convert this model to a continous-time model in the form of state-variable equations and then draw a block diagram of the system and evaluate its characteristic polynomial.

SYSTEM DESCRIPTION

Some of the people in a region are afflicted with a hereditary disease that generally goes undetected until it reaches an advanced stage, at which time it is fatal. With proper facilities, personnel, and educational programs, it is possible to detect the disease at an early stage by screening people. Those who are found to have the disease can be treated, and there is a high probability of their being cured.

As indicated in Figure 18.16, three state variables are defined as follows:

n_u, number of people with the disease undetected

n_t, number of people undergoing treatment

n_g, number of people without the disease

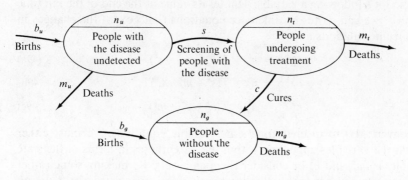

Figure 18.16 State variables and variables causing population changes.

People enter and leave these populations by the mechanisms indicated in the figure. The values of n_u and n_g are increased by births, and each of the populations is reduced by deaths. A person who is screened and found to have the disease moves from the population with the disease undetected to the population undergoing treatment. Those who are being treated and become cured move from the treatment population to the group without the disease. The variables denoting the rates of movement into and out of the populations, all expressed in units of people per month, are

b_u, rate of births of people with the disease undetected

b_g, rate of births of people without the disease

m_u, rate of deaths of people with the disease undetected

m_t, rate of deaths of people being treated

m_g, rate of deaths of people without the disease

s, rate at which people with the disease undetected are screened and correctly diagnosed

c, rate at which people under treatment are cured

The unit of time used to define these rates at which people enter and leave the three populations is arbitrary—e.g., second, day, month, or year. We shall use a 30-day month as the basic unit of time rather than the SI second, because the model coefficients and the times of interest will have more manageable values.

DISCRETE-TIME MODEL

Taking the time interval as T months and adopting the convention that a superscript k following a variable denotes its value at the end of the kth time interval,* we can write the difference equations that govern the changes in the three populations as

$$n_u^{k+1} = n_u^k + T(b_u^k - m_u^k - s^k) \tag{49a}$$

$$n_t^{k+1} = n_t^k + T(s^k - m_t^k - c^k) \tag{49b}$$

$$n_g^{k+1} = n_g^k + T(b_g^k + c^k - m_g^k) \tag{49c}$$

To convert (49) to discrete-time state-variable equations, we must either express the variables appearing within the parentheses in terms of the state variables n_u^k, n_t^k, and n_g^k or consider them as inputs. By making some rather arbitrary assumptions, we shall express each of the seven rates of population change in terms of the state variables, thereby obtaining a model with no inputs.

The expressions to be developed for the rates of population change will involve the probabilities of certain events occurring, such as the probability

---

* Recall that in Chapter 7 we used $q^{(k)}$ and $q^{(k+1)}$ to denote $q(kT)$ and $q[(k+1)T]$, respectively, where T was the step size being used for Euler's method. Also note the similarity of (49) to (7.18).

of a death within an interval of time. Such a probability is taken as the average number of occurrences of the event during a specified time interval divided by the number of people in the population. Policies and programs will influence the responses of the variables by modifying some of these probabilities.

Assume that the rates at which people are born are

$$b_u^k = P_b(n_u^k + n_t^k) \tag{50a}$$

$$b_g^k = P_b n_a^k \tag{50b}$$

where P_b is the probability of a birth expressed in units of persons per month per person, or $(\text{month})^{-1}$. It is assumed that all people with the disease, including those being treated, contribute to the birth of diseased babies, and that diseased babies are born only to parents having the disease.

The rate at which people in each population die is assumed to be a constant times the number of people in the population. Thus

$$m_u^k = P_{mu} n_u^k \tag{51a}$$

$$m_t^k = P_{mt} n_t^k \tag{51b}$$

$$m_g^k = P_{mg} n_g^k \tag{51c}$$

where P_{mu}, P_{mt}, and P_{mg} are the probabilities of a death in the respective populations, expressed in units of $(\text{month})^{-1}$. It is reasonable to expect that these mortality probabilities satisfy the inequalities $P_{mu} > P_{mt} > P_{mg}$.

The rate at which people with the disease undetected are screened and are correctly diagnosed for treatment is subject to a variety of assumptions and is dependent on the policies and programs implemented by the medical authorities. For simplicity, we assume that

$$s^k = P_s n_u^k \tag{52}$$

where P_s is the probability that a person with the disease undetected is screened and properly diagnosed within a 1-month period. The units of P_s are $(\text{month})^{-1}$. Because the people in the populations denoted by n_u and n_g are not identified on an individual basis, it is likely that those people to be screened will be drawn from both populations, perhaps with different probabilities. It is also possible that people without the disease might be diagnosed incorrectly as having it and given treatment, though this situation is not incorporated into our model.

Finally, the rate at which people undergoing treatment are cured and thus enter the group without the disease is assumed to be

$$c^k = P_c n_t^k \tag{53}$$

where P_c is the probability of a person's being cured during a 1-month interval, expressed in units of $(\text{month})^{-1}$. The value of P_c will depend strongly on the medical facilities and policies followed. Although this possibility is not included in the model, a person who has undergone treatment might be incorrectly thought to be cured. When treatment stops, that person would return to the population having the disease undetected, rather than to the population without the disease as indicated by (53).

When we substitute (50) through (53) into (49), we can write the discrete-time model in state-variable form as

$$n_u^{k+1} = n_u^k + T[P_b(n_u^k + n_t^k) - P_{mu} n_u^k - P_s n_u^k] \tag{54a}$$

$$n_t^{k+1} = n_t^k + T[P_s n_u^k - P_{mt} n_t^k - P_c n_t^k] \tag{54b}$$

$$n_g^{k+1} = n_g^k + T[P_b n_g^k + P_c n_t^k - P_{mg} n_g^k] \tag{54c}$$

If numerical values or functions of k were available for the probabilities and the interval size T selected, it would be a relatively routine task to implement a computer solution of (54). However, since we have not considered the solution of discrete-time models, we shall convert (54) to an approximate continuous model in the form of a set of differential state-variable equations. This process can be viewed as the reverse of the steps we used in Chapter 7 to develop Euler's method for the numerical solution of differential equations.

CONTINUOUS MODEL

To convert (54a) into a differential equation, we first rewrite it as

$$\frac{n_u^{k+1} - n_u^k}{T} = (P_b - P_{mu} - P_s) n_u^k + P_b n_t^k \tag{55}$$

The left side of (55) is an example of the constant-slope approximation to a derivative that was introduced in Section 7.1. Hence,

$$\lim_{T \to 0} \frac{n_u^{k+1} - n_u^k}{T} = \frac{dn_u}{dt} = \dot{n}_u$$

where the units of $\dot{n}_u$ are people per month because the probabilities appearing on the right side of (55) are based on 1-month intervals. To retain the second as the unit of time, we would have to divide the numerical values of each of the probabilities by the number of seconds in a month $(2.592 \times 10^6 \text{ s})$.

When we replace the left side of (55) by $\dot{n}_u$ and drop the superscript k from the state variables on the right side, then we have the desired differential equation. Carrying out the same steps for (54b) and (54c), we can write the state-variable model in differential-equation form as

$$\dot{n}_u = (P_b - P_{mu} - P_s)n_u + P_b n_t \tag{56a}$$

$$\dot{n}_t = -(P_{mt} + P_c)n_t + P_s n_u \tag{56b}$$

$$\dot{n}_g = (P_b - P_{mg})n_g + P_c n_t \tag{56c}$$

where the initial conditions $n_u(0)$, $n_t(0)$, and $n_g(0)$ must be specified in order for us to obtain a solution.

BLOCK DIAGRAM AND CHARACTERISTIC EQUATION

Provided that the probabilities are constant, (56) is a set of fixed linear differential equations. Transforming the equations and taking into account the initial conditions, we obtain

$$sN_u(s) - n_u(0) = (P_b - P_{mu} - P_s)N_u(s) + P_b N_t(s) \tag{57a}$$

$$sN_t(s) - n_t(0) = -(P_{mt} + P_c)N_t(s) + P_s N_u(s) \tag{57b}$$

$$sN_g(s) - n_g(0) = (P_b - P_{mg})N_g(s) + P_c N_t(s) \tag{57c}$$

For specified initial conditions, we can solve (57) for any of the three transforms $N_u(s)$, $N_t(s)$, and $N_g(s)$ as a rational function of s. If we do this, the numerator of the transform will be a polynomial whose coefficients depend on the three initial conditions, and the denominator will be the characteristic polynomial

$$D(s) = [s^2 + (P_{mu} + P_{mt} + P_s + P_c - P_b)s + (P_{mu} + P_s - P_b)(P_{mt} + P_c) - P_b P_s]$$
$$\times [s + P_{mg} - P_b] \tag{58}$$

A block diagram of the system model can be drawn from (57) and is shown in Figure 18.17. The diagram indicates that we can consider the system as the series combination of a second-order subsystem involving the state variables n_u and n_t and a first-order subsystem with the state variable n_g. The effect of this structure is evident in the characteristic polynomial $D(s)$, which appears in (58) as the product of a quadratic function involving the parameters P_{mu}, P_{mt}, P_s, P_c, and P_b and a linear function involving P_{mg} and P_b.

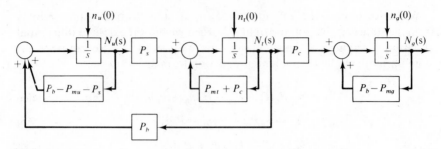

Figure 18.17 Block diagram of the continuous model.

RESPONSE OF THE MODEL

We shall consider the response for two situations that we can analyze easily.

No screening or treatment   In the absence of a screening and treatment program, we have $P_s = 0$ and $P_c = 0$. Equation (56b) reduces to

$$\dot{n}_t = -P_{mt} n_t \qquad n_t(0) = 0$$

which has $n_t = 0$ as its solution. Using this fact, we can rewrite (56a) and (56c) as a pair of uncoupled first-order equations, namely

$$\dot{n}_u = (P_b - P_{mu}) n_u$$
$$\dot{n}_g = (P_b - P_{mg}) n_g$$

which have the solutions

$$n_u = n_u(0) e^{(P_b - P_{mu})t} \tag{59a}$$
$$n_g = n_g(0) e^{(P_b - P_{mg})t} \tag{59b}$$

where t is expressed in months. If $P_b > P_{mu} > P_{mg}$, then both populations will increase exponentially with time. While this type of behavior may exist over a period of finite length, it clearly cannot hold indefinitely. In practice, a more accurate population model would involve nonlinearities that we have not considered. For example, decreases in food supply would tend to increase P_{mu} and P_{mg} and might tend to decrease P_b. Both of these effects would tend to limit population growth. Hence, if (56) is considered to have constant parameters, we should think of it as a linearized model that is valid only in the vicinity of a nominal set of constant populations.

Initiation of screening and treatment program   If $P_{mu} = P_b$, the solution of (59a) is a constant, which implies that the population of diseased people will be stable in the absence of a screening and treatment program. Now assume that such a program is initiated at $t = 0$, such that P_s and P_c become positive constants and $n_t(0) = 0$. From (56), the system model becomes

$$\dot{n}_u = -P_s n_u + P_b n_t$$
$$\dot{n}_t = -(P_{mt} + P_c) n_t + P_s n_u \qquad (60)$$
$$\dot{n}_g = (P_b - P_{mg}) n_g + P_c n_t$$

where $n_u(0)$ and $n_g(0)$ are positive and $n_t(0) = 0$. Without carrying out an explicit solution of (60), we can gain some insight into possible responses by considering the characteristic polynomial. Rewriting $D(s)$ with P_{mu} equal to P_b, we first get

$$D(s) = [s^2 + (P_{mt} + P_s + P_c)s + P_s(P_{mt} + P_c - P_b)][s + P_{mg} - P_b] \qquad (61)$$

Making the reasonable assumption that $P_{mg} < P_b$, we see that the root of $D(s) = 0$ that is at $s = P_b - P_{mg}$ is positive. This implies that the general population will grow without bound. The behavior of the state variables n_u and n_t depends on the locations of the two roots corresponding to the quadratic portion of (61). If the constant $P_{mt} + P_c - P_b$ is positive, then both these roots will be negative and both n_u and n_t will decay to zero as t increases. Hence, if the linear model were valid, the disease would eventually be wiped out. If this constant is negative, however, which would be the case if $P_c = 0$ and $P_{mt} < P_b$, then one of the roots of the quadratic will be positive and the other negative. In this case, the number of people under treatment would grow without bound. Nobody would be cured, but the diseased persons who are undergoing treatment would have their mortality rate reduced from P_{mu} to P_{mt}. Obviously, an effective overall program must involve both screening and treatment, in which case both P_s and P_c will be positive.

EXTENSIONS

The fixed linear third-order model represented by (56) and the block diagram shown in Figure 18.17 comprise only the most rudimentary approach to modeling a complex sociological system. An obvious extension is to improve the modeling of the birth rates b_u^k and b_g^k given by (50). Another extension is to include time-varying and nonlinear relationships. If we do this we can carry out computer solutions, but analytical analysis of the nonlinear time-varying model will not be feasible. We can develop a model in terms of incremental variables and inputs, however. Recalling the discussion of Section 5.4, we know that time-varying parameters such as P_s and P_c would lead to incremental inputs in such a model. In this fashion, changes in policies would be the model inputs and the resulting changes in the populations would be the outputs.

We have made no attempt to take the ages of the people into account. Certainly the various birth and death rates are age-dependent. To incorporate this effect, we could divide each of the three populations n_u, n_t, and n_g into several age groups. If we used three groups for each of the three populations, the model would be ninth-order. One could expect more realistic results than before, but the increased complexity would inhibit analysis. Fortunately, computer simulation remains feasible even for very high-order nonlinear and time-varying models.

In order to investigate the effects of alternative policies, we must revise the very simple models for the rates at which diseased people are screened and cured, as given by (52) and (53), respectively. For example, the effectiveness of the screening procedure depends on the number of people screened and on the accuracy of the diagnosis. With a fixed amount of money available per unit of time, either of these two factors could be improved at the expense of the other. It would be desirable to determine the combination of screening rate and diagnostic accuracy that results in the maximum number of diseased people being correctly screened per unit of time for a fixed funding rate.

We could expand the modeling of the treatment program to allow us to consider the effects on the cure rate of changing the number of people being treated and the length and quality of the treatment. Finally, we could investigate the effect of shifting funds from screening to treatment, or vice versa.

With a fully developed model, we can study in a quantitative manner the options for modifying the screening and treatment programs. To do this, we can implement on a computer optimization algorithms that are presented in more advanced books.

APPENDIX A ———————————
UNITS

We use the *International System of Units*, abbreviated as SI for *Système International d'Unités*. For an excellent comprehensive reference, see Ackley. In Section A.1, we list the seven basic SI units, of which we use only the first five in this book. A supplementary unit for plane angles is the radian (rad).

In Section A.2, we give those derived SI units that we use. In addition to the physical quantity, the unit, and its symbol, there is a fourth column that expresses the unit in terms of units previously given. For example, $1 \text{ N} = 1 \text{ kg·m/s}^2$.

A.1 BASIC UNITS

Physical Quantity	Name of Unit	Symbol for Unit
Length	meter	m
Mass	kilogram	kg
Time	second	s
Electrical current	ampere	A
Thermodynamic temperature	kelvin	K
Luminous intensity	candela	cd
Amount of substance	mole	mol

633

A.2 DERIVED UNITS

Physical Quantity	Name of Unit	Symbol for Unit	Expressed in terms of other units
Force	newton	N	$kg \cdot m/s^2$
Energy	joule	J	$kg \cdot m^2/s^2$
Power	watt	W	J/s
Electrical charge	coulomb	C	$A \cdot s$
Voltage	volt	V	W/A
Electrical resistance	ohm	Ω	V/A
Electrical capacitance	farad	F	$A \cdot s/V$
Inductance	henry	H	$V \cdot s/A$
Magnetic flux	weber	Wb	$V \cdot s$

A.3 PREFIXES

The standard prefixes for decimal multiples and submultiples of a unit are given in Ackley. The only ones used in this book are kilo (k), milli (m), and micro (μ). The terms in which they appear are as follows: $1 \text{ k}\Omega = 10^3 \ \Omega$, $1 \text{ mA} = 10^{-3} \text{ A}$, $1 \ \mu\text{F} = 10^{-6}$ F.

APPENDIX B
LAPLACE TRANSFORMS

B.1 TRANSFORMS OF FUNCTIONS

Time functions	Transformed functions
$\delta(t)$	1
$U(t)$	$\dfrac{1}{s}$
A	$\dfrac{A}{s}$
t	$\dfrac{1}{s^2}$
t^2	$\dfrac{2!}{s^3}$
t^n for $n = 1, 2, 3, \ldots$	$\dfrac{n!}{s^{n+1}}$
ϵ^{-at}	$\dfrac{1}{s+a}$

$t\epsilon^{-at}$	$\dfrac{1}{(s+a)^2}$
$t^2\epsilon^{-at}$	$\dfrac{2!}{(s+a)^3}$
$\sin \omega t$	$\dfrac{\omega}{s^2+\omega^2}$
$\cos \omega t$	$\dfrac{s}{s^2+\omega^2}$
$\epsilon^{-at}\sin \omega t$	$\dfrac{\omega}{(s+a)^2+\omega^2}$
$\epsilon^{-at}\cos \omega t$	$\dfrac{s+a}{(s+a)^2+\omega^2}$
$\epsilon^{-at}\left[\dfrac{C-aB}{\omega}\sin \omega t + B\cos \omega t\right]$	$\dfrac{Bs+C}{(s+a)^2+\omega^2}$
$2K\epsilon^{-at}\cos(\omega t + \phi)$	$\dfrac{K\epsilon^{j\phi}}{s+a-j\omega}+\dfrac{K\epsilon^{-j\phi}}{s+a+j\omega}$

B.2 TRANSFORM PROPERTIES*

Time functions	Transformed functions
$f(t)$	$F(s)$
$af(t)$	$aF(s)$
$f(t)+g(t)$	$F(s)+G(s)$
$\epsilon^{-at}f(t)$	$F(s+a)$
$tf(t)$	$-\dfrac{d}{ds}F(s)$
$f(t/a)$	$aF(as)$
$[f(t-a)]\,U(t-a)$	$\epsilon^{-sa}F(s)$
$\dot{f}(t)$	$sF(s)-f(0)$
$\ddot{f}(t)$	$s^2F(s)-sf(0)-\dot{f}(0)$
$\dddot{f}(t)$	$s^3F(s)-s^2f(0)-s\dot{f}(0)-\ddot{f}(0)$
$\dfrac{d^n f}{dt^n}$ for $n=1,2,3,\ldots$	$s^nF(s)-s^{n-1}f(0)-\cdots-sf^{(n-2)}(0)-f^{(n-1)}(0)$

* Restrictions on the use of these properties are not included in the table but may be found in Chapter 12.

$$\int_0^t f(\lambda)\,d\lambda \qquad\qquad \frac{1}{s}F(s)$$

$$f(0+) \qquad\qquad \lim_{s\to\infty} sF(s)$$

$$f(\infty) \qquad\qquad \lim_{s\to 0} sF(s)$$

APPENDIX C
MATRICES

This appendix is intended as a refresher for the student who has had an introductory course in linear algebra and is studying Chapter 17 on matrix methods. It is not a suitable introduction for the reader who has not had formal exposure to matrices. The appendix emphasizes only those aspects of the subject that are used in Chapter 17. References to more complete treatments appear in Appendix D.

C.1 DEFINITIONS

A *matrix* is a rectangular array of elements that are either constants or functions of time. We refer to a matrix having m rows and n columns as being of *order* $m \times n$. The element in the ith row and the jth column of the matrix $\mathbf{A}$ is denoted by a_{ij}, such that

$$
\mathbf{A} = \begin{bmatrix}
a_{11} & a_{12} & \cdots & a_{1n} \\
a_{21} & a_{22} & \cdots & a_{2n} \\
\vdots & \vdots & & \vdots \\
a_{m1} & a_{m2} & \cdots & a_{mn}
\end{bmatrix}
$$

A matrix having the same number of rows as columns, i.e., $m = n$, is a *square matrix* of order n. A matrix with a single column is a *vector*. Examples of vectors are

$$\mathbf{b} = \begin{bmatrix} b_1 \\ b_2 \\ \vdots \\ b_m \end{bmatrix} \qquad \mathbf{c} = \begin{bmatrix} 4 \\ -1 \\ 3 \end{bmatrix} \qquad \mathbf{x}(t) = \begin{bmatrix} x_1(t) \\ x_2(t) \\ \vdots \\ x_n(t) \end{bmatrix}$$

A matrix that consists of a single element, i.e., that has only one row and one column, is referred to as a *scalar*. A square matrix of n rows and n columns that has unity for each of the elements on its main diagonal and zero for the remaining elements is the *identity matrix* of order n, denoted by $\mathbf{I}$. For example,

$$\begin{bmatrix} 1 & 0 \\ 0 & 1 \end{bmatrix} \qquad \text{and} \qquad \begin{bmatrix} 1 & 0 & 0 \\ 0 & 1 & 0 \\ 0 & 0 & 1 \end{bmatrix}$$

are the identity matrices of order 2 and 3, respectively. A matrix all of whose elements are zero is the *null matrix*, denoted by $\mathbf{0}$.

We obtain the *transpose* of a matrix by interchanging its rows and columns such that the ith row of the original matrix becomes the ith column of the transpose, for all rows. Hence, if $\mathbf{A}$ is of order $m \times n$ with elements a_{ij}, then its transpose, $\mathbf{A}^T$, is of order $n \times m$ and has elements $(\mathbf{A}^T)_{ij} = a_{ji}$.

C.2 OPERATIONS

We can add or subtract two matrices that have the same order by adding or subtracting their respective elements. For example, if $\mathbf{A}$ and $\mathbf{B}$ are both $m \times n$, the ijth element of $\mathbf{A} + \mathbf{B}$ is $a_{ij} + b_{ij}$. We can form the product $\mathbf{C} = \mathbf{AB}$ only if the number of rows of $\mathbf{B}$, the right matrix, is equal to the number of columns of $\mathbf{A}$, the left matrix. If this condition is met, we find the $i\ell$th element of $\mathbf{C}$ by summing the products of the elements in the ith row of $\mathbf{A}$ with the corresponding elements in the ℓth column of $\mathbf{B}$. If $\mathbf{A}$ is $m \times n$ and $\mathbf{B}$ is $n \times p$, then $\mathbf{C}$ is $m \times p$ and

$$c_{i\ell} = \sum_{j=1}^{n} a_{ij} b_{j\ell} \qquad \text{for } i = 1, 2, ..., m \qquad \text{and} \qquad \ell = 1, 2, ..., p \tag{1}$$

For example, if

$$\mathbf{A} = \begin{bmatrix} 2 & -1 & 3 \\ 1 & 0 & 4 \end{bmatrix} \qquad \text{and} \qquad \mathbf{B} = \begin{bmatrix} 1 & -1 \\ 2 & 3 \\ -2 & 1 \end{bmatrix}$$

then

$$c_{11} = (2)(1) + (-1)(2) + (3)(-2) = -6$$

$$c_{12} = (2)(-1) + (-1)(3) + (3)(1) = -2$$

$$c_{21} = (1)(1) + (0)(2) + (4)(-2) = -7$$

$$c_{22} = (1)(-1) + (0)(3) + (4)(1) = 3$$

giving

$$\mathbf{C} = \begin{bmatrix} -6 & -2 \\ -7 & 3 \end{bmatrix}$$

We obtain the product of a matrix and a scalar by multiplying each element of the matrix by the scalar. For example, if k is a scalar, the ijth element of $k\mathbf{A}$ is ka_{ij}.

Some properties of matrices are summarized in Table C.1, where $\mathbf{A}$, $\mathbf{B}$, and $\mathbf{C}$ denote general matrices and where $\mathbf{I}$ and $\mathbf{0}$ are the identity matrix and the null matrix, respectively. Some of the properties that hold for scalars are not always valid for matrices. For example, except in special cases, $\mathbf{AB} \neq \mathbf{BA}$. Also, the equation $\mathbf{AB} = \mathbf{AC}$ does not necessarily imply that $\mathbf{B} = \mathbf{C}$. Furthermore, the equation $\mathbf{AB} = \mathbf{0}$ does not necessarily imply that either $\mathbf{A}$ or $\mathbf{B}$ is zero.

Table C.1 Matrix Properties

$$\mathbf{A} + \mathbf{B} = \mathbf{B} + \mathbf{A}$$
$$\mathbf{A} + (\mathbf{B} + \mathbf{C}) = (\mathbf{A} + \mathbf{B}) + \mathbf{C}$$
$$\mathbf{A}(\mathbf{B} + \mathbf{C}) = \mathbf{AB} + \mathbf{AC}$$
$$\mathbf{AI} = \mathbf{IA} = \mathbf{A}$$
$$\mathbf{0A} = \mathbf{0}$$
$$\mathbf{A0} = \mathbf{0}$$
$$\mathbf{A} + \mathbf{0} = \mathbf{A}$$

C.3 THE DETERMINANT AND INVERSE

Associated with any square matrix $\mathbf{A}$ is a scalar quantity called the *determinant* and denoted by $|\mathbf{A}|$. For the 2×2 matrix

$$\mathbf{A} = \begin{bmatrix} a_{11} & a_{12} \\ a_{21} & a_{22} \end{bmatrix}$$

the determinant is defined as

$$|\mathbf{A}| = a_{11}a_{22} - a_{12}a_{21} \tag{2}$$

One way of evaluating the determinant for larger matrices is in terms of the determinants of some of its submatrices, as indicated in the following discussion.

The *minor* of the *ij*th element of an $n \times n$ matrix **A** is the determinant of the $(n-1) \times (n-1)$ submatrix we obtain by deleting the *i*th row and the *j*th column of **A**. It is denoted by M_{ij}. The *cofactor* of the *ij*th element is denoted by C_{ij} and is

$$C_{ij} = (-1)^{i+j} M_{ij} \tag{3}$$

Thus the *ij*th cofactor is identical to the *ij*th minor if $i+j$ is an even integer and is the negative of the *ij*th minor if $i+j$ is odd. The *cofactor matrix* of an $n \times n$ matrix **A** is another $n \times n$ matrix whose elements are the cofactors of **A**. For the 3×3 matrix

$$\mathbf{A} = \begin{bmatrix} 2 & 1 & 0 \\ 3 & -2 & 1 \\ -1 & 2 & 1 \end{bmatrix} \tag{4}$$

the minors of the 1,1 and 2,1 elements are

$$M_{11} = \begin{vmatrix} -2 & 1 \\ 2 & 1 \end{vmatrix} = (-2)(1) - (1)(2) = -4$$

and

$$M_{21} = \begin{vmatrix} 1 & 0 \\ 2 & 1 \end{vmatrix} = (1)(1) - (0)(2) = 1$$

The corresponding cofactors are $C_{11} = -4$ and $C_{21} = -1$. Evaluating the cofactors of the remaining elements of **A**, we find the cofactor matrix to be

$$\mathbf{C} = \begin{bmatrix} -4 & -4 & 4 \\ -1 & 2 & -5 \\ 1 & -2 & -7 \end{bmatrix} \tag{5}$$

We can evaluate the determinant of **A** by selecting any one row or column of **A** and summing the products of the elements a_{ij} and their respective cofactors. Expanding along the *i*th row gives

$$|\mathbf{A}| = \sum_{j=1}^{n} a_{ij} C_{ij} \qquad \text{for } i = 1, 2, ..., n \tag{6}$$

while expanding down the *j*th column gives

$$|\mathbf{A}| = \sum_{i=1}^{n} a_{ij} C_{ij} \qquad \text{for } j = 1, 2, ..., n \tag{7}$$

For example, consider the matrix **A** defined by (4) and its cofactor matrix given by (5). Expanding $|\mathbf{A}|$ along the first row, we have

$$|\mathbf{A}| = a_{11} C_{11} + a_{12} C_{12} + a_{13} C_{13}$$
$$= (2)(-4) + (1)(-4) + (0)(4) = -12$$

If we had chosen to expand the determinant down the first column of **A**, the result would have been

$$|\mathbf{A}| = a_{11} C_{11} + a_{21} C_{21} + a_{31} C_{31}$$

$$= (2)(-4) + (3)(-1) + (-1)(1) = -12$$

For a 4×4 matrix, each cofactor is a determinant of order 3×3, which can in turn be evaluated in terms of three 2×2 determinants. We can always continue this process for any square matrix until only 2×2 arrays are left, and we can use (2) to evaluate each of these 2×2 determinants.

The *adjoint* of the square matrix **A** is denoted by adj[**A**] and is the transpose of the cofactor matrix; that is,

$$(\text{adj } [\mathbf{A}])_{ij} = C_{ji}$$

For the matrix **A** in (4)

$$\text{adj } [\mathbf{A}] = \begin{bmatrix} -4 & -1 & 1 \\ -4 & 2 & -2 \\ 4 & -5 & -7 \end{bmatrix}$$

If a matrix **B** can be found such that $\mathbf{AB} = \mathbf{BA} = \mathbf{I}$, then **B** is called the *inverse* of **A**, written $\mathbf{A}^{-1}$. Thus

$$\mathbf{A}\mathbf{A}^{-1} = \mathbf{A}^{-1}\mathbf{A} = \mathbf{I} \tag{8}$$

If the inverse exists, it is unique. A necessary but not a sufficient condition for **A** to have an inverse is that **A** must be square. A square matrix **A** of order $n \times n$ has an inverse $\mathbf{A}^{-1}$ if and only if $|\mathbf{A}| \neq 0$. If $|\mathbf{A}|$ is nonzero, it can be shown that

$$\mathbf{A}^{-1} = \frac{\text{adj } [\mathbf{A}]}{|\mathbf{A}|} \tag{9}$$

For the matrix **A** in (4), $|\mathbf{A}| = -12$ and

$$\mathbf{A}^{-1} = -\tfrac{1}{12} \begin{bmatrix} -4 & -1 & 1 \\ -4 & 2 & -2 \\ 4 & -5 & -7 \end{bmatrix} = \begin{bmatrix} \tfrac{1}{3} & \tfrac{1}{12} & -\tfrac{1}{12} \\ \tfrac{1}{3} & -\tfrac{1}{6} & \tfrac{1}{6} \\ -\tfrac{1}{3} & & \tfrac{7}{12} \end{bmatrix} \tag{10}$$

You can readily verify that (4) and (10) satisfy both of the relationships in (8). Although the methods described here for evaluating determinants and inverses are satisfactory for hand calculations with small matrices, other methods exist that are better suited for computer use with large matrices. The interested reader should consult Conte or Hamming for introductory material on the subject.

C.4 CHARACTERISTIC VALUES

Any square matrix of order n has associated with it a *characteristic equation* which, when written in terms of the variable s, is

$$|s\mathbf{I} - \mathbf{A}| = 0 \tag{11}$$

where $\mathbf{I}$ is the $n \times n$ identity matrix, and where

$$s\mathbf{I} - \mathbf{A} = \begin{bmatrix} s-a_{11} & -a_{12} & \cdots & -a_{1n} \\ -a_{21} & s-a_{22} & \cdots & -a_{2n} \\ \vdots & \vdots & & \vdots \\ -a_{n1} & -a_{n2} & \cdots & s-a_{nn} \end{bmatrix} \tag{12}$$

To evaluate $|s\mathbf{I} - \mathbf{A}|$, we can apply either (6) along any row or (7) down any column of the right side of (12). Regardless of the choice made, $|s\mathbf{I} - \mathbf{A}|$ is a unique polynomial of degree n in the variable s, and it is called the *characteristic polynomial*. Thus

$$|s\mathbf{I} - \mathbf{A}| = s^n + \alpha_{n-1} s^{n-1} + \cdots + \alpha_1 s + \alpha_0 \tag{13}$$

where the coefficients $\alpha_0, \alpha_1, \ldots, \alpha_{n-1}$ depend on the elements of $\mathbf{A}$.

Because the characteristic polynomial is of degree n, the characteristic equation given by (11) will have at most n distinct solutions, known as the *characteristic values* (*eigenvalues*) of the matrix $\mathbf{A}$. Should (11) have fewer than n distinct solutions, one or more of its solutions must be a multiple solution such that the total number of solutions, counting each according to its multiplicity, is exactly n. The characteristic values may be real or complex, but since the coefficients in (13) are real, any complex characteristic values must occur in complex conjugate pairs.

To find the characteristic values of the matrix $\mathbf{A}$ in (4), we form

$$s\mathbf{I} - \mathbf{A} = \begin{bmatrix} s-2 & -1 & 0 \\ -3 & s+2 & -1 \\ 1 & -2 & s-1 \end{bmatrix}$$

and then evaluate its determinant. Expanding along the first row, we find that

$$|s\mathbf{I} - \mathbf{A}| = (s-2)\begin{vmatrix} s+2 & -1 \\ -2 & s-1 \end{vmatrix} - (-1)\begin{vmatrix} -3 & -1 \\ 1 & s-1 \end{vmatrix}$$

$$= (s-2)[(s+2)(s-1)-2] + [(-3)(s-1)+1]$$

$$= s^3 - s^2 - 9s + 12$$

Since $|s\mathbf{I} - \mathbf{A}|$ is a cubic function of s, we resort to a digital computer to find that the three characteristic values are $s_1 = 1.4316$, $s_2 = 2.6874$, and $s_3 = -3.1190$.

APPENDIX D
SELECTED READING

This appendix suggests a number of books that are suitable for undergraduates. Some provide background for specific topics or can be used as collateral reading. Others extend the topics we treat to a more advanced level or offer an introduction to additional areas. The works cited here are included in the list of references that follows. This appendix and the references provide typical starting points for further reading, but these lists are not intended to be comprehensive.

BACKGROUND SOURCES

Introductory college physics textbooks describe the basic mechanical, electrical, and electromechanical elements and some other components as well. They also illustrate simple applications of physical laws to such systems. One pair of good references is *Resnick and Halliday Part I* and *Halliday and Resnick Part II*.

Purcell is a good example of a general college mathematics book. Somewhat more advanced books are *Boyce and DiPrima* and *Wylie*. For background in linear algebra and matrices, such as would be useful for Chapter 17, consult *Dorf* (1969) or *Nobel* (the first of which is in a programmed format suitable for self-study).

A comprehensive treatment of units is given in *Ackley*. In addition to definitions of the basic and derived SI units, it includes an extensive list of conversion factors from English units to SI units and a number of physical constants.

BOOKS AT A COMPARABLE LEVEL

The following four introductory systems books have an orientation toward physical devices. *Doebelin* (1972) emphasizes physical components and element laws. *Cannon* offers an extensive treatment of control systems. *Shearer, Murphy, and Richardson* makes heavy use of graph notation, while *Seely* uses graphs and topological formulas.

More mathematically orientated books include *Athans, Dertouzos, Spann, and Mason*. A related pair of books with a formal, concise, and abstract approach is *Polak and Wong* and *Desoer*.

BOOKS THAT EXTEND THE MATERIAL

More extensive discussions of algorithms for numerical solutions and estimation of errors are found in *Conte* and *Hamming*. *Bennett* treats analog simulation and includes a survey of digital simulation languages.

A number of books have a more general and advanced treatment of transform methods and state-variable techniques. These include *Schwarz and Friedland, Swisher, Timothy and Bona*, and *DeRusso, Roy and Close*.

Many books dealing with modeling and analysis are restricted to a particular discipline. Standard books on electrical circuits include *Van Valkenburg, Cruz and Van Valkenburg, Close*, and *Hayt and Kemmerly*. *Smith* (1973) gives an introductory treatment of transistors and electronic circuits, while such books as *Millman and Halkias* provide more comprehensive coverage of this area. Vibrations and machine dynamics are covered by *Thomson*.

Books such as *Smith* (1971) and *Del Toro* treat devices and applications in both electrical and electromechanical systems. *Fitzgerald and Kingsley* is an example of a book restricted to electromechanical machinery. *Doebelin* (1975) covers a wide variety of mechanical, fluid, and thermal devices. Textbooks dealing with thermal, hydraulic, and chemical systems include *Buckley, Tyner and May, Franks*, and *Hougen*.

It is natural to follow up your study of this book with books on general dynamic systems at a more advanced level that extend the principal techniques we have developed. Good examples include *Frederick and Carlson, Perkins and Cruz*, and *Lathi*.

Dorf (1974), *D'Azzo and Houpis*, and *Melsa and Schultz* concentrate on feedback systems. *Huelsman, Calahan*, and *Gupta, Bayless, and Peikari* emphasize computer solutions of electrical systems.

STARTING POINTS FOR TOPICS NOT COVERED

There are a number of important types of systems that we have not treated, except for passing mention or as a single example. A good treatment of discrete-time systems can be found in *Cadzow*.

Books are available that extend system modeling and analysis techniques to transportation, health, economic systems, and similar applications. Examples are *Helly*, *Catanese*, and the two books by *Forrester*.

Distributed systems are typically the subject of a chapter in books on such particular disciplines as acoustics or fluid systems. The treatment of stochastic systems requires some knowledge of probability theory. Though written at a rather advanced level, *Papoulis* combines probability theory and stochastic processes.

REFERENCES

Ackley, R. A. *Physical Measurements and the International (SI) System of Units*, 3d ed. Technical Publications, San Diego, Calif. 1970.

Alley, C. L., and K. W. Atwood. *Electronic Engineering*. 3rd ed. Wiley, New York, 1973.

Anner, G. E. *Elementary Nonlinear Electronic Circuits*. Prentice-Hall, Englewood Cliffs, N.J., 1967.

Athans, M., M. L. Dertouzos, R. N. Spann, and S. J. Mason. *Systems, Networks, and Computation: Multivariable Methods*. McGraw-Hill, New York, 1974.

Bennett, A. W. *Introduction to Computer Simulation*. West, St. Paul, Minn., 1974.

Boyce, W. E., and R. C. DiPrima. *Elementary Differential Equations and Boundary Value Problems*, 3d ed. Wiley, New York, 1977.

Buckley, P. S. *Techniques of Process Control*. Wiley, New York, 1964.

Cadzow, J. A. *Discrete-Time Systems*. Prentice-Hall, Englewood Cliffs, N.J., 1973.

Calahan, D. A. *Computer-Aided Network Design*, revised edition. McGraw-Hill, New York, 1972.

Cannon, R. H., Jr. *Dynamics of Physical Systems*. McGraw-Hill, New York, 1967.

Catanese, A. J. *Scientific Methods of Urban Analysis*. The University of Illinois Press, Urbana, Ill., 1972.

Close, C. M. *The Analysis of Linear Circuits*. Harcourt, Brace & Jovanovich, New York, 1966.

Conte, S. D., and C. de Boor. *Elementary Numerical Analysis*, 2d ed. McGraw-Hill, New York, 1972.

Cruz, J. B., and M. E. Van Valkenburg. *Signals in Linear Circuits*. Houghton Mifflin, Boston, 1974.

D'Azzo, J. J., and C. H. Houpis. *Linear Control System Analysis and Design*. McGraw-Hill, New York, 1975.

Del Toro, V. *Electrical Engineering Fundamentals*. Prentice-Hall, Englewood Cliffs, N.J., 1972.

DeRusso, P. M., R. J. Roy, and C. M. Close. *State Variables for Engineers*. Wiley, New York, 1965.

Desoer, C. A. *Notes for a Second Course on Linear Systems*. Van Nostrand Reinhold, Princeton, N.J., 1970.

Desoer, C. A., and E. S. Kuh. *Basic Circuit Theory*. McGraw-Hill, New York, 1969.

Doebelin, E. O. *System Dynamics: Modeling and Response*. Charles E. Merrill Books, Columbus, Ohio, 1972.

Doebelin, E. O. *Measurement Systems*, rev. ed. McGraw-Hill, New York, 1975.

Dorf, R. C. *Matrix Algebra*. Wiley, New York, 1969.

Dorf, R. C. *Modern Control Systems*, 2d ed. Addison-Wesley, Reading, Mass., 1974.

Fitzgerald, A. E., C. Kingsley, Jr., and A. Kusko. *Electrical Machinery*, 3d ed. McGraw-Hill, New York, 1971.

Forrester, J. W. *Urban Dynamics*. M.I.T., Cambridge, Mass., 1969.

Forrester, J. W. *World Dynamics*, 2d ed. Wright-Allen Press, Cambridge, Mass., 1973.

Franks, R. G. E. *Modeling and Simulation in Chemical Engineering*. Wiley, New York, 1972.

Frederick, D. K., and A. B. Carlson. *Linear Systems in Communication and Control*. Wiley, New York, 1971.

Gupta, S. C., J. W. Bayless, and B. Peikari. *Circuit Analysis*. Intext Educational Publishers, New York, 1972.

Halliday, D., and R. Resnick. *Physics, Part II*, 3rd ed. Wiley, New York, 1978.

Hamming, R. W. *Numerical Methods for Scientists and Engineers*, 2d ed. McGraw-Hill, New York, 1973.

Hayt, W., and J. Kemmerly. *Engineering Circuit Analysis*, 2d ed. McGraw-Hill, New York, 1971.

Helly, W. *Urban Systems Models*. Academic, New York, 1975.

Hougen, J. O. *Measurement and Control Applications for Practicing Engineers*. Cahners Books, Boston, 1972.

Huelsman, L. P. *Basic Circuit Theory with Digital Computations*. Prentice-Hall, Englewood Cliffs, N.J., 1972.

Kuo, F. F. *Network Analysis and Synthesis*, 2d ed. Wiley, New York, 1966.

Lathi, B. P. *Signals, Systems, and Controls*. Intext Educational Publishers, New York, 1974.

Melsa, J. L., and S. K. Jones. *Computer Programs for Computational Assistance in the Study of Linear Control Theory*, 2d ed. McGraw-Hill, New York, 1973.

Melsa, J. L., and D. G. Schultz. *Linear Control Systems*. McGraw-Hill, New York, 1969.

Merriam, C. W., III. *Analysis of Lumped Electrical Systems*. Wiley, New York, 1969.

Millman, J., and C. C. Halkias. *Integrated Electronics: Analog and Digital Circuits and Systems*. McGraw-Hill, New York, 1972.

Noble, B. *Applied Linear Algebra*. Prentice-Hall, Englewood Cliffs, N.J., 1969.

Papoulis, A. *Probability, Random Variables, and Stochastic Processes*. McGraw-Hill, New York, 1965.

Perkins, W. R., and J. B. Cruz, Jr. *Engineering of Dynamic Systems*. Wiley, New York, 1969.

Polak, E., and E. Wong. *Notes for a First Course on Linear Systems*. Van Nostrand Reinhold, New York, 1970.

Pugh, A. L., III. *Dynamo User's Manual*, 5th ed. M.I.T., Cambridge, Mass., 1976.

Purcell, E. J. *Calculus*. Prentice-Hall, Englewood Cliffs, N.J., 1972.

Resnick, R., and D. Halliday. *Physics, Part I*, 3rd ed. Wiley, New York, 1977.

Schwarz, R. J., and B. Friedland. *Linear Systems*. McGraw-Hill, New York, 1965.

Seely, S. *Dynamic Systems Analysis*. Reinhold, New York, 1964.

Shearer, J. L., A. T. Murphy, and H. H. Richardson. *Introduction to System Dynamics*. Addison-Wesley, Reading, Mass., 1967.

Smith, R. J. *Circuits, Devices, and Systems*, 3rd ed. Wiley, New York, 1976.

Smith, R. J. *Electronics: Circuits and Devices*. Wiley, New York, 1973.

Swisher, G. M. *Introduction to Linear Systems Analysis*. Matrix Publishers, Champaign, Ill., 1976.

Thomson, W. T. *Vibration Theory and Applications*. Prentice-Hall, Englewood Cliffs, N.J., 1965.

Timothy, L. K., and B. E. Bona. *State Space Analysis: An Introduction*. McGraw-Hill, New York, 1967.

Tyner, M., and F. P. May. *Process Engineering Control*. Ronald, New York, 1968.

Van Valkenburg, M. E. *Network Analysis*, 3d ed. Prentice-Hall, Englewood Cliffs, N.J., 1974.

Wylie, C. R., Jr. *Advanced Engineering Mathematics*, 4th ed. McGraw-Hill, New York, 1975.

System/360 Continuous System Modeling Program User's Manual, 4th ed. (H20-0367-3). IBM, White Plains, N.Y., 1969.

INDEX

$$47.7$$

$$57.11 + \quad = 80$$

$$\begin{array}{r} 77 \\ +85 \\ 95 \\ \hline 257 \end{array}$$